Globalization and Papua New Guinea: Ancient Wilderness, Paradise, Introduced Terror and Hell

Falk Huettmann

Globalization and Papua New Guinea: Ancient Wilderness, Paradise, Introduced Terror and Hell

Springer

Falk Huettmann
-EWHALE lab-, Department of Biology
and Wildlife, Institute of Arctic Biology
University of Alaska Fairbanks (UAF)
Fairbanks, AK, USA

ISBN 978-3-031-20261-2 ISBN 978-3-031-20262-9 (eBook)
https://doi.org/10.1007/978-3-031-20262-9

This Springer imprint is published by the registered company Springer Nature Switzerland AG
The registered company address is: Gewerbestrasse 11, 6330 Cham, Switzerland

Author's Devotion
When Tim Flannery wrote 1994[1] *in his dedication*

"I dedicate this book to Jim-Bob Mofett, his successors and all the other CEOs of mining companies with interests in Melanesia, in the hope that, through reading it, they will understand a little better the people whose lives they so profoundly change"

one cannot do and say more. This book belongs to the people of Papua New Guinea and of Melanesia. We owe it to them, and we owe much of our future to them too.

---

[1] Flannery T. (1998) Throwim Way Leg: Tree-Kangaroos, Possums and Penis Gourds. Atlantic Monthly Press, New York.

Preface

The strong smell of fire, burned wood, brings me back to Papua New Guinea's (PNG) legacy and deep times of Mother Earth. Tim Flannery calls it '*An Infinity Before Man*'. It's raw, authentic and pure. It's free of industrialization, hardly has a centralized feudal society, but all with ancient habitats and a species diversity and abundance one can now only dream of. It raises the curiosity of what was it like, how does it link with our modern world and own upbringing and what unites us, regardless of place, time and culture? Biodiversity, wilderness and how conserved well with an actual 'negative' global footprint?

PNG can actually tell you, and so can its people. Their legacy tells easily those tales to everybody who just wants to know and who is ready to listen and to inquire.

Travel and field work in PNG tends to rely on many things, a strong will to be there in the first place, and to get away from 'The West' and its enslaving money world, value systems and norms, breaking the borders of either system to engage locally and see that PNG is the window to the earth, to the ocean, to the universe and to ourselves.

For those who have the ability and privilege to work, travel, live and study in PNG, a field season year can look something like this for a biologist: Be in the field elsewhere during late summer, teach and research in fall, be in PNG over Xmas and New Year, teach and research again 'back home,' be in another field project elsewhere in spring, write and work things up, be in summer in PNG again, have 2–4 conferences in between and get all funding and book-keeping in order.

Repeat…

(Well, just the funding part is not as clear like described!). It's an intense spiral and can bring you in into the ancient times (and eat up your usual Western live; a traveler between the spheres under the universe).

Anyways, PNG remains very deep—and scary at times—but always entertaining and enlightening, intellectually and physically demanding. Clearly a 'whole body' experience. I found all encounters in PNG very friendly. I was lucky and I was able to dive deeper into PNG, into some of its society, 'the bush' and the villages and the species and habitat connections. Other than the rain, the mud and some critters, the PNG bush can easily be managed. Well, if you have support people. I was always treated with courtesy and respect in PNG. I am embarrassed I was not always able to come back more often; sometimes the odds were against us. Global COVID, universities or funding entities and airplanes are not always working for you, nor do they back home to 'get' PNG done the right way.

I like projects going "*in-and-out*," such type of fieldwork research gives you energy, reflection and much data to work on and to clean up. Much research here in PNG was done 'on foot' walking through ancient paths, rainforests and landscapes. It's a paradigm, similar to Gonzo-Science. Catching a ride—land or sea—as one could and being engaged and embedded, living it. So rather than being at one place for a few months—as many people do in PNG and try to describe new species to name from a field camp—I rather prefer to be nomadic; I see, sample and learn more that way. It's a natural approach for good scholars, instead of simply being pinned down to a place, with much gear, and being totally exposed under some plastic roof canvas from Walmart, for what? I really find intense and established field camps widely useless, a neurotic place of political irrelevance for good research, and instead I prefer to explore things by myself, rather than being in the footsteps of some big and richly funded guys who set up the well-infrastructured water-proof field camps (as most PNG is still widely male dominated; glad there are exceptions now).

When I started this work and book, I was aware of PNG's deep legacy and connection with Mother Earth. But what I was not aware, and learned intensely during this project, is that I am actually to track a global colonial and subsequent environmental crime story with industry that reaches way beyond PNG and which brings me back to Canada, Australia, Germany, the EU and similar so-called civilized, modern and advanced nations where people tried to educate us with an agency template of industrialized schooling. I wanted the get away from that, but in PNG I got sucked into it again instead. From all accounts, Australia and its industry operating in the region stands out as the big neighboring orchestrating bully. I see just now that this book tackles that very problem, with PNG as the action board and the wilderness biodiversity as the lost prize.

Now, 'that's really wild' (A phrase I learned from Alexandria Ortazio-Cortez (ABC) when she was asked how she felt about the January 6th implosion and tried-coup of a world democracy; the U.S.).

And accordingly, some traditional and conservative colonial minds—taxonomists and species collectors—who certainly will be reading this book critically will ask why the title, why the content and the facts when the core question is 'just' about biodiversity wilderness conservation?

Well, this dry question was answered for them already many decades ago and culminated in a well-cited phrase by Chico Mendes:

> *Ecology without Class Struggle is just Gardening*
>
> Chico Mendes (in Ludlam 2021)

This book is also for those ones who are still in doubt that the colonial approach is just widely off, that we are modern now, that western research is unbiased, that it would be perfectly fine to count beans and plants but ignoring the wider context and suffering, and that neocolonialism is just a global conspiracy thing. This book is for them and so that PNG can now move forward fine in its own, adjust and progress in globally sustainable ways with us all on board. I am grateful for that opportunity.

Papua New Guinea, it's really out there.

Fairbanks, Alaska, USA
March 2023

Falk Huettmann

Acknowledgements

"Thanks" to everybody who made after all these years this work and its message a reality. This book is just possible due to over nine field campaigns and seasons of visits and research inquiry during app. 20 years of overall PNG work; in the field, at sea, at the desk, from Alaska, online or during my other travels in Asia and remote field work. I owe it to the very helpful S. Oppel for the initial contacts, D. Wright and A. Mack for being the first and repeatedly kind hosts, the WCS team, L. Dahbeck and colleagues for model work and discussions over 3 years, M. Steiner for the great co-authored work, and all the many local PNG citizens, villages and travel friends. There are too many to be named here, but you know who you are. I owe it to them and thus far I can only say "THANK YOU"; this work is for you, and I hope we can do more!

I am also thankful to Becky from Santatravel for getting me incredible connections, for the pilots connecting us back, for my walking partners back to Goroka, and to guide Nupre of course and his grand village family in Sepik and along forest trails toward Madang. Then I thank the funders, including the Commerzbank Uelzen, Germany, for their great credit card scheme and unique account concepts to allow them access to my accounts even without my consent, as well as Mt. McKiney Bank, Fundus Nonexistencia and Grantis Nullis for reality support. Alaska Airlines was always helpful in this project and very flexible; airmiles work in the bush even. The internet in PNG was always available when needed and much faster than in Alaska; thanks to those who make it happen (the others may work harder). Thanks of course to my publisher and the publication team in various offices over the years.

Finally, I thank the large and super team around Chrome and the late Thor ("stunning", guys and gals), Arne and Sophia, and—of course—the incredible Hazel and Ela ground crew to keep Sherman Rd running free and wild. I am humble, blessed and in deep gratitude for all of this. THANK YOU!

Contents

About the Authors

Falk Huettmann is a 'digital naturalist' concerned with global sustainability. Being a well-rounded and published Professor in wildlife ecology employed with the University of Alaska Fairbanks in his EWHALE lab, he has worked in over 60 project locations worldwide, including Papua New Guinea during a time period of app. 20 years. Falk worked on pelagic seabirds in the North Atlantic, grizzly bear habitat in the Canadian Rocky Mountains, Pacific Northwest old-growth forests and marbled murrelts, snow leopard data issues for the Hindu Kush Himalaya, Matschie's Tree Kangaroo in Papua New Guinea and over 30 species in the ocean surrounding Antarctica, including one of the world's largest and remote marine protected area (MPA): the Ross Sea. His machine learning/AI modeling work features over 2000 species worldwide, including diseases. He teaches nationally certified online classes and international workshops, and his expertise is in rapid assessment methods such as open access digital data, geographic information systems (GIS), landscape ecology and disease perspectives all footed in ecological economics, predictive risk models and associated governance schemes. He appears in the world's media and has published over 300 publications, including high-impact books about the tropics, the three poles (Arctic, Antarctic and Hindu Kush Himalayas), oceans, Landscape Ecology, Machine Learning as well as Squirrels of the World, open access data and open source code with ISO-compliant metadata.

Moriz Steiner is a young scientist with multinational research experiences focused on small mammals, specifically squirrels. He acts now as a squirrel expert of the IUCN small mammal specialist group and the IUCN species survival commission. Most of his research projects use digital (=machine learning (ML), geographic information system (GIS), and artificial intelligence (AI)) approaches supported by field work seeking to develop state-of-the-art conservation ecology models and methods to better conserve our world's nature for future generations. Based on his global models with FH, Moriz supported much of the ecological niche model and data work for Papua New Guinea in the Australian and SE Asian context.

Chapter 1
Papua New Guinea (PNG)…The New Nation and a Global Experimental Sustainability Success of a Renewed Tribal Governance: Insights from a Widely Misunderstood Modern Control Study

Do not travel to Papua New Guinea due to COVID-19, crime, civil unrest, health concerns, natural disasters, and kidnapping. Some areas have increased risk.
Google online, 21st May 2021
Cannibalism was originally deemed as an abnormal or unnatural behavior in the animal kingdom (Dawkins, 1976; Maynard Smith & Price, 1973). *However, this view has changed dramatically over the past few decades, and it is now recognized to be adaptive, phylogenetically widespread, and relatively common in nature* (Elgar & Crespi, 1992; Hausfater & Hrdy, 1984; Hrdy & Hausfater, 1984; Polis, 1981)
Soler et al. (2022) *(citations within)*

Abstract Receiving its independence just in 1975 Papua New Guinea (PNG) is still a 'new' nation. While PNG as a 'nation construct' received its international status it was way too late. PNG is a 'rural' country and second-largest island in the world, but it actually is a nation of world records and of global relevance. Already the topography, biodiversity, marine life and the very diverse human society are stunning; humans can be traced back to PNG for easily over 47,000 years. But PNG is not short of problems and challenges. And according to many western nations and their (industrial) performance metrics, PNG gets classified as a failed nation, e.g. by Australia, business people and many colonial scholars alike. But then, PNG is essentially a western experiment that got pushed for Australian cost-savings from the stone age via colonial times straight into globalization and neocolonialism within less than a generation. PNG is modeled with an underlying Westminster governance scheme but lacks a solid tax base and computing power, and is not a typical industrial power house. PNG is a brainchild of the western world, namely colonial nations and adjacent Australia mandated by the UN, and those minds are to blame for any subsequent failure of PNG's design and PNG itself in the wider 'Pacific Theater.' Instead, in PNG a widespread and quite well-to-do society and successful community governance policy is found: tribal rules and the Wantok system. At minimum, it's resilient, has a proven deep-time sustainability record and actually keeps the nation afloat while most western measures—led by the former colonial powers, development aid and global powerhouses like Australia, UK and the U.S. and now also China—show environmental bankruptcy, including financial and social exploitation. Trying

F. Huettmann, *Globalization and Papua New Guinea: Ancient Wilderness, Paradise, Introduced Terror and Hell*, https://doi.org/10.1007/978-3-031-20262-9_1

for over 300 years, the western world still cannot comprehend PNG well and the international business community has not progressed well in PNG while the deep times still rule and survive, including sorcery.

Keywords Papua New Guinea (PNG) · Failed States · Political experiments · Independence · Globalization · Global governance

1.1 Introduction

Concepts that are found and described around Papua New Guinea -including the persisting colonial attitudes—are really wild.

PNG is many things to many people. It's a country consisting of over 1000 nations. It's thus very easy to be biased on the topic, and it's almost certainly that one is incomplete and behind.[1] To me, I find Papua New Guinea (PNG) is a very lush biological region north of the equator and a vibrant nation with a deeply-linked history of global sustainability relevance. PNG is part of globalization and Mother Earth as we know it. PNG is a global celebrity and certainly a megadiversity country (e.g. Beehler & Laman, 2020, https://en.wikipedia.org/wiki/Megadiverse_countries). It's a world civilization and global culture (e.g. Diamond, 2011a, 2011b). PNG offers geologically active areas, and it is one among the fasting rising ranges in the world. It's part of the Ring of Fire and has volcano islands like Karkar, Manu, etc. (Figs. 1.1 and 1.2).

Due to the combined deep geology, oceanography, biogeography, biodiversity and human aspects so readily accessible in PNG, PNG allows us to see, connect and experience with deep earth, live and the wider universe. Beehler and Laman (2020) refer to it as Earth's grandest island. It also has some three karst landscapes and those are listed as World Heritage Sites. Further, PNG is among the largest intact and contiguous old-growth forest area in Pacific-Asia (Beehler & Laman, 2020) if not perhaps even in the world. PNG is also full of wetlands, lakes and rivers. The largest freshwater lake in New Guinea is Lake Murray, in PNG it's Lake Kutumbu. The Fly river is PNG's largest, by flow (in the Fly headwaters the infamous Ok Tedi mine is located. It made world news when the toxic residues from a broken tailing

[1] Note from the author: It's virtually not possible to comprehend or cite, or even find and know, all articles and work on Papua New Guinea (PNG). That's because so many people from all over the world have worked on PNG during the last 300 years, collected data there but much is lost, not shared or hidden across the many languages and libraries of this world, that includes Asian and East European ones beyond English. The reader is advised that the same applies to this book. A good compilation and oversight of this problem is given in Hayes (2003) for instance. The author appreciates feedback from the readers for references to be included and considered in the text and for an improved interpretation. In the meantime, in the absence of a wider and consistent coverage, sometimes sources from The Guardian's Pacific Project—supported by The Judith Neilson Institute—are used. While some people dismiss news article citations in books like done here, in reality, there is no other, and hardly a better source for PNG. That's because those are fact-checked, come with citations and verified standards from specifically devoted investigations. They are as valid, likely less biased, than works from research or ones that are funded/supported by mining entities such as presented by Beehler and Laman (2020) as a typical and recent example 'creating' what the reality in New Guinea is to be.

Fig. 1.1 Mysterious Papua New Guinea

dam polluted this major river and its fish with lead, arsenic, mercury, cadmium, etc. (Cousteau & Richardson, 1999; Beehler & Latham, 2020). The Sepik river can be navigated by cruise ships making it an international industry, e.g. entertained by indigenous dance and penis gourd shows (Papua—part of Indonesia—has most of that culture though, less PNG; Flannery, 1998). Being so close to the equator, PNG is affected by the monsoon climate with elevational rain and temperature gradients as well as rain shadows, e.g. in the capitol Port Moresby. And PNG certainly is a strategic key area in the '*Pacific Theatre*' affecting world peace (Cousteau & Richardson, 1999 and references within). That holds for the past as well as now (Lockyer et al., 2021 for navy and submarines; see also Babbage, 1987). But the western construct of PNG as a nation is still in a very young state (Hawksley, 2006) and the nation got its independence just in 1975, after it was essentially ruled 1884 onward by three outside nations (Holland, Germany and England, helped by Australia and with a UN oversight). Whereas Spain and Portugal played a role in exploring New Guinea, they lost PNG early on, as these two empires lost many of their explorations. And colonial France was virtually not present in PNG still struggling then with North America, Africa and itself. Italy was not much present in that area, but several explorers and anthropologists operated in PNG later (Figs. 1.3 and 1.4).

Arguably, PNG was directly mismanaged by western powers for easily over a century (e.g. Hawksley, 2006). PNG was discovered by the western world through Spanish and Portuguese explorers, usually related to spice islands—and by some

Fig. 1.2 What sits behind the fascinating coastline and its mountains, reef to ridge? A question that occupies people to this very day

demand for Bird of Paradise skins (Beehler & Laman, 2020; Laman & Stoles, 2012)—and it centered around the Molluccas and western Papua (Bird's Head and Bird Island). Jorge de Meneses is credited for putting the main island of PNG on a western map, app 1526. Northern PNG (an area given by 1870s to the Germans and their companies) was explored by Ynigo Ortez de Retez in 1540 (see Beehler & Laman, 2020, p. 40). The Spanish navigator Luiz V. Torres sailed into Milne Bay (Southeast PNG). By then, '*Papua*' was referred to by the Portuguese and Dutch as the 'land of the fuzzy hairs' (Cousteau & Richards, 1999). During changing ownerships, Japanese occupations and world wars, it took until July 1, 1949, when the construct of PNG was founded and in 1975 it was pushed into independence to the world scene (Baraka, 2001; Hawksley, 2006; Somare, 1975). This was arguably done widely too late, setting PNG on a bad path it virtually never can ever catch up (Baaka, 2001; Gosarevski et al., 2019). Many groups in PNG itself did not even wanted independence, but it was cost-effective for Australia (details in Barak, 2001).

PNG is an island state, and it would compare with patterns found in other large area islands such as Japan, Borneo, Malaysia, Greenland or Iceland let's say, but PNG remains very unique. On the second-largest island area in the world, PNG features at least eight major geographic regions (Sepik, Huon, Border Range, Eastern Ranges, SE Peninsula, Trans-Fly. and islands; Flannery, 1990, Alcorn et al., 1993, Beehler & Laman, 2020; see also Wood, 1976). And PNG administrates them as 19 provincial

Fig. 1.3 Much of Papua New Guinea comes out of the deep forest

administrative units covering world-class biodiversity in a rugged and remote terrain with more than 700 languages (e.g. Wurm & Hattori, 1983). Which other nation in the world has such a complexity and can realistically be governed well, e.g. all done in one operating language and still being effective across public tenure (Baraka, 2001)? The nation that comes close perhaps is Nepal, and Nepal is widely driven by Maoist politics coming from India, and with a strong Chinese oversight from the North (Vanaik, 2012, 2015).

As there was no real struggle for independence, PNG is actually relatively benign on the political ideology of things, and it remains to be very sustainable with a next-to-nil ecological footprint. The PNG governance is widely debated for its performance, it's widely judged to be on the failing side, e.g. Gosarevski et al., (2019), Kurer et al. (2019). But relatively speaking, the biodiversity conservation and its achievements are among the best resources and role models in the world left. That presents a great unresolved conundrum though, because the best-claimed organized nations tend to entertain a rather poor environmental and sustainability record; not so PNG. Clearly, being well administered is not in favor of remaining wilderness (see justification with Ecological Economics, Daly & Farley, 2010; examples provided in Huettmann & Young, 2022).

Fig. 1.4 Pristine forests, clouds and rugged terrain - with a few trails- dominate much of Papua New Guinea and its culture

Eventually, PNG was to be given a governance structure on the international stage so that the global '*business as usual*' could be continued for income (Baraka, 2001). The natural resources in timber, fish and mining lay pretty bare in PNG—to be 'developed' (=extracted for market)—but the new construct allowed for further and template-style exploration all falling back directly and by coincident to the colonial powers, essentially seen as their investment right. PNG then roughly followed a federal state structure, similar to the Ancient Greek ideal but with a rather modern, British and Australian, spin to it. It allows to look good but keep exploitive structures (Baraka, 2001) while complex PNG as a nation was hardly viable (Gosarevski et al., 2019). Like applied in other places and at that time (Baraka, 2001), PNG was a simple, quick and dirty policy template—Westminster style for British Dominions—imposed onto over 47,000 years of a remote civilization. This template was used worldwide but celebrating nothing but destruction, neoclonialism with a 'resource curse' (Davis & Tilton, 2005; see Rich, 1994 for monetary policies), knowingly. It further comes with a monotheist Christianity in different, but closely related, fashions and forms—sects—that simply wipes out 1000 s of co-evolved tribal cosmologies (Beehler & Laman, 2020). And that's all done because PNG had many resources for the global community to obtain (see Streseman (1923) for '*Ausbeute*'; German term for ruthless exploitation of any kind; see Beehler & Laman, 2020 for modern and ongoing examples with millions of specimen collections). And thus a debacle was once again in the making. It just followed a repeat from other places in the world (see for instance the Massai in East Africa, Hughes, 2002). The colonial resource model failed so much elsewhere (Czech, 2002; Lines, 1999 for Australia), and here it did again.

And so in the 1970s PNG was part of the ongoing global wave of setting up so-called independent nations, e.g. in Africa, in nearby Fiji, etc. And thus, after 60 years Australia eventually had to let go of its former 'colony'—a protectorate-, as well as its protective attitude and culture toward its now fully recognized 'independent nation neighbor.' Officially, PNG is now run by the Queen of England as part of the British Commonwealth, but with decisions made '*by the people of PNG*' in a democratic parliament in the capitol of PNG, Port Moresby. Big Brother Australia remained the local super power though; it did that through power tools such as 'Australian Aid' (https://www.dfat.gov.au/geo/papua-new-guinea/development-assistance), policing, advisers (Gosarevski et al., 2019), use of science and by steering or 'educating' and training future leaders (examples for the PNG Prime Minister with Micheal Sumare and the biography of Chan, 2016; see subsequent chapters in this book).

Textbox 1: Federalism versus Tribalism in modern PNG

The western world, for a global governance, hails 'Federalism' as the golden standard. It's meant to be a success story in democratic governance so that everybody gets involved and gets an equal share of power with a great outcome. Nations—as we know them now—are usually designed in a federal structure, with states or provinces within, smaller regions, embedded in an overarching federal agency framework, as a set up and as supported by the supreme court and laws. It's meant to be a partnership between the federal entities and the states/provinces, with a supreme court overlooking the process and settlements. Most nations are set up and organized that way.

The royal courts got sidelined, and a parliament—power to the people—is essentially to drive all decisions. This is fueled by state funding, almost always based on a taxation scheme, e.g. income and/or sales tax. In some instances, a bas governance income comes from industry, e.g. oil, mining. That sets up the western world and its own understanding of law and order, the world governance primarily driven by extracted and manufactured resources.

Such a concept is to assure human rights and fairness for everybody; a bright outlook to a better life and world.

As the Federated System is widely based on power sharing, usually arranged between the state and the federal governments, there is an intended friction zone. At this friction zone there is meant to be a debate and creativity, to resolve the current problems that occur between the state and the federal governments in a mutual way. In reality, those are usually addressed through court cases though, the debate might initiate a topic (so does civil disobedience), but till this is resolved—and resolved well—it might take 20 years, or longer. Many problems by then have moved on, main actors are dead, and the win might be rather empty-handed with other more pressing problems elsewhere.

There are also several flavors and cultures of the Federated System as such. For instance in France, the Federal Governance is much dominated by a central federal governance—in Paris -, relayed top-down to the rural areas. In the U.S., the states instead are rather powerful though controlling policing, death penalty and drivers' licenses (which links to taxation identification, surveying, etc.).

The role of Supreme Courts is equally of relevance as they are to resolve any problems; but not all issues are even addressed in the Federal System. A typical example of failure would be cross-border issues like air pollution or migratory species and human immigrants.

Arguably, the federal system is about power sharing and power balance. How well that is achieved and done remains an ongoing discussion and is somewhat unresolved. Perhaps the western world currently faces a deep crisis on that matter.

What is certainly not unresolved is the topic of resources available in a finite state. All those resources, including land, are meant to be shared—democratically—with all its citizens, but in reality are not. The inequity of wealth, or of land, or access to human rights speaks to that problem clearly. This puts major constraints on the performance of the Federated System and resources available for all citizens.

Arguably, there are other forms of governance. Most dictators promote those views, and indigenous people have other views, so have other cultures. The French culture for instance promoted a reform away from a royal court—as made famous by the French Revolution. Many more examples and nuances exist in the over 200 nations of the world.

One may argue how close this concept of democracy really is to the ancient Greece (many variations exist in ancient Greece and its nations also, nor have the Greek really invented it in the first place; good but very old governance examples are found in China and Vietnam or in Latin America and Africa for instance). Clearly, the nation of Greece itself moved away from democracy, having had a royal family, a civil war, and now embedded in the EU framework but with a huge debt- load dominated by Germany. If Greece is such a great democratic role model, why is it not in the lead or a showcase?

Federalism and governance get even more interesting and problematic with modern developments such as globalization, and international corporations acting 'like a nation' but not approved by national citizens.

This matters for PNG because it was built under the construct of Federalism. It's meant to be a new modern nation. While the tax system was less strong—PNG is not much aware of a tax registration and subsequent income to be devoted to the tax and nation (e.g. Baraka, 2001; Chan, 2016)—other income streams were to be used, e.g. mining, forestry and some fisheries. Australia was to help and guide, the Prime Ministers were to carry it out. However, the reality after over 50 years, 2022 onward shows us the ugly truth of such a designed Federalism; it widely fails (see Peck, 2014 for such a role model and its problems).

Virtually all nations running a federal scheme of governance are in big trouble, socially, economically and certainly environmentally. It's a scheme that eats resources, adds to climate change, and it impoverishes people. The Federated System by the west primarily relies on a pension plan to award life-long job performance for stability, but which underlying resources are widely used up by most nations, social and capitalist alike. When the pension plan and associated concepts fail sustainability is virtually not possible,

In recent times, arguably, most powerful nations are not sustainable. They obtain their resources from the outside. Such federated governance systems are essentially then just subsidized from the outside; they live on the cost of others. This provides an impression that the federate system works, where it actually is not. Typical examples can be found with cheap labor to make an economy work, e.g. provided by immigrants or undocumented workers. The economy in the U.S. relies on such labor, so does Germany as the leading nation and powerhouse in the EU; see German Greencards for IT Specialists from India (Werner, 2002) or industrial "Gastarbeiter". Other examples can be seen with the use of energy products, e.g. oil or coal, which create a great nation that otherwise would hardly exist. Once more, the EU is a good example for it as it lacks most rare earth or any oil. For most of its time, Norway never was as rich as it became through oil of the last decades.

In such a world and system, the claimed pursuit of liberty and happiness is not achievable for all. As a matter of fact, many federated governance systems do not work well, and are highly inefficient, as can easily be detected when looking at metrics like inflation, unstable currency, taxation details or human happiness.

While democracy, and western style governance has been heralded as a sometimes frustrating but only realistic governance scheme, this must be put into doubt with man-made climate change (see Melick 2010 for PNG). And many of the global problems that citizens of this world are facing now after 60 years of an unprecedented feast eating up all resources available in the world. What the future has to offer remains dubious, but unlikely the western democracy can cater all of it; the track record speaks for itself as the largest destruction of natural resources, of live, of mother earth, happened coincidentally just in the last 60 years or so.

The so-called federated governance will stand out over time as an inherently poor scheme to achieve sustainable and subsequent happiness.

Instead, tribal governance—in all its facets including sorcery (Beehler & Latam, 2020)—prevailed in PNG for millennia (see Hillison, 1993), ~98% of its time and the nation remained sustainable. The set up of PNG as a federal state is described by Prime Ministers Somare and Chan (2016) as two of its architects and with first-hand perspectives. It's covered further in this book, and authors are asked to see details there on the PNG experience.

Time will tell, but as currently practiced, federalism looks pretty grim for the age of enlightenment, century of reason, of sustainability, of science and the online society. While new sustainable governance options are few, tribalism remains.

Textbox 2: The Bilum: Better than any plastic bag can be

'The Bilum' is many things. The first bilum I saw in action—realizing its value—was somewhere on a remote trail in the Bismarck Range. Initially, it looked to me more like a bad rucksack from ancient times …Well, until my own slick rucksack failed after a day and the bilum easily lasted for the entire trip under very rough conditions 'just fine.'

Bilums are used as a bag. They can carry sweet potatoes to a remote market, as well as camping gear to a homestead. Imagine to do so in a muddy place, 100 km away from the actual market and the village, in a rainforest during rain carried by kids, women and a man along the trail.

The bilum is a perfect tool for a world that has little material wealth. You can then fit all your belongings into a bilum. Imagine such a life; simple (and totally sustainable because the bilum is usually made of natural materials, e.g. tree bark!).

But a bilum can be more than just a day pack, as it can be a piece of pride and wealth. Done through weaving, e.g. using best suitable local plant bark or root, or now plastic fibers intertwined, bilums turn easily into art. Every locality has a certain bilum style. It's not described well nor well tracked and documented. Bilums live, they sit in the hands of the people, literally. Bilums are usually made by females and children. People in PNG do not have sheep or coats, and thus no wool really. Bilums are local and truly sustainable. There is no plastic waste, and no marine turtles or whales are killed from 'bilum waste,' nor is there any chemical plastic contamination or outwash, no micro-plastics neither. A global problem solved.

Bilums are usually tasteful colored. It makes for well-liked souvenirs, and should be an alternative to your common plastic bag now found all over the world, including the oceans where they often end up as a dump. Arguably, if you have 'less stuff' you will need less to carry and to move around, and a bilum shall do.

Consider the bilum a local adaption to the conditions in PNG, and a well-tested life item over millennia. It easily outcompetes any plastic bag I know of having an environmental impact next to nil.

Photo of Cool Dudes with their Billum: Photo by Author

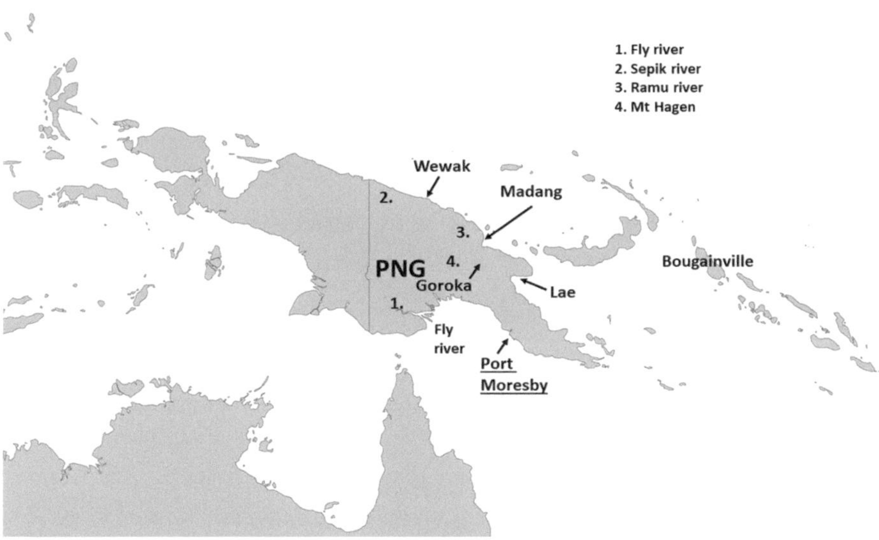

Still, PNG is of world interest, as presented by Flannery (1998, 2002), Diamond (2011a, 2011b) and others (e.g. Matthiessen, 1987, Cousteau and Richards 1999; see also movie 'First Contact' (Highlands Triology) https://www.youtube.com/watch?v=2Y5rC7kDx3o).

Already the exploration and initial state of PNG is quite well known, documented and a globally appreciated narrative, e.g. the HMS Rattlesnake Journey to Papua New Guinea 1849 (State Library New South Wales, 2022), the famous and influential Leahy expeditions (Leahy, 1994), the 11 Archebold expeditions (Cookson, 2000) or some less-known German, Japanese, Italian and British ones, including modern ones (e.g. Hoffmann, 2015; Salak, 2001). And so the '*mudmen of the highlands*' – among others - became a global icon for PNG (Cousteau & Richards, 1999): a globally recognized link with the deep past of the Earth for a virtually not understood human culture and its proceed. The Fly and Sepik rivers offer those links readily accessible, but it got exploited for mining and for tourists alike. Or see Flannery (1900, science photos p. 119, 134, 126, 216 with nakedness and penis gourds) exploited for fame, also shown in Martin (2005, p. 13). Such icons of humanity got widely perceived by the western onlookers as a primitive, if not even dirty but certainly very scary and savage representation of mankind and a certain inferiority (Hoffmann, 2015; Levi-Strauss, 1966; McFarlane, 1889; Matthiessen, 1987; Pratt & Pratt, 1906; Richardson, 2005)—they all add to the public PNG image and global perception what the new nation of PNG is to be. Visiting their Dominion, the British Crown still struggles with presentations of bare-breasted females during rituals when in PNG (examples can be found from the Queen to Prince Charles and Kate Windsor up to Meghan and Harris). But in that presented public image, nobody really saw within the 'big men' and females from the tribal and Wantok society in rural PNG acting today in parallel, and actually being alive and overruling modern globalization (see for instance Harari, 2015; Stiglitz, 2003 for globalization ignoring most people, 'the tropics' and PNG). That thought did not enter the media yet but it's all dominant in PNG. PNG remains widely tribal.

Beyond Facebook, in PNG the deep times remain in the driver seat stronger than ever before (Cousteau & Richards, 1999, p. 206; Beehler & Laman, 2020).

1.2 From Deep Time Until Just Very 'Recently': Over 47,000[2] Years Passed by and not Much Happened to PNG, or to the Earth!

PNG is part of Melanesia (Diamond, 2011a, 2011b; Beehler & Laman, 2020). The PNG part of the New Guinea island is part of the wider Sahul region (Flannery, 2002) and it has actually been settled for over 47,000 years (O'Connel & Allen, 2007, 2015).

---

[2] The term and concept of '47,000 years' old is used here and throughout this book in the sense of the Sahul region (Australia and PNG) which is likely being occupied by humans overall for over 60,000 years (Flannery 2002, Beehler and Laman 2020). Clearly, not all of PNG has such a time record but people did move and many waves of human settlement and cultures were found in PNG, many remain unnoticed to this very day still. It's a generic term and concept for 'deep time', with human generations and usually much older than the 'modern' European history (which often refers back to the Ancient Greek, c. 3000 years ago).

That makes then for more than 9% of the time before the western world contact (~1526 A.D. officially; Flannery, 2002; Beehler & Latam, 2020, p. 40 for more details). But in earnest, most of PNG just got closer exposure to 'the West' as late as during the 1930s (see Cousteau & Richards, 1999 for Australian Leahy brothers, etc.; Wilson, 2019 for first-hand account). While this is not taught much in public schools and science centers of the western world, for most of the human history on earth PNG had already a major civilization in place that operated in the region for a very long time and in a more or less globally sustainable way (Diamond, 2011a, 2011b; Flannery, 2002), transferring much of wild PNG into a multi-use garden state.[3] '*Gardening*' kept PNG a livable place for millennia (Diamond, 2011a, 2011b; Flannery, 2002), mostly with a co-evolved landscape and species set up that we experience today! Likely the PNG society removed trees from the highlands, planted new ones and turned it into grasslands and sophisticated gardens (Cousteau & Richards, 1999; Beehler & Laman, 2020 for examples). Tree kangaroos for instance got wiped out in adjacent Australia but less so in PNG (Dabeck et al., 2020, also discussed in Beehler & Laman, 2020). Human society in PNG simply lived its live for millennia, but for what western explorers, colonialists and writers called a '*stoneage*' (Hoffman, 2015; Matthiessen, 1987; Rick, 2020) and '*primitive*' (Mead, 2002 and citations within). And it centered around the practice of gardening, fishing, bartering, as well as a nomadic but sophisticated lifestyle around seasonal resources, and associated peace and warfare (Diamond, 2011a, 2011b). Warfare received a lot of attention in PNG Anthropology, and fighting was often a way of life for them (Flannery, 1998, Cousteau & Richards, 1999, Richardson, 2005). But it did live quite fine; at least until the last 300 years, or so, when the western people arrived in 'bulk' and made top-down changes to this lifestyle, including introducing 'foreign' food species like potatoes, and later, coffee (Flannery, 2002; see Baraka, 2001 for 'cash crops' impacts). Betel nut, pigs and dogs were likely introduced a bit earlier, likely from Asia (Beehler & Laman, 2020 and citations within).

Many things in PNG's colonial history went wrong and it continues to do so (Connel, 2005; Gosarevski et al., 2019). Already the name '*Guinea*' is based on a misconception by the earlier explorers equaling an Asian place with African Guinea (a location the Portuguese and Spanish were familiar with due to proximity with their Atlantic ports, but which is literally on a totally different continent and ocean; Beehler & Laman, 2020 for details). And so the misunderstandings continue on that wrong footing… with nature conservation paying the price (the faulty concepts of protected areas and national parks are discussed for PNG in subsequent book chapters).

The Western World simply cannot really wrap its mind around nations like the one of PNG, but which they essentially created as an administrative governance unit to save and make them money. And thus, as a matter of fact, just the last 50 years—when a western-driven form of globalization dominated PNG and steered by Australia, the

[3] Compare with the New Jersey 'the Garden State' in the U.S. https://www.state.nj.us/nj/about/facts/nickname/ and its poor environmental status. Such western approach to garden landscapes are far from sustainable whatsoever.

Fig. 1.5 A forestry trail at the Huon Peninsula that was likely in use for over 47,000 years connecting coastal areas with the highlands (Figs. 1.6, 1.7 and 1.8)

western powers—and later Asia- were among the most devastating ones for PNG (as will be shown later, the colonization and WW2 had a major impact, already when just judged by diseases; Diamond, 2011a, 2011b). The very inability and set up of the 'PNG construct' destroys PNG and other aspects alike; it's not viable much in itself (Gosarevski et al., 2019) and it even affects the global climate further. Needless to say that the colonial powers have a long track record of wiping out their own resources with virtually no legacy of a sustainability trace (see Diamond, 2011a, 2011b), or 'sustainable development' (Rich, 1994) as it is now widely claimed and promoted by the UN via Sustainable Development Aichi Targets (SDGs https://www.cbd.int/aichi-targets/) but without relevant progress beyond the wording itself, e.g. global climate change or the world's ocean crisis. It's fair to say that whatever those colonial and western nations got in their hands to administer, they tend to destroy it (Rich, 1994, see Limburg et al., 2011 for a fisheries example, and Cockburn, 2013 for the U.S. as one of the richest nations in the word). It's done by the governance and its leaders, including their mindset and religion (as part of that very governance scheme). It comes as a package and it's what the western societies does: 'destroy'. The western legacy on those matters is wide and deep (Aceomoglu & Richardson, 2013; see also Diamond, 2011a, 2011b; Blum, 2014; Harari, 2015). There is little sustainability to show and to see (Table 1.1).

Far from the '*noble savage image*' (Flannery, 2002), several events in PNG changed the initial so-called '*stoneage balance of millennia*' (i.e. the social, the environmental and the economic balance and co-evolution). We do not really know

Fig. 1.6 Sustainable rural village life as a core pillar for Papua New Guinea for over 47,000 years

much of the impacts of contact from earlier times and other cultures, e.g. Lapita and other waves Asia entering coastal New Guinea and northern PNG islands (Beehler & Laman, 2020 and citations within). But likely those contacts were resulting into some specific farming skills and certain improved nets and fisheries, assumingly changing the coastal and ocean set up (Jackson et al., 2001). Documented changes

Fig. 1.7 A typical Papua New Guinea Garden, the essence of sustainability on a global level

came after the contact from the sixteenth century onward, the increased appearance of the white people and their society last 400 years from several royal courts, introduction of diseases, potatoes by the Portuguese, missionary structures fighting indigenous believes and introducing further goods, aided by the onset of new but aggressive business models in the colonies, in PNG and with the global market overall (Diamond, 2011a, 2011b). This global market actually started many hundred years ago, likely before Christoph Columbus, certainly before James Cook and was enforced by the French court and its mercantilism and subsequent industrialization and colonies, e.g. in Africa, North and Latin America and India. And by now, the advent of the computer and the WWW have not markedly changed it, or made it more sustainable, but made it more intense and worse (e.g. Bowers, 2011).

Beyond the colonial contact and subsequent impacts, a real game changer then followed with a 'modern' governance, the federal nation set up, and its laws and courts, as well as its tried deviation from tribalism. It was tried to move into 'modernity' trying to abandon the 'Wantok' tribal systems and its village courts (see Gillison, 1993 for field examples). Whereas, nowadays, tribalism is still in the driver seat (Beehler & Laman, 2020) and unlikely to go away any time soon. With that, PNG is essentially independent by itself and quite resilient to globalization and its governance (see West, 2006a; Chan, 2016 for many examples) (Figs. 1.9, 1.10 and 1.11).

Fig. 1.8 A group of great friends

Table 1.1 Selection of Colonial Nations to PNG and their own state of the environment and sustainability, indicating poor and humble eco-leadership *(sensu* Ludlam, 2021)

Colonial Nation to PNG	Environmental metric	References	Comment
UK	Bird decline	Maclean (2010)	Birds are a flagship species group in the UK and vastly recognized as declining, also promoted by D. Attenborough and others
	Mammal extinction		Many large mammals like the Giant Elk, Moose, Aurochs or Walrus or many Whale species disappeared already 1000 years ago
	Fishing	Hutton et al. (2008)	Ongoing fishery crisis in the EU and beyond
	Climate Change	Stern report (2010)	A classic study, global outlook
Germany	Whales	Tønnessen and Johnsen (1982)	No large whales anymore in Germany and the North Sea bight
	Riparian Forest		Most of the 'modern' water management has resulted into managed water ways, usually ignoring value of natural flood plains. Those forests are gone for over 200 years already
	National Park buffer zone size		Germany can virtually not fulfill the IUCN size requirement for National Parks and their buffer zones
	Genocide	Moses (2008)	Mass killing of jews, roma, synti, etc Neonazis are still strong in German society and represented in the German parliament, including EU parliament seats
	Climate Change		IPCC failure, a global issue but driven by major industrial nations

(continued)

Table 1.1 (continued)

Colonial Nation to PNG	Environmental metric	References	Comment
Holland	Forest Cover		Holland is heavily populated, and the impervious surface is on the rise; forests have little space left
	National Park size		Holland can virtually not fulfill the IUCN size requirement for National Parks and their buffer zones due to size and people constraints
	Genocide	Weiner (2014) Anderson (2015)	West Papua, Indonesia and Timor have a very violent past and political aspects, as part of the Dutch governance and colonies
	Climate Change		IPCC failure
Australia	Forest Cover	Lines (1999)	This topic received a fresh debate with i recent forest fires
	Genocide	Ludlam (2021)	A major item in Australia's history and q widely discussed topic for betterment
	Climate Change	Ludlam (2021)	See Forest Cover Man-made climate change is widely debated and little progress is made to reduce greenhouse gas emissions, including methane from sheep
Portugal	Civil war in Portugal and former colonies	Strippoli (2016)	Often referred to as a colonial war, e.g. on Mozambique and Angola
Spain	Civil war and separatist movements (e.g. Basque, Barcelona)	Richards (1998)	A major part of Spain's history and still ongoing
	Fisheries crisis	Raakjær (2009)	Much of the EU Fisheries is driven by Spain, this includes efforts within the EU and outside, including PNG and the Pacific

Fig. 1.9 Local food processing of pandana nut in the bush; perhaps it contributed to Papua New Guinea's widely celebrated but mis-perceived cannibal legacy?

Fig. 1.10 Building and architecture in Papua New Guinea; those adobe buildings are earthquake resilient; consider PNG and its geology as part of the fastest rising ranges in the world

1.3 The Zigzagging Path of 'Modern' Development Toward Progress, Chaos or Death?

Papua New Guinea lived quite well for millennia and then became a divided colony of three western nations (Holland, Germany, UK and helped by Australia) for over 150 years. PNG literally brought very remote nations together, e.g. Australia was to share a border with colonial Germany! Portugal and Spain played initial roles also and were important for subsequent island treaties. This power game got perturbed, interrupted and eventually ended with the world wars. For any onlooker—and certainly PNG citizen—that must be confusing to watch and how it all unfolded on them. While WW2 was the stronger, more bloody, war propelling PNG into the industrialized western world, WW1 was actually the one that broke up the hundred-year-old colonial structures in PNG and re-arranged and unified it (there are many interesting facts about WW1 itself as it was 'predesigned,' quite naïvely perceived initially and meant to be the BIG WAR and 'the one and only' that ends all wars and within just a few months, conveniently to be over by winter; see for instance Effenberger & Macgregor, 2016). Initially WW1 was not so much considered to be even of relevance for PNG or of long-lasting global repercussions. But that view changed quickly…). Despite being already in existence for over 47,000 years, PNG was 'founded' after

Fig. 1.11 Bartering, e.g. for local garden products as shown here for betel nuts, are a core activity in Papua New Guinea. Betel nuts act as a currency throughout most parts of the country

the WW2, on 1st July 1949. Here then comes the attempted western construct and set up to PNG for a big misunderstanding, if not failure (May, 2006).

While New Guinea was Portuguese and Spanish, the eastern part developed a bit later with the colonialists. While that part was then initially claimed by the Dutch, and the northern islands and coast by Germans, the British still had some oversight and the approval power. The Germans developed a strong grip on Northern PNG, Finschaven, Bougainville, Solomons, etc. (Knoll & Hiery, 2010). A German New Guinea company set up shop and had a firm organization on the local villages. The capitols of 'German New Guinea' were Finschhafen, Madang, Herbertshöhe and Simpsonhafen. Catholic and Lutheran Missions unfolded in parallel. The island coconut colony cult by a German 'Hippie'—August Engelhardt—became world famous (for public knowledge see https://en.wikipedia.org/wiki/August_Engelhardt). But it was an artifact, all just fueled by national economic interests serving the emperor and spear-headed through Bismarck running the copra and coconut trade, with area claims of an emperor-backed German company, wrapped into religion as well as even the German vegetarian movement and its society (Bernard, 2017, see also Kracht, 2015). Back home, the core port for that exclusive German PNG trade was mostly Hamburg (less Bremen, and Kiel at the Baltic side)—as part of the established Hanseatic shipping culture. And to no surprise the islands of Northern PNG were quite shippable and due to Spanish-German treaties then became German territory; the proximity to

Fiji and Hawaii, with a German base (see also for 'Queen' Emma (Forsayth) and Paul Kolbe on nearby Kokopo island, New Britain), were in full support of that new empire.

Like with many of the colonial leftovers, the German footprint in PNG can be seen in many examples till this very day: *Mount Wilhelm, Weyland Mountains, the Bismarck Sea,* the '*King of Saxony*' paradise bird (*Pteridophora alberti*) and *Matschie's* Tree Kangaroo, etc. (see Beehler & Laman, 2020 for such naming biases and how done in taxonomy to please royals for getting funds and fame). But while the Germans essentially sat at the coast with easier access to resources, the actual inland and interior remained widely untouched by them and was widely unexplored by westerners overall still. Some thought it to be unpopulated, until the Leahy expeditions showed it otherwise (Leahy, 1994) and so did subsequent British and U.S. expeditions in the areas in the 1930s and 1950s (Hoffman, 2015 for Rockefeller; see Beehler & Laman, 2020 for high-density population centers, likely due to introduced pigs, trees and sweet potatoes several hundred years ago).

The Ramu river played a role for contact, but along most of that coast only the frequently used trails between the highlands of interior PNG and the coastal zones allowed for some western influence there! Indeed, a narrow but rather relevant trading trail with the 'outside' world (Figs. 1.4 and 1.5).

By the time of western contact for New Guinea, the Dutch had already assembled a world mapping center, e.g. Mercator (Zandfliet & Harvey, 1992). It allowed them to accumulate detailed strategic knowledge for navigation and exploration. And so, they had already strongholds in Java since the sixteenth century with the East India Trade Company (Rivers, 2016) and thus reached out rel. easily to PNG and its western

Fig. 1.12 Caved art from Papua New Guinea ranks very high, internationally

Fig. 1.13 Papua New Guinea grasslands are another cultural feature of the nation

parts and islands. Holland was well suited then with a ocean access and a Royal court in full support, and at that time it had some deep experience with international shipping, trade, with Asia and the tropics overall. At that time they were among the world leaders in such efforts. And thus, to this very day this spatial pattern of their exploits can be seen in the cities, ports and boundaries of Irian Jaya, Indonesia. The Dutch Trading Companies were actually the first truly worldwide operations and somewhat started western globalization, as a business model and with royal court authority (as they paid them a share to obtain that power; a typical royal scheme). And New Guinea, with the PNG part, was made an integral part of that scheme, by land and by trading route. The driving Dutch Royal Kingdome at that time was very sophisticated, but again, it was imperialistic overall and well thought out and planned that way. There was little consideration of PNG people and forced labor was the concept of the day; theft of land and resources was the other.

Europe lived well from the resources abroad. While the PNG colonizers changed somewhat, for PNG not much has changed in those globalization dynamics to this very day.

When the WW1 broke out, those colonies crumbled and were essentially handed over to the winning world powers. The winner takes it all, whereas PNG and its people had virtually no say in it, again. They were bystanders in their own nation and in a war that they had not started, not agreed upon and nothing to do with really; instead, the core disputes sat in Europe and with their related royal courts and families. And

PNG citizens could not have a say or intervene at those courts, as PNG was so far removed from what was said and where it was said, and how it was said. Core zones in that set up were located in London, with the English, a few European capitals and nations, their culture, their laws and its inbred clique of royalty, of self-declared emperors and subsequent ministries, agencies, expert advisors, the science and the money and subsequent banking powers (Effenberger & MacGregor, 2016).

The legacy of those colonial powers can be found still today. The case of the lost explorer Michael Rockefeller in northern PNG—as described in Hoffmann (2015)—showed clearly that the Dutch coastguard still played a relevant role in oversight and rescue for PNG and associated waters, e.g. due to independence struggles of Indonesia (Colonial Holland let Indonesia go independently in 1962 as a wider colonial debacle; Penders, 2021).

And to no surprise, the borders of PNG were set accordingly by the international community, following a 'silly' rational and arrangements (Die WELT 2021 for an example and global review). As found elsewhere in the world such divisions are the direct outcome of the imperial model, of imperial family deals within Europe, and those have their global effects to this very day. Once set in stone, how would one ever come out of it in a good way? An ordinary PNG citizen cannot, nor can the concept of voting in a democracy. And people tried…

A typical example for the world's 'silly' boundaries are found in PNG, with the islands (e.g. Bougainville), and with its border to Indonesia. Ecologically, New Guinea is one unit (Beehler & Laman, 2020). But the boundary to PNG is a straight line, convenient for some, but with a 'tweek' along the Fly river. While one might say a river boundary makes sense, it really does not. Already the location of the straight line in a river can easily be debated; it's arbitrary and ignores human realities and waterflow dynamics over time. And as for the river line, in such boundary settings one will always have troubles with river crossings, with bridge locations, with river floodings and changes, with erosion and river islands, and with the actual location of the boundary, e.g. located on the other side of the beach, at the water line, or the mid-river line. River boundaries, in real live, remain a headache to enforce and to agree on. That's a global experience, e.g. between U.S.-Canada, or China-Russia, or with the international administration of the Amazon (an international-law shipping route), etc.

In New Guinea that has a real-world outcome. People living in the Fly river region and the border zone between PNG and Indonesia come initially from the same and identical ecosystem and forest even. But they are prosecuted on one side of the border, however, not on the other side. They can escape from Irian Jaya into PNG and turn into refugees on their own land (see The Guardian, 2019 for updates and details). Such boundaries are simply western artifacts and create suffering and refugees that can easily be avoided. It's a colonial leftover and reality.

In that context, of special relevance for PNG are laws of the sea (UNCLOS) and the Exclusive Economic Zone (EEZ). That's because PNG has virtually over a million of islands, and those carry their own territories and seafloor geography and trenches. As will be shown later in other chapters, this matters for fisheries and sea floor mining, besides other things, e.g. trafficking, taxation, citizenship and refugees (e.g. Dastyari and Sullivan (2016), smuggling, navy strategy and offshore trading with fishing nations.

By luck of location/geography, Australia became the assumed beacon of civilization for PNG (Baraka, 2001). PNG was to excel out of the stone age into modernity; all just in a few years (Chan, 2016; Kiki & Cheshire, 1969; Wilson, 2019). And based on their assumed expertise with the adjacent land and aboriginee Australia was trusted with the entire PNG enterprise through a UN mandate (Bakara, 2001). That was conveniently done as part of the British Dominion and its funding machine (= 'The City'; London, which can take-in 'dirty money') linked with international aid and development donors. And thus, the real power agents never truly changed for PNG and PNG, and its people remained delivered to those for good. However, considering that Australia had to apologize to the aboriginal community for most of their wrongdoings (Cortassel & Holder 2008; Mellor et al., 2007; Reuters, 2008)—a scheme that became obvious already over the last 100 years—it shows us already that it was not a good foundation to start with. Considering Australia tried to get rid of PNG as a taxpayers cost item when PNG was the Protectorate (Baraka, 2001) through the U.N. it was not really to be trusted with PNG. And thus, to no big surprise, PNG went downhill from there for a struggle (Windybank & Manning, 2003).

PNG as a nation was offered the world stage way too late (see also Baraka, 2001). It's an ongoing struggle and with problems that can be seen to this very day (see example for a high-level Australia-PNG Scandal in The Sydney Morning Herald, 2010; The Guardian, 2022), e.g. police oversight by Australia for PNG and its court system. The 'Australian Arc of Terror' unfolds (May, 2006; see Lasslett 2012 for Bougainville terror; see subsequent chapters). And from my wider discussions in PNG, I know just a few PNG citizens who liked this blunt interference from abroad.

Bringing this home to the biodiversity conservation of PNG, this nation is the most-species rich island of the world and it provided many gifts to the human society. It keeps attracting taxonomists and specimen collectors, including a vast virtually unregulated trade of such species and specimen (details in Beehler & Laman, 2020). As it will be shown later, this global free-frawl is already easy to track and see in the biology and forestry sector, and an equal situation applies to fisheries, certainly in mining and oil & gas production. There will be little disagreement that the efforts and science works have not at all helped with conservation, humanity or land questions, rather vice versa. So-called 'Experts' like Beehler and Latam (2020) for instance state to this very day that the tree-living Birds of Paradise are not threatened, despite massive forest losses in recent decades (Mongabay 2022) and massive harvest/poaching pressures for those bird skins, and all other species of such sort being overharvested and extinct or close to it.

1.4 PNG as an Experimental Case of a Remote Ancient Society and Nation Thrown into the International Community and its So-called Free Trade Market Scheme: How Does the Control Site Fare?

Papua New Guinea was a problem case either way. Initially seen by Europe as a marginal and remote site, it was a left over and just later got precious for its resources to be used and developed. Soon some record profits were found or promised (e.g. Kare Mountain, Ok Tedi mine and sea floor mining; Henton & Flower, 2007, Kirsch, 2014). And a global demand for those raw products was there and could be built further. It was pure exploitation though of raw resources, and of labor (the labor was from PNG but also from outside, namely China—see Chan, 2016 for details—with expert labor like engineers, helicopter pilots and CEOs coming from Australia, etc.) Worth to mention here is for instance the dynamite license needed for mining and road projects, which is not in the hands of PNG but usually provided by Australian and foreign engineers. That's the classic definition of a Third World country, and of designed abuse top-down where the south pays the prize and provides resources for the (global) north to operate.

And so why then not giving PNG quickly a new structure, one that instead of a colonial dominance leftover rather looked like 'The West'—like a democracy and a federal system in its latest national design—and which then fits into the British Dominion concepts and looks good in the public eye? Money can be made in that process (Baraka, 2001), and liability then sits with the voting citizens of PNG, not the colonial designers and funders or with the Commonwealth. And then one can operate under the western premise of a promoted win–win situation: PNG gets happy and The West still gets the resources and benefits cheaply! In the meantime, the Australian tax payer has not to cover the costs for the protectorate (Baraka, 2001). This makes PNG an entire nation free for the taker. All of it, with Australia as its shining star of leadership of PNG into a modern world delivering the goods, a stable society?

And exactly that's what was tried, part of the wider global experiment: In the 1970s PNG now got a constitution, a court system, ministries, a central bank and a tax system, including a scheme for international engagement (Miller, 1973), a national airline (app 5 airplane routes, some planes are "hand-me downs" from Iceland, etc.; a scheme widely operated by Australia). The university can be added and its library (Lutton, 1981). This process was breathtaking for the speed of such a 'progress' while expertise and a grown culture and experience around those schemes was widely lacking (Chan, 2016; Wilson, 2019). PNG never really asked or fought for independence. Many of those subsequent governmental agency buildings now in place simply come from the mid-1970s, built within same years, and they are found next to each other on the same street—Magani Cres; a rather short street in Pt Moresby, arguably it's all a bit in decay by now (see "***PNG Building***": That is a building in Pt Moresby—actually it has a double meaning because PNG is a building and PNG is still being build—https://en.wikipedia.org/wiki/National_Parliament_House,_Port_Moresby) (Fig. 1.14).

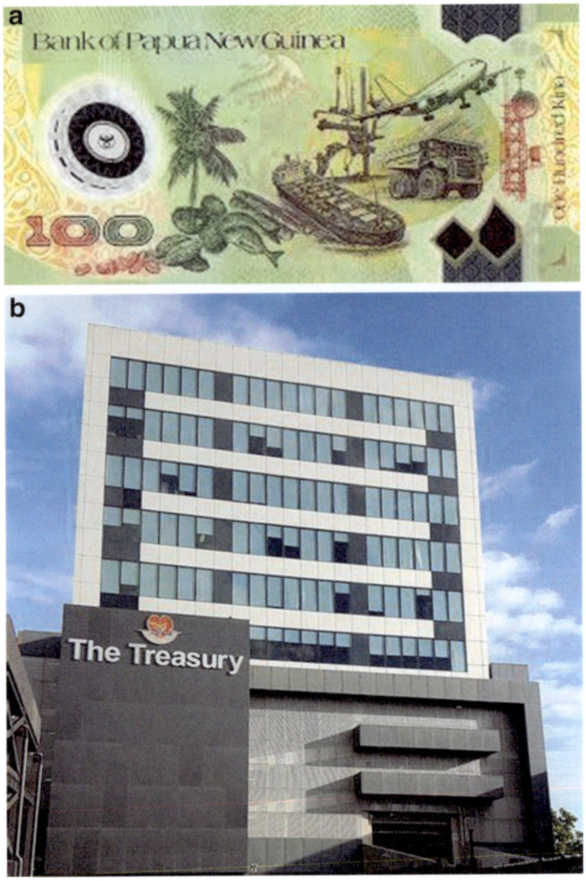

Fig. 1.14 **a** Papua New Guinea's currency, **b** the center of the modern PNG operation: The Treasury in Port Moresby

Because most of PNG's people never really asked for independence or a nation (Baraka, 2001), some deviations from a typical western, free and independent nation existed in PNG though. The PNG currency was tightly linked to the Australian Dollar; a reality to this very day (see Fig. 14a for the design of the actual bills following an Australian design; see Fig. 14b for the PNG Treasury). It was even printed in Australia; can one have a stronger statement of non-independence? PNG also got its military supplies and army training primarily from Australia, in exchange for good money to be paid by PNG to the supplier. And the PNG markets got heavily controlled by the Australian and Western nations -including Japan- and their actors anyways, by design. Already the airline and tourist markets are good examples for that. And now many of those features are controlled by Asia. PNG hospitals lacked the expert doctors and nurses and thus got them supported, trained and supplied from…again Australia and New Zealand (Chan, 2016). Other nations are also happy to jump on that development and aid train because it makes them look good; it's a low-hanging fruit and presents a typical colonial power structure top-down (see Rich, 1994 for

The World Bank). And schooling - as a key industrial scheme - presents us with a similar matter: The schools remained under strong control by the church; a certain conglomerate of Australian and Christian faiths and hardcore sects from the U.S. and the international scene (Beehler & Laman, 2020 for details).

Over the years, each time when in PNG and Port Moresby (POM), I looked hard for the Papua New Guinea Academy of Science, but found none though. So what a society is Australia really set up to build, and for whom really?

And just looking at the public media in PNG will follow similar patterns (Rooney et al., 2004; see also Foster, 2002).

It's easy to see that PNG lacks many essential aspects of a 'modern' nation; certainly the success aspects. Arguably, the independent PNG was not independent whatsoever to start out with, and it was not up for a good start, and did not recover from it well neither (see references for a failed nation, e.g. May, 2006; Tulkiewicz, 2006). PNG is not to blame for it, nor its people. It's not fair on them. PNG is primarily just a set up by western design to look good (Filer, 2004), a façade with much veneer for the global audience and its funders to obtain certainty and resources from PNG while the global public would not recognize it; PNG citizens are to be happy and not involved. It's window dressing at its finest. There are very few of those efforts to check and to expose reality (assessment by Gosarevski et al., 2019; see Henson & Flower, 2007; and see The Guardian with regular publications in their Pacific Project and supported by The Judith Neilson Institute; many citations are used in this book in support of evidence known by the author).

1.5 PNG as a Proclaimed Failed State, and Thus, a Failed Western Experiment

In modern terms, PNG's economy now roughly consists of agriculture (35.2%), industry (38.3%) and services (26.4%); truly precise numbers are hard to obtain. But in PNG these numbers speak little about PNG life and realities. PNG is far from an industrial nation, and it lived essentially a double-live: Set up now as a western modern (federated) nation concept—Westminster style—and then, an increasingly left-alone and ignored (but thus independent) PNG way of life, which is primarily a tribal live similar to what it was with early contact and if not even before (Diamond, 2011a). Most readers might not know that rural people only obey really the word of the '*Big Man*' (Cousteau & Richards, 1999, p. 206; see PNG president, Chan, 2016); and sorcery and shamans still play a big role in decision-making (Beehler & Latam, 2020; examples in Gillison, 1993). So what is the value of a federal government then?

The latter situation is the PNG of today, and there is little outlook for a change. The western experiment might appeal to some people, for modern nation-makers and leaders of the federal theory as well as missionaries perhaps, but it's not really and truly what happens on the ground in PNG and its rural society.

PNG's modernity simply stalled after a few years of PNG's nation creation by the west (stated first hand by its leader Chan, 2016). It might have started well with the PNG Central Bank, but then this stagnation and 'halt' moved into the PNG parliament and leadership, and from there into the university and the public school system, into Port Moresby municipality and into PNG overall. Australia witnessed and saw it first hand for decades, trying to help with aid, development and band-aid approaches like sending judges and policing efforts (e.g. Gosarevski et al., 2019). But the modernity is not so obvious in most of PNG.

Whatever The World Bank promotes (https://www.worldbank.org/en/country/png), PNG remains to be classified as a failed state (May, 2006; Tulkiewicz, 2006). It was not really able to pull it off, and people realized it—in PNG and from the outside (Fowke, 2006; Gosarevski et al., 2019). Of course that made biodiversity conservation and sustainability virtually impossible. It was obvious for a long time; faculty left the university in Port Moresby as a clear indicator (academics are to leave bad political systems first and subsequentially expose the case; a common global phenomenon, see for instance the classic case called "*Goettingen Seven*" by Strack 2017 with an example). The university library essentially moved to Perth (Lutton, 1981).

PNG cannot be a better version of Australia itself, or of the U.S. democracy for that matter, or beat the global market dynamics. Most nations in the world, certainly the western ones, have a widely failed Central Bank policy themselves (Stiglitz, 2003, e.g. Italy or Luxemburg, or U.S. itself). The so-called PNG failure (Duncan & Chand, 2002; Laurence et al., 2011; May, 2006) is nothing but a failure of the self-promoted western world and its institutions and attitudes itself, namely colonial and imperial minds and faked progress and a widely failed modernity and globalization. Just because a nation wins most Olympic gold medals and flies to the moon, like Russia, U.S. or China, it does not mean it's a great and world-leading nation. As a matter of fact, western style imperialism and its business model does not breed so well in PNG, hardly anywhere else (De Soto, 2000; see Elvin, 2008 for the Chinese experience and its environmental legacy, another megadiversity nation with a global dominance in high debate).

1.6 PNG as the Happy and Willing Resource Provider and Delivery Platform to the Outside World: Just another Failed Idea, Market and Mindset

While wild and vast, PNG and its world-class biodiversity reservoir (Beehler & Laman, 2020) and mining resources are not like Russia, U.S., Congo or South Africa, the Tibetan Plateau or Australia for that matter. Island PNG is sensitive, not endless, has little political power given, and instead its resource development and exploitation come with a prize. Most of the land is in public lend tenure, as the case for most of Melanesia (Lea, 1997). To operate well in PNG—on a finite space—takes expertise, money and skill as it happens here on a small land mass with a deep fabric of live (see Daly & Farley, 2010 for Steady State Economics as an applicable concept for PNG). People feel entitled to the land and will fiercely defend that right (Baraka, 2001). Sustainability is a science, and one that is hardly mastered by any modern nation or agency I know of (see Taber & Payne, 2003 for U.S. and Canada). As can easily be seen with man-made climate change, virtually none of the western and industrialized nations are sustainable, not locally not globally and not into the future (Stiglitz, 2003).

PNG is an island nation with various landmasses and ocean components, but it certainly is finite and cannot supply the entire world with its resources. Virtually all natural resource endeavors in PNG ended; and they ended rather quick and unpleasant, leaving warfare, destruction and death behind (examples provided in Hayan, 1990; Mack, 2014; West, 2006a; Kirch, 2014). Long-term sustainability is not a thing in PNG's modern industry endeavors shaped by The West. Boom-and-bust is the rule instead; the scheme seems to be 'grab as much as you can and ignore the wider clean-up costs'. Industrial projects that were actually developed with the local people, by the people and made local people happy and kept what was promised are far and few (see Cousteau & Richard, 1999 for negative examples, e.g. Japanese-Australian Forestry, fishing or mining). The colonial and industrial society approach to PNG was never sustainable, hardly successful yet. As will be shown later, certainly mining is not really among those, nor any modern agro-farming businesses, forestry, fisheries, urban projects or port projects. The resource curse (Ross, 2015) applies globally, and to PNG for sure. Most of the money for PNG comes from the outside, and it often well entailed bad business, as can be seen in casino business (The Guardian, 2021) or sea floor mining pushed via Canada which then ended in bankruptcy letting PNG pay bills (Earthworks, Deep Sea Mining Campaign et al., 2015). PNG has no relevant car industry, and no computer industry, or software market (other than some mobile phone towers), or military trades. Such income streams are left to the outside world and just to a few selected nations; most happen to speak English (or Chinese); and such a link is far from new. PNG is excluded.

1.7 What 'Science-Based' Policy Controls (Before-After-Control-Impact; BACI) Tell Us About the Western World: Clear Evidence for the PNG Case

Papua New Guinea is a control site for whatever the western and industrialized world has done and promoted during the last 300 years. Ideas like Industrialization, Marxism, Capitalism, Neoliberalism or Globalization come from the outside, usually with an English language wrap and origin. Instead, modern PNG shows us well what is ongoing outside of PNG, and how it comes into PNG and what it destroys there, and how it gets destroyed by local forces. The list of lost projects and their underlying

Table 1.2 Examples of western/global projects and thought that failed in PNG

Project	Metric	Comment
Communism	Wealth distribution, health and education metrics show no progress	There was little socialism and communism movement in PNG, but one may argue that the rural PNG society covers many of those aspects in its very fabric
Capitalism	PNG—with capitalism as its scheme—is on the verge of existence, was labeled a failed state and has major issues of debt and corruption	With a lack of registered citizens and no subsequent taxation scheme PNG can hardly run a capitalistic society Capitalism remains a global story of failure with a superficial global narrative of success, even happiness, e.g. The World Bank (https://www.worldbank.org/en/publication/trust-fund-annual-report-2019/trust-fund-achievements)
Neoliberalism	Any socio-economic metrics in PNG show lack of a win–win scenario, hardly progress	Neoliberalism is a bad idea (MacLaren & Kelly, 2014)
Missionary	PNG is one of the major christianized nations in the world (Beehler & Laman, 2020). But none of the church groups were able to establish themselves long-term, are really present in the bush 'in person' and the socio-economic metrics for PNG speak for themselves in parallel	A worldwide belief system in a global crisis
Globalization	The socio-economic metrics for PNG speak for themselves	Discontents published and expressed globally, e.g. Stiglitz (2003)
Chinese Management	China offers a longer history in Asia and its nations, also in PNG. China's development differs from the western model. However, it's widely acknowledged that China totally exploits forests and mines of PNG and offers little other relevant and sustainability support for PNG (e.g. Beehler & Laman, 2020)	China's export efforts focus on the Road and Belt initiatives (Connolly, 2020)

concepts is very long, and decay can be found all over PNG (see Table 1.2 for a selection).

It makes for a classic case of 'before and after impact' (BACI); seeing the PNG cases it's very difficult to design a better policy study in real life. Using science as the basic principle for inference on the evidence, one easily can apply BACI (Before-After-Control-Impact; e.g. Smith et al., 1993) here. It's not Rocket Science. If done for PNG and its modern nation building process last 60 years, a rapid assessment

Fig. 1.15 Pristine estuaries, coastal and mangrove forests are declining worldwide, also in Papua New Guinea

easily concludes pretty clean: The western nations, with Australia at its lead and supported by a U.N. mandate and global oversight, failed PNG dramatically in the set up (Duncan & Chand, 2002; May, 2006). China did not enter much in this mix but remained on its own, and it hardly looks better in PNG. This remains the key message from PNG leaders and prime ministers Somare (1975) and from Chan (2016) alike (Figs. 1.15 and 1.16).

1.8 PNG Forward, Back to the (Rasta) Roots

Papua New Guinea has settled with the west, with Asia and with itself. It's a solid nation *per se* (see Cousteau & Richards, 1999) and did well with itself and others for over 47,000 years. It has also settled with globalization, with Australia and with itself. PNG constantly defends itself, in a tribal Wantok style. The PNG lifestyle is heavily rural and thus, tribal; and likely it will stay there for a long time. This tribal lifestyle is extended with mobile phones, associated telephone towers and calling cards; some cars can be added to that mix, even fewer computers, but lots of music; Pacific songs and reggae music the most. Bob Marley—'the rasta' (Rastafari; Toynbee, 2007)—is still alive in PNG. Bob Marley is one of the few who can reach most PNG citizens and unite them; he has a message, e.g. peace, fairness, spirituality, and world citizenship,

Fig. 1.16 A beach in Papua New Guinea: PNG is a coastal nation affected by man-made climate change

free of its masters. Port Moresby—as the capitol—makes for a mélange of the modern PNG but widely fueled and driven by, rural PNG (which is with 86% the vast majority of the nation and over 95% is in public ownership).

That is the current PNG model, and it runs as good as it gets; it's sustainable for PNG, and as long as the framework allows for it, it gets even better.

References

Acemoglu, D., & Robinson, J. A. (2013). *Why nations fail: The origins of power, prosperity, and poverty paperback*.

Anderson, K. (2015). Colonialism and cold genocide: The case of West Papua. *Genocide Studies and Prevention: An International Journal, 9*, 5.

Alcorn, J. B., Beehler, B. M., & Swartzendruber, J. F. (1993). *Papua New Guinea conservation needs assessment* (Vol. 1). Biodiversity Support Program.

Australian Conservation Foundation (ACF). (2006). *Bulldozing progress: Human rights abuses and corruption in Papua New Guinea's large-scale logging industry*. Australian Conservation Foundation, Carlton, Australia.

Auty, R. (1993). *Sustaining development in mineral economies: The resource curse thesis*. Routledge.

Babbage, R. (1987). Australia and the defence of Papua New Guinea (PNG). *Australian Journal of International Affairs, 41*(2), 87–94.

Baraka, R. (2001). *The meanings of independence in post-colonial Papua New Guinea*. A Plan B Research Paper Submitted to the Faculty of the Center For Pacific Islands Studies in the Candidacy for the Degree of Masters of Arts. School Of Hawaii, Asia and Pacific Studies. The University Of Hawai'i At Manoa.

Beehler, B., & Lamann, T. (2020). *New Guinea: Nature and culture of Earth's Grandest Island*. Princeton University Press.

Bernhard, Z. (2017). *The Curious Case of August Engelhardt*, Leader of a Coconut-Obsessed Cult Gastro Obscura 30th November 2017. https://www.atlasobscura.com/articles/august-engelhardt-coconut-cult#:~:text=From%20his%20newly%20purchased%20island,transubstantiation%20of%20God%27s%20very%20flesh. Accessed July 3, 2022.

Blum, W. (2014). *Killing hope: US military and CIA interventions since World War II*. ZED Books.

Bowers, C. A. (2011). *Let them eat data: How computers affect education, cultural diversity, and the prospects of ecological sustainability*. University of Georgia Press.

Chan, J. (2016). *Playing the game: Life and politics in Papua New Guinea*. University of Queensland Press.

Cockburn, A. (2013). *A colossal wreck: A road trip through political scandal*. Verso Publishers.

Connell, J. (2005). *Papua New Guinea: The struggle for development*. Routledge.

Connolly, P. (2020). The belt and road comes to Papua New Guinea: Chinese geoeconomics with Melanesian characteristics? *Security Challenges, 16*(4), 41–64.

Cookson, M. (2000). *From the archives: The Archbold expeditions to New Guinea: a preliminary survey of archival materials held at the American Museum of Natural History*, New York City.

Cortassel, J., & Holder, C. (2008). Who's sorry now? Government apologies, truth commissions, and indigenous self-determination in Australia, Canada, Guatemala, and Peru. *Human Rights Review, 9*, 466.

Czech, B. (2002). *Shoveling fuel for a runaway train: Errant economists, shameful spenders, and a plan to stop them all*. University of California Press.

Dabeck, L., Valentine, P., Blessington, J., & Schwartzet, K. (Eds.). (2020). *Tree Kangaroos: Science and conservation*. Academic Press.

Daly, H., & Farley, J. (2010). *Ecological economics: Principles and applications* (2nd Ed.). Island Press.

Dastyari, A., & Sullivan, O. M. (2016). Not for export: The failure of Australia's extraterritorial processing regime in Papua New Guinea and the decision of the PNG Supreme Court in "Namah". *Monash University Law Review, 42*(2), 308–338. https://search.informit.org/doi/10.3316/informit.833786641621869

Davis, G. A., & Tilton, J. E. (2005, August). The resource curse. In *Natural resources forum* (Vol. 29, No. 3, pp. 233–242). Blackwell Publishing, Ltd.

De Soto, H. (2000). *The mystery of capital: Why capitalism triumphs in the West and fails everywhere else*. Basic books Publisher.

Duncan, R., & Chand, S. (2002). The economics of the 'Arc of Instability'. *Asian-Pacific Economic Literature, 16*, 1–9. https://doi.org/10.1111/1467-8411.t01-1-00001

Diamond, J. (2011a). *Guns, germs and steel: The fate of human societies*. Random House.

Diamond, J. (2011b). *Collapse: How societies choose to fail or succeed* (Revised Ed.). Penguin Books.

Die Welt. (2021). *250–000 Kilometer Absurditaet teilen die Welt*. 21.05.2021. https://www.welt.de/reise/Fern/article231236973/Grenzen-Die-Welt-ist-von-250-000-Kilometern-Absurditaet-eingeteilt.html. Accessed September 1, 2021.

Earthworks, Deep Sea Mining Campaign et al. (2015). *World's first deep sea mining proposal ignores consequences of its impacts on oceans* (28th September 2015) https://earthworks.org/releases/worlds_first_deep_sea_mining_proposal_ignores_consequences_of_its_impacts_o/. Accessed July 14, 2022.

Effenberger, J., & MacGregor, J. (2016). *Sie wollten den Krieg: Wie eine kleine britische Elite den Ersten Weltkrieg vorbereitete*. Kopp Publisher.

Elvin, M. (2008). *The retreat of the elephants*. Yale University Press.

Filer, C. (2004). A serious case of conditionality: The World Bank gets stuck in Papua New Guinea. *Development Bulletin, 65*, 95–99.

Flannery, T. (1990). *Mammals of New Guinea*. The Australian Museum.

Flannery, T. (1998). *Throwim way leg: Tree-Kangaroos*. Atlantic Monthly Press.

Flannery, T. (2002). *The future eaters: An ecological history of the Australasian lands and people.* Grove Press.

Foster, R. J. (2002). *Materializing the nation: Commodities, consumption, and media in Papua New Guinea.* Indiana University Press.

Fowke, J. (2006). Getting it wrong in Papua New Guinea. *Quadrant, 50*(12), 28–31.

Gillison, G. (1993). *Between culture and fantasy: A new Guinea highlands mythology.* University of Chicago Press.

Gosarevski, S., Hughes, H., & Windybank, S. (2019). Is Papua New Guinea viable? *Political Economic Bulletin, 19.* https://www.semanticscholar.org/paper/Is-Papua-New-Guinea-viable-Gosarevski-Hughes/6d71c70951c12fa82dea578a36b66923997f3f92. Accessed June 1, 2022.

Harari, Y. N. (2015). *Sapiens: A brief history of humankind.* Harper Perennial.

Henton, D., & Flower, A. (2007). Mount Kare gold rush: Papua New Guinea, 1988–1994. Published by Mt Kare Gold Rush, Brisbane Australia. ISBN 10: 0646482815; ISBN 13: 9780646482811.

Hawksley, C. (2006). Papua New Guinea at thirty: Late decolonisation and the political economy of nation-building. *Third World Quarterly, 27*(1), 161–173. https://doi.org/10.1080/01436590500370012

Hayan, D. M. (1990). *Road through the rain forrest: Living anthropology in highland Papua New Guinea.* Waveland Press.

Hayes, T. E. (2003). *New Guinea Bibliography*, October, 2003. https://nanopdf.com/queue/hays-2003-ng-bibliography_pdf?queue_id=-1&x=1656200487&z=MjYwMDozODc6NDo4MDM6Ojcy. Accessed June 1, 2022.

Hoffmann, C. (2015). *Savage harvest: A tale of cannibals, colonialism, and Michael Rockefeller's tragic quest.* William Morrow Paperbacks; Reprint edition.

Huettmann, F., & Young, B. D. (2022). The so-called modern 'Sustainable Forestry' destroys wilderness, old-growth forest landscapes and ecological services worldwide: A short first-hand review and global narrative on the use of 'growth-and-yield' as a destructive and even impossible goal. In *Forest dynamics and conservation* (pp. 53–82). Springer.

Hughes, L. (2002). *Moving the Maasai: A colonial misadventure.* Doctoral dissertation, University of Oxford.

Hutton, T., Mardle, S., & Tidd, A. N. (2008). The decline of the English and Welsh fishing fleet? *Advances in fisheries science: 50 years on from Beverton and Holt* (pp. 26–48).

Jackson, J. B. C., Kirby, M. X., Berger, W. H., Bjorndal, K. A., Botsford, L. W., Bourque, B. J., Bradbury, R. H., Cooke, R., Erlandson, J., Estes, J. A., Hughes, T. P., Kidwell, S., Lange, C. B., Lenihan, H. S., Pandolfi, J. M., Peterson, C. H., Steneck, R. S., Tegner, M. J., & Warner, R. R. (2001). Historical overfishing and the recent collapse of coastal ecosystems. *Science, 293*, 629–637.

Kiki, A. M., & Cheshire, F. W. (1969). *10,000 years in a lifetime.* Cheshire Publishing.

Kirsch, S. (2014). *Mining Capitalism.* University of California Press.

Knoll, A. J., & Hiery, H. J. (Eds.). (2010). *The German colonial experience: Select documents on German rule in Africa, China, and the Pacific 1884–1914.* University Press of America.

Kracht, C. (2015). *Imperium: A fiction of the South Seas.* Farrar.

Kurer, O. (2019). *The Papua New Guinean malaise: From redistributive politics to a failing state.* https://openresearch-repository.anu.edu.au/handle/1885/157813

Lasslett, K. (2012). State crime by proxy: Australia and the Bougainville conflict. *British Journal of Criminology, 52*, 705–723.

Laman, T. & Scholes, E. (2012). *Birds of paradise: Revealing the world's most extraordinary birds.* National Geographic Books.

Laurance, W. F., Kakul, T., Keenan, R. J., Sayer, J., Passingan, S., Clements, G. R., Villegas, F., & Sodhi, N. S. (2011). Predatory corporations, failing governance, and the fate of forests in Papua New Guinea. *Conservation Letters, 4*, 95–100. https://doi.org/10.1111/j.1755-263X.2010.00156.x

Lea, D. (1997). *Melanesian land tenure in a contemporary and philosophical context.* Lanham.

Leahy, M. J. (Ed.: Douglas E. Jones). (1994). *Explorations into Highland New Guinea, 1930–1935*. Crawford House Press.

Levi-Strauss, C. (1966). *The savage mind*. University of Chicago Press.

Limburg, K. E., Hughes, R. M., Jackson, D. C., & Czech, B. (2011). Human population increase, economic growth, and fish conservation: Collision course or savvy stewardship? *Fisheries, 36*(1), 27–35.

Lines, W. (1999). *Taming the Great South Land: A history of the conquest of nature in Australia*. University of Georgia Press, Atlanta U.S.

Lockyer, A., Burke, J., Lim, Y. H., Smith, F. (2021). *Manus Island and the Lombrum Naval Base: Five options for Australia's geostrategic gateway*. Royal Australian Navy, Sydney Australia.

Lutton, N. (1981). From Port Moresby to Perth, 30–38.

Ludlam, S. (2021). *Full Circle: Power*, Hope and the Return of Nature. Black Inc.

Mack, A. (2014). *Searching for PekPek: Cassowaries and Conservation in the New Guinea Rainforest*. Cassowary Conservation & Publishing.

McFarlane, S. (1889). *Among the cannibals of New Guinea: Being the story of the New Guinea Mission of the London Missionary Society*. Presbyterian Board of Publication.

MacLaran, A., & Kelly, S. (2014). Neoliberalism: The rise of a bad idea. In *Neoliberal urban policy and the transformation of the city* (pp. 3–19). Palgrave Macmillan.

Maclean, N. (Ed.). (2010). *Silent summer: The state of wildlife in Britain and Ireland*. Cambridge University Press.

Martin, R. (Ed.). (2005). *Tree-kangaroos of Australia and New Guinea*. CSIRO Publishing.

Matthiessen, P. (1987). *Under the mountain wall: A chronicle of two seasons in Stone Age New Guinea*. Penguin Classics.

May, R. J. (2006). Papua New Guinea: Disorderly democracy or dysfunctional state? In *Australia's arc of instability* (pp. 151–170). Springer.

Mead, M. (Ed.). (2002). *Cooperation and competition among primitive peoples* (Vol. 123). Transaction Publishers.

Melick, D. (2010). Credibility of REDD and experiences from Papua New Guinea. *Conservation Biology, 24*, 359–361.

Mellor, D., Bretherton, D., & Firth, L. (2007). Aboriginal and Non-aboriginal Australia: The dilemma of apologies, forgiveness, and reconciliation. *Peace and Conflict: Journal of Peace Psychology, 13*, 11.

Miller, J. D. B. (1973). Papua New Guinea in world politics. *Australian Outlook, 27*, 191–201.

Mongabay. (2022). *Papua New Guinea*. https://rainforests.mongabay.com/deforestation/forest-information-archive/Papua_New_Guinea.htm. Accessed March 29, 2022.

Moses, A. D. (Ed.). (2008). *Empire, colony, genocide: Conquest, occupation, and subaltern resistance in world history* (Vol. 12). Berghahn Books.

O'Connell, J. F., & Allen, J. (2007). Pre-LGM Sahul (Australia-New Guinea) and the archaeology of early modern humans. In P. Mellars, K. Boyle, O. Bar-Yosef, C. Stringer (Eds.), *Rethinking the human revolution: New behavioural and biological perspectives on the origin and dispersal of modern humans* (pp. 395–410). McDonald Institute Monographs.

O'Connell, J. F., & Allen, J. (2015). The process, biotic impact, and global implications of the human colonization of Sahul about 47,000 years ago. *Journal of Archaeological Science, 56*, 73–84.

Overseas Development Institute. (2007). *What can be learnt from the past? A history of the forestry sector in Papua New Guinea*. ODI.

Peck, J. (2014). Pushing austerity: State failure, municipal bankruptcy and the crises of fiscal federalism in the USA. *Cambridge Journal of Regions, Economy and Society, 7*(1), 17–44.

Penders, C. L. M. (2021). *The West New Guinea debacle: Dutch decolonisation and Indonesia, 1945–1962*. Brill Publishers.

Pratt, A. E., & Pratt, H. (1906). *Two years among New Guinea cannibals: A naturalist's sojourn among the aborigines of unexplored New Guinea*. Seeley & Company, Limited.

Raakjær, J. (2009). *A fisheries management system in crisis: The EU common fisheries policy*. ICES.

Reuters. (2008). *Australia apologizes to aborigines*, Reuters, Canberra, February 14, 2008. https://www.reuters.com/article/uk-australia-politics-aborigines-idUKSYD9123020080213. Accessed May 8, 2022.

Rich, B. (1994). *Mortgaging the Earth: The World Bank*. Earthscan Publishers.

Richards, M. (1998). *A time of silence: Civil war and the culture of repression in Franco's Spain, 1936–1945* (Vol. 4). Cambridge University Press.

Richardson, D. (2005). *Peace child: An unforgettable story of primitive jungle treachery in the 20th century*. Regal Publishers.

Rick, A. (2020). *Walking with ghosts in Papua New Guinea: Crossing the Kokoda trail in the last wild place on Earth*. Skyhorse Publishers.

Rivers, C. (2016). *The Dutch East India Company and British East India Company: The history and legacy of the world's most famous colonial trade companies*. CreateSpace Independent Publishing Platform.

Rooney, D., Papoutsaki, E., & Pamba, K. (2004, July). *A country failed by its media: A case study from Papua New Guinea*. Paper presented at the Annual Conference of the Asian Media Information and Communication Centre, Bangkok. https://hdl.handle.net/10652/1494. Accessed July 3, 2022.

Ross, M. L. (2015). What have we learned about the resource curse? *Annual Review of Political Science, 18*, 239–259.

Salak, K. (2001). *Four corners: One Worman Solo's journey into the heart of Papua New Guinea*. Counterpoint Publisher.

Somare, M. (1975). *Sana: An autobiography of Michael Somare*. Niugini Press.

Soler, J. J., Martín-Vivaldi, M., Nuhlíčková, S., Ruiz-Castellano, C., Mazorra-Alonso, M., Martínez-Renau, E., Eckenfellner, M., Svetlík, J., & Hoi, H. (2022). Avian sibling cannibalism: Hoopoe mothers regularly use their last hatched nestlings to feed older siblings. *Zoological Research, 43*, 265–274. https://doi.org/10.24272/j.issn.2095-8137.2021.434.PMID:35238184; PMCID:PMC8920838

Smith, E. P., Orvos, D. R., & Cairns, J., Jr. (1993). Impact assessment using the Before-After-Control-Impact (BACI) model: Concerns and comments. *Canadian Journal of Fisheries and Aquatic Sciences, 50*(3), 627–637. https://doi.org/10.1139/f93-072

State Library New South Wales. (2022). *The HMS Rattlesnake Journey to Papua New Guinea 1849*. https://www2.sl.nsw.gov.au/archive/discover_collections/society_art/png/rattlesnake/index.html. Accessed April 26, 2022.

Steiner, M., & Huettmann, F. (2021). Justification for a taxonomic conservation update of the rodent genus Tamiasciurus: Addressing marginalization and mis-prioritization of research efforts and conservation laissez-faire for a sustainability outlook. *The European Zoological Journal, 88*(1), 86–116. https://doi.org/10.1080/24750263.2020.1857852

Strippoli, G. (2016). Colonial War, Anti-colonialism and desertions during the Estado Novo. Portugal and abroad. In *III International Conference Strikes and Social Conflicts: combined historical approaches to conflict. Proceedings* (pp. 430–444). CEFID-UAB.

Stiglitz, J. (2003). *Globalization and its discontent*. Norton Paperback.

Stresemann, E. (1923). Ornithologische Ausbeute im Stromgebiet des Sepik. Ein Beitrag zur Kenntnis der Vogelwelt Neuguineas. Archiv fuer Naturgeschichte ser. A.A., 89(7)–96(8)–92.

Taber, R. D., & Payne, N. F. (2003). *Wildlife, conservation, and human welfare: A United States and Canadian perspective*. Malabar, Florida: Krieger Publishing Company.

The Guardian. (2019). *Refugees on their own land: The West Papuans in limbo in Papua New Guinea*, 30th November 2019. https://www.theguardian.com/world/2019/dec/01/refugees-on-their-own-land-the-west-papuans-in-limbo-in-papua-new-guinea. Accessed August 1, 2022.

The Guardian. (2021). *PNG residents evicted a decade ago for casino development face violence and poverty: Report*. https://www.theguardian.com/world/2021/nov/12/png-residents-evicted-a-decade-ago-for-casino-development-face-violence-and-poverty-report. Accessed March 17, 2022.

The Guardian. (2022). *PNG prime minister denies connection to suitcase full of cash found as voting starts*, 6th July 2022. https://www.theguardian.com/world/2022/jul/06/png-prime-minister-denies-connection-to-suitcase-full-of-cash-found-as-voting-starts. Accessed July 6, 2022.

The Sydney Morning Herald. (2010). *The PNG report links Somare to Moti scandal*, March 3, 2010. https://www.smh.com.au/world/png-report-links-somare-to-moti-scandal-20100303-piod.html. Accessed July 9, 2022.

Toynbee, J. (2007). *Bob Marley: Herald of a postcolonial world?* Polity Publishers.

Tønnessen, J. N., & Johnsen, A. O. (1982). *The history of modern whaling*. Univ of California Press.

Tulkiewicz, M. S. (2006). *Papua New Guinea: Weak, failing, failed?: An examination of failed state theory and the usefulness of the Failed States Index*. https://ro.ecu.edu.au/theses_hons/1283. Accessed May 8, 2022.

Vanaik, A. (2012) U.S. dominance and diplomatic choices in developing nations: Achin Vanaik. In *Diplomacy and developing nations* (pp. 64–88). Routledge.

Vanaik, A. (2015) *Nepal's maoist in power. New Left Review Mar/Apr 92*. https://newleftreview.org/issues/ii92/articles/achin-vanaik-nepal-s-maoists-in-power

Weiner, M. F. (2014). Racing slavery: Racial Neoliberalism, Social Forgetting, and Scientific Colonialism in Dutch primary school history Textbooks1. *Du Bois Review: Social Science Research on Race, 11*, 329–351.

West, P. (2006a). *Conservation is Our Government Now: The Politics of Ecology in Papua New Guinea*. Durham, NC: Duke University Press.

Werner, H. (2002). The current 'Green Card' initiative for foreign IT specialists in Germany. *International Mobility of the Highly Skilled, 321*, 1–12.

Wilson, M. (2019). *A true child of Papua New Guinea: Memoir of a life in two worlds*. MacFarland Publishers.

Windybank, S., & Manning, M. (2003). *Papua New Guinea on the brink*. No 30 - 12 March 2003. https://www.exkiap.net/articles/cis20030312_png_on_brink/ia30.htm

Wood, A. W. (1976). Rivers. In D. King, & S. Ranck (Eds.), *Papua New Guinea Atlas. A nation in transition* (pp. 98–99). Robert Brown Assoc.

Wurm, S. A., & Hattori, S. (Ed.). (1983). *Language Atlas of the Pacific Area*. Part I. New Guinea Area, Oceania, Australia. Australian Academy of Humanities, Canberra, Australia.

Zandvliet, K., & Harley, J. B. (1992). Art, science, and power in sixteenth-century dutch cartography. *Cartographica: The International Journal for Geographic Information and Geovisualization, 29*(2), 10–19.

Chapter 2
The Land of Papua New Guinea: A Quick Conservation Reader for What CEOs and Directors Ought to Know

Large-scale resource development in Eastern New Guinea never happens without fight
(*Beehler & Latam*, 2020, p. 353)
We are in the beginning of a mass extinction and all you can talk about is money and Fairy tales of eternal economic growth
(Greta Thunberg To the UN Climate Summit September 2019)
PNG is among the few nations in the world where ordinary people by virtue of birth can claim to secure access land
(*Baraka*, 2001, p. 8)
To destroy willfully the rich cosmologies of more than 1,000 ethnic groups across the Largest island on earth should be considered a crime against traditional humankind
(*Beehler & Laman*, 2020, p. 326)

Abstract Papua New Guinea (PNG) is a precious place; it carries a globally recognized civilization of public land tenure that is over 47,000 years old and globally sustainable. But a typically found conflict between western economy and the environment on a finite land mass and ocean destroys what PNG was and what it can be and contributes. This conflict began with colonialism and is ongoing for over 60 years. PNG is widely out of hands of democracy and its citizens running a western 'resource curse.' This scheme is led by international managers, CEOs, corporate administrators and politicians usually implying there is a win–win, or a tech-way out to make everybody happy and wealthy in a 'trickle-down economy.' As CEOs are placed to PNG by their respective companies with a working visa, they just tend to use superficial and simplistic narratives applied globally but failing locally. All too often it was shown that those leaders simply balance and increase budgets and subsequent maximized profits but ignore the rest of their trusted leadership tasks, or even frankly do not have the skill and expertise to account better and wider, to act more holistically, and to safeguard future generations and their trusted resource and the world. There has been no good partnership whatsoever between actors, and the 'public license' was not achieved well. Driven from the outside, politically correctness was left at the wayside, and bankruptcy with a boom-and-bust dominates. Such conflicts of sustainability become violent quickly on either side and leave '*bad blood*' for nothing. Based on field work and public data and open access information here some basic overview of PNG, of PNG principles and how it can be handled and lead better by CEOs that

F. Huettmann, *Globalization and Papua New Guinea: Ancient Wilderness, Paradise, Introduced Terror and Hell*, https://doi.org/10.1007/978-3-031-20262-9_2

operate in PNG are initiated with a discussion in order to achieve a better sustainability on many accounts for all actors involved and for future generations as global role model to go by.

Keywords Papua New Guinea (PNG) · Landscape ecology · Natural history · Finite resources, Mining, CEOs, Unsustainable development, Economic growth impacts

2.1 Introduction

Papua New Guinea has vast resources—biotic, abiotic, cultural and spiritual (Sullivan 2015); it's equatorial, oceanic, of global relevance and very sophisticated featuring a sensitive fabric to be aware of, specifically the concept of public land tenure. App. 97% of the PNG land is in the hands of traditional land owners but no corporations or governments dominate (Baraka, 2001; Beehler & Laman, 2020). Private land ownership on a larger scale does hardly exist in PNG and can interact with ancient souls, e.g. in sacred burial sites; Holzknecht (1999), Beehler and Laman (2020). In PNG people care for the transition of the soul into a good future. Public consent (aka '*Social License*') is widely required for any project, certainly when it is extractive, and open pit and from abroad (see Henton & Flower, 2007 and Macintyre, 2007 for experiences). Already the amount of gold and copper, beside other minerals, are found here and accessible in world-record amounts, and it represents money on the table for the outside world to grab. And it does not end there: PNG also features perhaps the largest oil and gas resources in the Pacific (Beehler & Laman, 2020, p. 74). And while it will just last a few years (Beehler & Laman, 2020), it's just a matter of time when those resources provided by Mother Earth are to be developed (see The World Bank promoting PNG to investors https://www.worldbank.org/en/country/png).

It thus attracts many companies and their leaders, CEOs. And it's the CEO that can steer many good or bad decisions for PNG and its people (see for personal CEO examples with Jim-Bob Moffet in Flannery 1998, with Denis Reinhardt in Henton and Flower 2006, and several Australian examples on aboriginee land in Ludlam, 2021, p. 231). A CEO is the face of a company, accumulates all maters in one person, can also report, and influence, the 'board' and stakeholders of the company. A CEO comes with a large colonial hinterland. In PNG terms, a CEO is the '*Big Man*' of their capitalism and industrialization (Figs. 2.1, 2.2 and 2.3).

In earnest, the western world history with PNG is just app. 300 years old—perhaps less than 98% of the 47,000-year-old human history of PNG, but it has been one of influx from abroad, specifically a dominance from Australia in relation to gold mining (Nelson, 1982; Beehler & Laman, 2020), beyond Asian people migration, colonial and global input. With that come many outsiders to PNG, taking over and making decisions about this nation and its people; not all of them are done well, good, sustainable or with a relevant vision. In the absence of good governance, e.g. in forestry (Beehler & Laman, 2020) or mining (Henton & Flower, 2007), the CEO decisions are dramatic to PNG; it has already inflicted terror on PNG (see Bougainville leading to civil war and national crisis; Chan, 2016), a nation that was

Fig. 2.1 The grass dance: a deep-rooted link of people, nature and cosmology exist in Papua New Guinea, also used to appease and welcome foreigners

co-designed and 'aided' by Australia. That is simply the legacy to consider and to know; and so any new activities on that front should acknowledge it and take that fully into account.

PNG as a nation is to c. 86% rural (Baraka, 2001; Beehler & Latam, 2020; Mongabay, 2022). PNG has one of the least dense road systems in the world. Roads and bridges are needed for making mining, etc. feasible; often in PNG it includes airports also (Sinclair, 1978); impacts of those can be rather devastating (e.g. Hayan, 1990). In the rural tribal world, life is nature-based instead. The amount of internationally informed, highly educated and urbanized effluent citizens is lower than in the western world. And thus it appears as relatively easy to be a manager or CEO from abroad, be in the driver seat, and sit in places like the established 'Papua Club' in Port Moresby (a well described 'home' for Australians and British for the last decades 'far away from home'; Cousteau & Richards, 1999, p. 205). Many of such spin-off clubs are found, e.g. the yacht clubs and resorts. From personal experience, its here were PNG policy is made and discussed, almost unconstrained, often based on hearsay 'from the bush' and certainly without relevant input of PNG citizens (who tend to act in such establishments as barkeepers and cleaners; the local business management there is usually in Asian hands). Traditionally, females are widely excluded in such localities and not allowed to enter even (Cousteau & Richards 1999).

Fig. 2.2 Spirits and ghosts of Papua New Guinea remain in the driver seat

But as the gap between rich and poor increases, there is a strong difference between the international investors—lead by CEOs and Directors—and the locals in PNG (often lead by Big Men). The ancient view of nature in PNG came from the belief that the phenomenon of nature actually were the very origins of the human beings (Cousteau & Richards, 1999, p. 211; see also Flannery, 2002 and Beehler and Laman 2020 for underlying and entrenched concepts of shamanism, spirituality and sorcery; examples shown in Gillison, 1993). The ancient and dominating view in PNG is that nature and its embodyment still houses the ancestors, and so the villagers have often a moral obligation not to harm or create suffering for them. Nature is an indivisible whole where humans are embedded within! A widespread view in most indigenous societies (Suzuki, 1993), but this model and concept does not much exist in the western world and their investors (example in Henton & Flower, 2007), rather vice versa: The more narrow/parsimonious and less connected the better for them (as per economy textbooks; see Daly & Farley, 2010 for alternatives; compare with Demeulenaere et al., 2021 for instance for the wider value, role and appreciation of trees and habitats in Melanesia).

PNG offers us with a civilization of world-relevance, and it does so already for over 47,000 years (Diamond, 2011a, b); it's a place where one can see and connect first-hand with all what earth is about. This society actually lived sustainable for over

Fig. 2.3 Tourist artifacts present a link with deep culture and skill now available for international sale

98% of its time, turning the PNG highlands and other areas in gardens and fishery grounds (Cousteau & Richards, 1999 for a coastal example with clams). Beyond what the '*Noble Savage*' image portrays (Ellingson, 2001), PNG offers us with one of the best global sustainability that mankind can produce, anywhere. Beehler and Laman (2020) refer to it as the grandest island in the world for a reason (Figs. 2.4, 2.5 and 2.6).

After Greenland—located in the Northern Hemisphere—PNG actually is the second largest island in the world (by area), and is actually the highest island in the world (by elevation). Beehler and Laman (2020) present many biodiversity aspects to justify it to be earth's grandest island. The PNG nation consists of thousands of island. Its geology links with the Australian mainland (craton; but which in Australia essentially is quite flat, apart from a few coastal areas; see Flannery, 2002 for effects, Beehler & Laman, 2020 for topography and biogeography). Many CEOs operate in PNG because the earth plates in PNG bring the interior resources to the surface. That's where the wealth sits and the resource curse starts (Molden, 2019; Ross, 2015). Located at the 'Ring of Fire,' PNG has the fastest rising ranges in the world with active volcanoes and sometimes with several tremors and earthquakes a day. Sink holes can be found.

Located at the Australian craton, PNG is part of the Sahul region (Flannery, 2002), which is the wider connected area between Australia and PNG (as its nearest

Fig. 2.4 Ridge to reef: all is connected in Papua New Guinea wilderness landscapes making environmental impacts severe

neighbor); its stems from a time when sea levels were between 55 and 130 m lower (Beehler & Laman, 2020).

One may say that Australia is dry and somewhat temperate, whereas PNG is essentially tropical and wet. PNG is known for its super wet zones making it difficult to operate in industrial conditions the 'The West' is used to. PNG shows some monsoon but with rain shadows, e.g. in Port Moresby making the capitol not so representative for PNG. By elevation, temperature cools down by app 0.65 °C/100 m. Dry events of El Nino can affect many ecosystem stresses; tsunamis are common. The operating business in PNG can change quickly.

2.2 (Foreign) Profits and the PNG Set up Mix Poorly

Whatever is said and for profits (see for instance The World Bank https://www.worldbank.org/en/country/png) PNG makes for a very complex (commercial) playing ground to operate in and for a company to make profits. The history shows colonial failures. The way how PNG is structured—a large main land, highlands, many steep and isolated valleys, a vast surrounding coastal area, few estuaries and thousands of islands—result into the situation that PNG is not really homogenous or unified.

Fig. 2.5 Tough but not impossible trails in Papua New Guinea

Instead it has one of the highest density of languages (over 700; e.g. Kulik, 2019), and many local tribes (see WIPO, 2006 for genetic theft and copyrights of indigenous people in PNG, Hagahai tribe) and customs; already the notion of 'one land' and 'one nation state' gets easily challenged in PNG (Chan, 2016; see Wilson, 2019 for the description of a double life). PNG is a country consisting of 1000 nations…PNG is not so united and it 'bubbles' all the time, politically and otherwise. Instead, Western nations often center around homogeneity; a singular tribe and a push for it with enforcements. Whereas PNG is virtually the opposite, a patchwork of tribes tried to maintain as a single unit trying to cater global expectations. Struggles, and fighting between tribes was part of the history (Matthiessen, 1987; Flannery, 1998 for first-hand accounts) and remains part of politics (Figs. 2.7, 2.8 and 2.9).

A good management and business theory for PNG has not been developed yet. Western models will not work well. Clearly, differential views should drive the management as there is no singular black or white. But to be considerate and to

Fig. 2.6 Crossing water is the rule of the day for working in Papua New Guinea landscapes

engage local community in a shared and mutual fashion that's what the international corporations are not so good at, as per track record and their legacy (examples seen in Henton & Flower, 2007; Kirsch, 2014). The western model trains CEOs in a different way, e.g. neoliberal (Duménil & Lévy, 2011). In PNG, the Masters of Business and Administration (MBA) comes easily to a full conflict on several grounds (Ludlam, 2021 for tightly related Australian experience on mining and indigenous people). Just think of insurance or liability, and then think of public land tenure, besides other factors (social aspects and global banking and stakeholder expectations; overview provided in Banks, 2008, with details in Mack, 2015; Beehler & Laman, 2020). Rural land is rarely sold or bought; it's not really thought of as a commodity, or something to be developed and used, or operated by foreigners and corporations abroad; it's fought for fiercely (Baraka, 2001). Many companies and their CEOs envision the world very differently though, pushing for profit for their stakeholders and economic growth in an assumed trickle-down economy instead and operating in a homogenous 'westernized' top-down law-and-order society where land is private (see Asian Development Bank ADB 2012 for economic growth without relevant development; Coleman 2015 for relevant farming sectors etc.).

In contrast to PNG, many nation states in the western and Asian world achieved the national homogeneity—and subsequent institutions—as a wider process of nation-building. This process is based on a longer history of peoples moving in and out of an area, and then settled by a certain and long string of warfare, land annexations, diplomacy, power top-down events and agreed settlements of land claims with required citizen taxation; it involved usually a fight for the nation. And some of those

Fig. 2.7 Muddy roads can make operating in Papua New Guinea very difficult for everyone

rather tragic and traumatic (see examples provided by Diamond, 2011a, b for North America, etc.). Instead, the culture and nation of PNG is a function of geography and topography, and then, of the subsequent evolutionary history over millennia with sea levels determining islands, remoteness and boundaries. People lived and fought within that. This is how PNG got started, as a region, with recent global political events then assigning the actual PNG nation in just the last 50 years or so (Manning, 2005; Gosarevski et al., 2019). PNG as a nation is young, somewhat naive and fresh with a nice outlook (e.g. Miller, 1973), but the negative industrial footprint is already very obvious and corrupting minds of 'politics' (Chan, 2016; The Guardian, 2019, 2022). A balance was hardly achieved, yet.

PNG is not really fitting the western industrial set up, an economic theory, anthropological social science narrative, or a classic management model, even if now forced into via globalization (see Stiglitz, 2003 for a factual review). In PNG most people are not really registered, hardly full-time employed with a social security or pension system, they use additional currencies than money, and industry and taxation is not a main income model. Health care is to be 'free' but of very low performance. Salaries and debt might not be a main motivation for coming to work every day.

Surprisingly, and like in most parts of the world, corporations do not participate in a suitable tax model for the nation neither. This makes for an odd relationship unfolding in PNG with international working visas etc.

PNG is not only complex politically, but first and foremost geologically complex and dynamic; the newly defined PNG nation construct never was a stable entity, or much homogenous (Chan, 2016; see Beehler & Laman, 2020 for the wider New Guinea 'unit'; Flannery, 2002 for the Australasian/Sahul region). And time and space

Fig. 2.8 Open water travel in basic unsecured conditions is reality in Papua New Guinea

are obviously linked and need to be seen as one unit! It's holistic and to remain intact. All other diversity comes from that and should be taken into account for nation-building by design[1] (as it was applied in PNG) and to operate there.

Already the mentioned geology will easily throw a curveball to any modern 'western-style' operation (see for instance Reuter Oil & Gas, 2018). PNG has a rich soil due to Pleistocene volcanoes. There are over 100 volcanoes of which app. 35% are capable to erupt (Cousteau & Richards, 1999 for examples). As a bad example stands the city of Rabaul, as one of the most geologically active places in the world. Volcano outbreaks in just recent times killed already many people; PNG is very unstable in that regard (see Chan, 2016 for experience and rescue operations). PNG is one of the most geologically active areas in the world affecting virtually all forms of business operations.

Further, PNG features a unique biodiversity and is located south of the Wallace Line and affected by the Weber Line which puts much of our western mind to a challenge also. That's because the PNG biotic resources of the nation—the biodiversity—is high, ancient and endemic, but it certainly is affected, and driven, by the

[1] Many other non-western nations suffer from similar problems, such as Pakistan and its former part of Bangladesh, or the Austrian-Hungarian Empire, former Czechoslowakia or Yugoslavia. None of those nations were inherently grown but just became a quick political western construct used for the global disposal. When the western demand stalled the entire new nation fell apart, often major bloodshed followed. That 'new nation' served its purpose, but nobody was really helped, certainly not the environmental aspects.

Fig. 2.9 Not all trips end well; flying in wilderness areas can be deadly

geographic set up. This in turn affects the people who live there, and how (classic descriptions made by Diamond, 2001; Beehler & Laman, 2020) (Figs. 2.10, 2.11 and 2.12).

This unique diversity sets the stage for the independence conflicts within the PNG nation, and the administrative headaches inside of PNG and its institutions, but also for outside of it. And so, PNG as a typical nation state, just as Spain or the British empire let's say, and with such minds of administration and economy, cannot work well (it hardly works well for Spain or England neither (e.g. https://www.britannica.com/topic/Western-colonialism/Decline-of-the-Spanish-and-Portuguese-empires), but PNG has it much more intense. And with over 700 languages—Wantok—what should the unifying language be? Melanesian Pidgin, Motu (as one of the dominating tribal languages), or English (and with the English tongue, there are at least four options to choose from: Oxford/Cambridge UK Standard, the Australian Standard, American English, or Indian English with India as the location in the world of the highest density of English speakers anywhere. And the New Zealand English is another option, so actually is Japanese, and also Chinese with two languages within: Mandarin and Kantonese)?

Clearly, if a local language dies, so goes the lifestyle that comes with it, often one that was sustainable for over thousands of years with a unique local adaptation (Kulik, 2019). Language is survival (K. Lester in Ludlam, (2022, p. 155).

So what do the CEOs tell us to do?

Fig. 2.10 People and their future sit at the core of the Papua New Guinea enterprise

Clearly, the profit and GDP metric that CEOs push for is pointless in the PNG setting, for the environment and beyond (Czech, 2019). It does not trickle-down well and does not feed people much in PNG (Madison, 1997). It creates unrest.

And what unified singular cultural model would such a diverse geography really follow for a good and robust governance scheme? Why is a western governance model needed really, and for what and for whom? While this might all sound theoretical to some, for a manager, director and CEO operating in PNG it plays a very practical role because already the baseline to start from is difficult to define, so what is the impact from that? Whom to hire, and who runs the strategic key positions and how to make it just, fair and inclusive in such a diverse human universe? How not to exclude? Simply looking at Australia, it was not well resolved there, yet.

And so the business model unfolds while there are many more fragments and smaller homogenous landscape units within PNG such as:

- The world-famous highlands of PNG are unique and quite fertile. They are well-placed being located just between the coastal people and islanders versus the forest people. When compared to the Western World industrial society, the Highlanders initially were never that abundant, wealthy, rich or influential (Matthiessen, 1987; Flannery, 1998); but that has changed, e.g. due to the introduction of the potato (Diamond, 2011a, b; Beehler & Laman, 2020) by the Portuguese providing a relatively large income to the local tribes after it got bartered up from the coast.

Fig. 2.11 Domestic violence and abuse are a severe problem in Papua New Guinea

Nowadays, other goods grown in the highlands add to that wealth too; it's a big culture and society.

- The other separate unit in PNG might be the coastal zone; it is famous for its marine link and similar to the islands mentioned below. But the coastalscape of PNG is pretty diverse of course far from any homogeneity, a universe of its own, e.g. Cousteau and Richards (1999).
- The islands then are another large universe of their own. It consists of a myriad of islets and small land masses connected with the ocean and coral reefs. Like the coastal zone, it has a direct link with the rainforest as well; many of those island forests are mountainous and easily present their own universe once more.

Many more landscape units exist, of course (see Lonely Planet Guide https://www.lonelyplanet.com/papua-new-guinea). Thus, when all is taken together, one will understand that 'PNG as a single nation hardly exists,' it actually cannot really exist

Fig. 2.12 Food reality in Papua New Guinea for many working people

and likely never will any time soon. From that, PNG cannot be managed well, and certainly not with a singular western world concept. PNG is a juggling act by design, hopefully without much trauma or bloodshed, as so many other colonial nations and CEOs experienced though. As a nation state it simply is a rushed and quick imperfect western construct that PNG citizens hardly asked for (Baraka, 2001); although PNG would likely take it, if offered well. But then, PNG is left alone with such a suicidal set up, while exposed to the outside pressures and Australian-type western and Asian predators, always pretending that 'all is fine.'

But PNG can operate in itself, if left alone and if not overruled and over-powered by outside forces; as usually done though by the imperialism and neocolonialism. It's a view widely found in PNG. This already makes life hard for managers and CEOs from abroad. Earning trust is part of the game indeed.

Leaving things to PNG tends to work best because PNG has clearly operated by itself for millennia, so why not now? Well, the answer is simply,why are PNG matters get meddled with and through CEOs: PNG has resources to offer. It's money on the table to grab for the winner while PNG is not really asked.

PNG remains then in the tribal world though and settles it that way (see Beltran, 2000 for a wider native and indigenous perspective). In the meantime, the resources are exploited under the umbrella of development, of modernity and of all what comes with it. PNG gets bullied and thrown around (Chan, 2016). When looking at what that really consists of, it's a rather empty package that PNG gets provided but does not agree with much. Many examples are described what this package consists of, e.g. copyright violations (The Guardian, 2022), clean up costs (Cousteau & Richards, (1999, p. 28), tribal dispute and worse (Henton & Flower, 2007; Kirsch, 2014).

2.3 What Does the PNG Set up Mean for Modern Business? A Crash Course on PNG Etiquette Will 'Crash'

PNG is community-based (MacIntyre, 2007), and this is not easily invested in, nor is it easily and quickly be used for natural resource purposes; hardly for any other exploitive schemes. In PNG, all is deeply connected; it's holistic and Ecology in pure form (compare with Naess, 2009 or Rozzi et al., 2015). In PNG, tribes are in charge and they try to be empowered more and more from their former masters and colonial powers; so is the trend and reality of times we are in (Blazey & Perkiss, 2016; see Short, 2010 for a reality review of indigenous legacy and recent trends). 'Business as usual' will fail in PNG (e.g. De Soto, 2000). There can be no quick crash course of PNG, or what it means to operate there for doing business in PNG. From the track record, western industrial business in PNG means '*boom and boost*' and a very high rate of risk and failure (The Guardian 2016) for an example. Instead, PNG has taboos and it values them. Complexities, and thus costs, will go through the roof the more one operates there, so will the turn-over. PNG requires time, expertise and fluency of the many local rituals and procedures (see Lonely Planet https://www.lonelyplanet.com/papua-new-guinea for a starter; Beehler & Laman, 2020 for New Guinea overall). Although it's precisely the history of western engagement, PNG is not a place for boom-and-bust type of resource exploitation; the latter is though what the western market scheme all is about (Stiglitz, 2003). It is set up that way (Rich, 2014). The Western world and its society are currently far from any relevant sustainability, whatever they say and claim otherwise (See United Nation's Sustainable Development Goals (SDGs www/undp/org) and compared with reality, e.g. for the global forestry sector; Huettmann & Young, 2022). Thus, PNG will clash with the Western World and its attitudes once international business models get rolled out and scaled up. PNG remains the control site, for the other word that is out there.

Table 2.1 A short selected list of CEO's must-reads for operations in Papua New Guinea

Topic	Citation	Comment
New Guinea field experience	Flannery (1998)	A great introduction to New Guinea and with examples of 'development' and impacts, addressing CEOs
Perspectives on guns, germs and steel with a wider Papua New Guinea focus	Diamond (2011a, b)	A generic overview of Papua New Guinea topics and how they relate to modern global aspects of society
Generic Economic Growth impacts	Czech (2002), Daly and Farly (2010)	Limits of growth, in real live
Ok Tedi mine legacy, Mining business model	Kirsch (2014)	Specific example of mining gone bad in Papua New Guinea with one of the largest mines
Understand ecological and spiritual complexities intertwined for the land	Suzuki (1993) Gillison (1993) Baraka (2001) Demeulenaere et al. (2021)	Those are examples to introduce the subject; many more exist
First-hand account of Mt Kare gold rush and Bougainville/Solomon Island perspectives	Henton and Flower (2007)	Specific example of mining gone bad in Papua New Guinea with one of the first modern mines

As PNG also has 'mummies' and celebrates them those ghosts of the ancient times are still around; they watch us, and you must not annoy them (see Cousteau & Richards, 1999; Beehler & Laman, 2020). The use of land needs to be done accordingly (Table 2.1).

PNG factsheet (*Sources* Cousteau & Richards, 1999, Lonely Planet Guide PNG, Beehler & Laman, 2020, online sources; see text for citations).

- island size: area of c. 462,840 km^2 (overall app. 2400 km wide and 1000 km high).
- number of languages: over 800, and number of tribal cultures over 600.
- at least 47,000 years of human civilization.
- 20% of its imports in PNG is oil and gas.
- human population of c.7 millions.
- closest neighbor: Australia.
- former colonial nations: UK, Holland, Germany (Spain and Portugal), Australian Protectorate with a United Nations oversight; Japanese invasion and dominating U.S. + China influence.

As any workforce involves debates about working hours, salaries, hiring and working conditions disputes about labor, land and profits are resolved in PNG, in a PNG way. Foreigners operate in PNG under a visa, which creates complexities. That

Fig. 2.13 At the end of the day, water, land, cosmology and people all come together as 'one'. That is true for PNG and anywhere else

will easily come to the forefront, and it involves frequently an international legal basis, making the cases more complex. But in a PNG world, it can mean to receive a 'spell,' get involved with sorcery and loose or not receive a visa (Fig. 2.13).

2.4 Beyond Just a PNG Factsheet, Some Quick Lessons from PNG's Development

- traditional western views will hardly work, hardly be effective or successful in PNG.
- excluding local views and PNG views, Melanesia views, will fail.
- widely inclusive, equal, mutual and fair approaches must rule.
- mutual learning must be part of the project and be 'built in'.
- multidimensional and multivalue approaches are currently the rule.
- long-term sustainability considerations are the only way forward.

Textbox 1: 'Bush flights' simply remain very risky, so is boating: Some experiences from Papua New Guinea

Flying can be fun…if somebody else pays the ticket and if it's safe and all 'in-time'. If not, you probably have the other side of the experience. Like anywhere in the entire world, bush flights are quite dangerous, specifically in cloudy terrain and wilderness, where maintenance is not guaranteed and certified, or proper tools often not available. Wildlife Biologists have been victim to that situation last decades, and many actually crashed or died. PNG shows a long and deep lasting legacy of bush flying, usually done for industrial and exploitive purposes (Sinclair, 1978); the cargo cult speaks to that fact.

Personally, in my life, I avoid flying, certainly small planes, helicopters and bush flights. I prefer walking. Thus far, I was always lucky with flying and subsequent solutions, but got many experiences.

I used in PNG once an air service provided by the church (as most of those are). This started out nicely—as booked—but the missionary flight then got delayed on departure already for ... 1.5 days. So I just got a sunburn while sleeping at the air strip waiting; traveled back and forth to the air strip, and mid next day we went off. Only 3 people fit in the plane, and after crossing a mountain range we landed after c. 40 min navigating through the clouds, finding a tiny hole eventually to fly down and to land. Navigation was done by sight; the airstrip was fresh-cut grass and we were the usual sensation of the villagers; save and sound.

More exciting was later the flight back out 10 days later. As everybody got excited and tense for waiting, the plane actually never arrived. The airstrip operator told us from his broken radio we were to wait a day in the jungle, but the no- show remained. A longer discussion ensued and perhaps the flight comes next week? I know of cases where people waited for three weeks. As I had an international flight to catch, with invited speaking engagements abroad, what to do? We got caught.

Well, I walked out.

I had walked 'in' in an earlier case; so what's the big deal? But complexities of that situation went quickly over the roof. In the village I found a guide relatively quickly, and off we went. I had no money for the guide but promised to pay everybody upon arrival. Well, that was a good assurance to arrive safely (cases have been heard of where carries get paid and then walk off, you get dumped in the middle of nowhere). The actual walk-out was bone-breaking, involved a super steep hill with mud in the heat, but worked. We stumbled into another village during church time and spoiled their service. We were the sensation; white man walking steep hills; people touched my legs in admiration! Staying at another village house on the way was another great experience, as one gets easily dragged into village life (a burial was ongoing with singing all night). At the morning at departure there was a mis-understanding on the price of the overnight stay, but I did not have to leave my camera behind as a payment and we made it to a road eventually (some unsafe sections got crossed, rascals aware), and then were able to hitchhike to a market next to the airport we started, voila. 18 h later we were back home again; expedition completed. The bush plane was not even leaving for us yet, but instead I got my international flight connection out of PNG, went via Japan and had the guest presentations done abroad.

When in hell, keep walking.

I had another bush plane experience, similar style. As part of a funding scheme we were coerced into the template of bush flying (sounds great to administrators on paper, so they booked it), and thus—after months of preparation—we flew in. Fine (but I hate to be a village sensation and to be used that way).

But then, the flight out, again, to pick us up, did not arrive. The old story; we got trapped. With everybody watching, we were in the hands of an old Australian airplane company owner who knew the game; their game. I was able to 'text' from PNG to North America, and they relayed the message back to the US. Office in another time zone and then to a PNG field office '*we got stuck in the bush and the agreed pick-up plane did not arrive, please ask them.*' Well, this got pretty funny quickly as a single flight can easily cost $6000 …if available. But there was none available anyways. We heard a plane, but it went elsewhere (a coffee flight). We waited and saw in the evening another plane far away in the sky, but it went to 'Kaboom' instead, so we were told. And perhaps a passenger plane flew over us connecting Singapur with Sydney?

Waited over night in the village, and hoped for the best next day. The food ran dry, as we were at the very end of a strenuous field campaign; welcome to the jungle and true bush life. A few other passengers arrived from the bush trails (they knew that the plane was not to arrive the other day anyways). Well, that feels better…We saw a plane, but it flew over us. Apparently a hospital emergency flight that had priority. Eventually we got a plane, end of the day, and we were back 50 min after. Live can take crazy turns.

While bush flights are bad enough, the insane helicopter ordeals I was able to stay away from. But for readers interested, it's described in sufficient length by Beehler and Laman (2020), and I have 100% no wish to repeat those experiences with years of preparations, last minute change of plans, coerced budget revisions, international credit card payments in the thousands, pilots and anxieties. It's gambling in the air, stay cool.

Luckily, I had no bad stories really with buses and public transport PNG neither (we repeatedly appeased mysterious rascals by throwing betel nuts out of the window for them at an agreed road crossing; worked wonders), but those stories seem easily to exist manifold. However, to be complete, one should also speak about 'shipping'.

I took a ferry once to Sepik, it was a great experience. But just to find out there was no return ship from Wewak for the next 6 (!) weeks; all closed (in the western world I spent years doing research from ferries, research boats and vessels, and the service office was usually fantastic; well, not here). At least the domestic airline was running. The complication here was that it was not just a ticket needed for me, but for the field crew members as well…so they demanded. Thus we (=I) bought those tickets, Aussi style, and left from a hotel. But a travel starting 4AM in the dark was not safe, and then even the plane crew was scared. But we got out, and once in the air, it was pleasant. Great views of the Bismarck Range and much money spent during a flight time of 35 min…

We also got out and returned safely from all our coastal travels over the years. I did several of those trips, often as an alternative to road travels, or to walking, or due to intense village fightings in the region that was to be avoided. Entire landscape got blocked that way. And so I like those boating trips indeed as I can do marine surveys and learn about the marine aspect better, too. However, not having a life vest - or spare fuel- on open water is a rather bad (deadly?) idea. Seeing the captain drinking 8 cans of beer in 2 h could be scary. And running out of fuel several times along the way, on the open ocean is even worse and when huge waves come in with the night on the way. Parts of such a deadly ordeal have famously been described by Hoffmann (2015).

On a concluding and sad note, I learned upon safe return in North America that our good and very friendly Australian bush pilot had crashed and died during rescue attempts (the crashed plane was not to be found in-time in the remote bush and steep slopes. Media release published here: https://www.abc.net.au/news/2018-01-04/concert-pianist-turned-bush-pilot-david-tong-dies-in-png-crash/9301562). May our souls rest in peace.

Textbox 2: "*Hey Director, one cannot simply come here, take, grab, and then boost about the physical exploits for your own promotion*": Some business examples gone wrong in Papua New Guinea

Among colonial-style experts Papua New Guinea is traditionally seen as a prime example for a collector's heaven: It's exotic, it carries a dangerous wilderness reputation, it's colonial offering cheap labor and good visa and permits, and it allows for new taxonomic species descriptions to the investigator and curator; PNG is significantly undercollected, and much of such research occurs there (when compared to Papua due to research permits and repatriation legacies) (Beehler and Laman 2020). And so, locals are expected to be easily paid off for delivery and support. What's wrong with it?

Well, it's already wrong for geological surveys, as the strategic and wealth information can easily be used for the apparent exploitation and subsequent destruction of the very land everybody lives on. Can you take a naïve-looking rock out of the PNG country (which then upon closer lab inspection might show a rare metal for a subsequent mining exploration? Answer: Yes). It's also wrong when considering the national ownership of those local resources and the land. It's further wrong when the actual gain is made manifold ("*exorbitant*") outside of PNG but without locals included, acknowledged or even considered. In reality, there is a certain disrespect shown in such endeavors as outsiders just come and take; one will agree: it's intrusive. Beehler and Laman (2020) clearly state that in New Guinea overall, most of such collection research occurs in PNG (arguably because the permits and regulations there are very relaxed, hardly enforced, opposite to the Indonesian side).

This textbox addresses the Directors and CEOs specifically because those are often approving the international expeditions, collection trips, fund-raise for it, and use it for their own PR, reviews, tenure and promotion, also done for money and the economy (e.g. Blazey 2016). It's very clear that much extraction work like the one frequently carried out in PNG is widely done for self-boosting and self-promotion, contracting on the cost of PNG. Many photographers, writers, NGOs and contractors operate that way. Work by The World Bank achieves hardly any other (Rich, 2014). So why approving of it?

It actually has a long legacy due to the colonial history, which tried to make those remote sites attractive, boost the western population there, the income and get the areas 'organized and peaceful'. But see reality. Still, it's a long and established business model, but one that fools most people involved. Many examples exist, e.g. Germans that had to pay a lot of money for land parcels in Northern PNG (e.g. Metcalf, 2011 for Marquis de Rays scams for settlers, see also Spanish-German colonial property contracts, or the sales price for Queen Emma's palm plantation).

Starting with Australian gold seekers, and anthropologists and philanthropists, Streseman (1923) is a subsequent typical example for the colonial '*Ausbeute*' (German word for an exploit), with many thousands of PNG specimen 'collected and brought home,' with the many decade-long Archbold expeditions as an easy follow up by the WW2 winners (Hoffmann, 2015; Beehler & Laman, 2020 list overall many many thousands of specimen taken abroad. Those are hardly documented or made available for a global audience, or for PNG citizens!). It's an easy consequence from there to get to the permits and work of Exxon and Mobile, or Chinese Nickel Mining once such a culture is entrenched and resources are frequently shipped away and comsumed with global harm. What does PNG get, and how is global conservation, humanity and sustainability served?

References

Asian Developemnt Bank ADB, (2012). Papua New Guinea: Critical Development Constrains. Manila Phillipines

Banks, G. (2008). Understanding 'resource' conflicts in Papua New Guinea. *Asia Pacific Viewpoint, 49*, 23–34.

Baraka, R. (2001). The meanings of independence in Post-Colonial Papua New Guinea. In *A plan B research paper submitted to the faculty of the center for Pacific Islands studies in the candidacy for the degree of Masters of Arts*. School of Hawaii, Asia and Pacific Studies. The University of Hawai'i at Manoa.

Beehler, B., & Lamann, T. (2020). *New Guinea: Nature and Culture of Earth's Grandest Island*. Princeton University Press.

Beltran, J. (Ed.). (2000). Indigenous and traditional peoples and protected areas: Principles, guidelines and case studies (Vol. 4). IUCN-The World Conservation Union.

Blazey, T., & Perkiss, S. (2016). The empowerment of Papua New Guinea's Tribespeople: Overcoming the challenges of foreign investment projects. *Australasian Journal of Regional Studies, 22*, 206–2024.

Chan, J. (2016). *Playing the game: Life and politics in Papua New Guinea*. University of Queensland Press.

Coleman, D. (2015). *Development Ideas: A study in comparative capitalism*. A thesis submitted for the degree of Doctor of Philosophy at The University of New South Wales.

Cousteau, J. M., & Richards, M. (1999). *Cousteau's Papua New Guinea Journey*. Henry Abraham's Books.

Czech, B. (2002). *Shoveling fuel for a runaway train: Errant economists, shameful spenders, and a plan to stop them all*. University of California Press.

Czech, B. (2019). The trophic theory of money: Principles, corollaries, and policy implications. In *Journal and proceedings of the Royal Society of New South Wales* (Vol. 152, pp. 66–81).

Daly, H., & Farley, J. (2010). *Ecological economics: Principles and applications*. (2nd ed.). Island Press.

De Soto, H. (2000). *The Mystery of Capital, Why capitalism triumphs in the west and fails everywhere else*. Basic Books.

Demeulenaere, E. S., Yamin-Pasternak, D. H., Rubinstein, A., Lovecraft, l., & Ickert-Bond, S. M. (2021). Indigenous spirituality surrounding Serianthes trees in Micronesia: Traditional practice, conservation, and resistance. *Social Compass SAGE Journal*.

Diamond, J. (2011a). *Guns, germs and steel: The fate of human societies*. Random House.

Diamond, J. (2011b). *Collapse: How societies choose to fail or succeed* (Revised ed.). Penguin Books.

Duménil, G., & Lévy, D. (2011). *The crisis of neoliberalism*. Harvard University Press.

Ellingson, T. (2001). *The myth of the noble savage* (Vol. 1). University of California Press.

Flannery, T. (1998). *Throwim way leg: Tree-Kangaroos, Possums and Penis Gourds*. Atlantic Monthly Press, New York.

Flannery, T. (2002). *The future eaters: An ecological history of the Australasian lands and people*. Grove Press.

Gillison, G. (1993). *Between culture and fantasy: A New Guinea highlands mythology*. University of Chicago Press.

Gosarevski, S., Hughes, H., & Windybank, S. (2019). Is Papua New Guinea Viable? Political Economic Bulletin 19. https://www.semanticscholar.org/paper/Is-Papua-New-Guinea-viable-Gosarevski-Hughes/6d71c70951c12fa82dea578a36b66923997f3f92. Accessed June 1, 2022.

Hayan, D. M. (1990). *Road through the rain forest: Living anthropology in Highland Papua New Guina*. Waveland Press.

Henton, D., & Flower, A. (2007). *Mount Kare Gold Rush: Papua New Guinea, 1988–1994*. Published by Mt Kare Gold Rush. ISBN 10: 0646482815, ISBN 13: 9780646482811

Holzknecht, H. (1999). Customary property rights and economic development in Papua New Guinea. Google Books

Hoffmann, C. (2015). Savage harvest: A tale of cannibals, colonialism, and Michael Rockefeller's tragic quest. William Morrow Paperbacks; Reprint edition

Huettmann, F., & Young, B. (2022). The so-called modern 'Sustainable Forestry' destroys wilderness, old-growth forest landscapes and ecological services worldwide: A short first-hand review and global narrative on the use of 'Growth-and-Yield' as a destructive and even impossible goal. In *Forest dynamics and conservation* (pp. 53–82). Springer.

Kirsch, S. (2014). Mining capitalism. *In Mining Capitalism*. University of California Press.

Kulik, D. (2019). *A death in the rainforest: How a language and a way of life came to an end in Papua New Guinea*. Algonquin Books.

Ludlam, S. (2021). *Full circle: Power*. Hope and the Return of Nature. Black Inc.

Macintyre, M. (2007). Informed consent and mining projects: A view from Papua New Guinea. *Pacific Affairs, 80*, 49–65.

Mack, A. (2015). *Searching for Pekpek: Cassowaries and conservation in the New Guinea Rainforest*. Cassowary Conservation and Publishing.

Madison, M. G. (1997). "Potatoes Made of Oil": Eugene and Howard Odum and the origins and limits of American Agroecology. *Environment and History, 3*(2), 209–238. http://www.jstor.org/stable/20723041

Manning, M. (2005). Papua New Guinea 30 years on. *Pacific Economic Bulletin, 20*, 145–158.

Matthiessen, P. (1987). Under the mountain wall: A chronicle of two seasons in Stone Age New Guinea. Penguin Classics.

Metcalf, B. (2011). Utopian Fraud: The Marquis de Rays and La Nouvelle-France. *Utopian Studies, 22*(1), 104–124.

Miller, J. D. B. (1973). Papua New Guinea in world politics. *Australian Outlook, 27*, 191–201.

Molden, N. (2019). Conflict dynamics of the Bougainville Crisis: Rio Tinto, Australia and the Resource Curse. https://www.academia.edu/21815557/Conflict_Dynamics_of_the_Bougainville_Crisis_Rio_Tinto_Australia_and_the_Resource_Curse. Accessed June 30, 2022.

Mongabay. (2022). *Papua New Guinea*. https://rainforests.mongabay.com/deforestation/forest-information-archive/Papua_New_Guinea.htm. Accessed March 29, 2022.

Naess, A. (2009). *The ecology of wisdom: Writings by Arne Naess*. Catapult Publishers.

Nelson, H. (1982). *Taim Bilong Masta, The Australian involvement in Papua New Guinea*. Australian Broadcasting Commission, Sydney . ISBN 9780642975669, 0642975663. Google Books https://www.google.com/books/edition/Taim_Bilong_Masta/y2MhAAAAMAAJ?hl=en

Reuters Oil & Gas. (2018). Papua New Guinea LNG project resumes exports—data. 15th April 2018. https://www.reuters.com/article/papua-quake-lng/papua-new-guinea-lng-project-resumes-exports-data-idUSL3N1RT1K3. Accessed July 1, 2022.

Ross, M. L. (2015). What have we learned about the resource curse? *Annual Review of Political Science, 18*, 239–259.

Rozzi, R., Chapin III, F. S., Callicott, J. B., Pickett, S. T., Power, M. E., Armesto, J. J., & May Jr, R. H. (Eds.). (2015). *Earth stewardship: Linking ecology and ethics in theory and practice* (Vol. 2). Springer.

Rich, B. (2014). *Mortgaging the Earth: World Bank, Environmental Impoverishment and the crisis of development*. Routledge.

Short, D. (2010). Cultural genocide and indigenous peoples: A sociological approach. *The International Journal of Human Rights, 14*(6), 833–848.

Sinclair, J. (1978). *Wings of gold: How the aeroplane developed New Guinea*. Pacific Publications. ISBN 0-85807-033-2

Stiglitz, J. (2003). *Globalization and its discontent*. Norton Paperback.

Stresemann, E. (1923). Ornithologische Ausbeute im Stromgebiet des Sepik. Ein Beitrag zur Kenntnis der Vogelwelt Neuguineas. Archiv fuer Naturgeschichte ser. A.A. 89 (7)-96 (8)-92.

Sullivan, N. (2015). Introduction to the prospects for sustainable development in Papua New Guinea. https://www.academia.edu/4492757/Introduction_to_The_Prospects_for_Sustainable_Development_in_Papua_New_Guinea. Accessed July 3, 2022.

Suzuki, D. (1993). *Wisdom of the elders: sacred native stories of nature*. BoD–Books on Demand.

The Guardian. (2016). Papua New Guinea's Prime Minister, Peter O'Neill, resigns 26th May 2016. https://www.theguardian.com/world/2019/may/26/papua-new-guineas-prime-minister-peter-oneill-resigns. Accessed May 10, 2022.

The Guardian. (2019). 'I will eventually get killed': Meet Bryan Kramer, Papua New Guinea's anti-corruption Tsar. 10th August 2019. https://www.theguardian.com/world/2019/aug/11/i-will-eventually-get-killed-meet-bryan-kramer-papua-new-guineas-anti-corruption-tsar. Accessed January 15, 2022.

The Guardian. (2022). 'Our designs are stolen': The fight to keep sacred tapa in the hands of PNG's Oro people 6th May 2022. https://www.theguardian.com/world/2022/may/07/our-designs-are-stolen-the-fight-to-keep-sacred-tapa-in-the-hands-of-pngs-oro-people. Accessed May 18, 2022.

Wilson, M. (2019). *A True Child of Papua New Guinea: Memoir of a Life in Two Worlds*. McFarland Publishers.

WIPO (World Intellectual Property Organization). (2016). Bioethics and Patent Law: The Cases of Moore and the Hagahai People. WIPO 5. https://www.wipo.int/wipo_magazine/en/2006/05/article_0008.html. Accessed June 30, 2022.

Chapter 3
The Oceans Surrounding Papua New Guinea: A Habitat View for Sustainable Fisheries and an Acknowledged Lack of Carrying Capacity Knowledge of the Ocean Ecosystem

> *Cousteau divers found extraordinary scenes of beauty and diversity in many parts of Papua New Guinea's little-explored waters.*
> *Cousteau and Richards* (1999, p. 199).
> *Parliament has been told that 18 Papua New Guineans working as fisheries observers in seas around PNG have disappeared in the past five years.*
> *Radio New Zealand* (2018)

Abstract Papua New Guinea is part of Melanesia and thus consists of a landmass with thousands of islands; it features one of the most diverse cultures on earth. While often not fully recognized PNG actually is an island nation embedded in the sea and its saltwater. The Exclusive Economic Zone (EEZ) of PNG is of major and strategic relevance for the nation; and the ocean wealth is another reason for PNG's diversity and uniqueness. Here the ocean resources surrounding PNG are briefly described and how the geology and biology affects PNG culture and human society in the context of the Anthropocene, which includes colonial, WW1 and WW2 battles, as well as modern gear and subsequent fisheries wars all done with a western and neocolonial oversight. Already 18 fisheries observers in PNG waters disappeared without obvious trace; still a 'cold case' but fully known to the fishing nations in the region like U.S., Japan, and New Zealand. Arguably, the world features a global ocean and coastal crisis, including a crisis of 'The Freedom of the Seas,' and those aspects are reported here also, including seabed mining and decay of sea grass, coral reefs, fish stocks and their connected aspects and collapse.

Keywords Papua New Guinea (PNG) · Pacific Ocean · Melanesia · Coral reef · Marine protected areas (MPAs)

3.1 Introduction

While nations are defined by land, much of PNG is probably better defined by the ocean and its processes and set up; it needs such a mindset for a more proper and more inclusive and effective conservation governance; saltwater matters (Pittmann et al., 2021; Robinson et al., 2022). This is a common-sense concept applied to '*The Earth*'

F. Huettmann, *Globalization and Papua New Guinea: Ancient Wilderness, Paradise, Introduced Terror and Hell*, https://doi.org/10.1007/978-3-031-20262-9_3

also. 'The Earth' actually is more appropriately called '**The Ocean**' simply because its salt water covers easily over 75% of the globe, and thus it dominates the planet and subsequently most of its ecological processes. The dominating ocean concept certainly is true for the app 5,100 km coastline regions of PNG, as much as for the large Melanesian part and for the many estuary regions, e.g. Fly river, Sepik river, Ramu river, and their river/ocean plume systems. It easily applies to the southern sections of PNG connecting geologically under water with Australia (the craton; the connecting Torres Strait is just app. 30 m deep as it presents a flooded land bridge due to sea level rise linked with the world's ancient glaciers and global temperatures; see Flannery, 2002; Beehler & Laman, 2020 for biodiversity and conservation impacts and biogeography).

While tropical waters are kind of nutrient-poor, PNG can beat that in various ways and presents some nutrient-rich hotspots and dispersal systems, e.g. via plumes, sediment wash-outs, currents, ocean floor vents, upwellings and seismic activity all making nutrients more available. Tides in the equator regions tend to be relatively small (c. a 1 m), but they do play a role over time and when 'king tides' and outliers come to the beach (those can have surprising damaging effects on navigating reefs, coastal erosion and brackish water spoiling dirking water).

Broadly speaking, in PNG's seafloor, three major geological plates interact there: the Bismarck Plate, the Australian Plate and the Pacific Plate. And within that geology, PNG comes peppered with many hills above water and subsequent smaller river estuaries and reefs also; it's part of an underwater 'garden of eden' (Stone & Obura, 2013; see for Coral Triangle Atlas at http://ctatlas.reefbase.org/). And thus many of those coastlines are rather remote and just can be reached by 'water taxi' (or bush plane) between major villages or national airports. Traditional outriggers still play a big role; not all are so safe to use and when modern engines are employed (Hoffmann, 2015); life vests or emergency beacons for navigations are often absent (Fig. 3.1).

PNG has easily over 600 larger offshore islands with thousands of smaller ones—most of them located on the shelf. For millennia rural coastal families visit each other on homemade canoes and outriggers (examples shown in Cousteau & Richards, 1999). Much of PNG actually is a coastal ocean nation and islanders intimately understand the high seas and sailing; they can read the environment and navigate for 1000 s of years in that marine universe (see, for instance, Chan, 2016 for the upbringing of the former prime minister and shipping, reef and beach combing experience and subsistence) (Fig. 3.2).

While not given much credit that way, PNG can easily be called a kingdom of islands driven by the ocean-scape and saltwater with a somewhat remote and removed land capitol in Port Moresby. It's a typical setup for (Melanesian) ocean nations, but which is so diverse in many respects. It's a universe in its own rights and it defies western policy (land) minds and governance styles. Already the sea depths of PNG show us no other: Some waters are widely shallow and others are world-record deep and widely untouched (Cousteau & Richards, 1999), and others again are dynamically changing due to sea floor plate dynamics, tsunamis, coral reef successions and some tides. It comes all together connecting with small estuaries,

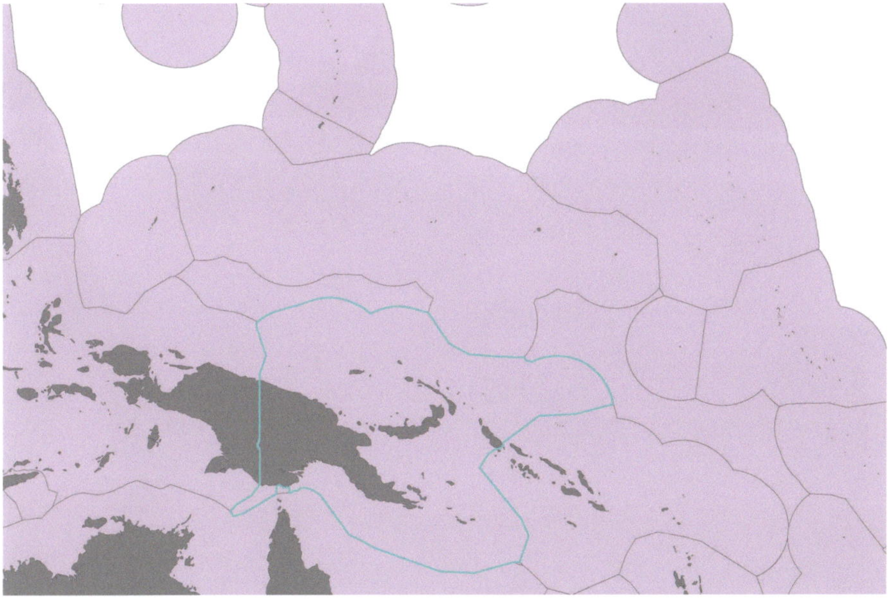

Fig. 3.1 Papua New Guinea Exclusive Economic Zone (EEZ; shown in blue outline). This map indicates what waters PNG has to administer and that there are direct national neighbors (EEZs shown in purple), whereas the international waters (in white) and their no-law zones are quite remote but still have an impact also

sandy and pebble beaches carrying palm trees and even covered with specific orchid species (Orchid Society of Papua New Guinea Inc., 2006). This is what makes already a large section of PNG: the coastline as the ocean–land interface. And not enough, the diverse seafloor extends out of the water and reaches on land up to 4509 m high above sea level (Mt Hagen). The concept of 'Ridge to Reef' is a true PNG feature (Carlson et al., 2019; see Fig. 3.3).

The oceans and coastal zones of this world are in a crisis state (as acknowledged in the UN declaration discussed in The Guardian, 2022a; Cousteau, 1979; Jackson et al., 2013) and so are most coral reefs (see Berzunza-Sanchez et al., 2013 for PNG). The waters of PNG still remain among the world's major destinations for pristine and diverse diving sites (Cousteau & Richards, 1999; Jackson, 2013). PNG is known for the soft corals; and tunicates and polyps are found in abundance. And the Gorgonian stalks are as mind-blowing as the groupers are, or the zebra sturgeon fish, surgeon fish, puffer fish or mother-of-pearls, sea cucumbers, sea shells and giant tridacna clams and their ancient management and human co-evolution (details in Cousteau & Richards, 1999; see also Hyndman, 1993; Jackson, 2013; for overfished mother-of-pearl and its statistics see Simard et al., 2022). But it's arguably the rainbow fishes that PNG is most famous for, and as highly seeked after species in the aquarium markets around the world.

Fig. 3.2 A typical coastal Papua New Guinea coastal set-up

What sometimes is forgotten in this context are the freshwater fish species. Those are often used in biogeography textbooks for their distribution patterns and offer interesting insights according to the Weber Line (Beehler & Laman, 2020 and citations within). On the landscape, the local community in PNG relies on eels for instance as an additional food source of fish found in the rivers, usually in valleys, those are caught in traps (see Fig. 3.4).

3.2 The Northern Ocean Side of PNG

This section of PNG is bordering to Indonesia (Iran Jaya—part of New Guinea), and the mainland coastline is dominated by the Sepik river and its estuary and plume (well described in Cousteau & Richards, 1999 for instance). There is a lot of exploration history in the Sepik river region with (German) colonialists and subsequent missionaries and industry active there for a long time in 'the remote bush' and along that coast. The UN tried the introduce 'exotic fish' in these watersheds but like elsewhere usually, not with good success wiping out endemic species and—in part—entire lifestyles and income systems (Beehler & Laman, 2020). Vanimo and Wewak are the main cities and regional hubs, with an airport and a ferry service for that region but essentially lacking a connecting road to Madang or Lae. There is a mountain range, and much interior forest landscape is often strongly harvested for its trees (The Guardian, 2015, photos shown in Beehler & Laman, 2020). Not surprisingly, several river headwaters are found there also which contribute the wider oceanic set up and freshwater inflow into this part of the PNG ocean. Some of those rivers are seriously affected by forestry and mining.

Nearby Manam Island has another regional importance as has the Kar Kar Island and Madang with a port. The Bismarck Sea is the central ocean waterbody in the region, surrounded by islands like Manus, more in the east are the New Ireland and New Britain groups. The infamous Wuvulu island described in Cousteau and Richards (1999) north of PNG represents a well-known marine biodiversity hotspot. Orcas are found there in abundance showing peculiar behaviors, apart from the major coral reef ecosystem at its finest. This island group had a diverse history in the colonial times with Spain, France, UK, Germany, and some smaller Swedish and Danish

Fig. 3.3 Ocean spirits and gods are everywhere and to be appeased

Fig. 3.4 Selection of diversity and wealth of a diverse and initially pristine ocean ecosystem

ownerships, also made famous by a Russian anthropologist (Nikolai Nikolajewitsch Miklouho-Maclay; see other chapters of this book). The islanders—who were there first and defending their land—were often described as 'fierce' (https://wuvulu.com/history.shtml#History) resulting into the diverse legacy. Wuvulu island was also famous for its palm tree plantations.

This wider part of PNG is truly ocean driven. Rare seabird species like the Heinroth's and Beck's Petrels (Flood et al., 2017) are found there, species like the Aleutian terns connecting from Russia and Alaska can be expected (F. Huettmann unpublished). The diving reports for that region confirm the underwater beauty. But also, fisheries destruction has already left havoc there once again. In those PNG waters local fisheries get used by Asian nations including Taiwan and Indonesia, and it is hardly controllable or enforceable (Cousteau & Richards, 1999, p. 215).

Since ancient times this ocean section connects with China and Asia, and with other parts of the Pacific Rim. This is well exemplified by the shark migration, e.g. Slavador et al. (2010), as well as sea turtles and seabirds (Block et al., 2011). From the human side, among other immigration waves, the relatively recent Lapita culture likely entered app. 3500 years ago along this corridor (see Beehler & Laman, 2020 and citations within) leaving wider impacts in PNG.

3.3 The Eastern Ocean Side of PNG Mainland

The eastern ocean side of PNG's mainland shows a very complex ocean seafloor. It features proximities to one of the deepest ocean trenches in the world (e.g. Challenger Deep c. 11 km deep; Mariana Trench). PNG waters are part of the New Guinea trench, app 10 km deep. Also, it actually is one of the world's most active areas with many sea earthquakes; often those are not so well recognized. Much study has been done there on those topics, e.g. deep ocean waves (Hollar, 2012 for the famous work on internal oceans waves by Walter Munk, etc.).

The coastline of this region is affected—in part—by the Ramu river, and this coast is referred as the Maclay coast with a longer colonial research and collection legacy (discussion in Ball, 1880 and citations within).

Further, the coral reefs and associated seas of this region are among the world's most fascinating, somewhat most pristine and still secretive waters (Jackson, 2013; see http://ctatlas.reefbase.org/ and Berzunza-Sanchez et al., 2013 for assessment).

Lae has a major port connecting with Australia and Asia. Other than that it's worthwhile to mention the lack of major hubs and city centers along the coastline and the Huon peninsula. While the area has ancient human occupation, it still lacks a 'modern' urban development and set up, making such areas attractive for explorers (see Cousteau & Richards, 1999; Hoffmann, 2015, etc.).

Milne Bay is a major section in eastern PNG and presents a bay with a large and deep surrounding culture. It was part of initial 'contact' and much international exploration, and later became a site of intense fighting during WW2 (details in Cousteau & Richards, 1999).

3.4 The Eastern Ocean Side of PNG: Islands

This part of the ocean includes the Solomon Sea and it is another part of true Melanesia, consisting of many small but remote islands, island groups and their diverse and unique cultures such as New Britain and New Ireland. There is a long colonial history for those regions. It's located in the southern part of the Melanesian islands including Bougainville that has virtually all aspects of the Solomon Islands. Just like the northern islands of PNG this part of PNG is truly ocean driven and has connections with Asia.

The Solomon Sea is most influential for this region, with the Trobriand Islands in the south, and Bougainville at the outer edge. New Ireland is just next to it, e.g. famous for coconut palm plantations and coral reefs. Other island groups are Schouten Islands, Vitiaz Strait Islands, and Milne Bay in the south, which includes D'Entrecasteaux Archipelago and Woodlark Islands.

This area is known for its deep human traditions also. Beyond many others, the giant clam is a fascinating species for sustainable management of the publicly shared 'clam gardens' (Lucas, 1988). Many of those ancient 'gardens' are now on the decline, e.g. due to overharvest and market demand in Japan and Asia. Another unique culture in that region is for instance the 'shark callers.' Shark calling is an ancient ceremony and includes coconut rattles to attract them. Once sharks are attracted and after entangling, the shark gets roped in with a vine noose and one brings the shark on board (details in Cousteau & Richards, 1999, p. 200). Another very interesting aspect of fisheries is found there using the plant-based nerve poison (rotenone) of derri roots; it's similar to what is used in Amazonia (details in Coustau & Richards, 1999, p.197). Those are essentially sustainable fisheries practices which the western world is rather short of (see Pauly et al., 1998 for harvesting down food chains worldwide but mostly carried out by western society and their institutions and agents).

Ecologically, this area is very complex and somewhat related to Guam (Micronesia; Cunningham & Beaty, 2001), and remotely, it connects with Hawaii even. However, albeit the indigenous people in this region are 'brothers,' the Polynesians that occupy these Pacific islands are not really so connected with mainland PNG and do not directly link with that section of Melanesia.

Bougainville, closely related to Solomon Islands, is another part of PNG part in this ocean section. It will be covered in subsequent chapters in more detail. But the politics and economics of such islands in the South Pacific are well documented (e.g. Filer, 1990).

All the islands north of PNG -northwest and northeast—have a deep and complex history including colonialism and during WW2 they were occupied and fought by Japanese troops, and then 'freed' by western allied troops, namely U.S. and Australia during fierce battles (ecological legacy described with photos in Cousteau & Richards, 1999 including subsequent dynamite fishing; first-hand accounts of the area are found for instance in Chan, 2016). The cold war legacy of those islands, including world-relevamt biological warfare experiments and impacts, affecting the ecological balance is described in depth by Rauzon (2016).

3.5 The Southern Ocean Side of PNG

The southern ocean side of PNG's seafloor is pretty 'flat' and shallow, and connected with Australia; it is part of the '*sockel*'—the Australian craton. In PNG overall it shows the same geological material than Australia, albeit much more rugged and steeper, even with potential snow coverage (now global warming makes it virtually impossible to occur anymore; as shown early on by Flannery, 1998 for New Guinea. The glaciers in New Guinea—located in the north—were very well and repeatedly studied already last 100 years ago, e.g. by British expeditions; see Beehler & Laman, 2020, p. 46 for overview and details). This part of the ocean—the northern Coral Sea—is often only 30 m deep, or so. There was a very relevant land bridge, app. 15 mio years ago due to Earth's glaciation history affecting the biogeography of the region, the seafloor and coastal zone. The southern section of PNG includes the capitol of Port Moresby (POM), and it's famous for some coral reefs accessible for tourists. POM is part of a major ocean travel route for PNG connecting with Australia and Asia, this includes crime and human migrants and trafficking.

3.6 The South Western Ocean Side of PNG

The western section of PNG oceans borders to Indonesia. It features the Arufa Sea in the west, the Coral Sea part in the east, and a prominent island group in the Torres Strait; the Torres Islands (e.g. Laffan, 1991). Those are of wider historic value and carry a deep human legacy (Flannery, 2002). By now, those are prime problem areas for climate change due to sea level rise (e.g. Green, 2006 for Torres Islands, and Edwards, 2013 for a PNG-wide example). Noteworthy is that this region is also part of an ancient, as well as a very modern, international travel route, many container ships use this part of the ocean to connect to India, Africa (Cape Horn or Suez Canal) and the Atlantic eventually. It's a certain bottleneck and an alternative to more pirate-infested shipping routes in SE Asia.

This area is a unique section because it has the Fly river estuary, the largest river in PNG (app 1050 km long). Relatively large wetlands are found there, e.g. attracting migratory shorebirds from Australia, Asia and the Arctic. The Fly river has received much attention by western explorers for centuries, including Sir Hillary after his Everest work and his subsequent tragic boating accident there (Gill, 2020). This region is also a source of the 'cannibal narrative' about PNG and of many other stories. Many western people obviously went there because it was quite accessible and it has been perceived as an ultimate adventure to be 'among cannibals.' Nowadays, this region is known for the river contamination caused by mining, e.g. Bolton (2009), Kirsch (2014) for Ok Tedi mine impact overview (Figs. 3.5, 3.6 and 3.7).

3.7 Overall Oceanography

PNG is part of the 'Ring of Fire,' and this also applies to its ocean and the plate boundaries rising and moving. PNG features parts of the Mariana Trench connection, the world's deepest seafloor allowing for many insights about earth and the environment itself (see Peng et al., 2018 for microplastics in the deepest and remotest section far away from humans). In such dynamic ocean seafloor areas therefore, ocean tsunamis are common and create problems for the local communities. The latter are well aware and have adjusted to such a life.

In addition, those 'seaquakes' cause active volcanism. Volcano outbreaks are part of PNG; some of them come straight from the ocean. Rabaul is one of those tragic outbreak legacy sites (see Chan, 2016 for a first-hand evacuation account). The direct impact on the ocean ecosystem remains little studied but the sea floor can quickly change, so does the coral reef ecosystem; the latter is now widely affected by climate change and coral bleaching. There is some good discussion how humans

Fig. 3.5 Handcrafted Papua New Guinea mask with kauri shells

also affect earthquakes, on land and at sea, due to melting glaciers, glacier weights and subsequently changing plate pressure dynamics causing ‘tensions and release’ between plates (Figs. 3.8 and 3.9).

3.8 A Very Special and Endemic Marine Set Up in PNG, Including Benthos

Papua New Guinea is an acknowledged world-class ocean resource and wilderness; endemism is exceptionally high. PNG is not only affected by the Wallace Line, but also by Weber’s line (Beehler & Laman, 2020) as shown by ichthyologists for many decades.

Fig. 3.6 Where precious ocean wealth now is used: cheap house gear. It remotely reminds of a former ocean diversity

PNG reefs are said to be often more diverse and still in a better condition than in Australia (Beehler & Laman, 2020, http://ctatlas.reefbase.org/; see Berzunza-Sanchez et al., 2013 for assessments). Over 2100 species of reef fishes can be found around New Guinea's reefs (Beehler & Laman, 2020). PNG further serves as a resource for endangered marine species, e.g. sharks (three species are typically found: Whitetip Reef Shark, Blacktip Shark and Bull Sharks; the latter even occurs in some freshwater lakes; see White et al., 2015 for river sharks). PNG also has sawfishes (Grant et al., 2021). It also has a very high ray fish diversity (Blaha et al., 2016; White et al., 2019). The historic size of the thuna stock—a predatory species in the food chain—was vast but now crashed due to overfishing (see for details Barcley & Cartwright, 2008), just as the case for whales (at least 15 marine mammal species were detected, Miller & Rey, 2021).

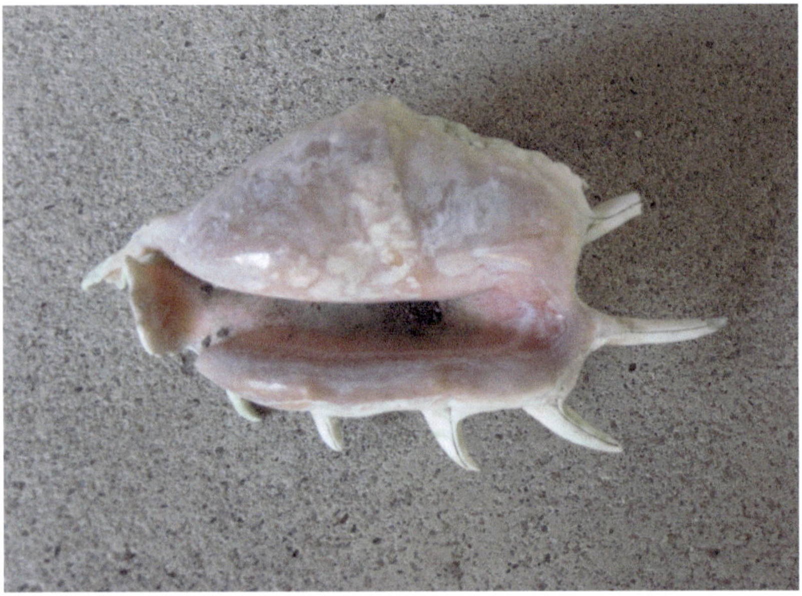

Fig. 3.7 Large shells as a precious ocean sustainability item found widely in-land, distributed on ancient-old trails in interior Papua New Guinea

Fig. 3.8 Sea turtles get fished, poached and offered on domestic markets for meat consumption

Fig. 3.9 Souvenirs for the tourist market showing manatees being hunted and penis bones being sold

Because PNG is so connected with all aspects of the universe, the earth's geology shows as quite accessible to humans, on the interacting surface. The seafloor and the trenches are a unique habitat and allow for a deeper earth insight. While the associated marine seafloor life remains a mystery in the abyss, another unknown, the actual benthos can be studied. While some benthos research has been done, it remains primarily—again—just a species list; much more is to be done and should (see Wei et al., 2011 for global benthos models to start from for PNG). It's a big deal for instance for sea floor mining questions (see chapters in this book and Steiner 2009, 2011) (Figs. 3.10 and 3.11).

3.9 Island Economics, Whales, Fisheries, Estuaries and Sustainable Oceans with MPAs?

The Pacific Ocean is vast, and it includes much of the world's fisheries (e.g. Dalzell et al., 1996). Islands of the Pacific are 'many' and each present their own microcosm. They are seen as a unit in (Island) Biogeography, over water and under water (see Mayr & Diamond, 2001; Steadman, 2006 for an applied example). These natural territories have been recognized early on by colonialists as precious and strategically relevant and they were intensely fought over, e.g. as part of the 'Pacific Theatre' ongoing today.

Fig. 3.10 Children's job in fishing communities consist of bait fishing

Fig. 3.11 Thuna on a local market, not a big one at all as those ones tend to be harvested out and used up by larger extraction industries

However, what is widely misunderstood is that those islands stand in their own, but as well as within wider, global dynamics. They are affected for instance by the Antarctic current, as well as by the Monsoon and the El Nino cycles all interacting in synergy, beside other impacts. Human and Polynesian intrusions and cultures played a big role for their (environmental) modifications. Expert collection of species on islands added further to the pressures. For Melanesian islands the specimen collection peak was achieved in the 1990s already (see in Mayr & Diamond, 2001); so why collecting more still, or even call PNG significantly undercollected as stated by Beehler & Laman, (2020)?

These islands are obviously not really independent or in a balance; see for instance Rauzon (2016) for ticks and invasive species on those Pacific islands deemed to be '*pristine.*' It's part of bioterrorism experiments during the Cold War and the Anthropocene, and it seriously affects the theory and inference for biogeography, and subsequently how we understand live, islands and PNG.

Like most of the world's ocean ecosystem (see UN Crisis Declaration in The Guardian, 2022a), PNG is not really sustainable anymore, nor should it be perceived that way or receive such a PR or management scheme with a static maximum sustainable yield (MSY) concept (Punt & Smith, 2001; which was never really achieved or exactly computed and aimed for anyways; Pauly et al., 1989; Pauly, 2019). Reaching a sustainable fisheries, in remote regions, with multispecies occurrences and interactions—including endemic and undescribed species—in times of climate change, is virtually impossible and almost never shown to have been achieved (Pauly, 1979; Pauly et al., 1989). The Aichi Targets (Sustainable Development Goals (SDGs; www/undp/org)) promoting a win–win for everybody are not helping on this matter as they just implement well-sounding neoliberal concepts and narratives and myths that have no good track record anywhere denying any relevant science for decades, e.g. it's a promoted growth scheme to operate on finite resources instead (see instead Limburg et al., 2011 for generic perspectives in a limited ocean; Safina, 2003 for reality perspectives).

Well familiar with PNG, already in 1979 Cousteau (pp. 411–414) stated the following problems with the oceans, summarized in just a few bullet points ('*The Planet is on Fire*'):

- man the waster.
- sowing less than we reap.
- waste of the world.

These succinct points speak very well to the problems we still have today and which we still have not resolved yet; far from it. It's easy to see that ocean research and marine ecology have not resolved that well for us in PNG (Rooney & Papoutsaki, 2004). The ocean governance in place, certainly for PNG, has not been much effective nor the institutions.

There has been a long, historical and deep shell trade in PNG from the south and coastal areas. It was quite sustainable. However, it went by the way side. The fisheries in PNG waters, or adjacent ones, simply is overharvesting the resource; computations of a sustainable harvest are either not existing, not possible, or not enforced, hardly monitored precisely. Robust institutions are widely missing in an area that is considered of many 'failing states.' Even basic items like sea cucumbers are not really manageable then (e.g. Purdy et al., 2017 for a required moratorium and impacts in PNG and markets). Relevant number of fisheries observers are not existing, but those who do join that profession get intimidated, or much worse (see The Guardian, 2022b for killed fisheries observers in Tonga; PNG has reports of over 18 dead fishery observers also; Radio New Zealand, 2018).

Shark finning, as a widely unsustainable fisheries method, is widespread also in PNG waters and done for markets in Asia, Canada and U.S. (Beehler & Laman, 2020), some shark products occur in the EU also (Fig. 3.12).

As a typical and still widely unresolved book-keeping problem in tropical nations and fisheries, the 'approved' numbers of fish harvested do not add up well. Note-worthy for most (tropical) fisheries here are the offshore trading quotas, illegal takes, and mislabeling of species on the global marketplace. For instance, PNG sharks are apparently found—and sold—in Greece within the EU and its subsidized efforts (Pazartzi et al., 2019; compare with White et al., 2018).

Instead, and like any ecosystem, PNG and its oceans are finite and delicate, and must not be overused (e.g. Limburg et al., 2011). This includes fish but many other ocean resources and ecological services also, including sea bed mining (Steiner 2009, 2011). Beyond catering international market demands and standards, PNG must look out for itself also: Sustainable ocean communities are real and they have existed fine for over 47,000 years (Hyndman, 1993)! This is clear for most island and ocean nations. Unless islands receive outside support, they struggle on their own for a decent survival while markets from abroad prey on them further till the entire system collapses. Virtually all islands are outcompeted by the bigger 'main land.' Islands become almost always political (e.g. McKenzie, 2021 for a South Pacific example). From deep time till now, the deep sea region of PNG was wilderness; virtually untouched due to sheer remoteness in the three dimensions. Back then, who knew about hydrothermal vents in the Bismarck Sea? Ecological services from the deep sea regions are excessive. Human pressures in that region were relatively small, compared to what we all experience today. This trend likely will change further with more sophisticated gear and deep sea mining plans (see subsequent chapters; Steiner 2009, 2011) while the bulk of the resource biomass just sits at the 0–60 m ocean depth and is already overexploited.

The ocean receives input from many sources. For examples of related river resource declines draining into the oceans, see invasive species in the Sepik river having resulted into a major human population shift due to the ecological collapse (Cousteau & Richards, 1999). For other rivers, the Fly river as the major river in PNG can be seen as a typical example: Fisheries there was affected via the headwaters of the Fly river where the Ok Tedi mine is located. When a tailing dam broke—as a typical and frequent problem in 'sustainable' mining—toxic residues polluted the natural resource with lead, arsenic, mercury and cadmium with ongoing concerns to this very day (Kirsch, 2014 for detailed descriptions). The real impacts are difficult to assess as the wider watershed and food chain is affected in PNG and in the Indonesian border zone as well.

Fig. 3.12 **a** and **b** Protein source for Papua New Guinea citizens straight from the processor and its fleet

And even worse, major game changers in PNG's history came with dynamite fishing; a destructive way of getting at the fish at all costs (Pauly, 1989). It's a leftover of WW2 dynamite—part of one of the largest bombardments in human history (as shown and documented in Cousteau & Richards, 1999). In parallel, the Asian fisheries fleet and their canneries wiped out major resources and devastated coastal zones and communities accordingly (Sullivan et al., 2003, 2011).

But before that, the steam engine and then the combustion engine left already major marks at sea, at land and in the atmosphere. The fishing styles changed over time subsequently. Already with an influx of new people from Asia 1000 s of years ago they brought with them different fishing styles affecting coastalscapes, oceans and PNG (e.g. highly populated highlands trading with the coastal communities). The impact of small engines and available fuel was vast. Major inaccessible remote fishing grounds now were accessible and got harvested without return. Having harpoons (for whales) available and adding never-ending monofilament plastic nets and global market demands did the rest; those net types are now all ubiquitous. And thus in parallel, the ocean decays with many fish stocks lost (Safina, 1999).

And the use of reliable but cheap outboarders changed the reach and intensity of fishing and access to fish stocks even further. More recently, the fisheries effort further increased, so did the gear used, helped with digital tools such as fish finders, sonar, GPS and the internet. All of this is easy to see and to detect in PNG as Asian fleets have almost perfected such skill set to get at the fish harvest and making money essentially at all costs for the fleet. Fisheries policies allowing such fleets now a selective access to PNG waters, e.g. for thuna, contribute dramatically to this problem.

Same can be said for 'winches,' and certainly, the all-dominant floating plastic net; the bigger the better, often measured now in 10 s of miles (= 'walls of death'). Spears are not that new, but gun-powered ones are, and ones that use rubber In PNG one can essentially buy them in any supermarket now (F. Huettmann pers. com.).

For many decades already local fishermen now routinely fasten slings of surgical rubber to their spears, which make it more powerful that way; it's very cheap and has impacts on fish (details in Cousteau & Richards, 1999, p. 205). Considering that in PNG fish and fish supply is provided by children and females (Beehler & Laman, 2020), those 'low tech' tools are now put at their free disposal, and they are ubiquitously used throughout the tribal society; the fish resource pays the costs.

Even worse, and as found in many parts of the world, including the U.S. National Parks, exotic fish got introduced. In PNG, already since 1949 the UN and other projects introduced exotic fish species in the Sepik river watersheds and elsewhere with the idea to improve food situation for the rural community; arguably a well-sounding idea but with very bad consequences (Beehler & Laman, 2020, p. 159). Overall, in PNG, at least 21 exotic species got introduced already for 'food security,' including species like karp and African tilapia (Beehler & Laman, 2022, p. 148). The Asian Development Bank projects did its share on 'fish ponds' (the author saw several of those start up projects, but that ran dry and afoul). Aquafarming and shrimp ponds are part of that scheme, with similar bad effects. A wild fish stock remains unbeatable; all else gets quite problematic one way or another.

It's difficult to deny that the fish stocks in PNG got compromised and plundered. PNG is part of the 'coastal collapse' described by Jackson et al. (2001) already a decade ago. And now that's all done in an industrial fashion, helped by outside nations, industries and embassies (the latter commonly provide visa support, legal and banking advise, contractual help, and intervene with vessel confiscations and labor questions). It is also clear that the major fish spawning habitats—reefs, mangroves, and seagrass beds—are on the vast decline, that sustainable management is widely absent, and that it continues to dominate in the absence of the western-made and -shaped political nation and EEZ PNG construct (Fig. 3.1). It's another outflow of colonialism gone bad, ignoring science and best professional practice.

Clearly the PNG coast guard cannot police all of the mandated EEZ. The seagrass beds alone are wild and vast, hardly accessible (due to being in very shallow waters). They feature stunning biological insights though as they seem to be often 'one plant' (= perhaps the largest plant in the world; Edgeloe et al., 2022). It is here where not only the framework, set by the new PNG national construct, but as well as by the international oversight, the UN mandate and NGOs have widely failed PNG's environment, ecology and its people, and sustainability at large. Food security in real life has a grim outlook there and consequently refugees are on the rise (Quayle et al., 2019).

PNG waters are certainly affected by humans, and have been so, for millennia (Hyndman, 1993). The human footprint existed but was benign, e.g. due to lack of access to remote areas and depths, and using just soft gear. The status of the giant clam gardens speaks to that situation in powerful terms, so does the recent sea cumber moratorium or widely increased harvest statistics of mother-of-pearls, and the generic decline of sharks and thuna (Fig. 3.13).

In the Western World, ocean zoning is now often practiced, based on marine protected area (MPA) optimizations. A MPA network is to be assessed and designed (e.g. Hamilton et al., 2009), while it's not so clear how it is to be funded, enforced and effective even (Klein et al., 2015), and by when it comes truly into action (e.g. Green et al., 2007). See for northern PNG for instance the Bismarck Sea Task Force (https://www.marinemammalhabitat.org/portfolio-item/bismarck-sea/). MPAs are not always providing the best protected biodiversity solution because they are biased by the input and contractor NGO. An example found in WAITTS (Ocean Protection Waitt Foundation and its aims to balance the environment AND the economy but catering economy the most (https://www.waittfoundation.org/ocean-protection). Something good and sustainable is to be done about PNG oceans, and here is some action then: The role of sperm whales and killer whales are used as criteria for protection, and it looks good in public. However, it's not very effective when considering that those are moving and migratory, virtually unsurveyed, and they are not so well indicating the small pocket endemics which are famous in PNG. Using indices is always an indirect and coarse approach missing many details. Arguably, PNG waters are more complex and bigger than just handling a few marine mammals. Similar to National Parks, the MPA idea is a brain child from The West, heavily supported now by Australia, perceived as the last 'cry' in ocean management (see, for instance, Halpin et al., 2022 for nearby ocean areas and Gadlflies. Those are endangered seabirds and are to be 'zoned out' into areas that are to be protected, but ecologically cannot), another typical western approach to ocean resources and failing. It's widely behind trying to protect anything, and anything relevant with a meaningful concept while we are in a world and PNG ocean crisis of a public good embedded in deep ecology (Figs. 3.14, 3.15 and 3.16).

In the meantime, the MPAs are usually just political compromises, and primarily consist of biologists circling areas on maps (as stated Beehler & Laman, 2020) in various shapes and forms, often based on aided optimization algorithms driving it (see Halpin et al., 2022, see Figs. 3.8, 3.9, 3.10, 3.11 for examples and realities in PNG waters). Whereas, the real issues like poverty, economic growth promotion, profit motive and good governance are not mentioned certainly not reached or resolved and almost never part of the 'modern' conservation equation.

Fig. 3.13 Shark calling and shark fishing is a long but sustainable practice in Papua New Guinea; but now taken over by commercial Asian efforts

Fig. 3.14 Inland fishing adds to fisheries views: photo of a fish trap

Last but not least, the biggest issue now comes from man-made climate change as the ultimate outcome of those synergies that were not addressed early on. Aquafarming will not fix it (Fig. 3.17). There is little good outcome from this situation—no winners and just a few losers—because the world's ocean currents, the water quality itself and coral reefs are all affected and the climate refugia are breaking down (Dixon et al. 2022) with implications beyond ocean ecosystem but affecting human well-being, globally. Without ocean, PNG as we know it, is 'toast.'

Textbox 1: Why is being a fisheries observer in PNG a deadly job, with 18 bodies missing and still counting? Troubles in the EEZ and with Freedom of the Seas

On the one hand, one cannot answer that question easily. While many fisheries observers report intimidation during their job, PNG fisheries observing would not differ from any other fisheries, observing or assessment job. But on the other hand, knowing that PNG is a dangerous place for crime to start with, why would the fisheries be any different?

It simply differs because PNG has the largest thuna fisheries area in the Pacific, and the PNG fisheries observers are treated by foreign vessels; their lives are essentially in their hands, out there in the wider PNG EEZ.

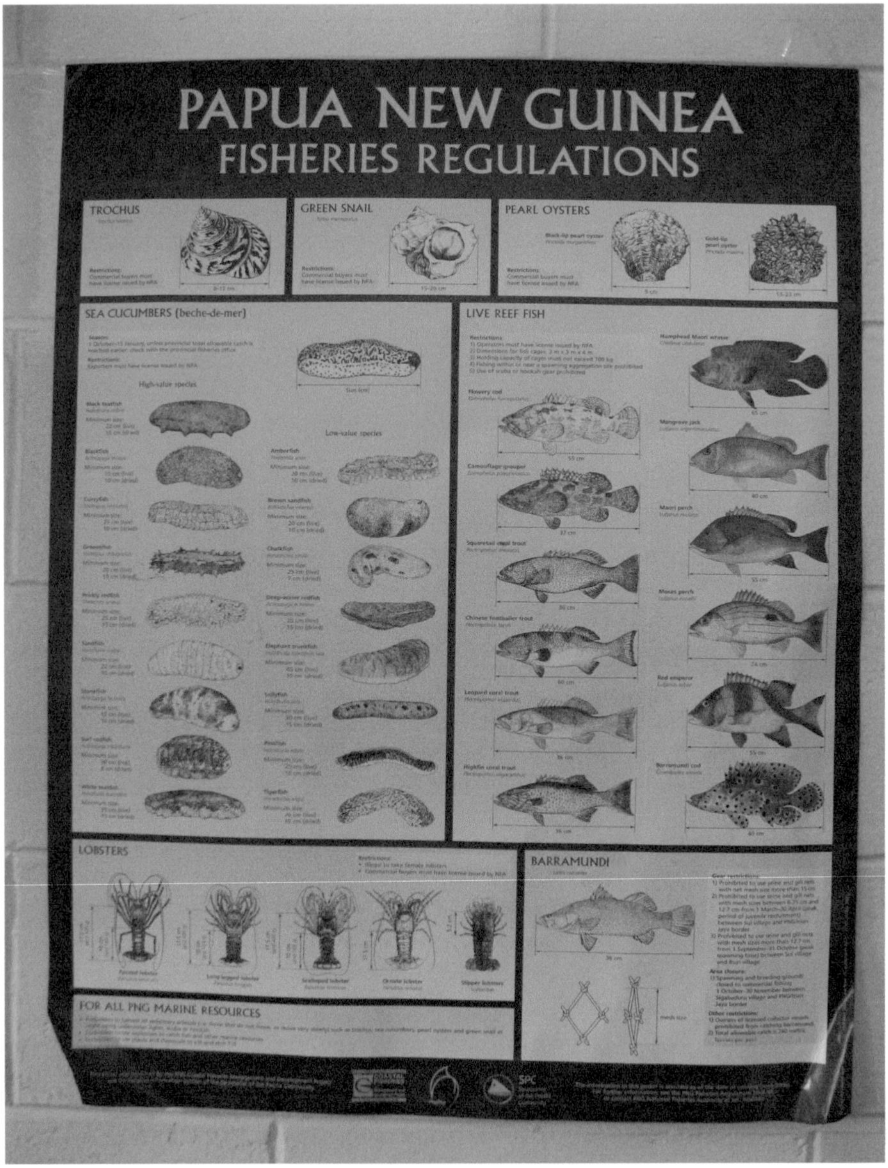

Fig. 3.15 Papua New Guinea's fisheries regulations for practitioners

PNG grants fishing access to resources in PNG waters of the EEZ through international agreements with Taiwan, Korea, the Philippines and China, as well through a multilateral treaty with the USA. Fisheries observers are then

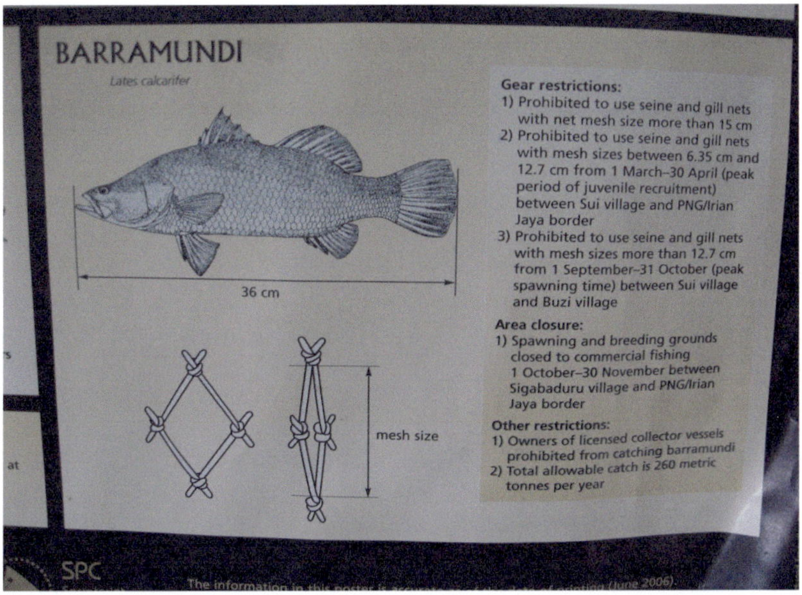

Fig. 3.16 Some details on Papua New Guinea's fisheries regulations for Barramundi

Fig. 3.17 Aquafarming as the solution to what?

affected by the laws of those nations; not 'just' PNG legislation. Certainly the U.S. customers of fish are very sensitive by now.

Globally speaking, fishing observers have a tough job and might not necessarily be well liked for their task by some exploitive industry members. After all, they can expose bad reality details and their reports can shut down the entire fishing task at hand; it's money off the table for some

Similar to whistle-blowers (Kennedy & Cuomo, 2000), fisheries observers do indeed make fisheries better and can help to show problems that otherwise would not be really visible, e.g. (Smart et al., 2016).

However, PNG seems to be specifically deadly and a warfare rages at sea centered around fishing observations. Radio New Zealand (2018) - not Australia or U.S. that usually have often the PNG oversight - exposed 18 missing PNG citizens in such a job without solution or good answers by the PNG government (see PNGAttitude, 2019 for more details and lack of progress).

The EEZ is supposed to serve anybody in PNG and allow for national protection and jurisdiction. The Coast Guards provide such a service…Well, if they are around, if their gear exists and is adequate, if all items line up well, and if a case can be made. Realities in the world fisheries show us a different picture though.

The EEZs tend to be too vast and too complex, with just a few coast guards—ships and helicopters—to really be on the ground and to enforce 'anything.' The EEZs are a real security and surveyance problem, certainly for PNG. Reality is, outer and wild areas of the EEZ - where some of the precious fish actually is located - are not enforceable.

In addition, the international sea, where the 'Freedom of the Ocean' sits, remains an utterly absurd concept: In times of legal regulation and governance, why should unregulated, criminal behavior be allowed at sea, anywhere? By now, the international seas are an international place of labor violations, of crime, a global place of war for everybody who can engage there. Money drives it, and always has.

From an environmental perspective, this applies specifically to fishing and following fishing regulations and contamination, e.g. bilge water cleaning (chronic oil pollution), drilling (oil spills), and ballast water (invasive species). Who wants observers there to expose cheap operations and bad practices affecting industrial income? While this dispute is usually fought in court, in many nations it's also fought on the ground, on the ships and with the fleets. The environmental outcome remains the same regardless: The ocean gets destroyed further, and that's precisely what we all see but should avoid if a claimed sustainability is the overall aim and best practice, the law!

References

Ball, V. (1880). Notes on the Papuans of Maclay Coast. *New Guinea. Nature, 21*, 251. https://doi.org/10.1038/021251b0

Barclay, K., & Cartwright, I. (2008). *Capturing the wealth from Tuna: Case stude's from the Pacific* (p. 268). ANU Press.

Berzunza-Sanchez, M. M., Cabrera, M. C., & Pandolfi, J. M. (2013). Historical patterns of resource exploitation and the status of Papua New Guinea coral reefs. *Pacific Science, 67*(3), 425–440. https://doi.org/10.2984/67.3.9

Blaha, M., Patoka, J., Kozák, P., & Kouba, A. (2016). Unrecognized diversity in New Guinean crayfish species (Decapoda, Parastacidae): The evidence from molecular data. *Integrative Zoology, 11*, 457–468.

Block, B. A., Jonsen, I. D., Jorgensen, S. J., Winship, A. J., Shaffer, S. A., Bograd, S. J., Hazen, E. L., Foley, D. G., Breed, G. A., Harrison, A. L., Ganong, J. E., Swithenbank, A., Castleton, M., Dewar, H., Mate, B. R., Shillinger, G. L., Schaefer, K. M., Benson, S. R., Weise, M. J., Henry, R. W., & Costa, D. P. (2011). Tracking apex marine predator movements in a dynamic ocean. *Nature, 475*(7354), 86–90.

Bolton, B. R. (Ed.). (2009). *The fly river, Papua New Guinea: Environmental studies in an impacted tropical river system.* Elsevier.

Carlson, R. R., Foo, S. A., & Asner, G. P. (2019). Land use impacts on coral reef health: A ridge-to-reef perspective. *Frontiers in Marine Science, 6*, 562.

Cousteau, J. (1979). *The ocean world.* Abradale Publishing.

Cousteau, J. M., & Richards, M. (1999). *Cousteau's Papua New Guinea journey.* Henry Abraham's Books.

Cunningham, L. J., & Beaty, J. J. (2001). *Guam: A natural history.* Bess Press.

Dalzell, P., Adams, T. J. H., & Polunin, N. V. C. (1996). Coastal fisheries in the Pacific Islands. *Oceanography and Marine Biology: An Annual Review 34*, 395–531.

Dixon, A. M., Forster, P. M., Heron, S. F., Stoner, A. M. K., & Beger, M. (2022). Future loss of local-scale thermal Refugia in coral reef ecosystems. *PLOS Climate, 1*(2), e0000004. https://doi.org/10.1371/journal.pclm.0000004

Edgeloe, J. M., Severn-Ellis, A. A., Bayer, P. E., Mehravi, S., Breed, F., Krauss, S. L., Batley, J. , Kendrick, G. A., & Sinclair, E. A. (2022). Extensive polyploid clonality was a successful strategy for seagrass to expand into a newly submerged environment. In *Proceedings of the Royal Society B*. 28920220538 20220538. https://doi.org/10.1098/rspb.2022.0538

Edwards, J. B. (2013). The logistics of climate-induced resettlement: Lessons from the Carteret Islands. *Papua New Guinea. Refugee Survey Quarterly, 32*(3), 52–78.

Filer, C. (1990). The Bougainville rebellion, the mining industry and the process of social disintegration in Papua New Guinea. *Canberra Anthropology, 13*, 1–39.

Flood, R. L., Wilson, A. C., & Zufelt, K. (2017). Observations of five little-known tubenoses from Melanesia in January 2017. *Bulletin of the British Ornithologists' Club, 137*(3), 226–236.

Gill, M. (2020). *Edmund Hillary—A biography: The extraordinary life of the beekeeper who climbed Everest.* Vertebrate Publishing.

Grant, M. I., White, W. T., Amepou, Y., Appleyard, S. A., Baje, L., Devloo-Delva, F., Feutry, P., Ibana, D., Jogo, D. J., Jogo, S., Kyne, P. M., Mana, R., Mapmani, N., Nagul, A., Roeger, D., Simpfendorfer, C. A., & Chin, A. (2021). Papua New Guinea: A potential refuge for threatened Indo–Pacific River Sharks and Sawfishes. *Frontiers in Conservation Science* 48.

Green, D. (2006). *How might climate change affect island culture in the Torres Strait.* Australia Commonwealth Scientific and Industrial Research Organisation.

Green, A., Lokani, P., Sheppard, S., Almany, J., Keu, S., Aitsi, J., Warku Karvon, J., Hamilton, R, & Lipsett-Moore, G. (2007). Scientific design of a resilient network of marine protected areas. Kimbe Bay, West New Britain, Papua New Guinea. TNC Pacific Island Countries Report No. 2/07.

Halpin, L. R., Mott, R., Clay, T. A., Humphries, G. R., Chatwin, T. A., Carlile, N., & Clarke, R. H. (2022). Predicting the foraging habitats of sympatrically breeding gadfly petrels in the South Pacific Ocean. *Frontiers in Marine Science*, 821.

Hamilton, R., Green, A., & Almany, J. (Eds.). (2009). *Rapid ecological assessment: Northern Bismarck Sea, Papua New Guinea*. Technical report of survey conducted August 13 to September 7, 2006. TNC Pacific Island Countries Report No. 1/09.

Hoffmann, C. (2015). *Savage harvest: A Tale of Cannibals, Colonialism, and Michael Rockefeller's Tragic Quest.* William Morrow Paperbacks, Reprint edition (March 10, 2015)

Hollar, S. (Ed.). (2012). *Pioneers in the World of weather and climatology*. Britannica Educational Publishing.

Hyndman, D. (1993). Sea tenure and the management of living marine resources in Papua New Guinea. *Pacific Studies, 16*, 16–16.

Jackson, J. (2013). *Dive Atlas of the world: An illustrated reference for the best sites.* IMM Lifestyle Books.

Jackson, J. B. C., Kirby, M. X., Berger, W. H., Bjorndal, K. A., Botsford, L. W., Bourque, B. J., Bradbury, R. H., Cooke, R., Erlandson, J., Estes, J. A., Hughes, T. P., Kidwell, S., Lange, C. B., Lenihan, H. S., Pandolfi, J. M., Peterson, C. H., Steneck, R. S., Tegner, M. J., & Warner, R. R. (2001). Historical overfishing and the recent collapse of coastal ecosystems. *Science, 293*, 629–637.

Kennedy, K., Cuomo, K. K. (2000). *Speak truth to power: Human rights defenders who are changing our world*. Umbrage Editions.

Klein, C. J., Brown, C. J., Halpern, B. S., Segan, D. B., McGowan, J., Beger, M., & Watson, J. E. M. (2015). Shortfalls in the global protected area network at representing marine biodiversity. *Science and Reports, 5*, 17539.

Kirsch, S. (2014). *Mining capitalism.* University of California Press.

Laffan, I. (1991). The Torres Strait and the environment. In *Proceeding of the Torres Strait Baseline Study* (pp. 451–460).

Limburg, K. E., Hughes, R. M., Jackson, D. C., & Czech, B. (2011). Human population increase, economic growth, and fish conservation: Collision course or savvy Stewardship? *Fisheries, 36*, 27–35. https://doi.org/10.1577/03632415.2011.10389053

Lucas, I. S. (1988). Giant clams: Description, distribution and life history. In J. W. Copland, & I. S. Lucas (Eds.), *Giant Clams in Asia and the Pacific* (pp. 21–33). ACIAR.

Mayr, E., & Diamond, J. (2001). *The birds of Northern Melanesia: Speciation, ecology, and biogeography.* Oxford University Press.

McKenzie, P. (2021). 'Is it to satisfy you or to satisfy us?' Why New Zealand's Pacific colony doesn't want independence. The Guardian. Accessed December 14, 2021. https://www.theguardian.com/world/2021/oct/30/is-it-to-satisfy-you-or-to-satisfy-us-why-new-zealands-pacific-colony-doesnt-want-independence

Miller, C., & Rei, V. (2021). Cetaceans of the Northern Bismarck Sea, Papua New Guinea. *Pacific Science, 75*(3), 395–406.

Orchid Society of Papua New Guinea Inc. (2006). *Spatulata Orchids Papua New Guinea.* Moore Printing Ltd. ISBN 9980-86-070-7

Pazartzi, T., Siaperopoulou, S., Gubili, C., Maradidou, S., Loukovitis, D., Chatzispyrou, A., & Imsiridou, A. (2019). High levels of mislabeling in shark meat–Investigating patterns of species utilization with DNA barcoding in Greek retailers. *Food Control, 98*, 179–186.

Pauly, D. (1979). Theory and management of tropical multispecies stocks: A review, with emphasis on the Southeast Asian demersal fisheries. ICLARM Studies and Reviews No. 1, p. 35. International Center for Living Aquatic Resources Management, Manila. ISSN 0115-4389

Pauly, D. (2019). *Vanishing fish: Shifting baselines and the future of global fisheries.* Greystone Books Ltd.

Pauly, D., Christensen, V., Dalsgaard, J., Froese, R., & Torres, F., Jr. (1998). Fishing down marine food webs. *Science, 279*, 860–863.

Pauly, D., Silvestre, G., & Smith, I. R. (1989). On development, fisheries and dynamite: A brief review of tropical fisheries management. *Natural Resource Modeling, 3*, 307–329. https://doi.org/10.1111/j.1939-7445.1989.tb00084.x

Peng, X., Chen, M., Chen, S., Dasgupta, S., Xu, H., Ta, K., & Bai, S. (2018). Microplastics contaminate the deepest part of the world's ocean. *Geochemical Perspectives Letters, 9*(1), 1–5.

Pittman, S. J., Yates, K. L., Bouchet, P. J., Alvarez-Berastegui, D., Andréfouët, S., Bell, S. S., Berkström, C., Boström, C., Brown, C. J., Connolly, R. M., Devillers, R., Eggleston, D., Gilby, B. L., Gullström, M., Halpern, B. S., Hidalgo, M., Holstein, D., Hovel, K., Huettmann, F., Jackson, E. L., James, W. R., Kellner, J. B., Kot, C. Y., Lecours, V., Lepczyk, C., Nagelkerken, I., Nelson, J., Olds, A. D., Santos, R. O., Scales, K. L., Schneider, D. C., Schilling, H. T., Simenstad, C., Suthers, I. M., Treml, E. A., Wedding, L. M., Yates, P., & Young, M. (2021). Seascape ecology: Identifying research priorities for an emerging ocean sustainability science. *Marine Ecology Progress Series 663*, 1–29. https://doi.org/10.3354/meps13661

PNG Attitude. (2019). Fishing's dark side: The cases of missing NFA observers 22nd January 2019. https://www.pngattitude.com/2019/01/fishings-dark-side-the-cases-of-missing-nfa-observers.html. Accessed July 22, 2021.

Punt, A. E., & Smith, A. D. (2001). The gospel of maximum sustainable yield in fisheries management: Birth, crucifixion and. *Conserv Exploited Species, 6*, 41–43.

Purdy, D. H., Hadley, D. J., Kenter, J. O., & Kinch, J. (2017). Sea cucumber moratorium and livelihood diversity in Papua New Guinea Coast. *Coastal Management, 45*(2), 161–177. https://doi.org/10.1080/08920753.2017.1278147

Quayle, A., Grosvold, J., & Chapple, L. (2019). New modes of managing grand challenges: Cross-sector collaboration and the refugee crisis of the Asia Pacific. *Australian Journal of Management, 44*(4), 665–686.

Radio New Zealand. (2018). PNG parliament told about fisheries observers who disappear. February 16, 2018. https://www.tunapacific.org/2018/02/16/png-parliament-told-about-fisheries-observers-who-disappear/. Accessed June 1, 2022.

Rauzon, M. J. (2016). *Isles of Amnesia: The history, geography, and restoration of America's forgotten Pacific Islands*. Hawai University Press.

Robinson, S., Bouton, E., Dolan, M., et al. (2022). A new framework for rapidly assessing national adaptation policies: An application to small island developing states in the Atlantic and Indian Oceans. *Regional Environmental Change, 22*, 37. https://doi.org/10.1007/s10113-021-01855-2

Rooney, D., & Papoutsaki, E. (2004). *Who is research for? An observation from Papua New Guinea*. Paper presented at the First Colloquium of the UNESCO Forum on Higher Education, Research and Knowledge, Paris.

Safina, C. (1999). *Song for the Blue Ocean: Encounters along the World's Coasts and Beneath the Seas*. Holt Paperback.

Salvador, J., Reeb, C. A., Chapple, T. K., Anderson, S., Perle, C., Van Sommeran, S. R., Fritz-Cope, C., Brown, A. C., Klimley, A. P., & Block, B. A. (2010). Philopatry and migration of Pacific white sharks. In *Proceedings of the Royal Society B: Biological Sciences* (Vol. 277, pp. 679–688). https://doi.org/10.1098/rspb.2009.1155

Simard, N. S., Militz, T. A., Kinch, J., & Southgate, P. C. (2022). Shocks within a Pacific island fishery: A historic study of events impacting the mother-of-pearl fishery in Papua New Guinea. *Marine Policy, 143*, 105173.

Smart, J. J., Chin, A., Baje, L., Green, M. E., Appleyard, S. A., Tobin, A. J., & White, W. T. (2016). Effects of including misidentified sharks in life history analyses: A case study on the grey reef shark *Carcharhinus amblyrhynchos* from Papua New Guinea. *PLoS ONE, 11*(4), e0153116.

Steadman, D. W. (2006). *Extinction and biogeography of tropical Pacific birds*. The University of Chicago Press.

Steiner, R. (2009). Independent Review of the Environmental Impact Statement for the proposed Nautilus Minerals Solwara 1 Seabed Mining Project, Papua New Guinea. Conducted for the Bismarck-Solomon Seas Indigenous Peoples Council Madang, Papua New Guinea.

Steiner, R. (2011). Establishing the Papua New Guinea (PNG) Deep Sea Mining (DSM) Citizens' Advisory Council (CAC) Richard Steiner, Professor, Oasis Earth.

Stone, G. S., & Obura, D. (2013). *Underwater Eden: Saving the last coral wilderness on earth.* The University of Chicago Press.

Sullivan, N. et al. (2003). Tinpis Maror: A social impact study of the proposed RD Tuna Cannery at Vidar Wharf, Madang. https://www.academia.edu/4492602/Tinpis_Maror_A_social_impact_study_of_the_proposed_RD_Tuna_Cannery_at_Vidar_Wharf_Madang?email_work_card=title. Accessed July 22, 2022.

Sullivan, N., Warkia, R. K., & Lewis, A. (2011). A social, economic and environmental impact assessment on the EU-Pacific economic partnership agreement for Eurothon. https://www.academia.edu/4492625/A_Social_Economic_and_Environmental_Impact_Assessment_on_the_EU_Pacific_Economic_Partnership_Agreement_For_Eurothon. Accessed July 2, 2022.

The Guardian. (2015). The incredible plan to make money grow on trees. November 24, 2015. https://www.theguardian.com/world/2015/nov/24/redd-papua-new-guinea-money-grow-on-trees. Accessed May 28, 2022.

The Guardian. (2022a). UN head declares 'ocean emergency' as global leaders gather in Lisbon|Oceans. 27th June 2022a. https://www.theguardian.com/environment/2022a/jun/27/un-head-antonio-guterres-declares-ocean-emergency-as-global-leaders-gather-in-lisbon. Accessed July 10, 2022a.

The Guardian. (2022b). 'It's a mystery for us': The puzzling death at sea of a Tongan fisheries observer. https://www.theguardian.com/world/2022/jan/04/its-a-mystery-for-us-the-puzzling-death-at-sea-of-a-tongan-fisheries-observer 3rd January 2022. Accessed May 12, 2022.

Wei, C., et al. (15 co-authors) (2011). A global analysis of Marine Benthos Biomass using Random Forest. *Public Library of Science (PLOS), 5*, e15323.

White, W. T., Appleyard, S. A., Sabub, B., Kyne, P. M., Harris, M., Lis, R., & Naylor, G. J. (2015). Rediscovery of the threatened river sharks, Glyphis garricki and G. glyphis, in Papua New Guinea. *PloS One*, *10*(10), e0140075.

White, W. (project leader) (2018). Project: Sustainable management of the shark resources of Papua New Guinea: Socioeconomic and biological characteristics of the fishery. CSIRO Australia. https://www.aciar.gov.au/project/fis-2012-102

White, W. T., Baje, L., Simpfendorfer, C. A., Appleyard, S. A., Chin, A., Sabub, B., Rochel, E., & Naylor, G. J. P. (2019). Elasmobranch bycatch in the demersal prawn trawl fishery in the Gulf of Papua, Papua New Guinea. *Scientific Reports 9*(1), 1–16.

Chapter 4
Deep Times of Papua New Guinea with Over 47,000* Years Nothing 'Relevant' Happening: Why It Matters for the World's Future as a Leading Sustainable Steady-State Economy Role Model and a Sustainability Super-Nation

We declare our second goal to be for all citizens to have an equal opportunity to participate in, and benefit from, the development of our country.
Papua New Guinea Constitution (https://www.parliament.gov.pg/constitution-of-the-independent-state-of-papua-new-guinea; *accessed 11th July 2022)*

The remarkable thing is that we Westerners absolutely do not believe in the efficacy of sorcery, whereas our counterparts in rural New Guinea absolutely believe in the workings of sorcery
Beehler and Laman (2020, *p.* 327)

Abstract There are just a few locations in the world thus far known to have humans consistently and sustainably on the landscape and seascape for over 47,000 years. Papua New Guinea (PNG) is among those, and it offers great tracks of humanity including the first cultivation of food items (farming) and an incredibly diverse human culture with over 700 languages. After contact and colonialization PNG was essentially forced to move from deep time to modern time in less than a generation. It's a world record in cultural adjustment. Instead of abandoning the PNG life style, here it is shown that this culture offers many contributions to the world for moving forward with a highly needed and widely seeked-after global culture of sustainability and well-being. PNG has already for millennia what most industrial nations are still looking for: benign sustainability for the land and seascape with a robust atmosphere and society.

* The figure of 47,000 years is widely used in this chapter and throughout the book as a generic reference. One may argue whether it is a bit more or a bit less. The archeological site records all rely on 'detection' of a site, just a few of those ones exist, and thus relying on those can easily be a hit or miss. But it's 100% clear that PNG is associated with the Australian craton and thus part of Sahul. For the wider Sahul region, one may safely assume people in the wider landscape for over 47,000 years. Flannery (2002) for instance refers to 60,000 years of human occupation in Australia. Genetically, one can link PNG with Siberia and that movement must apparently happened before the 47,000 years as a reference in PNG. It's also clear that not all of PNG was populated everywhere and all the time. While one may argue the validity of the figure of 47,000 years, it should not make a big difference for the sustainability argument presented here. Arguably, the 47,000 years figure will likely become a larger number once more science is done and more archeological sites and artefacts are found.

F. Huettmann, *Globalization and Papua New Guinea: Ancient Wilderness, Paradise, Introduced Terror and Hell*, https://doi.org/10.1007/978-3-031-20262-9_4

Keywords Papua New Guinea (PNG) · Archeology · Deep times · Ancient civilizations · Human history · Infanticide · Cannibalism

PNG presents us with a time machine in action (Kiki & Cheshire, 1969; Wilson, 2019). It is known to have hosted humans for over 47,000 years (Beehler & Laman 2020; O'Connell & Allen, 2007). Whereas the last glacial maximum (LGM) was app. 24,500 years ago. While a connection with Siberia can be shown in the DNA of PNG people, three waves of human immigration are known (Beehler & Laman, 2020, p. 241). The Lapita culture came from Asia c. 3600 years ago. Afterwards, more of the exotic species like dogs, timber trees and pigs occurred with a big impact on the ecosystem overall, but they just really had a short history in PNG thus far. Overall, due to such a complexity, the human stories about PNG run deep and diverse (e.g. Flannery, 1998, 2002; Gillison, 1993; Matthiessen, 1987), but most of them are reported by foreigners (Table 4.1), only a few really come from the native voice and with a wider context (e.g. Chan, 2016; Kiki & Cheshire, 1969; Wilson, 2019). Most of the reports come from 'white males' (exceptions are for instance Gillison, 1993 and Wilson, 2019). Even then, those records are necessarily incomplete and biased due to the very high cultural and ecological diversity in PNG, all brought back into a narrow framework of English-written science. So what is a representative scientific sample for PNG? There is no normal distribution hardly a sufficient sample size.

Table 4.1 Narratives about Papua New Guinea provided by major (western) sources (a selection)

Attribute	Author	Comment
Primitive	Mead (1930, 1932, 1935, 1967)	While this was used as a professional term, the meaning and misuse is clear
Undiscovered	Beehler and Latam (2020)	This means undiscovered by the western world, ignoring over 98% of the deep legacy
Bare-naked	Flannery (1990, 1998)	It's clear from the writing that Flannery supports indigenous people but the way how presented (photos and implied) is widely imperfect
Intelligent	Diamond (2011a, 2011b)	It's clear that this is used as a positive feature, however, it remains widely disputed as a concept
Cannibals	Pratt and Pratt (1906), Hoffmann (2015)	Arguably, there are no relevant modern cases of predatory cannibalism in PNG these days; however, the book title plays with this term as an attention seeker. Arguably, these references are very quite on PNG being a major vegan nation
Savage	Levi-Straus (1966)	Needs to be read in context of its time
Stone age	Matthiessen (1987)	Likely this term reflects the attitude of its time
Deep myths	Gillison (1993)	Mysterious deep ancient times unfathomed

And oddly enough, seen globally, there is virtually no internationally recognized PNG Nobel Prize winner, or PNG poet in Hollywood, an internationally recognized creative awardee or similar that gets listened to for his/her traditional ecological knowledge and sustainability skill (Suzuki, 1993; compare also with Bringhurst, 2011). App. 97% of the land is in the hands of traditional land owners; no corporations or government (Baraka, 2001; Larmour et al., 1981). Related conservation management customs are passed down as part of the oral tradition and belief system (Cousteau & Richards, 1999; Gillison, 1993, see West, 2016 for new directions). And so, the stories from PNG that are found for a global audience are usually based on hearsay picked up by foreigners for impact, half-broadcasted for the soundbites, often got lost in translation and have evolved by now; we just see a leftover of earlier times unknown (Gillison, 2002). Virtually design, the PNG information tends to be exotic, sensational or if not even sounding raw or barbaric to a western audience, promoted by directors for funding and promotion, and by publishers for sales to a widely uninformed audience on PNG issues (e.g. Hoffman, 2015) or with a personal agenda. To attract attention, PNG is simply to be presented that way: raw, wild, stone age, savage and primitive; ideally naked and with cannibals (Matthiessen, 1987). It's a PNG profile designed by 'the west' and that makes western people feel good, somewhat superior and civil, and provides them income. But PNG—the deep times—begs to differ though.

PNG indeed provides entertainment and stimulus for the global intellectual, keeping academia occupied, (e.g. Acemoglu and Robinson 2013, Diamond, 2011a, 2011b; Gillison, 1993; Mead, 1930, 1963; West, 2006) (Figs. 4.1, 4.2, 4.3, 4.4, 4.5, 4.6, 4.7, 4.8, 4.9, 4.10 and 4.11).

And then there is the endless promise of 'wealth' in wild areas that are to be 'developed'; much of the colonial history is driven by such greed. Gold is a classic driver for western society (see for instance Diamond, 2011a for Latin America): "That remote place we put in some much effort must be valuable," not? PNG lived for over 47,000 years in its own huge wealth. But as a very common scheme, from their earliest sightings of New Guinea, the Europeans assumed they would find gold ("Isla del Oro"; see Beehler & Laman, 2020, p. 41), at least spices and personal gain and fortune was to be found. All else what PNG had to offer was to be ignored.

The geography of the PNG landscape, rugged mountains, dense vegetation and the unknown interior usually convinced explorers that there was much treasure of a sort. For a while, spices took over the excitement for that area (in the Mollucas), nutmeg, masoi bark and dried sea cucumbers as well as crocodile kins, and soon also Birds of Paradise plumage as another item of desire and 'must have' (worn as a prestige by members of the royal courts in Europe and outside) (Laman & Scholes, 2012)! But finally, in 1890 a small amount of gold was eventually found on the Gira river as well as on the Yodda trail near Kokoda. But it took until the early 1900s for the more successful gold discoveries to be made and reported. The Markham River area became a center of attention and subsequent road, bridges and airports were developed there with cities and ports, e.g. Lae. It's the fabric of modern nation PNG driving business till today.

Fig. 4.1 Ancient times are everywhere

Instead, for thousands of years, the indigenous people of PNG knew very well themselves about their (geological) treasures and how to use them. Industrial extraction and use of gold had little value to them throughout most of the PNG history. Clay work was of equal or more value (e.g. from the Sepik-Ramu area; Beehler & Laman, 2020, p. 239 as an artefact shows on display for a small audience from Young Museum, Fine Arts in San Francisco[1]). Gold was locally known but not a major item to seek after really. Instead, PNG people have mined and bartered stone implements and ochre and also used clay to make pottery. Gold was first really seen and 'discovered' by Europeans in Papua New Guinea in 1852 as accidental small traces in pottery coming from Redscar Bay, Papuan Peninsula. A wider gold rush started from then on, spurred by Australian Leahy explorers in the 1930s for interior PNG. PNG then lost much self-determination in that process (Golub, 2014; Kirsch, 2014); just as it happened elsewhere in the world too (see Bélanger, 2019 for an entire 'Empire of Mining' driven by the British court and its spin-offs, e.g. Canada promoting it worldwide!). When mining enters the landscape, so usually do Australians and Canadians;

[1] This argument matters as it's widely claimed to take artefacts out of to own them globally, which is widely untrue. Neither PNG citizens nor most people in the world can enjoy them when shown in rich nations, at elusive places and where visa are required; public transport is hardly available. In the meantime, artefacts in (traveling) exhibitions can get damaged or stolen.

Fig. 4.2 A mumu: an ancient tradition feasting for the family and tribe with an esteemed 'earth oven'; a feast for Anthropologists

Fig. 4.3 Meeting a travel party on foot, just like done for ancient times

Fig. 4.4 Ancient craft to perfection

Fig. 4.5 Selling crafts now to tourists

Fig. 4.6 Tree fern, a very important and super-abundant plant fabric for millennia

Fig. 4.7 A very sustainable income and food system: gardening with pigs, dogs and coconuts:

Fig. 4.8 Cut wood and search for a food item of the wild: grubs

Fig. 4.9 A grub as a delicious food item of the wild: Who needs to starve?

Fig. 4.10 Modern intruders enter with global warfare beyond 'just' colonization

Fig. 4.11 Typical 'station': a hub of rural live in Papua New Guinea centered around a phone tower and a landing strip

this pattern applies to seafloor mining also (review in Earthworks et al, 2015). PNG got played with once more by outside nations, most of them colonial by nature and coming through the UN mandate.

And when a subsequent missionary program enters a landscape—a commonly linked scheme–all traditions get usually destroyed—at least modified[2]—and thus, a lifestyle and its cosmology that evolved for over thousands of years is lost (Beehler & Laman, 2020 for a critique). It's man-made extinction in action (Short, 2010). Has a soul been saved in that process?

Along with the lost lifestyles goes the loss of taboos. One may easily argue that bushmeat hunting creates a severe and widely unmanaged burden to wildlife and might even now lead to extinctions (Beehler & Laman, 2020, p. 267). But with (co-evolved sustainability) taboos, this is not to happen. It makes an easy argument favoring the need for a framework in the 'tragedy of the commons,' or that the commons have no relevant tragedy in ancient times; the spoilage instead comes from modern man itself and the feudal system; not? (For Nobel Prize winner Eleonore Ostrom on that entrenched discussion see Araral, 2014).

Research work in New Guinea villages has shown now that people have already started to loose their ancient link with the landscape and the deep times that way (F. Huettmann unpublished, Beehler and Laman 2020 for 'untouched' Foja mountains even etc.). It's also described as ongoing for over 30 years with the advent of cash crops, e.g. Baraka (2001). This ongoing loss of expertise must come to no big surprise (as described by Kulik, 2019 for instance) because that's precisely what the Leahy brothers and governmental patrols really wanted: to pacify, to civilize and to bring 'peace' in the bush, for all of the PNG nation while their own profits grow in parallel unchallenged. Australia forbid cannibalism in PNG, and thus it got abandoned by the 1960s. But some outcomes were different from intended (see metrics and conclusion from Gosarevski et al., 2019). It's the direct consequence of the imposed modern life, the western life system and funding streams during globalization. It comes with serious 'modern lifestyle diseases.'

Even in most remote bush villages, the dwellings tend to be now centered around squared air strips, as another western construct. Forest people and their villages now are organized into convenient units for governments and churches to administer (Having been in many of those myself, those are quite boring set ups and places). In earlier times, those have been family compounds instead dispersed widely in the forest and landscape for natural resources (Beehler & Laman, 2020). A church, another squared set up, is part of such a western-style governance culture, whereas the old but world-famous longhouses are mostly gone now.

---

[2] What usually happens in such cases—human nature—is that human society divides on the issue and keeps the ancient traditions, e.g. under cover, and blends them with the missionary ones. Examples are seen easily in the undercover voodoo cults ongoing after church times in the Caribbean or in coastal Brazil. The green Christmas tree, the Easter bunny and many of the pilgrimage trails (e.g. Campostella) show the identical pattern for western Chistianity; those symbols are part of an earlier pagan belief system in central Europe. For the non-believers of this statement, one may easily hint to the Vatican in Rome, a city and location which was initially driven by a heathen belief system and tried to outlaw and kill Jesus and Christianity itself for over century.

Still, PNG is alive in modern times and is trying not to go 'under,' just like most other nations and cultures these days. It's the usual struggle of live ongoing since mankind, globally, but now accelerated by globalization. In the meantime, sorcery remains widespread (Beehler & Laman, 2020, p. 326–327; Gillison, 1993, see for isolated cases of spiritual cannibalism to transfer the soul of a relative till the 1960s in Liberski, 2013; concept in Pedersen and Willerslev 2012.

In reality, in the human history though, PNG is no global exception in the struggle for survival and sustainability and as a tropical nation (see for instance Dublin & Tanaka, 2015). Life and societies go up and down (Heinsohn, 2021 for a global up and down, e.g. tenth century; see Diamond, 2011a, 2011b; Harari, 2015). Beehler and Laman (2020) discussed that process for PNG in some detail and Mack (2014) and Kulik (2019) showed it in wider reality for PNG linked with culture and language (e.g.Wurm and Hattori 1983). PNG is actually living in the modern world now. That's the reason why antiquated approaches on top-down Anthropology by so-called elite institutions fail us so much; it's a mismatch (and that mismatch comes from the scientists and their champions that have not realized in what world they are now living and what they study; PNG in the year 2022. It remains widespread) (Table 4.2).

PNG is located in the tropics near the equator closest to the sun, where most human live on earth is found in abundance for millennia. PNG's patterns and profiles are often nothing special really and found worldwide. Nor are kinship dynamics of a male-driven or maternal society, and even killings, crime, or eating the enemy and cannibalism (see reports in Flannery, 1998; Richardson, 2005 etc.) anything new or special really in human history. It is widely known and described worldwide, e.g. in Central Africa, Amazonia, in the polar regions and in the human record overall—ancient, medieval or modern European (see St. Petersburg siege during WW2; Table 4.3). King (2000) put this in a wider modern context and asked who are actually the cannibals; the western tourists? And if not killing infants for sustainability and population control and associated warfare, certainly abortion is known to have been practiced by females, families and communities throughout the world for millennia, as it is described for PNG also (Choudhary et al., 2017; see also Gillison, 1993). Cannibalism, which PNG gets accredited for so widely (Beehler & Laman, 2020; Hoffman, 2015), is nothing unique to this part of the world nor for mankind or civilization, and there is no truly and relevant reported predatory cannibalism case of PNG since the 1930s anyways (there were a few spiritual burial practices till the 1950s, or so, Liberski, 2013; compare also Mithen 2007 for other ancient records). So why does PNG get the bad and sensational press? See instead the concept of a New Cannibalism brought from the outside (Patience, 2012)!

Arguably, those reports from PNG tend to sound brutal and barbaric to the uninformed and when provided without wider context. It makes money, and it steers the public attention, e.g. Pratt and Pratt (1906). But then, so-called civilized nations are not free of barbaric events and actions neither, e.g. German Holocaust, Japanese war atrocities, or the throwing of nuclear bombs as a Genève Convention violation, declaring international wars without proper justification and evidence, using drones to kill innocent bystanders, food embargos for entire nations and their people? Virtually any 'highly civilized nation' with a royal court or similar can report those 'uncivil'

Table 4.2 Selection of cases of reported local cannibalism and related cases of widely debated inhumane treatments in the world (in alphabetic order)[a]

Nation/culture	Reported topic	Citation	Comment
Amazonia	Birth control	Hern (1992)	An ancient technique found globally
Arabs	Slavery	Gordon (1989)	A classic subject but found almost worldwide
Austria	Cannibalism	Der Spiegel (2007)	Modern issues, case relates to homelessness
Central Africa	Birth control	Eyong (2007)	Consumption on specific wild meat etc
France	Cannibalism in a prison	BBC News (2007)	A man-made feature and facilitated by institutions
Germany	Cannibalism	Daily Mail (2020)	Online case
Malaysia (Not New Guinea)	Human trophy (shrunken head)	Ginging (2007)	Also found at other locations
Indonesia (New Guinea)	Cannibalism	Beehler and Latam (2020)	
Ireland	Slavery	Rodgers (2007)	Widely done worldwide. The Irish slave trade in the tenth century connected with the Vikings and with Central Europe, widespread
Some North American Indian tribes	Scalping and Cannibalism	Bullock (1991), Abler (1992), see Whitehead (1984) for the adjacent Caribbean	Common features of most wars in the world
Papua New Guinea	Cannibalism	Pratt and Pratt (1906), Flannery (1998), Richardson (2005), Leberski (2013), Hoffmann (2015), Beehler and Latam (2020)	Two forms of human consumption were reported, eating the enemy, and eating a dead relative for saving its soul from decay by worms and beetles

[a] Many more can be found in public knowledge bases such as https://en.wikipedia.org/wiki/List_of_incidents_of_cannibalism

Table 4.3 Terms used by indigenous people when the white man discovered indigenous people ("contact") and started to take over

Location and tribe	Term used	Event	Citation	Comment
Interior Alaska Dene	World broke down	Mining	Ayers et al. (2009)	A commonly heard statement when white people entered
North America	Various experiences	Christoph Columbus contact	King (1999)	A diverse set of experiences on the continent, over time, usually the Europeans came better prepared and equipped due to wider and previous experiences worldwide
Polynesia, Samoa	Friendly contacts	Laperouse expedition 1722 onwards	Tcherk (2008)	A widely interpreted topic re. sexual behaviors among tribes
Highlands PNG	Believe that white people were ancestors, locals attempted to rub off the white skin	Leahy exploration 1930s	Griffin (1978)	A reported narrative

actions and collaterals as a known part of their actions, e.g. Saudi Arabia, Belgium (e.g. Hochschild, 1999 for Congo) and the English and Spanish royal family (as per name, 'concentration camps' were widely started and practiced by the Spanish and British governances). The report on cannibalism in modern Austria, a former kingdom empire, as well as in Germany, all facilitated with 'modern' tools of the internet makes that argument (Daily Mail, 2020). The western industrial lifestyle, governance and culture is far from benign itself (King, 2000). So who is to judge (see Patience, 2012 for a new form of cannibalism)? Instead, we are to live in a good and mutually agreed framework of human well-being.

One may see it that way perhaps: PNG's past and created reputation helps tourism, increases book publication sales, provides career boosts and institutional recognition, using the fear factor (Hoffmann, 2015; King, 2000; Salak, 2001, it's an old business plot in capitalism, e.g. Pratt & Pratt, 1906). It helps people to link with PNG, create a narrative and an association, albeit in a somewhat unfair and biased way. It puts PNG on a map in the landscape of your mind and in geography with a reputation designed by the west. But arguably, PNG is more than a nation that offers exotic stories for your bartender; it's a human place and an inherent part of the world with a deep earth link and geology located in a vast and deep wilderness (for relevance of wilderness globally and PNG's contribution see Pérez-Hämmerle et al., 2022) and coming from the eternal garden (that actually would make a link with the bible, but I am not sure the missionaries play up that card much.). One cannot simply say: PNG is 'out there' and not part of latest globalization and its society; rather vice versa. It's the mother of all, certainly sustainability. Like Nepal not being defined by Mt Everest, PNG is not defined by its bad stories narrated by the western colonizers; many are even outdated, reported third hand without relevant context and disputed (e.g. Foerstel, 1994 based on etc.; see also Silverman, 2012).

Arguably, PNG's stories are much more than drama or hearsay! They matter, in a way, because this type of society stayed sustainable, more or less (Flannery, 2002). It did so for thousands of years, certainly for over 47,000 years. That is not only a feat, but also a world record because no other nation, certainly not the modern nations, got it done well over time (see Elvin, 2008 for the Environmental History of China and what is like today in a Megadiversity Landscape). Most modern nations lack such a governance, culture, commitment and expertise—a real sustainable culture is absent—to operate on earth that way for thousands of year: running the big garden. Whereas the modern nations now seem to destroy even themselves, and many others, in just less than 100 years (Harari, 2015). Any visit to a zoo or herbarium will show you no other, showing many endangered species in captivity and/or with photos as their last resort; those are the 'zombies' and walking dead while wilderness habitat disappears. It exposes global environmental decay and wide lack of conservation, wilderness progress metric and even ethical disregard of ancient societies and diversity (see museum value discussion here https://www.theguardian.com/commentisfree/2018/apr/03/brooklyn-museum-white-curators-african-art-open-letter). Beehler and Laman (2020) have discussed the cultural decay in length.

It's a common scheme that the western society looks at non-western society as primitive, exotic and barbaric, (e.g. Levi-Strauss, 1966; Mead, 1932). In PNG, this was reported early on, e.g. killing is on the order of the day and it were the Australia miners and Christianity who changed that (Nelson, 2016). Whatever the brutality in the PNG society is (see Salak, 2001 for experiences traveling as a Solo Woman; see also domestic violence in Gillison, 1993; West, 2006 and experiences by the World Birder Phoebe Snetsinger, Snetsinger & Pratt, 2003), PNG remains a super sophisticated sustainability culture and nation in the global context, just like most of the human societies that operated well on earth for over 47,000 years. It cannot be any other. Like other nations, PNG perfected the use, governance and conservation of gardening as well as fisheries and a nomadic sustainable lifestyle. PNG did not really develop a feudal system across PNG. Already the topic of sustainable shark fishing (compare with White et al., 2018) partly achieved with shark calling (Cousteau & Richard, 1999, p. 197), or use of natural nerve poison for a long-term human-use fisheries speaks to that (Cousteau & Richards, 1999, p. 197). It stands beside the many other and unique but great policies and skills that PNG has co-evolved and developed (e.g. Diamond, 2011a, 2011b; Flannery, 2002).

It is in that context and culture that PNG used mummies, refers to sorcery and believes humans came from the very nature it was using; thus nature was 'holy,' part of human life itself. That is the taboo avoiding overuse of nature. PNG culture is a sustainable culture and inherently living with nature, and from nature. It has that 47,000 years long proven track record! The destruction of nature, loss of land, is not what is promoted in PNG and by its rural people (Cousteau & Richards, 1999; Flannery, 1998, 2002; see Demeulenaere et al., 2021 for Melanesian approach to trees).

Now how did the deep-time society in PNG really look like?

What was village life and PNG like before contact? Was it just sustainability by luck, by design or a drama, suffering was it warfare, or was it an outermost cruel and barbaric society, or was it a benign life like anywhere in most ancient societies mostly at peace with themselves? This very argument, that native tribes before contact lived in peace and harmony, is globally intensely debated (Ellingson, 2001 for 'Noble Savage' and Raymond, 2007 for Ecological Nobel Savage; see Bernhardt, 2017 German for a myth of a coconut cult). It would be too nice to be all true, but scholars report opposing views; see Matthiessen (1987) or Golub (2014) (see also Rauzon, 2016 finding that many 'remote' Pacific islands covered ticks, diseases and invasive species spoiling the harmony view and the biogeography balance). The PNG village life and the great Melanesian way is promoted by Narokobi (1975, 1983) as an essential pillar of the PNG nationhood!

But clearly, if the harmony would even just remotely be true, the western lifestyle would clearly be a debacle (as it is far away from any harmony, or from sustainability of any sort, nor does it promote it; it easily lacks the deeper cosmology of the Melanesian way, e.g. Gillison, 1993, 2002). But it certainly was a different live than what industrialization and the western world offers. One can find many hints in the literature and interview what indigenous populations experienced when the encountered the 'white man' or its society (see Tables 4.1 and 4.4).

Despite the discussion and debate ongoing, in the following, I would like to show some presumed robust metrics of Deep Time in (terrestrial) PNG:

Tribal structure: Many tribes roamed the landscapes in parallel with assigned boundaries, little overarching structures existed beyond bartering and reciprocity (Walton & Jackson, 2020).

Structure of family: Maternal or male-dominated societies were found, often just small groups (<15 individuals) (Beehler & Laman, 2020; Flannery, 1998, see Gillison, 1993 for some family details).

Village structure: Remote and loosely dispersed family housings existed, usually due to a semi-nomadic lifestyle centered around swidden-farming (slash and burn gardening by use of fire and hatchet/stone axe, farming moves determined by the speed when forest succession recovered; Beehler & Laman, 2020).

Population: Human density now increases but likely has been rather low for millennia (Beehler & Laman, 2020), even for the 'highly' populated areas in the highlands encountered by Australians in the 1930s.

Tools: Bow and arrow remained relevant. Decorative stone implements were widely found. Drums, carved out of wood as an ancient communication tool were used. The stone axe remained a major tool.

General biological species structure and set up: On a larger scale, big and low species got reduced, e.g. Eastern Long-beaked Echidna became extinct in

Table 4.4 PNG as a sustainability leader: Examples and topics

Topic	Practice	Citation	Comment
Swidden gardening	Burn, cut, plant and let re-grow, nomadic style	Beehler and Laman (2020)	Essentially a global practice
Forest conservation	Remote landscapes with little human populations and almost free of hunting		Low human populations allow more resources to access and to use
Local landownership tenure	Sophisticated laws and concepts of using products and services from the land, often overlapping	Narokobi (1983) Baraka (2001), Golub (2014)	Together with cosmology a core aspect for Melanesia
Live off the land, for medicine and health	Shamans using plants for treatments, sorcery etc	Narokobi (1983), Beehler and Laman (2020)	Like above
Personal property	Traditional PNG societies have very little individual and material wealth, just common wealth Wealth was measured in pigs, and only a few people had those. They used them to indebt and impress on others and often shared the wealth	Beehler and Laman (2020)	
Personal housing	Such houses naturally only lasted 7 years or so	Beehler and Laman (2020)	Often nomadic; temporary residences widely used. Personal dwellings not needed really

Australia and are widely reduced in PNG also, only found there now in remote places. Tree kangaroos were widely reduced, often extinct. Due to human hunting pressures, many mammals became rather cryptic and nocturnal. Some marine species must have been reduced also, but little evidence exist.

For Birds of Paradise (BoP), Beehler and Laman (2020) think that the BOP harvest has little to no impact, including the colonial and more modern ones. It would not affect the BOP conservation status. But that can arguably be disputed as these experts show no population numbers, trends and remain with a widely outdated 'surplus' argument and narrative that failed anywhere else (e.g. Hernandez et al., 2013 for a typical hunted species and textbook example managed under such an initial surplus 'spilled milk' assumption).

Overall, app. 88 species of wildlife became extinct in the Sahul region (that is mostly for Australia; Beehler & Laman, 2020 and citations within; Flannery, 2002; Hocknull et al., 2020; Johnson et al., 2016).

Landscapes in PNG have been modified by ancient humans, e.g. by fire and for hunting (Examples provided by Flannery, 2002; Beehler & Laman, 2020, p. 122). Further on a landscape scale, wild nuts and fruits, and some timber trees got planted and maintained. Earliest PNG inhabitants likely lived from yams, pandana nuts, game and fish (Beehler & Laman, 2020). As PNG is one of the oldest centers of agriculture (e.g. a c. 9000 years old taro field are found; Beehler & Laman, 2020 and citations within), it is likely one of the first areas in the world for cultivation of bananas, sugarcane, taro and greater yams. It can be said that sweet potatoes, timber and pigs entered the interior, and yams and fish were used at coastal areas.

On a microscale, village diseases are of interest such as hookworms, roundworms, tapeworms, giardiasis, tuberculosis, common cold, hepatitis A and diarrhea. Advanced staphyloccus infections and open sources, ulcers, can now be found. Malaria is widespread, but the latter is likely made more severe in modern times.

National structure: None really existed. PNG was not a nation state back then and a central/feudal island king did not exist.

Reference to the cosmos and the universe: A clear link is made with the wider universe and the relevance of a holistic view in which humans are just one part and are an inherent part of nature.

Instead, the PNG constitution, just done in the 1974 onwards and set up by western minds following Westminster-style federated governance, clearly does not—and cannot—achieve what the deep time society of PNG achieved (see also Stewart, 1983). It's simply not suitable, nor adjusted well and embedded in PNG's complex habitat. So where then is the progress last 60 years brought by modernity of a society that otherwise did well for over 47,000 years in a sustainability and governance framework that is virtually opposite to globalization?

Reality-reporting of modern PNG in globalization presents to us as if PNG people are lost in their own land and time; it might easily appear that way to the onlooker. Whereas, it's really vice versa, the western people are lost, confused and lost themselves in this world, certainly when in PNG but also in their own lands. They do not understand PNG much, hardly themselves. Many metrics show us no other, e.g. a record-high use of anti-depressants, drugs and social decay on the search who they are (e.g. Hemels et al., 2002). The Central Banks are consistently adjusting the policies for a stable currency, and the actions and arguments become absurd, widely divorced from the citizens (e.g. Stiglitz, 2003).

One may easily conclude that in PNG—like in most parts of the world—the last 60 years have been devastating and created turmoil of a nation that sat well on earth for over 47,000 years, done without a feudal structure, without colonial powers, without a federated system and without a Westminster legal style (Steward, 1983).

Who is to blame?

As a look into history and legacy shows, modern western society, its leaders and promoters, of the last 300 years, specifically last 60 years, are messing up well-entrenched and sustainable systems. For what? The colonial history of PNG was no fun ride, but WW2 and its aftermath was the worst. What currently gets promoted globally, on an Asian level (e.g. Asian Development Bank, 2000) or globally with the UN, does not work well neither; the underlying economic model and its concepts and laws lay at fault (see United Nations own assessment, as presented in The Guardian, 2022; see also Rodriguez-Labajos et al. 2019).

At best, the future will include a blend where modernity is merged with the ancient ways of the last 47,000 years. Many of such moves exist (e.g. Ludlam, 2021; Majumder, 2021); and what other options do we really have left? Such a style will certainly do some good to the tropical landscapes and global processes, e.g. Hannah et al. (2020).

It's from such perspectives that indigenization, decolonization and repatriation must be seen. What can we give back and how done best for a future? The achievements and outcomes should be rather clear from that.

Textbox : First-hand incidence reports from Papua New Guinea (PNG) regarding its brutality, rocks thrown, axing, machete injuries, wife bashing, claimed cannibalism and crime rate (experiences by FH)

I have been visiting and traveling in PNG for several years, and hear reports by co-workers about mass rock-throwing toward busses ("stoning"). I have also been called at 2 am by compound neighbors about a break-in and intruders still in the area. A close research colleague from the bush reports to me machete injuries by their husbands as the most common injury among females reported to western doctors. I further read and heard news reports repeatedly of axing, e.g. of the local judge or unpopular people in the community. Wife bashing is exposed as 'bad' in public posters with calls to end it and exposing much of the brutal reality of widespread domestic abuse and helping to avoid this calamity. A missionary reports to me about the break-in of his house with his kids being traumatized for years. Another coworker got dragged down by the bilum on the market, nobody came and helped (likely for reasons of fear and payback). Myself, I got chased by a local teenager (on drugs) with a machete, I tried to lock myself into my bamboo hut until I realized those walls are rather thin. And I drove with a public transport bus through the front lines in a village fight, we rerouted for that reason. Reports of such kind can be endless and filled with terror and fear. Clearly, the crime rates in PNG are high. But then, I came back each time fine and I was never harmed. As a matter of fact, more than once I was protected by my co-workers or other local people—including a police officer guiding me through the neighborhood—who saw a potentially dangerous situation to occur. I was safe!

For the western mind, that's a lot to swallow and to stand. But those who read The Secret Barrister (2018) will agree, the western civilization comes with aggressions and microaggressions, and it's not the harmonious and peaceful place many believes it is whatsoever.

References

Acemoglu, D., & Robinson, J. A. (2013). Why nations fail: The origins of power, prosperity, and poverty paperback.

Araral, E. (2014). Ostrom, hardin and the commons: A critical appreciation and a revisionist view. *Environmental Science & Policy, 36*, 11–23.

Asian Development Bank. (2000). A Pacific Strategy for the new Millennium. © Asian Development Bank. http://hdl.handle.net/11540/3043. License: CC BY 3.0 IGO.

Ayers, H., Harman, D., & Pennington, L. (2009). Arctic gardens: Voices from an abundant land. Boone, NC.

Baraka, R. (2001). The Meanings of Independence in Post-Colonial Papua New Guinea. A Plan B Research Paper Submitted to the Faculty of the Center for Pacific Islands Studies in the Candidacy for the Degree of Masters of Arts. School of Hawaii, Asia and Pacific Studies. The University of Hawai'i At Manoa.

BBC News. (2007). France probes 'cannibalism' case 5th January 2007. http://news.bbc.co.uk/2/hi/europe/6234097.stm. Accessed July 22, 2022.

Beehler, B., & Lamann. T. (2020). *New Guinea: Nature and culture of Earth's Grandest Island*. Princeton University Press.

Bélanger, P. (2019). *Extraction empire: Undermining the systems, states, and scales of Canada's Global Resource Empire, 2017–1217*. MIT Press.

Bernhard, Z. (2017). The Curious Case of August Engelhardt, Leader of a Coconut-Obsessed Cult. Gastro Obscura 30th November 2017. https://www.atlasobscura.com/articles/august-engelhardt-coconut-cult#:~:text=From%20his%20newly%20purchased%20islandtransubstantiation%20of%20God%27s%20very%20flesh. Accessed July 3, 2022.

Bringhurst, R. (2011). A story as sharp as a knife: The classical Haida mythtellers and their world (Vol. 1). Douglas & McIntyre.

Bullock, P. Y. (1991). A reappraisal of Anasazi Cannibalism. *Kiva, 57*(1), 5–16. https://doi.org/10.1080/00231940.1991.11758180

Chan, J. (2016). *Playing the game: Life and Politics in Papua New Guinea*. University of Queensland Press.

Choudhary, M., Rani, S., Sharma, P., Choudhary, N., & Budhwaar, V. (2017). Anti-fertility and abortifacient potential of hydroalcoholic leaves extract of Alstonia scholaris in female rats: An ethnomedicine used by Papua women in New Guinea. *Bulletin of Faculty of Pharmacy, Cairo University, 55*(1), 123–127.

Cousteau, J.M., & Richards M. (1999). *Cousteau's Papua New Guinea Journey*. Henry Abraham's Books.

Daily Mail. (2020). 'Friendly and polite' German cannibal who ate a man he met online has been allowed out of prison on day trips while wearing a disguise, police reveal. 12th May 2020. https://www.dailymail.co.uk/news/article-8310545/German-cannibal-ate-man-met-online-allowed-prison-day-trips.html. Accessed July 22, 2022

Demeulenaere, E. S. Y.-P., Rubinstein, D. H., Lauren Lovecraft, A., & Ickert-Bond, S. M. (2021). Indigenous spirituality surrounding Serianthes trees in Micronesia: Traditional practice, conservation, and resistance. Social Compass SAGE Journal.

Diamond, J. (2011a). *Guns, germs and steel: The fate of human societies*. Random House.

Diamond, J. (2011b). *Collapse: How societies choose to fail or succeed. Revised Edition*. Penguin Books.

Dublin, D., & Tanaka, N. (2015). Comparison between Satoyama and the Ainu way of life. *Kawistara Jurnal Sosial Dan Humaniora, Universitas Gadjah Mada, Indonesia., 5*, 221–328.

Earthworks, Deep Sea Mining Campaign, et. al. (2015). World's first deep sea mining proposal ignores consequences of its impacts on oceans. September 28, 2015. https://earthworks.org/releases/worlds_first_deep_sea_mining_proposal_ignores_consequences_of_its_impacts_o/. Accessed July 11, 2022

Ellingson, T. (2001). *The myth of the noble savage* (Vol. 1). Univ of California Press.

Elvin, M. (2008). *The retreat of the elephants*. Yale University Press.

Eyong, C. T. (2007). Indigenous knowledge and sustainable development in Africa: Case study on Central Africa. *Tribes and Tribals, 1*(1), 121–139.

Flannery, T. (1990). *Mammals of New Guinea*. The Australian Museum.

Flannery, T. (1998). *Throwim way leg: Tree-Kangaroos, possums, and penis gourds*. Atlantic Monthly Press.

Flannery, T. (2002). *The future eaters: An ecological history of the Australasian lands and people*. Grove Press.

Foerstel, L. (1994). *Confronting margaret mead: Scholarship, empire, and the South Pacific*. Temple University Press.

Gillison, G. (1993). *Between culture and fantasy: A New Guinea highlands mythology*. University of Chicago Press.

Gillison, D. (2002). *New guinea ceremonies*. Harry N Abrams Inc.

Gingging, F. A. M. (2007). I lost my head in Borneo: Tourism and the refashioning of the headhunting narrative in Sabah, Malaysia. *Cultural Analysis, 6*, 1–29.

Golub, A. (2014). *Leviathans at the gold mine: Creating indigenous and corporate actors in Papua New Guinea*. Duke University Press.

Gordon, M. (1989). *Slavery in the Arab world*. Rowman & Littlefield Publishers.

Gosarevski, S., Hughes, H., & Windybank, S. (2019). Is Papua New Guinea Viable? Political Economic Bulletin 19. https://www.semanticscholar.org/paper/Is-Papua-New-Guinea-viable-Gosarevski-Hughes/6d71c70951c12fa82dea578a36b66923997f3f92. Accessed June 1, 2022.

Griffin, J. (1978). *Papua New Guinea portraits: The expatriate experience*. Australian National University Press.

Hannah, L., Roehrdanz, P. R., Marquet, P. A., Enquist, B. J., Midgley, G., Foden, W., Lovett, J. C., Corlett, R. T., Corcoran, D., Butchart, S. H. M., Boyle, B., Feng, X., Maitner, B., Fajardo, J., McGill, B. J., Merow, C., Morueta-Holme, N., Newman, E. A., Park, D. S., Svenning, J.-C., et al. (2020). 30% land conservation and climate action reduces tropical extinction risk by more than 50%. *Ecography, 43*, 943–953. https://doi.org/10.1111/ecog.05166

Harari, Y. N. (2015). *Sapiens: A brief history of humankind*. Harper Perennial.

Hemels, M. E., Koren, G., & Einarson, T. R. (2002). Increased use of antidepressants in Canada: 1981–2000. *Annals of Pharmacotherapy, 36*(9), 1375–1379.

Heinsohn, G. (2021). The disastrous tenth Century: Cataclysm and collapse. *Academia Letters*. https://doi.org/10.20935/AL2958

Hern, W. M. (1992). Family planning, Amazon style. *Natural History, 101*(12), 30–37.

Hernández, F., Brennan, L. A., DeMaso, S. J., Sands, J. P., & Wester, D. B. (2013). On reversing the northern bobwhite population decline: 20 years later. *Wildlife Society Bulletin, 37*(1), 177–188.

Hochschild, A. (1999). King Leopold's ghost: A story of greed, terror, and heroism in colonial Africa. *The Wilson Quarterly, 23*(1), 103–105.

Hocknull, S. A., Lewis, R., Arnold, L. J., Pietsch, T., Joannes-Boyau, R., Price, G. J., Lawrence, R. A., et al. (2020). Extinction of eastern Sahul megafauna coincides with sustained environmental deterioration. *Nature Communications, 11*, 1–14.

Hoffmann, C. (2015). Savage harvest: A tale of cannibals, colonialism, and michael rockefeller's tragic quest. William Morrow Paperbacks; Reprint edition (March 10, 2015).

Johnson, C. N., Alroy, J., Beeton, N. J., Bird, M. I., Brook, B. W., Cooper, A., Bradshaw, C. J., et al. (2016). What caused extinction of the Pleistocene megafauna of Sahul? *Proceedings of the Royal Society B: Biological Sciences, 283*(1824), 20152399.

Kiki, A. M., & Cheshire, F. W. (1969). *10,000 years in a lifetime.* F. W. Cheshire Publishing.

King, J. C. (1999). *First peoples, first contacts: Native peoples of North America.* Harvard University Press.

King, C. R. (2000). The (Mis)uses of cannibalism in contemporary cultural critique. *Diacritics, 30*, 106–123.

Kirsch, S. (2014). *Mining capitalism.* University of California Press.

Kulik, D. (2019). *A death in the rainforest: How a language and a way of life came to an end in Papua New Guinea.* Algonquin Books.

Laman, T., & Scholes, E. (2012). *Birds of paradise: revealing the world's most extraordinary birds.* National Geographic Books.

Ludlam, S. (2021). *Full Circle: Power, Hope and the Return of Nature.* Black Inc. Publishers.

Larmour, P., Crocombe, R., & Taungenga, A. (OOs.) (1981). *Land, people and government.' public lands policy in the South Pacific.* Institute of Pacific Studies, in association with the Lincoln Institute of Land Policy, University of the South Pacific.

Levi-Strauss, C. (1966). *The savage mind.* University of Chicago Press.

Liberski, P. P. (2013). Kuru, the first prion disease: A travel back in time from Papua New Guinea to Neanderthals extinction. *Aktualności Neurologiczne, 3*(13), 217–229.

Matthiessen, P. (1987). *Under the mountain wall: A chronicle of two seasons in stone age New Guinea.* Penguin Classics.

Majumder, N. (2021). Paradigm war: A distinct indigenous vision. *Academia Letters*, Article 926. https://doi.org/10.20935/AL926

Mead, M. (1930). *Growing up in New Guinea.* Morrow Publisher.

Mead, M. S. (1935). *Temperament in Three Primitive Societies.* Part I. Morrow Publisher.

Mead, M. (1932). An investigation of the thought of primitive children, with special reference to animism. *The Journal of the Royal Anthropological Institute of Great Britain and Ireland, 62*, 173–190.

Mead, M. (1963). *Sex and temperament in three primitive societies* (Vol. 370). Morrow.

Mead, M. (1967). The rights of primitive peoples: Papua-New Guinea: A crucial instance. *Foreign Affairs, 45*(2), 304–318.

Mithen, S. (2007). *The singing Neanderthals: The origins of music, language, mind, and body.* Harvard University Press.

Narokobi, B. (1975). *Foundations for nationhood.* Papua New Guinea Press and Bookshop.

Narokobi, B. (1983). *The melanesian way.* Institute of Papua New Guinea Studies.

Nelson, H. (2016). Black, white and gold: Goldmining in Papua New Guinea 1878–1930. ANU Press, Google Books https://www.google.com/books/edition/Black_White_and_Gold/8M0PDQAAQBAJ?hl=en&gbpv=1&printsec=frontcover

O'Connell, J. F., & Allen, J. (2007). Pre-LGM Sahul (Australia-New Guinea) and the archaeology of early modern humans. In P. Mellars, K. Boyle, O. Bar-Yosef, & C. Stringer (Eds.), *Rethinking the human revolution: New behavioural and biological perspectives on the origin and dispersal of modern humans* (pp. 395–410). McDonald Institute Monographs.

Patience, A. (2012). The New Cannibalism: The international community and the problem of governance weaknesses in Papua New Guinea. *Asia Pacific Journal of Public Administration, 34*(1), 31–55.

Pedersen, M. A., & Willerslev, R. (2012). "The soul of the soul is the body": Rethinking the concept of soul through North Asian Ethnography. *Common Knowledge, 1*, 464–486. https://doi.org/10.1215/0961754X-1630395

Pérez-Hämmerle, K.-V., Moon, K., Venegas-Li, R., Maxwell, S., Simmonds, J. S., Venter, O., Garnett, S. T., Possingham, H. P., & Watson, J. E. M. (2022). Wilderness forms and their implications for global environmental policy and conservation. *Conservation Biology*, e13875. https://doi.org/10.1111/cobi.13875

Pratt, A. E., & Pratt, H. (1906). *Two years among New Guinea cannibals: A naturalist's sojourn among the aborigines of unexplored New Guinea.* Seeley & Company, limited.

Raymond, H. (2007). The ecologically noble savage debate. *Annual Review of Anthropology, 36*, 177–190.

Rauzon, M. J. (2016). Isles of Amnesia: The history, geography, and restoration of America's forgotten Pacific Islands.

Richardson, D. (2005). *Peace Child: An unforgettable story of primitive jungle treachery in the 20th Century.* Regal Publishers.

Rodriguez-Labajos, B., Yánez, I., Bond, P., Greyl, L., Munguti, S., Uyi Ojo, G., & Overbeek, W. (2019). Not so NATURAL an Alliance? Degrowth and environmental justice movements in the global South. *Ecological Economics, 157*, 175–184.

Rodgers, N. (2007). *Ireland, Slavery and Anti-Slavery: 1612–1865.* Springer.

Salak, K. (2001). *Four corners: One worman solo's journey into the heart of Papua New Guinea.* Counterpoint Publisher.

Short, D. (2010). Cultural genocide and indigenous peoples: A sociological approach. *The International Journal of Human Rights, 14*, 833–848.

Silverman, E. K. (2012). From Cannibal Tours to cargo cult: On the aftermath of tourism in the Sepik River, Papua New Guinea. *Tourist Studies, 12*(2), 109–130.

Snetsinger, P., & Pratt, H. D. (2003). *Birding on borrowed time.* American Birding Association.

Spiegel, D. (2007). Austria shocked by suspected case of cannibalism. 29th August 2007. https://www.spiegel.de/international/zeitgeist/brain-on-a-plate-austria-shocked-by-suspected-case-of-cannibalism-a-502792.html. Accessed July 22, 2020.

Stewart, W. J. (1983). Stone age and twentieth century law in the independent state of Papua New Guinea. *BC Third World LJ, 4*, 48.

Stiglitz, J. (2003). *Globalization and its discontent.* Norton Paperback.

Suzuki, D. (1993). *Wisdom of the elders: Sacred native stories of nature.* BoD–Books on Demand.

Tcherk, S. (2008). *First contacts in Polynesia: The samoan case (1722–1848) Western misunderstandings about sexuality and divinity* (p. 241). ANU Press.

The Guardian. (2022). Humans need to value nature as well as profits to survive, UN report finds: Focus on market has led to climate crises, with spiritual, cultural and emotional benefits of nature ignored July 11, 2022. https://www.theguardian.com/environment/2022/jul/11/humans-value-nature-survive-un-report-age-of-extinction. Accessed July 11, 2022.

The Secret Barrister. (2018). *Stories of the law and how it's broken.* Picador Publishers.

Walton, G., & Jackson, D. (2020). *Reciprocity networks, service delivery, and corruption: The wantok system in Papua New Guinea.* U4 Anti-Corruption Resource Centre, Chr. Michelsen Institute, Norway

West, P. (2006). Conservation is our government now. In Conservation Is Our Government Now. Duke University Press.

Whitehead, N. L. (1984). Carib cannibalism. The historical evidence. *Journal De La Société Des Américanistes, 70*, 69–87.

White, W. (project leader). (2018). Project: Sustainable management of the shark resources of Papua New Guinea: Socioeconomic and biological characteristics of the fishery. CSIRO Australia. https://www.aciar.gov.au/project/fis-2012-102

Wilson, M. (2019). A true child of Papua New Guinea: Memoir of a life in two worlds. MacFarland Publishers.

Wurm, S. A., & Hattori, S., (Eds.). (1983). *Language Atlas of the Pacific Area. Part I. New Guinea Area, Oceania, Australia.* Australian Academy of Humanities.

Chapter 5
Colonial After-Times in Papua New Guinea: Urgency for Repatriation, Biodiversity Compensation and Fair Treatment to Assure a Better Global Future

Never leave the law to the lawyers
Source unknown
...having been requested by the people of Papua New Guinea, through their Constituent Assembly, to become the Queen and Head of State of Papua New Guinea...
*PNG Constitution (*https://www.constituteproject.org/constitution/Papua_New_Guinea_2016.pdf?lang=en*)*

Abstract From the fifteenth century onwards, just a handful western nations and their royal courts simply divided the entire world into their colonies and resources for self-interest. After the discovery by Portuguese and Spanish explorers, Papua New Guinea (PNG) was essentially part of the Dutch, British and German colonial commercial enterprise, Japanese intrusion, and then after WW1 and WW2 widely handed over to Australia as its protectorate, supported by the so-called League of Nations and later by the United Nations (UN) through a mandate essentially to this very day. It‘s part of the 'Pacific Theatre' of the U.S., now also affected by Chinese and ongoing EU interests. The 'independence' of PNG, declared in 1975—and arguably a status obtained way too late and simply done to save money for Australia—resulted into a Dominion recognition in the British Commonwealth headed by the late English Queen to be a 'modern' nation construct and for being set up for globalization and subsequent resource extraction again by others. PNG itself was hardly involved nor really requesting it even. Arguably, PNG cannot deliver to those international demands well and is not really set up for doing so by its former colonial powers. PNG lies bare for the taker, e.g. corporations of mining, oil and gas, timber and fisheries industries, including the science enterprise collecting specimen in a biodiversity wilderness environment that is still perceived -by some- as 'significantly *undercollected.*' Repatriation remains an effort to pay back PNG in a fair way for a better set up. But that has not really happened yet, or has not happened well, e.g. from Australia and most colonial powers, including Japan and Germany. While PNG has lived quite well for over 98% of its time by itself it's the western influence, Christian missionary nations and sects and a similar globalization approach to live that destroys in PNG a lifestyle, a culture, a land- and seascape, a sustainability governance and parts of the atmosphere, and it leaves dramatic global repercussions in the universe while the Melanesian way remains quite resilient.

F. Huettmann, *Globalization and Papua New Guinea: Ancient Wilderness, Paradise, Introduced Terror and Hell*, https://doi.org/10.1007/978-3-031-20262-9_5

Keywords Papua New Guinea (PNG) · Colonialism · WW1 · WW2 · Australian protectorate · British Dominion · Repatriation · Colonial injustice

5.1 Introduction

In the wider universe, our planet is a rather small, unique and finite place with limited resources, and always has been that way. In the solar system, it's actually the only planet known with life, and again a relatively small one when compared to other planets out there. The layer that really has 'live' is actually less than 9 km above the sea level and c. 10 km below = app. a 19 km thickness overall. It's only here, in certain concentrated areas like coastal zones and valleys where most of mankind lives and makes money; it's the only space assuring human well-being and happiness, all under the wider universe (Naess, 2009; Buckley & Lama, 2021; see Gillison, 1993 for an example of self-understanding of people in PNG). And within that narrow strip of life (in reality just 60 m in the ocean and 3 km in altitude), areas like PNG really stand out but get now exploited dramatically. That's what at stake here; globally relevant biology hotspots, and associated deep human cultures, their habitats and ecological services, sustainability practices and best governance cultures.

Globalization is nothing new really; it's often linked with colonial powers and their expansion (see Morris, 2018 for relevance of geography in that pattern). The global hunger for more resources started already several centuries ago. The migrations into Europe 1000 of years ago, or Easter Island historyshow that as much as the Mongolian invasions into China etc. (e.g. Diamond, 2011a). Europe took a more recent and dominant approach to it, e.g. Hanseatic trade, British Commonwealth, empires of Holland, Spain and Portugal, leaving a devastating footprint; the Dutch East India Company was one of the first global enterprises and affecting New Guinea directly, while 'mercantilism' was initiated by the French court.

Ynigo Ortiz de Retez reached New Guinea and landed there first in c. 1545 (Beehler & Laman, 2020), it was followed later by the Dutch and others, and then setting up consistent ports like Sorong and Port Moresby (named after the English Capt. J. Moresby) for instance; the hope was to find wealth; gold. Spices took over the curiosity for a while but gold then was found eventually and set the direction for Australia being a dominant cultural outside power for PNG on the ground. Aspects of typical colonialism were found in PNG (see for instance Fitzpatrick, 1980 for slavery-like concepts applied). In the northern PNG section, Germany was later able to establish itself further. It tried building a form of a new but oddly placed Germany ("*Deutsch-Neu-Guinea*"). Finschhaven served as a capitol of German New Guinea from 1885 to 1892; other cities followed till 1918, all based on the New Guinea Company (Ohff, 2008). But as a matter of fact, the Germans did little to develop Morobe region; they hardly made relevant money (Firth, 1972). Still, they left a company structure and subsequently strict village organization and culture that differed from other colonial areas. After the global WW1 changes the colonial structures got modified, as the winner takes it all and writes history. WW2 and the Japanese invasion turned things around once more. Australia used force to defend during WW2 what it perceived as its own U.N. -given protectorate (Mackenzie, 1941) and what was perceived for them to come; this is described as '*heroic*' and used then as a contributing argument for a commercial claim over much of PNG—as per UN mandate—including the ‘modern’ PNG nation and Australia’s Arc (Rumley, 2006; Henton & Flower, 2007, p. 293, see also Gosarevski et al., 2019). Already the labor indenture system that got applied throughout PNG between the 1880s and the 1960s caused dramatic disruption and decay of PNG’s social systems (Baraka, 2001). While Christianity preaches empathy and equality, the missionaries did not change such society problems in PNG, but made them worse because the co-evolved interaction with the local environment—sophisticated cosmologies—developed and co-evolved over millennia got outlawed and wiped out within years and were to be replaced with ‘Jesus’ as the prophet (Beehler & Laman 2020; a religion and concept totally foreign to this part of the world). The London Missionary as well as German Catholic and Lutheran Missions were major players on that front but hardly compensated for their bad impacts (while Jesus was never in the UK or in Germany; it’s make-believe). Instead, PNG remains one of the nation with the highest missionary effort in the world (Beehler & Laman, 2020); a complete artefact and an absurd colonial leftover in the ‘hope’ of fixing western governance problems with a make-up religion primarily catering political power (but not serving the poor, as per promoted Christian values).

Subsequent ‘modern’ development in PNG got stuck in its bad path and roots; just as shown in the state of the world 2022 (Table 5.1). It is noteworthy for New Guinea that relevant economic development was done mostly 1950s onwards and mostly on the Australian side (= PNG). The Dutch did not do so well (Beehler & Laman, 2020), but both sides are equally struggling now; that is certainly true for wilderness and biodiversity conservation.

Table 5.1 State of the post-colonial and neocolonial world

Metric	Global status	PNG status	Comment
Pandemic	On the rise	Fully included in the global trend	COVID, Avian Influenza, Malaria
Destruction of indigenous society	On the rise	Fully included in the global trend	Example loss of cultures and languages
Gap between rich and poor	On the rise	Fully included in the global trend	PNG has few but very rich and dominating families, e.g. in Milne Bay
Efforts for an equal distribution of wealth	Little	Promotion of Economic Growth and neoliberalism using unconsiderate extraction of natural resources	Seabed mining, fisheries, forestry, development aid
Wilderness and old-growth forest	On the vast decline	Fully included in the global trend	Promotion of fast-growing tree plantations as well as private forest land—all as promoted in PNG by Australia, etc. confirm this trend

Due to their own colonial legacy most of those western nations are simply poorly governed and struggle with their initial set up and burden dumped onto them (Boerhringer & Giles, 1977; Windybank & Manning, 2003; Hawksley, 2006; Bougainville terror as a typical example after being a German colony handed over as an Australian protectorate and its Australian mining conflict; https://www.britannica.com/place/Papua-New-Guinea/Attempts-at-secession). A selection of modern impacts of development are already shown in Foster (1999, 2002) also recognizing the role of the media to move into a 'modern society' but which happens to stand against the Wantok system and public land tenure and resource ownership of PNG. Feudal society, the nation state, remains an odd concept for PNG.

Europe and its royal courts and family clique arrangements had vast global repercussions. Simple local power, family and debt considerations in Europe decided the outcome for many colonial areas and their people worldwide, e.g. Seewald (2022) for a tropical example involving three nations and a royal and commercial debt settlement using indigenous grounds. The Westphalia Peace Treaties set a stage and wider world order, including for remote PNG. When the world's colonies were predetermined and divided at the Treaty of Tordesillas (1524), the Westphalian Peace Treay (1648), the Treaty of Paris (1898), Berlin (Congo) conference (1884–1885), and the Brussels Conference Act 1890, it had global repercussions long-term; millions of people got affected and many thousands subsequently died. Often the impacts are seen to this very day and the pain remains for many citizens of this world (Hinton et al., 2014; Suzuki, 1993) including border disputes, warfare and refugee crisis (Jenkins & Schmeidel, 1995). A typical and tragic example can be found with warfare around resources in such nation constructs and with displaced migrants. In the Pacific and for PNG, Manus island and its dreadful state for immigrants (Dastyari & O'Sullivan, 2016) makes for a good point in time. And there is hardly a change in sight from this global century-old policy of conquer-and-divide and to dominate in a top-down fashion. It comes and remains in many fashions, despite 'changes'.

I find, looking just at Wuvulu Island and its legacy, history and ownerships—national and private—reflects the European influence onto people and land ownership very clearly (https://www.wuvulu.com/history.shtml). The history and naming of 'New Holland' (now Australia), New Britain, New Ireland and Solomon Islands including Bougainvillc shows us no other.

For PNG, it resulted in the wider PNG nation set up (as there is some change over time and a subsequent intense debate what the PNG area really is and should be, Baraka, 2001; see Beehler & Laman, 2020 for New Guinea and Indonesia). The PNG area consisted of a Dutch part, a German part and a British part with their oversight (Fig. 5.1). As a result of WW1, the Germans had to give up their sectors quickly and after WW2 PNG became an British approved Australian-assigned protectorate within a UN mandate overall while in reality the UK Dominion aspect still prevails till today in all its bad side aspects, e.g. the late queen is the head of the state, and money runs through London and related financial links and outlets (Hongkong and Singapore). This structure even got more enforced after 'independence' (Baraka, 2001; and see the recent visit by Meghan Markle etc. and its racial discussion, EU Observer, 2011). The UK cannot really let go of its colonies, certainly not one that is so close to Australia and New Zealand offering a great political block aligned with US and Europe (referred to as '*Australia's Arc*', e.g. Dinnen et al., 2006; Rumley, 2006).

Fig. 5.1 Thousands of years of human civilization speak from that carved statue, note the relevance of the crocodile!

Unless there is war and subsequent debt settlement, those royal powers tend to stay in power for a long time, as enforced by tradition and its ministries, constitutions and employees. It's 'culture' and widely written in stone through the pension plans and general 'tradition' too. Whereas, in reality, PNG is as far away as it gets from UK and Europe, hardly a meaningful representation for a Dominion governance based in Europe. That's because PNG never had an island-wide feudal-style top-down government (much different from Hawaii etc.). PNG's association with the western world and its governance simply comes from colonial times and is a bad link. PNG did not ask for it and does not embrace it, much (Fig. 5.2). But here are the realities of the world we all live in.

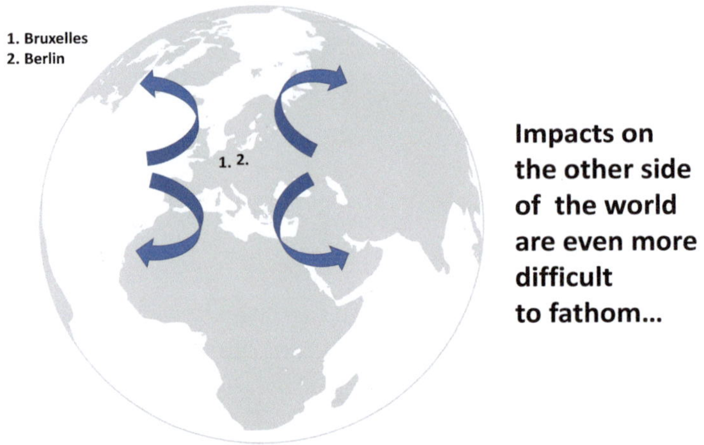

Fig. 5.2 Brussels and Berlin Treaties and global map and boundary impact; other impactful treaties in Central Europe set the stage; for details see text and citations within

Although widely and typically ignored in many modern economic assessments for PNG and its metrics, already this historical set up and its legacy directly affects many perspectives about colonialism, protectorates, postcolonialism and neocolonialism (e.g. Hawksley, 2006) all done in times of the ongoing globalization and the Australian, western and now Chinese sequential dominations, acting in synergy. This has then many wider implications for the Pacific region, and globally, including PNG as a member nation of the Antarctic Treaty System (ATS; https://www.ats.aq/index_e.html) and similar set ups and political maneuvers, e.g. in the UN (see https://www.scoop.co.nz/stories/WO1009/S00517/png-moves-closer-to-contributing-1st-blue-helmets.htm for Blue Helmet participation of PNG). Clearly, PNG is not so modern then, hardly free, an equal UN contributor, or has a good outlook under such a political lock up. It simply got caught, innocently and as a victim, in the western global colonial set up and it cannot really break out in a good and sustainable way. And that's all; it's the life, landscape and biodiversity we now see in PNG (Fig. 5.3).

Textbox 1: Papua New Guinea and the Antarctic Treaty System ATS

The world has many wilderness places, and some got accessible or became relevant just recently. Usually, that's due to intensified resource completion and competition driving prices up of the resource at that remote place. Papua New Guinea is one of those cases, but so is Antarctica. Both sites play a relevant role on the world agenda and for world peace. And both places were part of the colonial discovery game starting with James Cook and similar, e.g. the Russian World Expeditions (lead in part by Baltic-German Captn F.G.

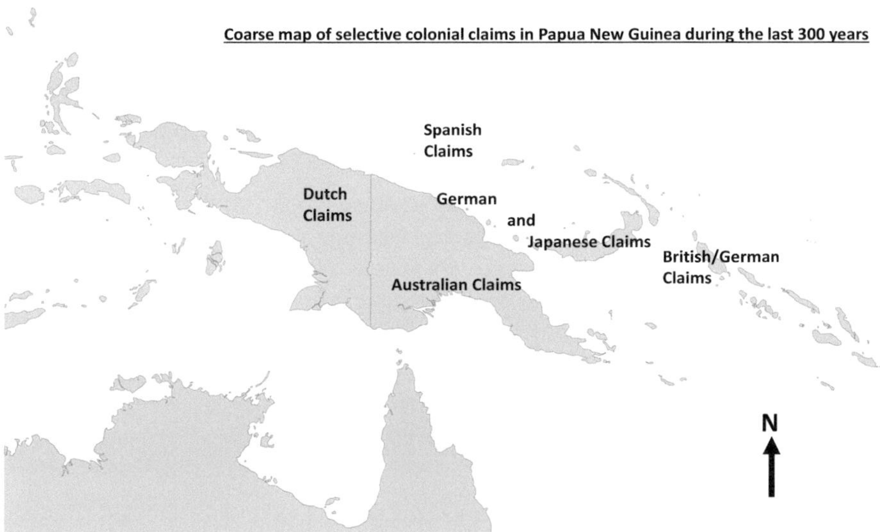

Fig. 5.3 Papua New Guinea's colonial division pie; a coarse map of politically very complex, involved and impact claim zones. To this very day, those nations operate in those zones (but hardly in other nation ones, unless asked. Essentially, a colonial scheme still in operation)

von Bellinghausen) that explored both sites. From that effort, there were even—mistakenly—three penguin species assigned to PNG (e.g. *Pygoscelis papua*—the Gentoo penguin—a species that 100% does not live in Papua New Guinea; other details in Beehler & Laman, 2020).

The modern nation construct of Papua New Guinea is part of the British Commonwealth, mandated by the UN through Australia, and thus, it usually aligns with the 'Western Block' eventually. Antarctica during the Cold War became a pawn in the East–West conflict. As Australia is a dominant player in Antarctica with research stations there and internationally assigned areas to operate, PNG as a former Australian protectorate was a logic option for a strategic power-swap and voting partner to join the western bloc of the Antarctic Treaty System (ATS; Kimball, 1988). It looks good for global democracy. The ATS is meant to be a model of international co-existence and co-operation (but totally fails PNG and other nations - and the world - by ignoring man-made climate change while Antarctica holds most of the world's snow and ice for the world and its oceans and currents)

Thus far, PNG never played a large role in the ATS but it's present there and somewhat strengthens the involvement of southern and tropical nations for Antarctica. Other tropical and southern nations in the ATS are Malaysia for instance, and then South Africa, Pakistan, India and of course many Latin American nations such as Brazil, Argentina, Chile, Uruguay and Equador.

While the PNG membership in ATS might sound a bit absurd to the outsider, the ATS actually comes peppered with outside and other interests. One may argue and ask what Germany has to do with Antarctica, or why Norway (a relatively small nation of just 5.3 mio people) invested so much in its exploration of the southern pole? Apart of fisheries, Antarctica is simply part of a wider power balance and complicated global fabric, also related to global nuclear weapons, world peace, world's cruiseship industry, whaling and fishing, and potential mining interests on land and seafloor. Climate change now plays a major role also.

One may look at PNG's membership in the ATS as an antiquated post-colonial power construct, but one may also look forward as a more equal and more fair involvement of PNG and its people in modern topics of global relevance and its resources (temperature included).

Now, PNG as a modern nation is a construct that got essentially created (Baraka, 2001 for overview); it's to look good and smart. It never was fought for by PNG itself; there was virtually no struggle in PNG for that independence. But instead it's essentially a quick self-convenient construct designed by other, more powerful, political systems in the remote western world for themselves. Much has to do with its geography, natural resources and proximity to the powers in charge. Just like with Nepal or Alaska, it was remote, marginal, difficult to access and perceived as a not so worthwhile leftover place that was not so interesting then to the global powers, poorly explored, remote to get to with exotic stories and diseases reported, and while other locations are 'better' and more 'worthwhile'—as judged in the context of its time. The stories and attributes that came with it go accordingly: crocodiles, man-eaters, rugged and wild (Flannery, 2002; Diamond, 2011a). The BBC—a service known for its bias and political influence (Mills, 2020)—similar to the National Geographic contribute further to the story, as does even the own PNG media (Foster, 2002; Rooney, 2004). For Australia, as the assigned power over the PNG protectorate, it became a public cost-saving effort to have PNG run as an independent nation (more details in Baraka, 2001).

Arguably, that assessment of the PNG value and what it is worth has a bit changed by now from the 1970s. But in earnest, PNG remains a certain artefact mostly created by 'the global west', affected by Asian powers, couched between Australia and Indonesia within the '*Pacific Theatre*' dominated by the U.S. with a geography set by the ocean and sea level, formerly supported by UK and the US eventually in Australian hands, now heavily affected by China, and some globalization.

5.2 The World's Colonial Set up Provides Ongoing Patterns of 'Modernity' for a Tripple Whammy

Another colonial leftover angle in PNG and its structure is how it is governed, internally. PNG follows a top-down concept that was adjusted and fine-tuned by the colonial powers (Holland, Germany[1] and Britain all came from virtually the same monarchy style, even royal family tribe, similar religion, they are also strongly blood-related (Ceballos & Álvarez, 2013; O'Neill, 2020) and then PNG got adjusted to suit an Australian model and latest governance schemes from the UN to be 'modern' (Chan, 2016; Wilson, 2019 for first-hand descriptions).

In reality though, most of those do not work well, include much conflict, e.g. union battles, coming poorly funded, having no sustainable business model, and failing while PNG then is left with its own way of doing business. PNG lives in its own internal world, AND it has to compensate for western damages, locally, nationally and internationally, now and in the foreseeable future

PNG got hit by a Tripple Whammy it currently cannot get out of by itself due to a global lock-in!

Many of such businesses operating from abroad in PNG are not even living up themselves to industry standards, and tax-payments, as the western world promotes it. PNG is one of the few places in the world where there is no railroad, as otherwise found as a typical infrastructure feature and governance tool. PNG uses bush planes instead (a widely unsustainable and non-local practice heavily cashed in by outsiders, e.g. from Australia and missions; see Sinclair, 1978 for legacy, pilots and organisation style). Even a national road system is widely incomplete, without a valid business model (Beehler & Laman, 2020). It has many potholes and erosions, and gets frequently entirely blocked by bridge closures, local warfare or plain crime (e.g. rascals; Huettmann pers. com. for first-hand experiences). It may be stated here—once more—that the generic western governance model is widely unsustainable by itself, ecologically, economically and socially. That is specifically true for its transportation systems that are highly inefficient, and energetically wasteful and debated (see frequent train re-re-privatizations or airline bancruptcies). It's also true for the colonies as well as for the British 'motherland' itself.[2] The rail privatization under M. Thatcher and the ongoing BREXIT situation and its governance crisis speak to that fact, similar applies to the EU and its peace and currency.

PNG is a unique place featuring landownership tenure, but which now stands in conflict with 'modern' land developers and their land grabs destroying the 'usual' lifestyle (Beehler & Laman, 2020). A land grab now comes down to PNG, and land that is in public land tenure will be seen as underdeveloped, which is off and a neocolonial land grab once more affecting the land rent/tax. PNG gets the 'double whammy and worse', again.

Clearly, a key question in this situation centers on the topic of people's wealth: taxation, income and distribution. It is meant to be sustainable, and mostly has been in 98% of the PNG lifestyle itself, e.g. when compared to other nations and globalization. PNG—historically—was globally sustainable for over 98% of its time

[1] Germany has a longer and very bad colonial history, e.g. with Venezuela as early as 1528 (Seewald, 2022).

[2] This concept easily goes back to Australia, UK and Spain for instance. Imposing the western model onto PNG may be seen as setting up another death spiral for PNG and its people. That is shown widely elsewhere already, e.g. Africa and Latin America. And PNG is to follow that trend with open eyes. The path of sustainable development has left us with no other assessment (e.g. Rich, 1994).

(Flannery, 2002; Diamond, 2011a. See Wilson, 2019 for a first-hand description). It actually still is, at least when judged by carbon sequestration and carbon offset per capita. PNG helps the world by essentially being a carbon sink of world importance. And here is where it gets pretty perverse then and even more abusive: Like the Torres Islands or adjacent Kiribati, PNG did no harm but instead pays the bill for climate change and environmental problems, many of them created by the western world. PNG gets more than the 'double-whammy': It comes out of colonial times to recover and then got a broadside from the western world via their climate change emissions.

Seen from that angle, New Guinea had the back luck of being discovered by 'the west.' After contact, PNG essentially got enslaved into colonialism, then it even supported WW1 and WW2 and fought there with many victims and war heroes (Cousteau & Richards, 1999 for numbers, and Chan, 2016 for first-hand descriptions of island-life of that time), and then got modernized and is to pay all the bills full force, for itself and for others, for the global community that is subsidized by PNG through its resources (just like many other tropical nations do, it comes as a concept; Rich, 1994. See Belanger, 2019 for the 'mining empire' and Kirsch, 2014 for a PNG example).

No wonder then that PNG and its citizens are poor; how can one not be poor and aggressive in such an imperial construct? "*A hungry mob makes for an angry mob,*" a famous lyric by Bob Marley. If one lives for 47,000 years in a sustainable set up, why not being angry when this lifestyle gets destroyed?

It must be surprising that the question of compensation, repatriation and fairness did not really pop up earlier, e.g. for former colony-holding nations (see Knoefel, 2019, 2021a/2021b for German colonies[3]; it's already a wider topic in Holland and UK though, as well as U.S. where many of those cases get debated and legally settled; usually with an African focus). Voices that show more of the reality are still far, few and insufficient, e.g. Wilson (2019), compare with Laman and Scholes (2012) simply focusing on 'beauty' of PNG *sensu* Bandura (2007). Beauty comes with a context and framework. And in PNG that is a colonial framework, which is rather abusive (Gosarevsky et al., 2019) (Table 5.2).

Textbox 2: Science data, the need to have and to understand, lack of open access, ecology and repatriation arguments

There seems to be the misunderstanding that (western) humans must touch, investigate and take home everything they can get their hands on in the universe (see Everest, deepsea floor, moon and mars explorations). The widely heard and promoted worldview goes: "*we need to understand everything first before we can act and manage well*", "*understand the mechanism and causation, not just correlation*" (=a set of arguments and reasoning widely disproven for decades as humans never really truly understand the underlying molecular interactions nor do they act on it (see for instance '*three body problem*';

[3] See Schneppen (2001) for German researcher of East Africa, founder of the Society for German Colonization, who got deprived of the imperial high commission for Kilimanjaro in 1897 due to misuse of official power in his treatment of the Africans. Similar cases are now also investigated for Oceania and Papua New Guinea, e.g. Muehlhahn et al. (2014).

"*there is more to life than what meets the eye*", underlying particle physics (e.g. neutrinos and quarks), complex ecology or the universe and 'big bang'. In the meantime, it's surprising that the resource exploitation goes on widely uninformed and in brutal force, if not by the West than by others, e.g. China. Typical examples are provided for the Ross Sea in Antarctica where an '*experimental fisheries*' without proven concepts or track records—hardly any numbers—operates on an industrial scale wiping out endemic species without knowing even the most basic life history metrics; after the massive fish stock removal the area—widely depleted of whales and fish—is now declared a conservation marine protected area (MPA)). Papua New Guinea is just another example of many key areas to show how wrong, dangerous, self-serving and self-enriching that mentioned concept is; we need to know first and collect the specimen...

There are many aspects where the open access argument of internationally shared specimens from PNG fails, e.g. "*we collect specimen for other people to see*" (for the large failure of that argument one just needs to try to match the vast exploration efforts in PNG with actual specimen shown and their data made available in GBIF.org as a mandatory data repository most colonial nations agreed by law. The impossible exposure of timber harvest, or fishing tracks and harvests is another good example, just like interview data of the anthropologists, or recordings! Some tree kangaroo data examples on this topic are shown in Huettmann, 2020a, 2020b, 2020c; see also Zuckerberg et al., 2011; Huettmann, 2015). For artefacts, typical examples of that mismatch for PNG are found with sacred mass carvings, and pottery from Sepik area, which are now dispersed all over the world and sometimes shown there in colonial-style museums and art galleries for an entry fee. But there

Table 5.2 Examples of compensation for colonization and other intrusions

Selected example	Nation	Item details	Comment
Iraq	UK	Looted artefacts	Many of such examples can be named involving many nations
Nigeria	Germany	Artefacts	Ongoing and wider discussions now
Holocaust	Germany	Various, e.g. art pieces, loss of live, trauma	
German Northwest	Namibia	Art pieces	
Aboriginees	Australia	Loss of live, trauma	A widely discussed topic, equal or worse to Apartheid
Space	U.S., Russia, China, India etc.	Moon travel, Mars travel, satellites	A topic that has not hit mainstream media yet

is hardly a known central source, or synergetic analysis done, ever. Likely it will never be possible even with 100s of years of PNG science efforts wasted? That's where the international science progress argument falls easily apart, thus far.

For conservation, science, economy and social aspects, the specimen collection concepts promoted by Beehler and Laman (2020) failed already for long time in Europe and North America, e.g. with colonial expedition collections lost during world wars there, and before (UK collections re-house and sold to the U.S., others lost or burned in fires. So why would it work in the tropics, or in PNG?

Throughout centuries, there were major expeditions who did nothing but collect specimen and artefacts for themselves (details provided in Beehler & Laman, 2020). The bird work by Mayr and Diamond (2001) and Diamond (2011a and citations within) for instance was widely based on shooting birds as evidence ("*must be collected for science*", but not on statistically-sound survey work, research design, or detection transects, as usually done in modern ornithology as a science. Non-intrusive research methods also rank high these days on the agenda and can be done instead). But despite recent developments in bird surveying and inference, Beehler and Laman (2020) go further and state that PNG would actually be significantly undercollected! Authors - featuring support from extractive but widely critiqued industries and corporatations—then consequently critizize for instance in their book (p. 57) the science xenophobia and overregulation of field work in New Guinea (by islamic Indonesia as well by PNG. PNG's regulations seem to be less stringent and thus invite more collectors, but the high crime rate and violence puts a limit to that for its attraction). One would partly agree with difficult handlings and hostilities with the administration, but it comes for a reason. By now, Indonesia got rather aggressive for repatriation of samples. The authors' claim that "*all collections and data are globally shared and that it benefits wider mankind*" is just 100% not true and not applicable (meaningful data and specimen are not well shared and accessible, see Huettmann, 2020b for examples on tree kangaroos; and artefacts held in San Francisco and Fine Art exhibits do not reach a global audience at all, nor for free. Remember that the average global citizen earns less than $5 a day and does not live in the U.S. or in San Francisco, hardly can get a visa to travel there), nor that much is learned.

Even museum and collection experts like Mayr and Diamond (2001) stated clearly that Melanesian islands received the collection peak for its species etc. already in the 1990s; so why further collecting, killing animals and spending money on it in a monopolizing effort with royal and national science academies and institutions, and private entities for pure self-justifications?

And the subsequent ecology lessons made by those collectors and their specimen remain bleak, and the actual insights from those 'scientific' collection trips, e.g. killing trips lacking efficiency, good research design, hypothesis, mutual community buy-in, and information provision to the global public, including PNG citizens, is arguably minor. What was learned from the hundred of thousands of specimens, or million of critters killed when it comes to insects? Use of modern analysis methods for those data and non-intrusive approaches like Machine Learning and AI has been absent for decades, despite that method being in good use in wilderness research for long time, 1970s onwards (e.g. Ohse et al., 2009 for an example; Humphries et al., 2018 for overview).

Overall, and until modernized, the concept of specimen collections should be embargoed until a good vision and progress is mutually agreed on, and while all existing data got analyzed well first, and made publically available for a better environmental decision making and outlook. Thus far, it just remains a grim topic exposing a wide failure of the modernity - western society - when compared to ancient lifestyles that remained much more sustainable for over 47,000 years than what we see today. It's puzzling, at best.

That matters because there is already a big discussion in African nations and elsewhere (media, including Hollywood; whereas 'Bollywood' and its funders remain widely free of such concepts) on colonial compensation for slavery etc., (see Hughes, 2002 for the Massai); it reaches into war damages also and all want to be compensated for their suffering and losses.

And there is a similar question—in concept and in reality—with 'artefacts' (pieces of art, archeology etc. One may happily include here sound files and anthropological interview recordings, e.g. Niles, 2012). Based on excessive collection efforts in PNG, many collector's items are parked in museums, in their basements for decades (some unwrapped and unprocessed even; Germany has some of those for instance) and those are often distributed abroad (The Pledge Times, 2019; for PNG see for instance Beehler & Laman, 2020, p. 239 for artefacts held in 'Fine Arts' collections of San Francisco; or for Vatican in Rome see for instance https://news.artnet.com/art-world/pope-francis-canada-2151805, Patriangeli, 1982; Metzler, 1990). PNG has its own museums (see LonelyPlanet guide https://www.lonelyplanet.com/papua-new-guinea), but the big ones with high-paying audiences and patrons are often located in big nation cities and capitals, e.g. of the former colonial powers, like London, Paris, Berlin or in Amsterdam let's say, and now, in the U.S. Visas are required and PNG citizens are widely excluded. Like most big nations, Russia is also involved in such collections and Russia did have Anthropologists on the ground in PNG (for Nikolai Miklouho-Maclay see Webster & Webster, 1984; Shelton, 1994). Whereas the Japanese efforts in PNG remain even more cryptic for most global citizens to join.

And where does that obtained money go from those efforts? Overall, I am also not aware of many PNG curators being employed on PNG artefacts in museums abroad; I am sure many of them are not even aware what is out there (see discussion why curators are mostly white The Guardian, 2018; The Guardian Nigeria, 2018).

While the world is now waking up to such of its wrongdoings in (African) colonies (for artefacts in U.S. see for instance The Guardian, 2022a, 2022b. For Germany see Der Spiegel, 2013; Knoefel, 2021a/2021b, 2019), the level of awareness and discussion there has not really reached yet PNG and its specimen, much (but see WIPO, 2016 for DNA). The holdings and ongoing actions by the German Lutheran and Catholic Missions should be taken into account also (https://www.gutenberg.org/files/55819/55819-h/55819-h.htm). Beehler and Laman (2020) stated the wide humane destruction of local cultures when missionaries enter the landscape and operate there. Many other nations got confronted with those details earlier but have not acted yet.

Similar can be said for expedition information and tracks, interviews done, photos and videos taken and digital data (where it now all sits; Huettmann, 2015; see Huettmann, 2020b for Tree Kangaroos, Huettmann, 2020c; Suwal & Huettmann, 2020 for similar tropical nation examples).

In this regard, a global key decision might be pending with slavery compensation cases in the U.S. The U.S. has already for a long time reacted and supported Africa, specifically Liberia and Ghana, let's say. But similar action has not happened with the other colonial nations yet, certainly not PNG. With the rise of awareness after Trump'ism, those issues now come to the forefront again, and as a repercussion the repartition might be a relatively easy case and one that helps to 'safe face' for the colonial powers. It's essential though for an equal and fair treatment of international partners and for a science-based interaction and subsequent wilderness biodiversity conservation!

These efforts are specifically useful in the light of new powers entering the global agenda, e.g. China, India and Russia, playing big roles in the global scene now and coming with their own ideologies. If the west would get its own act cleaned first, the new cases by others can easier be resolved for good progress and for a good world! It is here where we see again a failed opportunity provided by 'the west', thus far.

5.3 Any Fairness by the Winners? What Court, What Judge and What Culture of Fairness? What Future Culture?

The public argument on post-colonial fairness seems rather clear: "*We took the resources from them, mistreated the colonized nations, cashed in on it, and thus, need to pay compensation and adjust those actions now, all seen in a modern context for a good future outlook.*" Most people will agree with those concepts (unless some voices want to argue that colonizing other nations was the norm of the day and thus not judgeable today, and thus get ignored). The notion of '*we came and brought them law and order*'—as shown by Nelson (1982; see Beehler & Laman, 2020 for governmental bush patrols to pacify uncontacted peoples, done with Australian help in PNG till the 1970s) is widely and easily disputed (e.g. Chan, 2016; Wilson, 2019 for a local perspective). PNG is not better off before such efforts started by mining.

But once corrected, one can set a good precedence for the future, sensu Rossi et al. (2015; Huettmann, 2020a, 2020b, 2020c).

Well, this makes all sense, but these details are not always possible to get correct over time as these concepts are based on simplistic concepts, rejecting liability (The Guardian, 2022a, 2022b for a typical example in Sahul), and need good evidence/data to hold up with scrutiny. For one, one cannot really just pay money for a lost live to bring it back, and then all is well. It never gets well, as any family member and relative can tell you. The person is dead and gone, pain and trauma remains for very long time and healing needs time and 'love'.

Secondly, the amount of the damage done in the past is difficult to assess, and so is the amount of compensation to be paid. Consider this on a global scale! These settlements are more of a cultural agreement, but which are difficult to do between nations, e.g. PNG versus UK, or globally. Arguably lump sum payments are not achieving such questions well, nor a few hyped up public media cases (see WIPO, 2016nfor DNA questions in PNG as an example).

Third, as many of those wrong-doings were done in the past; they are virtually impossible all to claim and to settle in the absence of good documentation and facts, out of context.

Fourth, many nations will claim they fully acted within the law and approved by governments elected by democratic principles and 'the people' which back then was very different than today on those matters. So why to settle in the first place?

And in theory, there should be a Global Court which settles all those things between parties. But this is a tricky concept and with many cultural drawbacks. Evidence on this is showing all over the world (e.g. Pomerance, 1996; see The Secret Barrister, 2018 for realities of the legal approach and profession itself). The concept of one global and unified legal justice, of supreme courts, gets relatively easily perverted and finds its limits quickly. That's because it is fully embedded in the western style of democracy, of colonial structures, and global power politics, which relies on resource and funding input, and its democratic governance (see Ziegler, 2013 for how the UN fails its missions and drifts toward dictators). So now what? (Fig. 5.4).

5.4 Enforced Industrial Unfairness Brought by Globalization and Its Neoliberal Market Capitalism Ongoing and into the Future: It Must Be Improved!

And so, while the legal system gets bought out during industrialization/capitalism and fails its very own idea—science and truth (The Secret Barrister, 2018)—enlightenment and globalization comes forward and gets pushed into the last vacant and

Fig. 5.4 Water, forest gardens and children: that's a core aspect of the past, and for the future to come

accessible spaces of the planet, and its atmosphere and the universe. Fueled by industrialization consuming natural resources at a record rate, injustice increases, just as fast as man-made climate change rises on a finite space.

And that injustice has left us with a record rate of problems, with minorities put on the cutting block (see Suzuki, 1993; Stiglitz, 2003 for examples worldwide). Certainly, PNG is quite powerless in such a scheme and when it has to deal with corporations like Google, Facebook, Yahoo or the globally acting Mining and Shipping Corporations coming from Australia etc. Those are the real Future Eaters (Flannery 2002). Patience (2012) calls it the new cannibals! (Fig. 5.5)

In the meantime, Asia is now super dominant in PNG and elsewhere, and simply judged by demographics, India and China and their ideologies will be the powerhouses and in the driver seat for fairness and ethics. As the colonial powers have not well settled their cases yet, in PNG, India and China, one must not expect Asia for doing any better. The West can hardly intervene here and has not much to show anymore. It's out of power. Those western efforts will likely relatively easily be dismissed, and as part of the power politics PNG gets exposed even further and stripped off its resources, again. That's a real-world outlook for this region.

Fig. 5.5 People and their unity ("One People One Nation"): for PNG it's more than a slogan on a plastic bag (=commercial version of a Bilum)

References

Bandura, A. (2007). Impeding ecological sustainability through selective moral disengagement. *International Journal of Innovation and Sustainable Development, 2*, 8–35.

Baraka, R. (2001). The meanings of independence in post-colonial Papua New Guinea. A Plan B Research Paper Submitted to the Faculty of the Center for Pacific Islands Studies in the Candidacy for the Degree of Masters of Arts. School of Hawaii, Asia and Pacific Studies. The University of Hawai'i at Manoa.

Bélanger, P. (2019). *Extraction empire: Undermining the systems, states, and scales of Canada's global resource empire, 2017–1217*. MIT Press.

Beehler, B., & Laman, T. (2020). *New Guinea: Nature and Culture of Earth's Grandest Island*. Princeton University Press.

Boerhringer, G. H., & Giles, D. (1977). Criminology and neocolonialism: The case of Papua New Guinea. *Crime and Social Justice, 8*, 58–63.

Buckley, M., & Lama, D. (2021). *The fragile planet: His holiness the Dalai Lama on the environment*. Sumeru Books. ISBN 978-1-896559-73-5.

Cousteau, J.M., & Richards M. (1999). *Cousteau's Papua New Guinea Journey*. Henry Abraham's Books.

Ceballos, F. C., & Álvarez, G. (2013). Royal dynasties as human inbreeding laboratories: The Habsburgs. *Heredity, 111*(2), 114–121.

Chan, J. (2016). *Playing the game: Life and politics in Papua New Guinea*. University of Queensland Press.

Dinnen, S., McLeod, A., & Peake, G. (2006). Police-building in weak states: Australian approaches in Papua New Guinea and Solomon Islands. *Civil Wars, 8*, 87–108.

Der Spiegel (U. Knoefel and O. Celik). (2013). Museum wars: Ankara demands artifacts from Berlin, 14th March 2013. Der Spiegel, Germany. https://www.spiegel.de/international/germany/dispute-heats-up-between-germany-and-turkey-over-contested-artifacts-a-888398.html. Accessed May 11, 2022.

Dastyari, A., & O'Sullivan, M. (2016). Not for export: The failure of Australia's extraterritorial processing regime in Papua New Guinea and the decision of the PNG Supreme Court in "Namah" (2016). *Monash University Law Review, 42*(2), 308–338.

Diamond, J, (2011a). *Guns, Germs and Steel: The Fate of Human Societies*. Random House.

EUObserver. (2011). Meghan Markle, royal racism and the 'European Way of Life. 17 March 2011. https://euobserver.com/opinion/151250. Accessed July 1, 2022.

Firth, S. G. (1972). The New Guinea Company, 1885–1899: A case of unprofitable imperialism, 361–377.

Fitzpatrick, P. (1980). Really rather like Slavery: Law and labour in the colonial economy in Papua New Guinea. *Crime, Law and Social Change, 4*(1), 77–84.

Flannery, T. (2002). *The future eaters: The future eaters: An ecological history of the Australasian lands and people*. Grove Press.

Foster, R. J. (1999). TV talk in PNG: A search for a policy in a weak state. *Pacific Journalism Review, 5*(1), 53–79.

Foster, R. J. (2002). *Materializing the nation: Commodities, consumption, and media in Papua New Guinea*. Indiana University Press.

Gillison, G. (1993). *Between culture and fantasy: A New Guinea highlands mythology*. University of Chicago Press.

Gosarevski, S., Hughes, H., & Windybank, S. (2019). Is Papua New Guinea viable? *Political Economic Bulletin, 19*. https://www.semanticscholar.org/paper/Is-Papua-New-Guinea-viable-Gosarevski-Hughes/6d71c70951c12fa82dea578a36b66923997f3f92. Accessed June 1, 2022.

Hawksley, C. (2006). Papua New Guinea at thirty: Late decolonisation and the political economy of nation-building. *Third World Quarterly, 27*, 161–173.

Henton, D., & Flower, A. (2007). *Mount Kare gold rush: Papua New Guinea, 1988–1994*. Published by Mt Kare Gold Rush, Brisbane Australia. ISBN 10: 0646482815; ISBN 13: 9780646482811

Hinton, A. L., Woolford, A., & Benvenuto, J. (2014). *Colonial genocide in indigenous North America*. Duke University Press.

Huettmann, F. (2015). On the relevance and moral impediment of digital data management, data sharing, and public open access and open source code in (Tropical) research: The Rio convention revisited towards mega science and best professional research practices. In F. Huettmann (Ed.), *Central American biodiversity: Conservation, ecology, and a sustainable future* (pp. 391–418). Springer.

Huettmann, F. (2020a). Spirituality beats it all: A quick conservation overview, self-organization and the great value of (indigenous) religions for Hindu Kush-Himalaya landscapes, its geo-parks, species, ecological processes and watersheds. In *Hindu Kush-Himalaya watersheds downhill: Landscape ecology and conservation perspectives* (pp. 283–311). Springer.

Huettmann, F. (2020b). Chapter 24—Investigating Matschie's tree kangaroos with 'modern' methods: Digital workflows, big data project infrastructure, and mandated approaches for a holistic conservation governance. In D. L. P. Valentine, J. Blessington, & K. Schwartz (Eds.), *Tree Kangaroos: Science and conservation* (pp. 379–391). Academic.

Huettmann, F. (2020c). Chapter 25. The forgotten data: A rather short but deep story of museums and libraries in HKH and similar information sources in support of the global biodiversity information system (GBIF.org) and model-predictions for improved conservation management. In G. R. Regmi, & F. Huettmann (Eds.), *Hindu Kush-Himalaya watersheds downhill: Landscape ecology and conservation perspectives* (pp. 497–520). Springer Gland.

Hughes, L. (2002). Moving the Maasai: A colonial misadventure [Doctoral dissertation, University of Oxford].

Humphries, G., Magness, D. R. & Huettmann, F. (2018). *Machine Learning for Ecology and Sustainable Natural Resource Management*. Springer.

Jenkins, J. C., & Schmeidl, S. (1995). Flight from violence: The origins and implications of the world refugee crisis. *Sociological Focus, 28*(1), 63–82.

Kimball, L. A. (1988). The Antarctic treaty system. *Oceanus*, 14–19. Summer

Kirsch, S. (2014). *Mining capitalism.* University of California Press.

Knoefel, U. (2019). A hidden legacy: Time for a new look at Nazi art. Der Spiegel, 14th August 2019. https://www.spiegel.de/international/zeitgeist/hidden-legacy-time-for-a-new-look-at-nazi-art-a-1281602.html. Accessed May 11, 2022.

Knoefel, U. (2021a). *Streit um Kolonialkunst: Grüne Ministerin legt »Roadmap« zur Rückgabe von Raubkunst vor.* Der Spiegel, 23rd April 2021a. https://www.spiegel.de/kultur/theresia-bauer-gruene-ministerin-legt-roadmap-zur-rueckgabe-von-raubkunst-vor-a-fd1a889c-2267-41b1-8f81-cfae8930f7a0. Accessed May 11, 2022.

Knoefel, U. (2021b). *Debatte über koloniale Raubkunst: Schluss mit der Verharmlosung.* Der Spiegel, 28th May 2021b. https://www.spiegel.de/kultur/koloniale-raubkunst-schluss-mit-der-verharmlosung-kommentar-a-338f7afd-8f91-4da2-98e6-272cec7cf2e1. Accessed May 11, 2022.

Mayr, E., Diamond, J. M., & Diamond, J. (2001). *The birds of northern Melanesia: Speciation, ecology and biogeography.* Oxford University Press on Demand.

McFarlane, S. (1889). *Among the cannibals of New Guinea: Being the story of the New Guinea mission of the London missionary society.* Presbyterian Board of Publication.

Mackenzie, S. S. (1941). *The Australians at Rabaul. The capture and administration of the German possessions in the Southern Pacific, official history of Australia in the war of 1914–1918* (p. 412). Angus and Robertson Ltd.

Mead, M. (1967). The rights of primitive peoples: Papua-New Guinea: A crucial instance. *Foreign Affairs, 45*(2), 304–318.

Metzler, J. (1990). The Vatican secret archives and their missionary holdings. *Mission Studies, 7*(1), 108–117.

Mills, T. (2020). *The BBC: Myth of a public service.* Verso Books.

Morris, I. (2018). *Geography is destiny: Britain and the world, a 10,000 year history.* Profile Publisher.

Muehlhahn, K., Berman, N., & Nganang, P. (2014). *German colonialism revisited African, Asian, and Oceanic experiences.* University of Michigan Press.

Laman, T., & Stoles, E. (2012). *Birds of paradise: Revealing the world's most extraordinary birds.* National Geographic Publishers.

Mayr, E., & Diamond, J. M. (2001). *The birds of northern Melanesia: Speciation, ecology and biogeography.* Oxford University Press on Demand.

Naess, A. (2009). *The ecology of wisdom: Writings by Arne Naess.* Catapult.

Niles, D. (2012). The national repatriation of Papua New Guinea recordings: Experiences straddling world war II. In *Ethnomusicology forum* (Vol. 21(2), pp. 141–159). Taylor & Francis Group.

Nelson, H. (1982). *Taim Bilong Masta, The Australian involvement in Papua New Guinea.* Australian Broadcasting Commission. ISBN 9780642975669, 0642975663. Google Books https://www.google.com/books/edition/Taim_Bilong_Masta/y2MhAAAAMAAJ?hl=en

Ohff, H. J. (2008). Empires of enterprise: German and English commercial interests in East New Guinea 1884 to 1914 [Doctoral dissertation, Adelaide University, Australia].

Ohse, B., Huettmann, F., Ickert-Bond, S., & Juday, G. (2009). Modeling the distribution of white spruce (Picea glauca) for Alaska with high accuracy: An open access role-model for predicting tree species in last remaining wilderness areas. *Polar Biology, 32*, 1717–1724.

O'Neill, C. (2020). Royal blood: Queen Victoria and the legacy of hemophilia in European royalty. Hektoen. *International Journal of Medical Humanities, 12*(2). https://hekint.org/2020/02/10/royal-blood-queen-victoria-and-the-legacy-of-hemophilia-in-european-royalty/. Accessed July 23, 2022.

Patience, A. (2012). The new cannibalism: The international community and the problem of governance weaknesses in Papua New Guinea. *Asia Pacific Journal of Public Administration, 34*(1), 31–55.

Pietrangeli, C. (1982). *The Vatican collections: The papacy and art.* Metropolitan Museum of Art.

Pomerance, M. (1996). *The United States and the world court as a "Supreme Court of the Nations": Dreams, illusions and disillusion* (Vol. 26). Martinus Nijhoff Publishers.

Rich, B. (1994). *Mortgaging the earth: The world bank, environmental impoverishment and the crisis of development*. Earthscan Publishers.
Rooney, D. (2004). Information for empowerment and development, why media is failing the people of Papua New Guinea. In *Paper presented at World Media Freedom Day Conference*, Divine Word University, Madang, PNG, 30 April.
Rozzi, R., Chapin III, F. S., Callicott, J. B., Pickett, S. T., Power, M. E., Armesto, J. J., & May Jr, R. H. (Eds.). (2015). *Earth stewardship: Linking ecology and ethics in theory and practice* (Vol. 2). Springer.
Rumley, D. (2006). Australia's arc of instability: Evolution, causes and policy dilemmas. *The Otemon Journal of Australian Studies, 32*, 37–59.
Schneppen, H. (2001). The case of Karl Peters (1856–1918). A German colonial civil servant on trial. *ZEITSCHRIFT FUR GESCHICHTSWISSENSCHAFT*, *49*(10), 869–885.
Seewald, B. (2022). Welser-Die-Deutschen-sind-schlimmer-als-die-wildesten-Loewen. In *Die Welt*, 27 March 2022. https://www.welt.de/geschichte/kopf-des-tages/article237787217/Welser-Die-Deutschen-sind-schlimmer-als-die-wildesten-Loewen.html. Accessed March 28, 2022.
Shelton, K. (1994). Nikolai Miklouho-Maclay, the forgotten anthropologist [Unpublished PhD thesis; Northern Illinois University].
Sinclair, J. (1978). Wings *of gold: How the aeroplane developed New Guinea*. Pacific Publications. ISBN 0-85807-033-2.
Stiglitz, J. (2003). *Globalization and its discontent*. Norton Paperback.
Suwal, M. K., & Huettmann, F. (2020). Chapter 26 A rather short story of the GIS data layers in HKH. In: G. R. Regmi, & F. Huettmann (Eds.), *Hindu Kush-Himalaya watersheds downhill: Landscape ecology and conservation perspectives* (pp. 521–563). Springer Gland.
Suzuki, D. (1993). *Wisdom of the elders: Sacred native stories of nature*. BoD–Books on Demand.
The Pledge Times. (2019). The 'loss' of an unusual dinosaur uncovers how rich countries exploit southern fossils (19th May 2019). https://pledgetimes.com/the-loss-of-an-unusual-dinosaur-uncovers-how-rich-countries-exploit-southern-fossils/. Accessed April 24, 2022.
The Guardian. (2022a). Not calling a massacre a massacre is ridiculous': A model of truth-telling at Myall Creek, 2nd May 2022a. https://www.theguardian.com/australia-news/2022a/may/03/not-calling-a-massacre-a-massacre-is-ridiculous-a-model-of-truth-telling-at-myall-creek. Accessed May 15, 2022a.
The Secret Barrister. (2018). *Stories of the law and how it's broken*. Picador Publishers.
The Guardian. (2018). Why are white curators still running African art collections? https://www.theguardian.com/commentisfree/2018/apr/03/brooklyn-museum-white-curators-african-art-open-letter. Accessed May 5, 2022.
The Guardian Nigeria. (2018). Outrage continues over white curator for Brooklyn Museum's African collection, 8th April 2018. https://guardian.ng/art/outrage-continues-over-white-curator-for-brooklyn-museums-african-collection/. Accessed May 5, 2022.
The Guardian. (2022b). The New York DA's office fighting to stop trade of looted antiquities, 30th May 2022b. https://www.theguardian.com/us-news/2022b/may/30/fight-to-stop-trade-of-looted-antiquities. Accessed June 1, 2022b.
Webster, E. M., & Webster, R. (1984). *The moon man: A biography of Nikolai Miklouho-Maclay*. Univ of California Press.
Wilson, M. (2019). A *true child of Papua New Guinea: Memoir of a life in two worlds*. MacFarland Publishers.
Windybank, S., & Manning, M. (2003). *Papua New Guinea on the brink*. Centre for Independent Studies.
WIPO (World Intellectual Property Organization). (2016). *Bioethics and patent law: The cases of Moore and the Hagahai people*. WIPO 5. https://www.wipo.int/wipo_magazine/en/2006/05/article_0008.html. Accessed June 30, 2022.
Ziegler, J. (2013). *Betting on famine: Why the world still goes hungry*. The New Press.
Zuckerberg, B., Huettmann, F., & Friar, J. (2011). Proper data management as a scientific foundation for reliable species distribution modeling. Chapter 3. In C. A. Drew, Y. Wiersma, & F. Huettmann (Eds.), *Predictive species and habitat modeling in landscape ecology* (pp. 45–70). Springer.

Chapter 6
Biodiversity of Papua New Guinea (PNG): Attempting a More Meaningful Conservation Description and Approach of Its Use, Co-evolution, Generic Status and Grim Outlook

> *Neither the government of PNG nor of Indonesia shows the necessary capacity or commitment to actively conserve mammal species and their critical habitat.*
> *Beehler and Laman (*2020*, p. 220)*
> *The most extraordinary and the most beautiful of the feathered inhabitants of the Earth.*
> *Edgar Wallace reference to Birds of Paradise (cited in New South Wales State Library,* 2022*)*
> *Bats can hear shapes. Plants can eat light. Bees can dance maps.* (This citation exist in several variations. And one may easily add infrasound, 3D and ocean issues, magnetic detection, and the feelings and counting skills by plants and trees etc. Overall, if all of those things are already occurring, what do we not know yet and how biased are humans in their knowledge, perception and subsequent conservation management? There is more than what meets the eye. PNG deeply entrenched in its cosmologies can open those pathways to the non-believer. Unfortunately, I know of many PNG experts that are not among those people but who remain with conservative, parsimonious descriptions of the Western Society the most, instead of seeing the wider, telecoupled, more holistic picture that PNG offers us and that its citizens know for millennia already.)
> *@CryptoNature in Ludlam (*2021, p. 272*)*

Abstract Papua New Guinea (PNG) is one of the few megadiversity nations in the world. It's essentially an ancient species engine and hosts world-relevant populations of wildlife and plants, with a co-evolved human society for over 47,000 years. Many of the species in PNG are among the oldest in the world and endemic providing unique DNA and lessons of evolution for global mankind and well-being. One can see in PNG the more original set up of species in the tropics, and the world. Many international and grand expeditions, collections and research were done in PNG but with virtually little sustainability success. Most data remain not available, hardly known. However, while PNG was forced to engage in a global commodity market during colonial times and subsequent globalization the conservation status for most of those species and habitats in PNG remains poor and with little relevant action or vision presented. It follows a loose laissez-faire model from Australia added with neocolonial attitudes and Asian input. PNG remains a 'feast' for the global enterprise. It's

F. Huettmann, *Globalization and Papua New Guinea: Ancient Wilderness, Paradise, Introduced Terror and Hell*, https://doi.org/10.1007/978-3-031-20262-9_6

shown that the current nation set up and governance for megabiodiversity nations like PNG and wider Melanesia results in the wholesale destruction of otherwise globally relevant world wilderness, species, ecological services and sustainability.

Keywords Papua New Guinea (PNG) · Biodiversity · Multispecies · Conservation · Wallace Line · Endemism · Ecology · Global Biodiversity Information System (GBIF.org)

6.1 Introduction

Like many tropical nations, Papua New Guinea (PNG) is known for its unique biodiversity; much of it is deep, ancient and endemic but virtually all of it co-evolved with a low density of humans (Beehler and Laman, 2020; Diamond, 2011; Flannery, 2002). By now it's probably public knowledge that PNG is part of the 17 megadiversity nations in the world (https://en.wikipedia.org/wiki/Megadiverse_countries). But despite being a 'tropical' nation PNG's equatorial biodiversity stands in wide contrast to the widely heralded and promoted biodiversity patterns and concepts in Central or Latin America, Africa or other parts of Southeast Asia for instance. PNG is not much 'packed with species' like Costa Rica is (Huettmann, 2015 and citations within), and the species and diversity densities tend to differ also. PNG is south of the Wallace Line, which creates a schism in Southeast Asia; paradoxically PNG is virtually free of monkeys and squirrels (other than a few introduced species etc; see subsequent chapters in this book). It's also affected by Weber's line, driving fish species distributions (Beehler & Laman, 2020 and citations within). In an evolutionary earth history sense, birds and mammals of PNG are usually younger, whereas plants and insects are the older groups. But it's the insects and plants that crossed the Wallace Line.[1]

PNG sits at the Australian craton—essentially a geologically ancient rock connecting Australia directly with PNG that still moves geologically—and thus PNG belongs to the wider Sahul region, initially all part of a unified Godwanaland (Beehler & Laman, 2020; Flannery, 1990, 2002). PNG is essentially a rugged landmass with many island fragments surrounded by saltwater located between Australia and Southeast Asia. And therefore it was isolated for quite a long time with an associated human co-evolution (a selected visualization of those aspects is shown in Figs. 6.1, 6.2, 6.3, 6.4, 6.5, 6.6, 6.7, 6.8, 6.9, 6.10, 6.11, 6.12, 6.13, 6.14, 6.15, 6.16, 6.17, 6.18, 6.19, 6.20, 6.21, 6.22, 6.23, 6.24, 6.25, 6.26, 6.27, 6.28, 6.29, 6.30, 6.31, 6.32, 6.33, 6.34, 6.35, 6.36, 6.37, 6.38, 6.39 and 6.40).

Islands are known to present us with 'biological labs of evolution,' somewhat independent experiments that unfold over time in various directions. But while PNG as a nation is facilitated by a large island land mass, it's also defined by a myriad of islands and islets and by all of its interactions as a wider whole (part of Melanesia,

[1] There is a good question where the Wallace Line and Weber Line actually are located, and how that is determined. In reality, one needs to be prepared to argue those details and to handle a grey zone instead.

Fig. 6.1 Papua New Guinea is home to the oldest fig trees. It's where much of the forest cover and its co-evolution with the biotic and abiotic world starts

the Pacific and beyond). The ocean plays a big role for PNG. Most of the islands are covered with closed canopy vegetation. Non-forested areas in PNG are relatively species-poor but they still carry unique species, e.g. specific snakes in the grasslands (of which some are rather poisonous).

Biogeography processes relate to barriers and dispersal (MacArthur & Wilson, 1963); this makes saltwater, barriers and mountains a prime topic to study, just as it applies still to PNG (see at Beehler & Laman, 2020 for a New Guinea perspective). The discipline of biogeography studies typically distant islands, island size effects, comparable high and low islands, land bridges, highly movable species, corridors and mountain-top 'islands.' Elevation effects are dominant in biogeography, as described by Alexander von Humboldt for South America two centuries ago and also found by Wilson and MacArthur (1967) and Diamond (1976) for PNG, besides many others (not all are given credit). Further, for PNG Species-Pair Competition and Ring-Species were also described (e.g. Mayr and Diamond 2001). Other effects shaping biographic patterns are the harshness of the environment. In PNG that's for

Fig. 6.2 Large orb-web spiders: a good welcome to Papua New Guinea's biodiversity

instance, ancient glacier effects and the El Nino dry events affecting fire occurrence and subsequent species set up for land and the ocean.

PNG features most aspects of the classic Island Biogeography theory from Wilson and MacArthur (1967); this theory is essentially a common sense approach published from the 1960s though. It actually falls quite short on accounting for human impacts, and it is not much based on modern data and the latest assessment methods (e.g. best-available globally compiled data sources at hand, DNA evidences, survey detection statistics, habitat data, disease information and computing-intense analysis for better inference. The knowledge about evolution also progressed dramatically since then).

But in addition to those research topics, PNG has many additional dimensions foreign to short-term outside observers, one of them is 'deep time' another one is, year-round effects as well as cosmology (Baraka, 2001). As a living place PNG makes for a unique set up that the Western World has a hard time to understand and to grasp with, or to manage well for sustainability. PNG is not like Europe or North America

Fig. 6.3 Ancient beetles

(but where most of the theory-driving investigators, actors and publishers for PNG currently are coming from). And PNG is just one of many places in Asia; the Pacific Rim is indeed a big place, truly connected with deep earth, and that the Western World has a difficult time with and fails frequently (for evidence of this true'ism just see Myint-U, 2006 for Burma experience; Glavin, 2008 for ocean explorations, Rauzon 2016 for Pacific Island set up, or Krishna et al., 2022 for rabies as a wider landscape disease).

As stated by Beehler and Laman (2020), PNG is a great species generator for plants and animals. And so in that regard, PNG compares easily with the much larger Amazon basin and Congo. PNG provides us with a gift of species and human co-evolution on a global level. That is specifically so due to the 'young' geology making a universal link across all forms of live we can see and experience, land and sea.

And PNG's biodiversity is not only shaped by a 'set of isolated islands' located south of the Wallace Line and the Weber Line allowing us to see ancient species, e.g. for insects and plants. Due to the elevational gradient, the diversity of species increases and then falls after c. 2500 m (Beehler & Laman, 2020), and associated endemism is highest in mountain areas (Beehler & Laman, 2020). As well, PNG acts as a North–South species conduct, and thus it is part of the exchange corridor between Australia and Indonesia, South East Asia, with a peak biodiversity from both regions, as it is the case for insects and plants, let's say (whereas other species groups mixed and moved less, e.g. mammals and birds). All of those details matter

Fig. 6.4 'Walking stick' species

for the species set up found in PNG and its human co-evolution. But this matters more for some species than for others. It gets complex…

PNG is a geographic species bottleneck and acts as a biogeography textbook in action; and much of that book's chapters have not even been written yet. In the meantime, one hopes PNG's nature remains in a good shape so that its study and subsequent conservation can be achieved. That is unlikely though. The current outlook looks rather grim (details in the following chapters of this book).

While many of PNG's species are usually described as ancient, more simple and primitive—in the evolutionary sense—but they are not less fascinating as they allow us to see species concepts that are already gone. Often those are simply 'unique' and among the first species we know of that existed or that are leftover for humans to experience. In PNG, we can see into Earth and Earth's history, under the sea surface, what it was like and where we ourselves come from, and how we connect to the universe overall. In PNG, we can truly understand ourselves!

A great example for that are the ancient figs and the orchid diversity of PNG, or some large butterflies (= largest in the world; see Beehler & Laman, 2020 for such species and their trade).

From that any scholar, student and naturalist easily can conclude that PNG species can open your mind and understanding of what species and ecosystems consist of, what earth really is, how we evolved under the wider universe, and where species might have come from. A typical example for that are the mammals, where in PNG it's

Fig. 6.5 Butterfly collection from Papua New Guinea; many species are bred and exported, alive of dead—for the world's butterfly lovers

the only place where all major clades of living mammals coexist: monotremes (egg laying mammals), marsupials (pouch-raising mammals) and placentals (mammals with a placenta as essentially found worldwide and often seen, wrongly though, as the quintessential mammals); e.g. Flannery (1990), Martin (2005). Many species here co-evolved with the environment because PNG is a few of the world's unique places where species had much time to do so, all done without much initial pressure from human densities. Endemism is consequently high in PNG; we currently just see what is left of it. Due to this unique set up, many PNG species now carry a conservation concern (because species that are simple, slow and big get eaten and collected fast…)! The Sahul region has already experienced such a species loss over time (details further below; see also Flannery, 2002; Martin, 2005).

6.2 Birds: The Cheapest and Effective Conservation Platform Left Widely Unused

PNG has one of the most diverse avifauna in the Pacific, c. 900 species of birds, c. 9% of the world's species (app. 9,750 species worldwide; but with more DNA research, political lobbies and Birders pushing for species splits this number is to increase soon whereas most of the actual species and their habitats on earth are more

Fig. 6.6 Butterfly cocoon in the wild

threatened than ever). Arguably, a sole species focus—such as Birds of Paradise—ignoring habitat issues and context is misleading for conservation, as can easily be shown for PNG.

Like in many British colonies worldwide, birds from those places to be settled in the British Empire got promoted for a long time. They are presented and get promoted in paintings, by artists and writers (e.g. Cocker, 2006 for a typical style) and are made 'special,' e.g. with narratives and through paid 'experts' (whose assumed job it might well be to show the citizens of the kingdom how special and thus worthwhile the place and its species are, promoting such colonial constructs. One may easily argue that the work by David Attenborough—as part of the BBC—is an outcome of such a long tradition now entrenched as a global culture and in modern media, that otherwise was not so prominent in other colonies and royal courts, e.g. with the Spanish, Portuguese or French, let's say. Due to competitive collection expeditions to bring home 'curiosities' for the cabinet those colonial nations do still have large museum collections with assigned curators and experts; details can be tracked just partly in GBIF.org because not all data are not widely shared with the global audience; see also Beehler & Laman, 2020 for expeditions and institutions; Huettmann, 2020; Table 6.1 for more details). A quick GBIF.org search for PNG data shows available data from the 1940s onwards, hardly before (but where major specimen collections were actually made, likely in the hundred thousands; more details are below and in textboxes).

Fig. 6.7 An ancient species group with its own co-evolved species famous for Papua New Guinea

Like found elsewhere in the tropics, bird species assemblages can be a patchy occurrence in the (tropical) forest landscape; for PNG those can consist of over 30 species (see also Beehler & Laman, 2020). Having spent much time in PNG forests myself, I hardly see that many species though! And so it's nothing unusual for a bird watcher in PNG to encounter much less species, or for a long while none at all…Perhaps the birds have to find you instead!? Despite its fame, PNG remains a challenge for bird watchers as birds do move fast, are smart (e.g. corvids), are used to human prosecution and escape fast, and often operate in high and complex canopies. Detectability is 'a thing' in PNG (in my own work in PNGs' rainforests, I only see app. 20% of birds, whereas I get over 80% of species and individuals just by hearing. Arguably there are very few sound recordings and ID guides for PNG available making reliable bird surveys a true challenge to most people and incoming students. Doing mist netting, shooting, camera-trapping and telemetry does not overcome this problem of very low detection rates and thus bias. This puts an incredible confidence question and liability on bird work in PNG and one wonders how earlier scholar dealt with it, e.g. E. Stresemann, E. Mayr or J. Diamond? Metadata do matter for such works. For instance, the German colonial bird work presented by Stresemann, 1923 carries large issues around spatial accuracies and data availabilities, e.g. in GBIF.org for Germany as a signatory, lacking metadata; compare also with Beehler & Laman, 2020 on that issue of German and other colonial expeditions).

Fig. 6.8 A megapod (bird) mound: those breeding ovens include eggs which can be harvested sustainably over time

A noteworthy finding in the ancient forests of PNG is the ground nesters, as it's a common finding for remote and rel. undisturbed wilderness areas, such as PNG (see Freeman et al., 2013; Mack, 2014 for ground-living birds and findings). When stray dogs, cats or other mammals and predators come into a landscape (dogs came into PNG rather late, perhaps 2,500 years ago), this cohort of birds tends to disappear quickly. And PNG had dogs and specifically hunting dogs to aid that process (e.g. Flannery, 1990, 2002; Martin, 2005).

There are of course many other very interesting aspects in birds of PNG and their conservation (the latter subject is poorly tackled though for a meaningful policy and impact studies; see Richards, 2018; Richards & Whitmore, 2015 for examples and where rapid species lists dominate but accepted taxonomies, meaningful research design and statistics are side-lined for valid inference and meaning). In PNG vultures are missing, but Black Kites are now found in large abundance in human-changed areas and at forest fires and around garbage dumps, e.g. in the Ramu sugarcane area. Forest birds in PNG have often two eggs, sometimes spread out over several months. Switching of the ecological niche can be observed, such as with the pygmy parrot, operating in a 'nuthatch niche' (Beehler & Laman, 2020). PNG has the world's largest pigeon, the smallest parrot in the world and also rather large parrot (the Papan Vulture Parrot).

Fig. 6.9 Nest attractant of a bowerbird

Textbox 1: The ongoing fallacy of specimen collection: Pseudo-science without relevant research design and controls, Neocolonialism in museums and utter lack of conservation and habitat progress

Nature and its biodiversity are 'really big,' it's megascience. The associated complexities increase manifold when collecting over time for specific time periods, or years. One can easily collect oneself to infinity and get exhausted for live but still not get it all described or done well. As many collectors have experienced, it's virtually not fundable! Arguably, that creates a bias in itself. But it's virtually impossible to describe nature in objective terms and being 'complete' or representative, as nature is wider than what humans, or a group of humans, perceive. Putting those collections in an institutional

Fig. 6.10 Chick of a cassowary; those chicks are often caught in the wild and sold or kept. However, in captivity those can turn rather aggressive and in the first year get usually released or killed (= eaten) as those birds are pretty tasty. Worth to mention that those birds are cute but over time can be rather dangerous and can injure and even kill people

framework involving human lives, careers and 8 h-a-day employment makes is a formidable task indeed.

And so, many of the commonly encountered biodiversity issues worldwide are now also found in PNG—Earth's grandest island in many respects. Often those issues are of international relevance and carry a price tag to deal with. One of the typical stereotypes encountered center around the traditional style of collection expeditions carried out for 'remote' PNG. Virtually all aspects of endemic species from PNG are collected by museums, herbariums and zoos, usually with governmental approval, visas, permits and public or private

Fig. 6.11 Moss, a mystery species set for Papua New Guinea

funds- embassies included. Many of those examples can be found documented as data in GBIF.org, but many more not all (e.g. many records of the colonial times; see also Huettmann, 2020 for some PNG details). Insect collections, namely butterflies and beetles, as well as early plants, might lead that list of 'no shows' in the publically available realm (see Flannery, 1990; Martin, 2005 for paucity of records for mammals despite centuries of study efforts in PNG).

Beehler and Laman (2020) reported on science xenophobia and overregulation of field work (p. 57) making it difficult for the traditional sciences. In New Guinea, that can be based on religious beliefs, e.g. Islamic governance, indigenous cosmologies and/or bad century-long legacy of western efforts in that region. Australia dominates many of the biodiversity topics,

Fig. 6.12 Forest ecosystem with fungi at its finest in pristine Papua New Guinea; those wilderness forests and ecosystems, and their ecological services, are on the decline

e.g. taxonomic Delta keys (https://www.delta-intkey.com/www/overview.htm) or tree guide of PNG and all its field data get sent to Sydney and collection managers there (https://www.idigbio.org/wiki/images/e/e9/Guide_to_trees_of_Papua_New_Guinea.pdf). The claim made by Beehler and Laman (2020) that all such collections are globally shared is just 100% not true and not much applicable, nor that much is learned; as the missing conservation progress and crisis for PNG presents. The ecology knowledge from those collections remains bleak, as any literature search for PNG can easily show, and as stated by Beehler and Laman (2020) in their own words.

Fig. 6.13 Old-growth trees loaded with epiphytes: wilderness habitats at its finest and of disproportional relevance (e.g. Taylor et al. 2022)

Besides large decade-long collection expeditions in PNG shown in Table 6.1, many many more exist; not all are well known even. One may add here many local collection efforts such as the Wau Ecology Institute (https://en.wikipedia.org/wiki/Wau_Ecology_Institute) now in a funding crisis, if even operating effectively.

To convince in an argument, the western science world centers around 'evidence'; the smoking gun. For species presence and taxonomy, this is usually provided with a voucher specimen to refer to as 'proof'. Voucher specimen allow to confirm and investigate a species detail. '*One must have a bird in the hand*,' so to speak. However, in the year 2022 this is a very outdated concept when nowadays DNA records drive much of taxonomy and when statistics and online analysis are done 'in the cloud.' to actually make the case. This now sits at the core of knowledge production but is rarely fully available, certainly not understandable for the lay audience.

In the '*cabinet of curiosities*' mentioned for specimen by Beehler and Laman (2020) as the root of modern taxonomy to go from, certainly for PNG, the major (public) museums competed for exotic species and specimen and collected

Fig. 6.14 A typical epiphyte load in a canopy of old-growth trees

accordingly. The collection expeditions served virtually no other purpose. Institutions and their funders boosted against each other who has most (exotic) specimen and from which locations. It was as simple as that. And the royal courts and wealthy nobles were leading such vanity approach to science and seeking funding, as well described by Beehler and Laman (2020), Martin (2005). In the year 2022 one should move forward and beyond though, serving the public at large, e.g. Graham et al. (2004). Science does not need to be intrusive, unlikely should be (see Humphries et al., 2018 for options). Of course, all old material should be made available, fully described, in a good format, and usually be fully analyzed first before new ones get collected. One may have an embargo on research collections till then; why not?

Fig. 6.15 An orchid harvested from a tree canopy

While many people keep collecting, for the sake of collecting (e.g. '…*PNG is seriously undercollected*…'; Beehler & Laman, 2020, p. 123), the intense collection efforts made in PNG must appear dubious to many people. Many nations competed and participated, not all such expeditions were successful, many are not shared and a wide 'bycatch' exist—also in the social aspect of it for impacts. And if the data are shared, they are shared in a filtered form and incomplete fashion. A typical example is with the Bird of Paradise specimen and the tree kangaroo ones without proper locations, or even without a source (widely discussed in Flannery, 1990; Martin, 2005) or any standardized meta-data across languages and disciplines to understand for a global audience what was done (see Huettmann, 2020 for examples).

Fig. 6.16 Tropical flower beauty

The birds of PNG are arguably a global conservation highlight; many of the species are highly seeked after by most bird watchers (e.g. Snetsinger & Pratt, 2003), researchers and certainly by museums and their eager curators in the world who are paid and funded for collecting them (see for instance bird collections mentioned in Diamond, 2011; Mayr & Diamond, 2001, and as mentioned in Mack, 2014, and listed in Beehler & Laman, 2020 with taken numbers, easily in the many thousands), and with the Birds of Paradise (BoP) easily on the prime list (e.g. Laman & Scholes, 2012). Graduate student projects from many nations collecting hundreds of bird specimen can also be found. BoPs have actually been intensely collected for centuries by many actors, with 100,000 s of individuals exported over many years (see subsequent details in the BoP section below). Arguably, the collections intensified with colonialism and globalization, and they have not stopped; a recent international poaching upsurge has been observed due to apparent demands by Asia

And of course, nowadays people are also interested in other avian highlights like '*the bird with poison feather*s': The Pitohui (as stated in the western media; details in Beehler & Laman, 2020; Mack, 2014 with citations within). This bird represents the first chemical defense found in any bird really ...by Western people, as the locals knew this species and its 'bad taste' for long time. It was not tasty and simply avoided. There is another species of this sort, the Blue-headed Ifrit also has a light poisson and due to feeding ants that seem to produce such a toxic substance (Beehler & Laman, 2020). But this is 'just' another bird item to be hyped up about PNG while PNG

Fig. 6.17 Ferns growing on tree stems and branches; it's a large contribution to forest biomass and non-timber value

remains very deep and complex with selected bird narratives falling short of wider and earnest conservation overall, or of any western understanding of PNG well and appropriately. In the following I continue to present some more details about PNG's birds:

The 20 species of bowerbirds—famous for their nest attractants and using modern shiny features—are also of great ecological interest for various reasons, but they are much less recognized by a global audience, or their scientists. Most of such species are widely under-researched and actually lack relevant and serious conservation efforts.

Birds that I have seen a lot myself in the deeper jungle during many years of fieldwork are sulfur-crested cockatoos, and Blyth's Hornbill, but almost no Papan Vulturine Parrots. Was I just lucky, or unlucky or has that a wider conservation meaning? Arguably, those species are easy to detect by call and flight, and thus biased detections.

Further, PNG features ground doves as well as various parrots, but no real hummingbirds. The latter group is likely compensated by flowerpeckers and other species instead. Notably, but not new, are the specific elevational associations of avian endemics and ecological niches (as widely described by Diamond, 1973; see Beehler & Laman, 2020; Steadman, 2006 for citations) and then also the many island ranges as some species failed to colonize islands and sites across saltwater and islands vary in elevation.

Fig. 6.18 Field work is demanding: hiking and researching ancient forest trails connecting the coast with the highlands that were used for millennia

The bird biogeography described by Diamond (1973, 1976) onwards became of world fame. It set a template and consists of elevational gradients, island hoppers, island skippers and other peculiar cases. See also work by Mayr and Diamond (2001), Thibault and Cibois (2017) on islands, seabirds and the island declines (Steadman, 2006). Each island has its own set up and history; a microcosm of life.

Shorebirds—part of waterbirds—are also of interest for PNG but a bit overlooked, specifically the 'peeps' (sandpipers) at mudflats (see Long et al., 2021 for a PNG Ramsar wetlands affected by oil & gas industry), and Far East Curlews (see Fig. 6.36 for estuary habitats) as those come from Russian Far East, Australia etc., and appear on the decline now. And then also the other many long-distance shorebirds that migrate between Russia, Australia and landing in PNG's wetlands like the Ruddy Turnstones, and Rednecked Stints; some plovers might connect with Tibet and the Hindu Kush-Himalayas or Mongolia. Others might just extend from Australia across the Torres Strait, including gulls terns and some white-bellied sea eagles.

And many people might not know that PNG acts as a wintering ground for Russia's and Alaska's Aleutian tern species, many now on a dramatic decline. It's part of a wider species movements and similarly connects sea turtles and sharks from Hawaii and even from the Californian current and from parts of southern Alaska. Along the same lines, PNG is globally connected with marine mammal migration, e.g. humpback whales, sperm whales and associates. Birds are often part of such ocean communities and we can just see and comprehend the left-overs now.

Fig. 6.19 Matschie's tree kangaroos, in captivity (where most people see them; in the wild they are rather elusive in the deep old-growth canopy and remote and rugged forest areas)

Many biodiversity aspects actually do move and migrate indeed, often as a wider community; exact details are not well known yet and discoveries are still made. Thuna might play a role as their foraging follows fish flocks and brings them to the surface for other species like seabirds to feed and to join. It comes as a co-evolved wider migratory community, and that aspect is widely unstudied because polar environments are still not perceived—hardly studied—as directly connected with the tropics (Zoeckler, 2012), or with PNG for that matter. PNG gets still widely seen as a stand-alone unit, but is more than that and requires context. Research on avian influenza tends to support that due to virus exchange (e.g. Gulyaeva et al. 2020).

Already those migratory species carry with them many conservation issues from the outside. This can be zoonotic diseases or population questions and acting with a time lag; many of those aspects are not studied well and hardly known, certainly not acted on in valid management scenarios. Examples would be found with Zandri et al. (2009 and in Robbins et al. 2016).

Without doubt, PNG remains a bottleneck and deal breaker in case you want to enter the World Birding club (see public biography of one of the leading world record birder, the late Phoebe Snetsinger: https://en.wikipedia.org/wiki/Phoebe_Snetsinger; Snetsinger & Pratt, 2003), there are app. 103 endemic species to pick from in PNG. If you want to be a relevant world birder, PNG is a 'must see.'

Fig. 6.20 A scrap mark of a tree kangaroo on a tree; 'presence only' data in the wild

The latest 'rare' bird chases for PNG involve subspecies, as well as Beck's and Heinroth Petrels (e.g. Flood et al. 2017; see also Davis et al., 2018). It's another continuation and spin-off from the age-old western colonial bird pursuit, ticking, as an off concept for conservation but still heavily pursued by BirdLife International (a UK-based organization and funding platform), helped by such 'experts' and many amateurs from New Zealand and Australia, the funders of oil and gas, U.S. Fish & Wildlife Service and so on. Such minds and their projects never run short (see Long et al., 2021; Richards, 2018) while the actual conservation of birds, seabirds and their habitat remains an increasing worry and is a wide neglect towards absurdity. That is certainly true for PNG, and with few exceptions its citizens are usually excluded from such work on their own land and knowledge, e.g. on an author level or for being able to join. A quick science, funding and literature search on birds of PNG for PNG authors and PNG institutes will easily confirm that, see in Laman and Scholes

Fig. 6.21 Chewing and grazing signs from tree kangaroos that come down from the tree to feed; experienced 'rangers' (="Waldlaeufer") can read and intrerpret the forests like no others

(2012), Mack (2014) for BoPs. The few token authors from the region make that argument even further. 'Modern' Ornithology has a privilege and elite problem of untackled magnitude. In PNG that does not only include local PNG citizens and their absence, but also bird workers from nations other than UK, U.S., Australia and New Zealand. How many native African, Latin American and tropical-experienced bird researchers have worked and published in PNG?

In terms of 'using' birds for livelihoods PNG offers many examples. Cassowaries are frequently caught and young birds kept as pets (they tend to turn very aggressive after a few months and thus get eaten; Mack, 2014 for studies on those topics; see Gillison, 2002 for more photos). Pet cockatoos can be widely found in villages, and their feathers are used in singsings and for instance in the YAMS dance (Cousteau & Richard, 1999).

For ornithologists of the world, PNG is world famous for the work by Diamond (1973, 2011). It's an archetype of field work in exotic field camps. Indeed it provided great baseline data, concepts and put birds on the global agenda and international arena, asking 'interesting' questions, but often this was based on specimen collections, little statistical research design and likely had quite a local impact. It was essentially done with a shotgun approach and without statistical detection and modeling considerations a pioneering but blunt concept that still dominates in most

Fig. 6.22 Leaves are beautiful

museums and is pursued and done that way, e.g. with Burke Seattle Museum specimen holdings for all over the Pacific (https://www.burkemuseum.org/). In addition, this work by J. Diamond is not well-tested statistically, modeled, predicted and/or really updated and was never done with GIS and environmental layers, or with any relevant modern quantitative methods, and it awaits field-based counts and statistical detection surveys, or DNA assessments. Modern man-made Climate Change questions rule in their absence. Diamond's work was not digital and thus remained widely unprovable with modern data, with a quantitative rigor and confidence, hypothesis, valid research design and quantitative data and models (but as science is widely done now these days). As bird distribution data for the general public do virtually not exist for PNG—or is not publicly shared in a meaningful digital form with metadata to understand it—such an assessment has not happened. Instead, detailed multiyear nice-looking photographic efforts were done with narratives (see Gillison, 2002; Laman & Scholes, 2012) not adding much to the sciences and conservation that was recommended as best practice and that had been possible or was suggested. Opportunistic field work still rules in PNG, what is a research design, and for whom?

Consequently, the bird data world in PNG remains wide open (e.g. Freeman et al., 2013 for new findings; for digital study approaches see also Huettmann, 2020. All of this stands in contrast to what Jerry Diamond expressed, that birds of PNG are essentially all discovered now and studied, with robust and well-proven principles of biogeography; quite far from it). Simply dealing with detection bias and correction

Fig. 6.23 Leave structure in a tropical fall; PNG rainforests have little seasons and many biological events occur interwoven

factors for presence and abundances, or model-predictions and remote sensing habitat layers in PNG in the Anthropocene will likely make for a big change and update on what we know about birds of PNG, their range and biogeography. An ornithological culture is to change and modernize still.

Just because a few studies have done phylogenomics and many opportunistic point-location shotgun collections are carried out does not mean PNG Ornithology is in the twenty-first century; Beehler and Laman (2020) are far from correct on that assumption. Even after decades of study the BoP taxonomy is not agreed and well resolved at all, and no protection levels really exist, neither for harvest nor for habitats and forestry, or for fishery impacts and seabird bycatch. A good focus species on that would already be the Frigatebirds, or kites.

People that are actually experts on birds of PNG — including nest finding and bird calls essential to ID and detect them in the field, or on life history details—are far and few, and beyond specimen collections much of the research focus is not very deep, well-coordinated or well-published even. Bird collection, camera traps, mist netting, banding—and now geotagging—and an ongoing, easy but aged obsession with species descriptions, naming and taxonomic splitting, and some DNA works still drive most of the widely fragmented research agenda of PNG birds, hardly with a Melanesian-style conservation focus whatsoever. A bird banding recovery atlas or

Fig. 6.24 Plant specimen for herbaria. Virtually endless attempts have been made in PNG and elsewhere with (plant) specimens collected by international actors in PNG in the millions. But what has been learned, what was the research design and location, when done, where are the data, what synergy result, and how does the (plant) conservation fare in Papua New Guinea? In the meantime, many specimen collections fade away by themselves and remain totally understudied and insufficiently analyzed for their global sustainability value and contribution; PNG has little plant conservation whatsoever nor is that promoted or achieved by many botanist. In reality, PNG is a total global botanist society conservation failure

meaningful band/ring recoveries for PNG birds shared wider does not exist (A good example found with the shorebirds and waterfowls as studied along the flyway by Australia, New Zealand, China, South Korea etc). It's disappointing to see and when new scholars and students engage that way, e.g. via research proposals of their 'dream' project by their supervisors to be funded for them in PNG, all based on repeated and widely empty-handed ornithology narratives and claims handed down over time but achieving so little. As a reviewer, I have personally seen such applications all the time in various reviewing platforms, e.g. National Science Foundation, NGOs and with graduate, PhD students and postdocs. and their institutions, zoos included But what for? Birds of PNG are still not better off, likely will not be the next decade

Accordingly, while many birds of PNG migrate within PNG and by altitude, many details are not well known and shared; despite years of study. Some known intercontinental migration occurs, mainly shorebirds, waterbirds and seabirds. Some birds connect between Australia and PNG; noteworthy here is for instance the brolgar (essentially a Sarus Crane species), as well as some large gulls and shorebirds.

Fig. 6.25 A tropical species of begonia widely seen in Papua New Guinea's forests

While PNG supports a large number of avian fruit eaters and nectar eaters, the ecology of seed dispersal is widely not understood or untangled but remains very relevant for forests and forestry, let's say. One of the longest and best studies on that topic used cassowaries but conservation outcomes remain little (Mack, 2014). For instance, the connection to forest harvest, done in a science-based manner for sustainability, is widely missing in PNG. No wonder then that environmental impact assessments, e.g. done for mining or oil & gas projects, get totally lost in species lists, and thus, have no relevance to actually express 'the impact.' The World Bank, a large funder of forestry works in PNG, never truly used such ecological works for policy, hardly asked for them. Why is that? In my view, any bird research done in PNG now should have a conservation requirement before it gets carried out and done, not?

While it should be clear at least that forest canopy clearings, and when done in mid-elevation valleys, should result in bird changes, even this simple fact is not agreed upon, e.g. for Birds of Paradise (Beehler & Laman, 2020, whereas the expert publication by Laman & Scholes, 2012 remains widely silent on the topic all together, so does most of the work by Jarred Diamond. A lot of cutting happened in PNG from the 1950s onwards, certainly 1970s (see chapters on forestry in this book)).

Fig. 6.26 A 'touch-me-not' plant frequently encountered on trails in Papua New Guinea. Likely this species is spread by humans to 'beautify' the trail. A typical example of the 'PNG Garden' and the Anthropocene but in a style that was rather sustainable for millennia. PNG leads the way

Textbox 2: "*We sort'em post mortem*": Taxonomy *ad absurdum* without any relevant conservation perspective other than self-interest and money/greed?

In a classic sense, taxonomy is the study of species descriptions, names, and their evolutionary trees. The approach usually is based on a genus and a species name, all as promoted in the seventeenth century by Carl Linnaeus in Sweden, as the state-of-the-art back then. It reminds of first name and last name, applied to humans. This concept then got applied worldwide and even today 300 years later in times of DNA and digital approaches (Beehler & Laman, 2020). What has really changed ? Well, the human footprint expanded and many species declined or got extinct, and that is not a cycle but a trend and consequence from unconstrained human consumption.

This taxonomy is essentially a western classification approach to the world's environment, but which is now often flipped on its head. Instead of Sweden and Europe, the species cradles are often in the south, but which got described the last, but with a structure done from the North and with voucher specimen that

Fig. 6.27 Flower beauty in red

are actually evolutionary younger than many tropical ones where the species started to evolve.

For PNG, this is obvious in the concept of its ancient figs, insects and amphibians, including crocodiles (many of such species virtually did not exist in Europe but where many of the taxonomic experts are sitting and acting from the remote, including the IUCN with the U.N.).

When it comes to taxonomy from collected bird specimen, PNG has already left a global impact giving a name to a penguin that does not live there, the Gentoo penguin scientific name *Pygoscelis papua*. As a matter of fact, three penguin species got initially assigned to PNG (Mayr & Diamond, 2001). The authors provide longer list of taxonomic errors and impacts, see for instance Lowe (2004) for primates in Indonesia, and Steiner and Huettmann (2021) for squirrels and some global aspects. Just the language choice alone creates problems to describe species. While a numeric description for a species, a taxonomic serial number (TSN) makes it easier to compute and store, it does not resolve well the problem of the species question *per se*. Does it add to the confusion?

PNG has over 700 languages, Melanesia has many more, and Pisin English is a major language there but not well written down by its users, whereas

Fig. 6.28 A white flower beauty

colonial powers all have their own species names, changed them over time, and now use mostly English while taxonomy is based on Latin and ancient Greek terms. Arguably modern species taxonomy is very complicated! But who truly needs taxonomy?

It was a hope that taxonomy will get better over time and when using 100s of criteria, as it is suggested for some species to ID them 'correctly.' The use of software such as the DELTA Key from Sydney/Australia is to provide help and solve issues and ID species in a more objective fashion for PNG and beyond (https://www.delta-intkey.com/).

And then DNA methods are used to eventually provide 'the truth' to follow and to apply. And that 'train of reasoning' was widely followed the last 50 years and pushed throughout the science enterprise and its journals and paid editors. But more advanced methods came online, new lab machines frequently occurred, and those 'new methods' are not all in agreement neither; 'revisions' get frequently published and updated (typical examples are found with the late American Orithological Union AOU bird lists, e.g. shown here https://www.audubon.org/news/here-are-biggest-changes-aou-checklist-north-american-birds). Retractions are record high in the DNA disciplines. And then there are also species that cannot really be done with DNA

Fig. 6.29 A commonly found flower in village gardens of Papua New Guinea

methods well (e.g. Vasilita et al., 2022), or where DNA is 'not there yet' to do so. There are also experts and nations that tend to reject DNA methods and who stick with morphometric criteria instead, besides others.

Mis-identification also remains a rampant issue, even with specimen collections; see Mayr and Diamond (2001) for historic examples over time.

It's '*taxonomy ad absurdum*,' again: All what western institutions and their employees often know and do in PNG is to describe specimen, but no ecological insights or conservation are really possible from such collections and harvest expeditions, as shown and stated by Beehler and Laman (2020).

Man-made climate change provides once more a good example for the lack of progress in the collection-based taxonomy worldview, e.g. the concept of "*all needs to be known first before we can act*": The curious glaciers and snow areas of New Guinea have been studied for many decades, and they were a piece of research early on, e.g. for British explorations (as described by Beehler & Laman, 2020; see citations within). Mt. Hagen was climbed by many Alpine celebrities, including Reinhold Messner (famous for Mt. Everest etc climbs). Flannery (2022) even had referred to the changes of these areas early in the 1990s onwards due to global warming, but a relevant recognition or actions to stop climate change did not come from it to the very day. Historically, these

Fig. 6.30 A red flower with big volume

areas were heavily 'collected' though. But still, no relevant climate change actions have come from it, see The Guardian (2020) for public statements.

To the outsider, and as widely portrayed, taxonomy appears like a sound and robust science, based on respected institutions even describing new species to justify their administrative funding existence. But a closer look will easily reveal an institutional debacle of the endless kind. PNG shows that very clear with its 'penguin,' the Birds of Paradise (disputed for centuries), albatrosses, the super complex reef and marine species (virtually not well inventorized even in a century of effort; see also for cryptic species, e.g. crayfish, Blaha et al., 2016) and unnamed but occurring new species for insects, frogs and plants.

Fig. 6.31 White and sophisticated: PNG flowers are magnificent

Beehler and Laman (2020) actually described rather well how taxonomy and its naming schemes functions: running after European nobles to please them all done as a competitive race among scholars for greed and fame. Naming, the new species after some powerful 'august' or member of royal family to get support. It became a sport, with countries like PNG as an arena. For PNG, we see such concepts applied and now everybody needs to live with it in perpetuity, written in stone. Conservation efforts were ignored. For instance, it was tried to do so with Queen Carola of Saxony, bird of paradise (Beehler & Laman, 2020, p. 291). The science of taxonomy simply works that way, taking up western institutions. In the meantime, those species names can be rather confusing and

Fig. 6.32 Flower beauty in the microview

a whole barrage exists for the same species a group (see any example in Rhodin et al., 1980 for turtle names).

For PNG, with over 100 species of amphibians new to science coming online, but not officially described yet (details in Beehler & Laman, 2020) a naming frenzy is to happen further. Amphibian numbers can easily be expected to double. Who can pay for it, who can keep the books and who manages it all well for conservation, validity or quality? PNG insect numbers are also to increase, more bird species are likely getting split, and many mammals are still poorly described or known such as tree kangaroos. PNG easily puts western taxonomy *ad absurdum* while collectors keep collecting anyways; it's a pleasure. Done by helicopter and similar "*fly me in and out, please, …and others pay for it*" as described as a common concept by Beehler and Laman(2020, p. 293), but it has no real landscape and reality context of the place—and the people—where they work in and what they analyze.

One may easily call taxonomy a bad, failing and repeated book-keeping approach to nature (Alcorn, 1993; Faith et al., 2000; Beehler & Alonso 2001 for such compilations; see Wikramanayake et al., 2002 for a similar approach using Ecoregion clustering), done by dominating, money-hugging and power-grabbing western institutions and their employees and subcontractors (many

Fig. 6.33 A small orchid (PNG is the world's headquarter of orchid species; in comparison, Hawaii just has 3 original orchid species)

National Academies actually operate as contractees) and often paid by extractive industry. So where really is the science, and why based at public scientific institutions?

And it does not end there. To better describe species and their biology, one applies for help. Citizen science offers such shared workloads in excess of a curator, and para-taxonomists are now used worldwide, also in PNG. Many more species get described, more people are looking, but who sorts it out for overlaps and accuracy, or effective conservation? What is the truth and where located? Arguably, there is more work to come...but thus far conservation certainty remains elusive.

Fig. 6.34 Typical ornamental flowers in a village of the Huon Peninsula, Papua New Guinea

6.3 Widely Misunderstood Birds of Paradise (BoP) and Hyped World Records Lead the Way…into a Superficial World Audience Ignoring Conservation Reality?

Arguably, the 'Bird of Paradise' (BoP) is a headline for any naturalist coming to PNG. They have been called '*Birds from Paradise.*' This group of species still makes for a great discussion item at any cocktail party, for those people who feel ignored and seek the attention (and subsequent money or other fame). But whoever has pursued BoPs in the wild and watched the wider BoP scene will agree: These are far from easy birds to find and to see, or to photograph (Latam & Scholes, 2012 as acclaimed BoP experts and photographers; see camera-trap imagery and gear used within).

Fig. 6.35 Ferns are found in many shapes and forms in Papua New Guinea's rainforests

BoPs are smart like crows (as they are essentially 'tropical corvids'), social, fast, occur in the canopy individually or in small flocks, either hidden by the shadow or the sun glare, and are often found in remote steep-sloped areas; depending on the species. Seeing BoPs also usually means you actually need to be in rural PNG first of all; a trip most people do not do or dare (a few species occur in Australia and Indonesia also).

Because BoPs are part of the taxonomic corvid group (e.g. crows, ravens), they are similar to any other of the world's corvids (a group known for their intelligence). Depending on the taxonomy used, there are 40 BoP species to choose from—38 of them in PNG; their headquarters. Some people are 'obsessed' to see all their species or to photograph them, now done frequently with high-powered lenses (Gillison, 2002) or even better, camera traps (Latam & Scholes, 2012); it's done for their records, for status, money, fame and a certain cultural world dominance; PNG citizens, villages and the sophisticated landscape maintenance are widely excluded (Latam & Scholes, 2012). That's specifically the case in the western world, and in Europe where BoPs were promoted by the remote royal courts and by the powerful for their own agendas. For instance for Australia, John Gould—an 'obsessive' bird collector—promoted BoPs in PNG early on for the global audience (https://australian.museum/learn/collections/museum-archives-library/john-gould/). It does not come as a surprise really that virtually no PNG authors are included in any of the BoP literature. It's as PNG

Fig. 6.36 A beach in Papua New Guinea (a), staging site for migratory curlews (field sketch; (b)) (the latter species group is now of conservation concern along those type of habitats)

people do not exist in the BoP research, while they have been with those birds for millennia.

Perhaps one of the easiest BoPs to see in PNG is the Raggiana (e.g. see Gillison, 2002), also the iconic national bird of PNG. As a matter of fact, many locals in PNG clearly know the western obsession for this bird and track them well for that very reason, and for western observers. If not careful, one may easily become prey

Fig. 6.37 Stray village dogs found on shorebird and sea turtle nesting beaches; those are not good friends...

of the BoP and their culture and get looped into endless fund raising campaigns on such matters. Locals in the PNG bush build permanent blinds for people to encounter the BoP at a given time of the day (or to hunt them for the bird skin to be traded). But beyond a plain birding tick ("*I think I saw them, well must be*"), one can also buy a BoP skin (it's illegal in PNG but remains ongoing and widespread; details in Beehler & Laman, 2020. See also Flannery, 1990 for such trade around mining sites, whereas many mine sites make hunting illegal for impact reasons).

Clearly, the breeding displays of BoPs are the highlights for the bird watcher, and global audience, ideally with a great photo, or now with a YouTube video or to skype it to your relatives overseas—in-time. '*Why not being like David Attenborough with the BBC?.*' Arguably, that is not to happen for most people and birders 'though.' (the few seconds of BoP TV fame are coming from years of high-level effort, tries and gear; see Laman and Stoles 2012) But keep trying... People literally tried to build Eco-Lodges and entire helicopter tours around BoPs (see details in Beehler & Laman, 2020; Mack, 2014; West, 2006). An assumed win-win for everybody, and for the birds. But it turned into a triple whammy, and worse (West & Kale, 2015). Needless to say that those 'lek' matings are crucial for BoP reproduction and thus for BoP nesting, population maintenance and conservation. Saving the leks means saving the birds and their habitats. It's therefore not possible to loose forests at the current dramatic rate but BoPs would still be fine. BoPs have co-evolved over time in a

Fig. 6.38 Pigs found on shorebird and sea turtle nesting beaches in Papua New Guinea

complex setting, with humans being around in the landscape and pursuing them. That's likely where some of the mis-understanding comes from, that BoPs would simply be able to stand the human pursuit and habitat loss (doomed surplus, 'spilled milk' concepts from the 1930s but proven utterly insensitive and not applicable; see Karger et al. 2021 for global forest loss). BoPs cannot stand it, just like many other species can't in PNG and beyond.

The fascination with BoPs is part of a PNG culture for thousands of years. The western obsession with BoPs is not so new neither and also reaches back from the start of western contact with PNG. In the absence of gold found, BoPs reached high on that list to justify the colonial efforts. In addition to the spices, BoPs were seen as 'wealth from the colonies,' to get 'value' from those remote colony areas and to make it all worthwhile. One had to bring back something! It is believed that the first BoP skins came to Europe already in 1522 by the surviving crew members of the *Victoria* (the only ship that has completed Magellan's circumnavigation voyage around the entire globe; State Library New South Wales, 2022). In 1630, Rembrandt was one of their painters among many others and helped to bring them to world fame. A colonial PNG PR trick was born, ongoing til today (e.g. Laman and Stoles 2012, Beehler and Laman 2020).

As reported by, and in display with, the State Library New South Wales (2022) the aforementioned naturalist and painter John Gould (1804–1881) visited Australia in 1838. Gould became famous for his seven volume works *Birds of Australia*. But

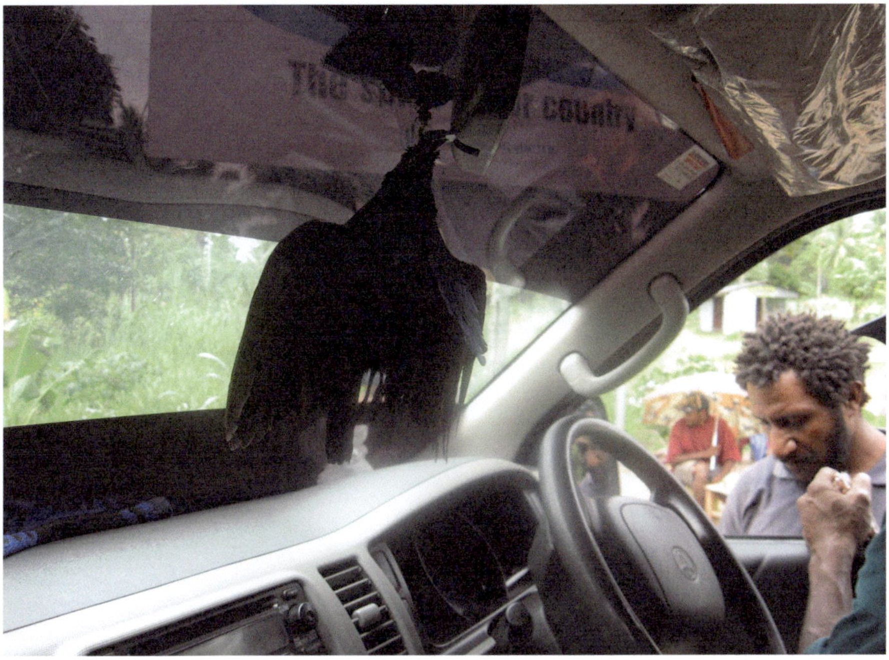

Fig. 6.39 Birds of Paradise skin in a bus to bring luck

later he was also engaged in compiling *The Birds of New Guinea and the Adjacent Papuan Islands* (finally completed after his death by Dr. R. B. Sharpe). This work then further contributed to the wide exposure and appreciation of BoPs and of PNG. But it also helped to set the foundation for Australia as a connoisseur, curator and thus control-agent of PNG and BoPs, and similar relevant items. It just came as another British fabrication in order to promote and dominate colonies (and their people) while PNG citizens, BoP habitats or BoPs themselves were mostly ignored. Traditional Ecological Knowledge (TEK) around BoP is described little, if at all (see Gillison, 1993 who studied nearby a BoP research hotspot).

Bird taxonomy is widely in flux. Depending on the taxonomies used, there are app. 38–43 species of BoP, but the BoP taxonomy remains dubious and done in English (Beehler & Laman, 2020), and splitting of species is not so clear, e.g. for the Superb BoP (https://www.birdsofparadiseproject.org/new-vogelkop-superb-bird-of-paradise-changes-up-the-old-song-and-dance/). There are long and severe disputes on BoP taxonomy and its phylogeny for many decades, and it is still widely unresolved! What's a subspecies in BoPs?

As PNG hosts all but two of the c. app. forty BoP species, there is a deep and co-evolved BoP culture in PNG on the ground. The tribal societies of New Guinea traditionally used BoP plumes in their dresses and rituals (Gillison, 2002 for rituals). Among many, men from the Yonggom tribe for instance were very knowledgeable about these birds and successfully hunted the Greater Birds of Paradise. Their feathers

Fig. 6.40 Close-up of a Bird of Paradise skin in a car; an image widely found in PNG

were used as adornment on the ceremonial headdresses. Those were worn during ritual dances (Latam & Scholes, 2012). The amount of skins and species used in those PNG 'sing sing' events can be easily in the hundreds (! Details shown in Latam & Scholes, 2012). Arguably, BoPs are heavily pursued.

BoP ecology is not well known, but these species operate as seed dispersers and they might create a high-quality seed rain. Already a single fig tree can have up to 40 bird species (Beehler & Laman, 2020). Still, details of such relevant forestry aspect are easily hindered by tree species and fruit identification, lack of an agreed taxonomy and study methods. How to conclude?

BoPs are also sexually dimorphic, males occupy 'fixed leks.' Despite their massive and fascinating, defying evolution and subsequent pursuit, BoPs only lay one egg though. Predators seem to be few. And while forests with BoPs are on the generic decline, one can spend a fortune to see and photograph BoPs. And a few people do

Table 6.1 Short selected list of major collection expeditions for Papua New Guinea

Expedition	Focus	Nation	Citation	Comments
Crane expeditions	Sepik region, done by a fish researcher	U.S.	Herre (1936) in Webb (1995, 1996)	Two years work with a photographic focus and record of many PNG details. Context is discussed by V. L. Webb but actual interpretation just comes many decades after the fact
Dutch colonial expeditions	New Guinea	Dutch	Holthuis (1949)	National Dutch explorations while Holland still had the oversight of its colony
Archbold expeditions	New Guinea	U.S.	https://www.archbold-station.org/documents/publicationspdf/lohrer_2019_11ArchboldExp.pdf	Famous and very long expedition work with the American Museum, 7 expeditions done directly in PNG, and a few others
British speleological expedition	PNG	British	Holthuis (1978)	One of many British expeditions for New Guinea (see Beehler & Laman, 2020 for more details)
Finisterre Gebirge expedition	Colonial exploration	Germany	Zöeller (1891)	Germany explored in its accessible areas a lot, but little information reached the mainstream research body of today, e.g. due to language barriers and efforts not designed well to be shared globally
Ok Tedi mine and Fly river Cambridge expedition	Fly river region with a focus on the Ok Tedi mine project	British	Boyden et al. (1975)	A classic for funding/support, 'objectivity' and elite research. See the rainbow fish species that carries Ok Tedi in its name (https://rainbowfish.angfaqld.org.au/Oktedi.htm). Such mind-sets still drive a lot of the research world and in PNG today

(continued)

Table 6.1 (continued)

Expedition	Focus	Nation	Citation	Comments
German botanical collections	Northern PNG (German colonial sectors)	Germany	Hiepko (1987)	Many of those specimen information are typically blurred, published with a vast delay, hardly in GBIF.org, carry no metadata and are debatable for (modern/international) taxonomies, etc.
German ornithological collections	Northern PNG (German colonial sectors)	Germany	Stresemann (1923)	Many of those data are not available in English for a wider audience, hardly for a PNG audience

There are many more collection expeditions and surveys done in Papua New Guinea, here a list with expedition years to be inquired more (sources found in Hays 1993 etc); they all have deep stories to tell with data underneath: Borgman (1960/61), Brass (1928–1939), Carr (1935), Clemens (1931–1936), Lederman (1914), Liditker and Ziegler (1968); see also Huettmann (2021) for tree kangaroos etc

spend it (see Latam & Stoles, 2012 for equipment and travel list; funders not much exposed).

But ask yourself, and after people having done it for over 300 years: What came from such intense pursuit? Did the people of PNG benefit, and the PNG nation, or the habitats and birds overall? So then beyond some photos and 'birding ticks' (Gentile, 2009), who really has understood a BoP and who really truly cares for them, or for their habitats? The conservation status of BoPs is supposed to be 'stable' with virtually no species having been assessed in demonstrated quantitative terms, and with a defendable conservation status even. What is the BoP management method and who does it, and does it follow any wildlife management principles, other than '*laissez-faire*'? Thus, an assigned 'status' as done by IUCN or BirdLife International etc sounds perplexing and it cannot really be 'stable' when forests are their prime habitats and when those are lost in a record-high rate and without any forest management or data whatsoever (see subsequent chapters of this book).

Beehler and Laman (2020) think that the BoP harvest, now and historically, has little to no impact. Laman and Stoles (2012) widely ignore the subject all together. That assessment one would judge as very doubtful. The authors provide no real data on their claim that BoP are of no conservation concern but rely essentially just on the '*surplus harvest argument*,' which has been disproven so tragically worldwide for many decades already (Williams et al., 2004) and which the authors use all the time in reverse on PNG's large mammal decline (e.g. high human pursuit levels). Because BoPs have little conservation and little shown and available data and management, they are in a precarious state with experts that do not notice it well. And we have been there before.

6.4 Trees of PNG: Precious, not Surveyed, not Understood, Mis-labeled, Unpreserved but Dramatically Cut-Down all with the Global Community Watching and Consuming

PNG is known globally for its virgin forests. Compared to other areas in the Pacific and worldwide, many of them are still widely spread, relatively pristine, diverse and gigantic, and even widely unexplored! Such forests are true retreats and the 'living room' for a remaining civilization of the last thousands of years. Indeed PNG has one of the largest blocks left of intact old-growth rainforest in the Pacific (Beehler & Laman, 2020).

Already the known tree diversity covers c. 600 tree species (http://www.pngplants.org/PNGtrees/). But with a deeper taxonomic inquiry over 3000 trees can likely be expected for PNG. A one-hectare plot can already support app. 70–200 species (Beehler & Laman, 2020). There are Antarctica beaches, 150 rhododendrons but just a few Diptocarp species (less than found elsewhere in SE Asia). Locals do know the relevant trees and what they have to offer; a certain symbiosis and co-evolution with humans and beyond were developed (e.g. for Melanesia see for instance Demeulenaere et al., 2021). Trees up to 45 m in height can be found, e.g. the Benuang (*Octomeles sumatrana*). So how to 'manage' it sustainably with a massive cutting ongoing?

This book has a specific chapter devoted to forestry, and a wider PNG context is presented there. But it's worth to mention here that forests are a core-livelihood feature for most people in PNG, and there are a great many noteworthy forest species and ecology aspects in PNG forests to mention, such as the oaks, tree ferns as well as ancient figs (with PNG as the cradle). Already the fig trees are important for many BoP species, and those can cluster around them. For BoP survival figs play an essential role! Many other aspects are not well known ye and certainly not managed.

'Swidden forestry' carried human society in PNG for many thousand years, people modified forests with fire and planting nut trees, and later, timber trees (those might have been introduced, such as Casuarina tree arriving app. 1200 years ago. Similar applies to betel nuts and their palm trees; Beehler & Laman, 2020). Further, in PNG people beautify their forest trails, e.g. with *noli-tangere* species due to their beautiful flowers, and fruit, and thus, 'garden species' can be found widely dispersed in the wild and remote bush. It makes for a unique, sustainable landscape feature. PNG locals are the great landscape gardeners, and it's quite sustainable and has been for over 47,000 years! So why now the destruction?

But despite forestry ministries, international experts, forest policy revisions, development aid, certified forest markets and sustainable forestry, industrial Australia as the next-door neighbor and advisor, one must say to this very day: There is no pixel-based PNG-wide forest inventory or a decent distribution map for PNG trees, nor is it really known how many tree species there are in PNG, their life history, or any approach on how to manage them.

As a profession, 'Forestry' as well as Botany and such sciences and institutions clearly served PNG poorly (examples are provided in Cousteau & Richards, 1999, pp. 206–207; see also Beehler & Laman, 2020).

6.5 Plant Overview: Power to the Flowers

Already a non-expert in plants will easily detect the poinsettia flowers that are planted all over the nation. They have leaves looking like actual flower petals and give a great testament to the fact that PNG is the great garden, man-made! PNG likes flower beauty. Much power sits within that.

Arguably, orchids, moss, some domesticated flowers, ancient figs and tree ferns might be dominating the discourse about PNG plants; but there is so much more. With perhaps 13,000–20,000 plant species in New Guinea overall, PNG is among a world record plant nation and island. The grasslands, usually carrying much less species than forests do, but add another relevant dimension (Beehler & Laman, 2020). While grass is very complex for a species group and its ecology, it's even less studied and known. Most grasslands are man-made, likely a certain effect of fires started by humans throughout the 47,000 long history for better access, farming and hunting (Robbins et al. 1976); and part of a swidden forest process (details for PNG in Beehler & Laman, 2020; Flannery, 2002). In the traditional absence of large (animal) browsers, it's a co-evolved 'grass-scape.'

Plants of PNG are described and collected for centuries by imperial-funded scientists (see Beehler & Laman, 2020 for overviews and expeditions). However, relatively little precise information and conservation maps came from it; and virtually no conservation efforts (unless it is of outermost commercial interest and invokes DNA copyrights, bioprospecting investments, international trade and consumption). That is certainly true for PNG. See for instance work and data in Hiepko (1987) for the decades-long colonial German collections but which are widely lacking international context, metadata and are hardly found and accessible in Genebank or GBIF.org with context to be used for scholars, with a research design or for progress. And then, see Paijmans (1976) for an actual PNG plant atlas (not in a digital format, and species taxonomies widely debated still). Nowadays, plants of PNG are covered in the ongoing and incomplete 'Flora Melanesia' (https://floramalesiana.org/new/) with all data to be shown in GBIF.org for a global reference and appreciation. However, no lay audience is able to really use and employ this information yet and species naming concepts are widely diverse across cultures and languages (PNG has over 700 languages, Melanesia has many more, and Pisin English is a major language but hardly written down, whereas the many colonial powers for the area have their own names, changed them over time, and now use mostly English while taxonomy is based on Latin and ancient Greek; a concept very remote from PNG but which had their plants known and used for over 47,000 years in a sophisticated fashion which is c. over 44,000 years before the Ancient Greece did. Arguably plant taxonomy is a complicated topic which will favor PNG but which currently botany or taxonomy are not doing!). Farming of many tropical food species likely started in PNG!

PNG carries many botanical world records, already its lichen diversity is the highest in the world; add the mosses. It further features the largest orchid species hotspot in the world (Montgomery & Bishop, 2006), with *Bulbophyllum* and *Dendrobium* (Spatulate orchids) being specifically species rich (Orchid Society of Papua New Guinea Inc., 2006). Some members of the species group of orchids can be immersed in ocean water and feed off salt spray. Most orchids are epiphytes, and app 85% have a high endemism (Beehler & Laman, 2020). Overall, New Guinea has app. 2850 orchid species (Beehler & Laman, 2020), and there are likely less in PNG (details remain unknown though while orchid trade and harvest are ongoing virtually unabated or managed; associated enforcements are virtually not heard of). In villages of PNG, some orchids are used for 'arm and wrist bands' and get planted for that reason in a certain domesticated fashion (Beehler & Laman, 2020).

While the study of the PNG flora remains in its 'youthful' stage (Beehler & Laman, 2020), studying plants should also help for a better forest management. However, thus far it has widely failed to do so. I spoke with botanists about this topic, and most just ignored it or just threw their arms in the air. It was clear that they ignored the subject and were even angry when I raised that question. Non-timber products seem not to be much *en vogue* with such botanists. But in reality, those are essential in any forested landscape; now more relevant than ever.

Textbox 3: Environmental Impact Studies in reverse: *Laissez-faire* in real life

Environmental impact studies sound like a great idea: Assess whether an industrial effort is harmful, and if so, it will be stopped, addressed and/or mitigated

Well, judged by the ever-increasing industrial footprint, that has rarely happened, and it has not happened effectively to stop bad industrial impacts, or to improve. Already just looking at man-made climate change and the release of CO_2 shows that clearly. Mostly, the legal argument of impact assessments are a smokescreen, a true greenwash and used for demagogy; nature is not given a real chance. For PNG, those questions get more complex and more sensitive because it involves people, funding and the fundamental national set up, as the case in Bougainville showed (see chapters in this book). Can one accept an impact for the wider public good?

To get at the actual impact, there are many questions what to measure, how done and by whom, and who pays for it, and whether that is even possible and meaningful?. Classic studies for PNG used for such purposes and with such questions are found with Boyden et al. (1975) involving Cambridge University, Ok Tedi mine and water supply affecting PNG and food security for a long time. A more complete picture on that impact can be seen in Kirsch (2014). Similar works exist and are frequently done, e.g. Richards (2018); see Earthworks,

Deep Sea Mining Campaign et al. (2015) for a review of seafloor mining impacts.

The binding framework how such impact studies are actually to be used matters also. Sullivan (2015) showed that those studies get often used as a '*pick and choose like cherries*,' which puts doubt on the objectivity, purpose, effectiveness of such efforts and on the legal use of such studies. The latter profession gets exposed for their actual 'impact'.

Taber and Payne (2003) show and discuss those failing concepts for North America—many more by now exist—and it's clear those steps are legally required, but hardly perform. Those assessments can turn into public soap operas, and many of those exist, while the economic development industrial efforts are steaming along either way, anyway; business as planned proceeds regardless then, globally (Czech, 2020).

So what value do such (rapid) assessments really have, when not done thoroughly, without thinking and reflection, and when not given a chance to truly show, critique and stop the development, often just favoring a strategic use and subsequent destruction of the natural resource? Impacts cannot be divorced from the process, or simply bought out. A de-coupling is not possible on a finite space and resource (Daly & Farley, 2010), e.g. in a sophisticated and interconnected island-state like PNG. This basic reality and the inherent flaw in such measures rarely comes to the forefront though, and those weak steps of policy are to be addressed and resolved better, if nature is to have a fair chance, or PNG for that matter.

So what should a botanic scholar really do, and write about PNG plants overall? Typical examples of those open questions for PNG are found with Webb et al. (2005) based on Harvard references and with CSIRO and NGO support. A concept that is sold to the indigenous people that way also for (western) relevance and authority. But it's narrow, not holistic, lacks conservation and one easily can do more and better, considering the forest landscape and its complexity is of global relevance but they simply get 'lost' at a record rate; conservation does matter.

While it appears to some people perhaps like an off topic, one may easily report first that (western) botany in itself is in a crisis. While often funded by commercial entities and wealthy donors, botanists and their institutes usually never clarify whether it should be a capitalistic botany, to make rich people richer, celebrate colonial efforts and insist on copyrights and privatizing the common good, plants and their DNA and clones, e.g. for farming or medical reasons? Most of the PNG botany done by the international community sees it no other: bioprospecting and providing steps to get there and set it up with species lists and some whereabouts; now all done online. Habitats do not matter in such a plant science. Much of the PNG botany stems from those types of questions, and while PNG remains deep and difficult to study it indicates that such botanists never really tackled the 'real' questions and the context among themselves for a good outcome of their doings:

> What really are the research questions in botany worthwhile, and for a sound and robust botany in Papua New Guinea with a Melanesian view point?

Just sticking to 'names' in various languages, digital or not, intransparent inventories without proper and agreed- upon ID keys or research designs, and 'just' then century-old naming of items collected is not objective, hardly useful or even science (e.g. what is the hypothesis tested, and how done, what research design and sample design, what sample size, how mapped and how assessed quantitatively with statistics and confidence, shared with the wider global public in a transparent and repeatable manner?).

And how are all the other many aspects of botany catered, including representative sampling, use of latest methods, environmental and social justice, inference for generalization, landscape scales, and PNG needs which are all so embedded in food and thus in plants (for PNG those are essential questions: Demeulenaere et al., 2021; Narokobi, 1975, 1983; see Heinemann et al., 2009 for Agriculture at the Crossroads report)? Botany can be pursued in so many directions. But the western route primarily took on the commercial DNA taxonomy and book keeping one. It was done through 'white mans' angle as the mainstream, a path of greed—with a bit of added ecology to ponder and justify complexity (which humans and experts have such a hard time to grasp). Staying in that mind-set, most western-style Botanists can only add here to the endless bias and write it into stone further with species lists while trying to hang on to their funding and employment positions. PNG plants are just a collateral that game, so are the people of PNG who rely on those plants.

For PNG, it took the collection, ecoclassification and bioprospecting angle; much of PNG's botany done by experts is simply stuck in collecting, documenting, botanical plots, and selling flowers and (genetic) plant parts abroad, breeding them, and at best, clustering and describing them, e.g. in photo books for the coffee table and creating income and 'slash funds' for professional societies (Orchid Society of Papua New Guinea Inc., 2006) but doing virtually nothing though on robust science, people, climate change, and effective conservation, let alone data for good use and progressing the nation, or plants for that matter. The latter two would be rather honorable goals; so why not pursued? Like with many tropical botanies, PNG botany still sits largely in the colonial mind-set; arguably it is widely dominated by Australian efforts and their agencies, helped by Europe and the U.S. - as the dominating science power in the world.

Botany work in PNG had over 400 years to do otherwise, but still it did not. It took the wrong exit. The wide fallacy of such a botany dominated by just a few driving entities can easily be seen in the very sorry and pity conservation state of tropical plants and their management and policy, see forestry itself; a world crisis. While evolution-wise, PNG has the world's oldest plants (likely millions of years ago), the taxonomy used actually still comes from remote Sweden, seventeenth century onwards. Our worldwide dominating taxonomic and evolutionary understanding was built up from the North, using Latin and ancient Greece, but this is far from the cradles of biodiversity, such as PNG is (examples discussed in Martin, 2005 for mammals). So how can such a system and its experts ever cope with PNG, its plants and do it good justice (details in textbox taxonomy; see also chapter in this book on Forestry)? One awaits for answers while the habitat angles got virtually ignored. Already moving into a Landscape Ecology perspective would be progress.

It's clear that so many more plants still await their research, discovery even, and that it can lead us into new directions eventually, but the conservation policy hardly exists to do so in a modern world for PNG or for the tropics. So what does that show and where does that leave us?

6.6 Insects: An Overview and Description is Next to Impossible

People not familiar with insects in PNG will still easily recognize the huge spider webs by the large orb weaver spider. But while PNG features actually a relatively low number of spider species overall, for people with sharp eyes, they will detect many other fascinating insects, e.g. the 'walking stick' (an ancient species group, many species exist within). And PNG is globally known for the bird-wing butterfly, Queen Alexandria's bird-wing (as the largest butterfly in the world highly seeked after by naturalists) but now highly endangered, subsequently also a highly prized collector's item. Values increase with conservation declines, and a high pursuit of 'world record species' will put pressure on them (many examples for that pattern exist, e.g. large beetles or for birds, the Ivory-billed Woodpecker in the U.S and Cuba Gotelli et al., 2012). Insects are often popular with collectors and 'insect lovers', e.g. done in PNG for the markets abroad through the Insect Farming and Trading Agency (https://www.pngyp.com/company/1513/Insect_Farming_Trading_Agency). Species get collected in the wild but are hatched in captivity for the export trade. Collections play a large role, as shown in Beehler and Laman (2020, pp. 141–142). As a matter of fact, selling butterfly cocoons for market export is a real business and one that strives in PNG (Beehler & Laman, 2020; Cousteau & Richards, 1999). Butterfly farms, such as in Bulolo, present a new form of cash crops. One may find it a bit dubious as those are used abroad for releases in weddings, parties, religious mercy releases and the like, e.g. in Australia and Asia (it remains unclear where the butterflies go from there after the release; presumably they are left to die). While collected in the wild, they usually hatch in domestic conditions and get sent off from there for a profit for such uses; PNG fully participates in such efforts.

New Guinea has app 300,000 insect species (Miller, 2007 in Beehler & Laman, 2020). That's app 5% of all insects in the world. Typical representatives are the mentioned stick insects, but also katydids, cicadas and crustacea. Further, app 10,000 species of moths are found in New Guinea. For beetles, PNG is a hyper-diverse site and the Giant Horned Beetle is another collector's item. In PNG some insects are also eaten, like grubs found in tree stumps.

As long as insects—the invertebrates—are not recognized with any animal care rights such trades will continue though and with a somewhat Australian oversight and approval, usually with Asia as a 'big market.' It's a bit strange because one of the world's leading animal right activists comes from Australia, Peter Singer (https://en.wikipedia.org/wiki/Peter_Singer), but no real progress is made for insects on that matter in the region. Whereas in contrast most indigenous and PNG people understand and know for millennia that we are all connected with the universe, and its animals, butterflies included. Traditionally, the pursuit and bycatch of insects, including their habitat loss and transition, has never been that high in human history, all helped now by colonial powers and the western world and globalization, e.g. in pursuit of economic growth on a finite space. What culture is sustainable, and which one is not?

6.7 Study Insects for What, How Done and for How Long?

Along the same lines, and close to the heart of western society and its sciences, sits the actual study of collected insects in PNG. For instance, tree canopies over 20 m high up to 45 m are widely understudied for insects in the wild and have an attractive appeal. Tree climbing or canopy cranes are often used to address that, e.g. there is a PNG canopy study site in the Baitabag Village 45 m high (in Beehler & Laman, 2020), and it really looks spectacular. But how does it conserve PNG and insects, considering record losses of rainforests have been realized in PNG without any relevant constrains? While describing insects and count them how is insect research effective on those essential (habitat) measures relevant for mankind?

Instead, insects are typically still studied with collections and get pinned in boxes for an assessment. DNA just adds a new tool in that limited portfolio ignoring effective conservation measures and policy. As Beehler and Laman (2020, p. 51) showed, 100,000s of insects get collected that way in expeditions and their repositories. Presumably the real number of collected and killed insect individuals is in the many millions, but to what outcome? In PNG, a research group—from Czech in Europe—actually dominates this approach and section of biodiversity work in PNG for decades already. The method of choice remains counting and obtaining the insect in the hand, e.g. done via 'fogging.' This essentially means one kills them off, with blunt insecticides done for 'science.' It has many scientific and other problems, one is unintended and widely uncontrolled 'bycatch.' In that method it's nothing unusual to kill off entire tree canopies, e.g. by cutting the entire tree down and fog the accessible canopy, and then count out the 'exotic' insects falling on the ground or accessible. It's essentially a textbook approach to studying insects worldwide for over 200 years ("*we sort'em post mortem*"). While this approach makes sense to some people, it's bare-bone human-old work, but now even paid to Czech and its researchers through the EU and by the National Science Foundation (NSF) of the U.S. Scientifically, it hardly allows for an individual-based research design or a shown wider inference. Other methods like non-intrusive predictive modelling or data mining of the insect data holdings are not on the table, that expertise is widely ignored for decades. The established researchers for PNG simply fell in love with their method...and got stuck there. But so are the outcomes. Ethically, nobody can really count, and identify, all species that live in canopies of such large trees and in old growth. Many new species can get detected; it comes with the job and is expected for taxonomists; it's expected of them. Thus, nothing special. Fogging can often just ID and study a tiny fraction of the insects killed; the rest goes to waste (as those individuals were fogged, it makes for contaminated waste!).

In reality, it's simply the 'agreed practice' in such professions for publications, power and funds abroad. It gives them a job, and thus it all continues that way. Usually, as just a fraction of insect species 'fogged' gets sampled, a large number of the caught species 'dies' and is widely left unaccounted for; it creates bias which puts pressure on the scientific rigor and validity of such a science in the first place. A research design would be critical, but is almost impossible to achieve for parametric inference. In the meantime, this work is labor-intensive and requires funds for students and field workers in a nation (PNG) where many people earn less than $4 day. Science for what and how done? It's certainly not done for conservation or to serve PNG and its people, as the progress on that issue shows and is 'tiny,' if even that. Is insect research colonial?

Considering that insects present a large chunk of biodiversity, so where then does that leave us for any informed and defendable discussion on biodiversity progress and insect ecology and conservation research for PNG?

While non-invasive methods exist, those are not much used or accepted nor is there a strong and enforced insect conservation policy, not even meaningful trade and export laws, as CITES underperforms wholesale even for its trade concepts, certainly for taxonomies and conservation (which it was not designed for). It's not well under control in Asia, certainly not in PNG; for details see annual TRAFFIC reports: https://www.traffic.org/. How can there be quantitative assessments done on species like endangered small beetles with a complex life history and grubs (hidden away in a tree stem or underground)? How can the ever-changing species taxonomies be checked for trading? And how are insect laws enforced and by whom? The Asian trade reviews on poaching, bushmeat and species trading show us an abysmal performance picture: Borders are virtually totally open in Asia to whoever wants to cross them for products; the tiger, elephant or rhino poaching situations show us no other (see also CITES websites https://www.traffic.org/ and https://cites.org/eng; see Dinnen, 2017; Supuma, 2018 for PNG).

Let's agree, and like with bird research (Mack, 2014) or the earlier discussed plants, insect research in PNG—as developed, overseen and promoted by Australia, colonial powers, Europe and North America—has not moved much forward beyond Neanderthal-style vanity collections, selective species naming abroad and information narratives, and that such institutions, funders and practitioners got stuck in such a mind-set cloning new graduates along those lines set for perpetuity. The use of DNA methods does not make it better. Most of this is well critiqued by Beehler and

Laman (2020) who refer to such collections as '*the cabinet of curiosities.*' It would be nice for PNG to move into progress regarding insect research; arguably, with the funded actors involved that is not to happen any time soon though.

6.8 Mammals of PNG: What This Species Group Really is About

App. 244 terrestrial mammal species are found in PNG (in New Guinea app. 293 overall; Beehler & Laman, 2020). Like with the other species groups, it's a classic and world-wide renowned but complicated research topic (Flannery, 1990; Martin, 2005). It remains widely unresolved for a good mutual agreement on species, taxonomy and abundances (Huettmann 2020 or tree kangaroos; see Chap. 28 for squirrel ecological niche discussion; consider the marine mammals also poorly inventorized).

While many endemic mammals are found in PNG, the geography and land mass of PNG is not really the typical island of mammals (as known from the 'Old World'). Already the 'Old World' mammals are not found there, e.g. primates or squirrels (but glidersetc., might fill that niche instead; see chapter in this book). PNG is not really short of mammal species as can also be seen below with marine mammals. But then, already the diversity of bats in PNG is to be reckoned with! Tropical nations generally tend to have a high species diversity of bats and associated species. But PNG features app. 31 bat species with microbats being a biological highlight, also including the flying foxes and the mentioned gliders.

Emblematic mammals of PNG are the echidnas (insect-eating, egg laying mammals that are found only in New Guinea and Australia) and marsupials, including wallabies, tree-kangaroos and the possum-like cuscus. Those are rather odd species indeed for Europeans and spark interest with any scientific-minded conservationist (those species are conservation poor; see Beehler & Laman, 2020; Flannery, 1990, 2002). Those must not miss in any mammalogy lecture because it will show the true diversity found in mammals, not just mammals of the Old and New World

Much is still to be learned about mammals and of PNG ones, including bandicoots, dasyurids, bandicots, striped possums, pygmy-possums, tree kangaroos and their subspecies, wallabies, cuscuses, rodents, ringtails, fruit bats, microbats and the marine species.

The advent of the (hunting) dog some 1000 years ago changed much of the PNG ecosystems and set ups (Flannery, 1990, 2000). With those ones, the cuscus and quoll can get widely seeked after for its fur (see details in Flannery, 1990). Mammals avoided humans for 47,000 years and thus are usually nocturnal and cryptic, unless a dog can find them. The decline of long-beaked echidna species and tree kangaroo species speak to that impact (see for instance Huettmann, 2020; Wildlife Conservation Society, 2009).

For the tree kangaroos, depending on the taxonomy used (Martin, 2005) 14 species exist worldwide but only PNG—as the only place in the world—has as many as 12

Fig. 6.41 Isolated and fenced Matschie's Tree Kangaroos in a Mountain Zoo in Colorado U.S.; what's here the conservation gain, for the species, for PNG and for the world?

species left. Australia lost them early on and just retained 2 species out of that pool. This species group has received major attention by Australians (Flannery, 1990, 2002; Martin, 2005), specifically in similar habitats than PNG, the Atherton Tablelands/Australia (but which got widely destroyed for its original old-growth forest; Ludlam, 2021; Martin, 2005)!

The archeology of PNG and the Sahul region overall is widely studied but actually leaves us with just few sites and little outcome for real-world conservation and practice; it's usually just based on a few point samples. The now extinct thylacine (marsupial wolf) received a lot of public attention and was for instance found as artifacts in caves of PNG until the Holocene. And a small sea cow was probably the first/oldest mammal species found in PNG's history (while the sea cow is extinct, the dugongs are in a rather poor shape now also). The question remains, again: What was really learned for conservation or mankind, e.g. Fig 6.41, 6.42?.

Fig. 6.42 Sign post for the Matschie's Tree Kangaroos in a zoo from Fig. 6.41 Is that all of the information we need and want? That's the justification and outreach for such a complex species and conservation subject? It's widely uninformative and totally underwhelms

6.9 Amphibians and Reptiles

App. 395 amphibian species are found in PNG (And app. 424 in New Guinea overall Beehler & Laman, 2020) with numbers actually expected to double (!) while new methods and more field work come forward. Over 100 species already just await publication and wider acceptance boosting New Guinea and its amphibian species list further (Beehler & Laman, 2020) but leaving protective efforts even more desolate because money for those 'new' species remains widely unavailable by agencies in charge. It's a typical example for inappropriate business models in nations like PNG, e.g. a trickle-down economy. In te meantime, many people who study and promote amphibians and reptiles flock to New Guinea and such regions to find new species and to be able to name them... is that not like describing the deckchairs on the Titanic? And why done and a good use of resources? And where are the people of PNG in all of this?

The first structured herpetology expeditions to PNG were made 1826 onwards. PNG features freshwater as well as an estuarine crocodile with six turtle species and over 100 snake species, some are deadly (e.g. the Papuan Taipan, Eastern Brown Snake and Papua Black Snake). The Papuan Monitor as the largest lizard on earth is up to 4 m long. With its tropical rainfall as a world record, PNG is an amphibian island: 90% of PNG frogs are endemic. The Spike-nosed (Pinocchio) Treefrog got famous, it just was discovered recently and described by an Australian researcher (e.g. https://www.smithsonianmag.com/smart-news/meet-newly-named-long-nosed-pinocchio-frog-180972385/). In the meantime, the toxic cane toad that got introduced to control the Sweet Potato Moth went *havoc* in PNG, as in other places in the world.

With that many frog species in Papua New Guinea, it's a given that many are 'new to science.' Conveniently for fame, this group was widely claimed by another research group from another nation, namely the Bishop Hawaii Museum (https://www.bishopmuseum.org/) of the U.S. But beyond the endemic species, and assigned national science topic budgets in the $100,000s the world of amphibians and reptiles is in crisis and will likely remain so for decades. Massive species loss is almost the rule, and fungi and invasive, in addition to habitat decay, habitat transitions, wholesale habitat transitions and climate change (warming–drying) tend to make it a grim world for amphibians to live in. And the many new species described to science will not change that and can provide an utterly wrong message when 'naming the deckchairs on the Titanic' while the Titanic actually sinks in full steam.

6.10 Marine Biodiversity

The oceans and waters around and in Papua New Guinea carry their own record; they have virtually no equal due to being so little explored (Cousteau & Richards, 1999 for details; compare with the Caribbean for instance; Huettmann, 2015 and citations within; see related chapters in this book).

This is well known and documented for over two decades by now (Jackson, 2013). But what is done about it to halt and improve the trend of ocean decay? It is here where the western model fully fails us all again; as shown for many years in any coastal ocean of the world and their status (e.g. Jackson et al., 2001).

Major fish groups found in PNG consist of—but are not limited to—sharks, sawfish, sting rays, herring and anchovy, catfishes, garfish, tuna, perches, grunters, snappers, biddies, croakers, porgies, archerfishes, mullets, blennies, gobies, gudgeons, soles, thuna and Barramundi as a major eaten fish species; world-renowned are the rainbow fishes.

But the freshwater fishes should also be mentioned for PNG; e.g. eels that are found in rivers. The freshwater fishes have received much attention as their distributions in lakes and watersheds show fascinating distribution patterns, e.g. Weber Line, and some are affected by mining effluents in rivers and estuaries, e.g. classic study by Boyden et al. (1975); see Kirsch (2014) for an applied example.

The largest freshwater lake in PNG, Lake Murray, is now exposed to invasive fish pushing out endemics (Beehler & Laman, 2020, p. 25). Like found in other tropical nations (e.g. Nicaragua, Australia) sharks are frequenting those brackish water sections, often very close to humans. Bull sharks are found in freshwater lakes such as Lake Yamur on Papua (Beehler & Laman, 2020, p. 25).

Coastal fish harvest is usually done by kids and women (Beehler & Laman, 2020). As it will be treated in another chapter, thuna and sharks are widely overfished, as stated by Barclay and Cartwright (2008).

PNG has all the general marine species and set ups found in the tropics. But then it has more; the marine mammals are one aspect of those. River dolphins as well as dugongs are found there in the estuaries and in the excessive seagrass areas; those are all now in decay. Same can be said for the river dolphins (Leatherwood, 1991). The dugongs in Torres Strait region carry a more tight management regime than in PNG (see for IUCN listings and details: https://www.iucn.org/resources/publication/dugong-status-reports-and-action-plans-countries-and-territories).

On the pelagic side, many whales, ocean dolphins and porpoises are also in PNG waters, at least 15 species. It's a minimum estimate because of taxonomic splits and the beaked whales (a group of app. 24 species worldwide) are not fully detectable and studied for their range and occurrences, and most whales tend to move and migrate—usually under water—long distances, and all are now affected by climate change in many ways. It is debatable to some experts whether the abundances for those marine species are declining due to bycatch of commercial fishing, microplastics or due to seismic shockwaves intruding into ocean depths due to seabed exploration and military and civil shipping and navigation, submarines included. But the marine resource declines are a generic scheme reported by locals (e.g. Marsh et al., 2015 for concepts). While PNG seriously lacks detailed marine data some surveys found rather low counts of whales. Marvae et al. (2021) documented at least 6 marine mammal species, with spinner dolphins and pantropical spotted dolphins both as the most frequently sighted and abundant species within that group (see also Hamilton et al., 2009).

Noteworthy to the western whale watcher are perhaps the pygmy killer whale and then the pygmy sperm whale as well as dwarf sperm whale. The regular sperm whales of PNG reached a sad world fame for their contamination (lead) loadings (Savery et al., 2014). Unique orca behaviors have been described for Wuvulu Island, northern PNG, by Cousteau and Richards (1999, p. 217).

Arguably the coral reefs are a highlight around PNG (Jackson, 2013), but those are also under threat (see also other chapters of this book). With climate change, there is no good outlook for those reef regions acting as the certain climate refuges (Dixon et al., 2022; a very misleading term considering Global Change comes with many dimensions and remains widely unhalted

And not short of records, PNG is part of the world's deepest water trench systems—e.g. Manus Trench, New Guinea Trench and nearby Challenger Depth—one of the deepest known, and its species await more exploration but while plastics are documented there already.

Crocodiles fared very well in PNG for millennia, and a deep culture and worshiping centers around them, e.g. in Sepik river region (Cousteau & Richards, 1999). It's a unique culture! But they also represent a real danger to people to this very day (just as they are in other tropical nations such as in Central America; e.g. Huettmann, 2015). However, the wild crocodile skin resource was overharvested by colonialists already a long time ago, certainly in the 1960s (Beehler & Laman, 2020; Burgin, 1980). Crocodile farms were set up, e.g. at Angoram, Sepik river, and crocodile skins are sold to EU and Japan; see Burgin (1980) for more details.

Fisheries remain another black eye for modern sustainability management in New Guinea and its waters (see Radio New Zealand, 2018 for 18 PNG fishery observer deaths just recently reported). PNG and Indonesia cannot really get their EEZs under control, and outside it's even worse for control and management. Modern (international) pirates do occur, in the fishing industry and outside

Information from undisclosed inside sources provided to the author go that way:

> "*The world's largest purse-seiner was operating north of New Guinea, within their EEZ, but much without permission. Spain is involved through the EU, in part. A Fishery Observer who was murdered came from further south, and possibly from a longliner. In parallel, the U.S. Coastguard apprehended two South American vessels fishing illegally north of here, in Micronesia, likely for albacore, but released them without penalty as instructed to do so by the State Department, who needed to be consulted for such cases. State said that these vessels were owned by a country which was 'friends" with the U.S. and thus released. Greenpeace had the details & photos of this vessel online at the time*".

6.11 Endemism and Its Centers

Arguably, not all species are distributed randomly or equally. Instead, centers of species diversity can be found, e.g. due to environmental harshness over time, other biogeographic factors mentioned above (as expressed in Biogeographical theory), or due to more recent human impacts. PNG is known to be an endemic center for spatula orchids, and for tree kangaroos—a global hotspot (Flannery, 1990; Martin, 2005; Montgomery & Bishop, 2006). But also for species like giant butterfly and most of the Birds of Paradise. And then, mountain tops and some specific islands serve as endemic centers (see Montgomery & Bishop, 2006 for cloud forests and tree kangaroos). Many of those centers are not even well known yet and DNA research might prove insightful for such assessment and for connectivities. Those known centers of endemism got highlighted in conservation priority reports and rapid assessments (e.g. Alcor, 1993; Hamilton et al., 2009; Richards & Whitmore, 2015 for marine areas). However, neither their locations nor their processes got really protected yet, have a budget, and modern management and wider global buy-in, nor is that a meaningful approach really when data are not well shared, analytical methods remain cryptic without documented open source code, a public land tenure exists and needs to be catered (Baraka, 2001). At minimum, national park-type protection efforts need to be reconsidered and to follow a wider community approach to be effective in PNG (e.g. Narokobi, 1983).

Where are the centers of endemism exactly? While this question needs a lot of science and data to answer, it's a bit of a dilemma then. After over 400 years of scientific collections and research, nobody can really answer that question yet with good certainty for all of PNG, or for parts of it even. One may be pragmatic and use some data at hand, but it will usually highlight existing hotspots, mountain peaks and areas with a large track of undisturbed rainforest lacking roads (and people; or mining for that matter). For a PNG conservation, any protected pixel currently counts though while a generic onslaught hit Melanesia either way, e.g. by climate change or by 'Mine Melanesia' (Kirsch, 2014).

6.12 The Unknown, the Ecology and Toward a Better Nature-Environmental Relationship in 'The West' and Worldwide

And there is (much) more to the biodiversity story than just rapid assessments, limited ecology and DNA-based species lists that are consistently expert-revised (= the western and dominating biodiversity effort in PNG paid by tax papers and industry). PNG's deep ecology remains widely unstudied; that is certainly true for forest ecology and for marine ecology. No need to think of species new to science or undescribed, as typically found for instance in the remote forest, ocean and deep sea when searched harder. It comes with the job at hand. In PNG new species come as a fact and are expected; while PNG citizens usually knew them for millennia, and in their own words. So what's new, and what a discipline worthwhile would that be; Biology?

But even more so, much of the co-evolution and its ecology is not well studied for PNG's species and habitats at all (while again the locals know the species, their interactions and their uses and meanings very well for millennia, including the ecosystem connections and a link with the universe and their own being, and how it's all managed sustainably). But then think of cryptic species and new DNA studies showing us once more newly revised phylogenies and new but endless new splits and re-arrangements of existing species groups and species. That occurs virtually each time a new analysis method and lab machine provided by industry occurs on the western market. So what's objective, unbiased and what is the point of all of this? Technology drives species descriptions, its science, and so does the money paying for such a technology! It gets circular then and self-serving quickly. But how is PNG served and better served than before?

Instead, what one should study more in PNG is the wider and deeper connection of the living world with the universe and its people, locally and globally, the cosmology, and all done in harmony with Mother Earth, across land and sea, and in a sustainable fashion for future generations, holistically. It matches what the elders of this world report (https://www.wisdomweavers.world/). Conservation matters, as one cannot live without it. PNG is fully embedded in it and has much to share, but the western academy is just not listening, nor studying it! It makes no money for them.

This aspect of the acclaimed works by Wilson and MacArthur (1967), or Diamond (2011), Flannery (1990)—or David Attenborough (The Guardian, 2017 for the Blue Planet) for that matter—is widely absent in the public discourse. Why is that?

Textbox 4: Ecology anybody?

Ecology is not new and just a western and institutionalized concept, and it is also to be applied a bit to Papua New Guinea (PNG). But as shown in the species sections of this book, ecology is hardly known or studied even for PNG, a relevant policy link is missing. Consequently, wider views and policy impacts are essentially ignored; and relevant ecological science sits in its infancies for areas like PNG. Ecology is widely left unstudied there; a typical example can be seen in Cassovary's role in rainforest rejuvenation (Mack, 2014), and when considering these birds distribute seeds and communicate with low-frequency sounds. With PNG having such deep cosmologies, all is connected and links with the wider universe, certainly within a complex forest ecosystem; but how linked, how studied, and how all connected and managed for ecosystem maintenance, sustainability for future generations? Nobody knows how done and how to apply it, unless the Melanesian way is followed more.

PNG is widely understudied, and under-researched; western humans never will understand 'everything' or apply it well anyways (just see man-made Climate Change where man-made CO_2 release is not achieved, after 300 years of international science). Arguably, the period of enlightenment and over 100 years of 'modern' research, in PNG one still got stuck in opportunistic collections and primarily species descriptions across a few competing outside actors, but which are not even agreed on much. The Australian scientists see the world very differently from the UK ones, or the U.S. Birds of Paradise taxonomy make for a good point in time; a topic widely disputed internationally for decades with fierce infights. Taxonomy of benthos, and the deep sea, a topic where PNG has much to offer, is hardly tackled yet. Needless to say, the PNG scientists and citizens have little to say in this whatsoever; they are hardly on the table even.

But when it comes to ecology, the advanced study of interactions, even less is known and done. How many cutting-edge ecology studies are done in PNG, e.g. using telecouping?

Most PNG ecology studies probably still relate to elevational gradients, which is an over two hundred years old concept from Alexander von Humboldt etc and conceptually nothing new for over a century, hardly exciting. Over 50 (!) years agao Diamond (1973) promoted already such works again for birds, islands and island peaks and corridors, and with little progress since. We all know that mountain tops differ in the climate and thus in their ecological and species set up; who would have guessed? Elevational bands are described for over 100 years globally, so is island hopping (or island skipping). Such type of studies, most just repeats, logic theoretical reasoning and study templates from elsewhere and then mindless applied to PNG with strawman questions ("would it apply to PNG also?)", are quite far from innovative thinking or the ecology definition and what it has to offer these days and how studied well, e.g. holistic concepts or telecoupling (Liu et al., 2018). Pollination is another one, but also got stuck in the descriptive co-evolution approach. Then there are some more high-tech studies such as the use of DNA trying to answer complex questions with complex methods. That has its own problem for inference. The notion of predators, ecology of fear, food chains or the Anthropocene have hardly entered PNG or its research practitioners. Instead one sees shallow studies that try to describe for a month or two what species eat, e.g. using indirect samples like feces.

In the meantime, the local PNG citizens know their living world since deep times and handle it accordingly. Who needs science, or ecology and an underlying taxonomy for that matter?

6.13 Biodiversity Conservation Anyone, Done for Purely Commercial Reasons Adding to Mining, Oil and Gas Explorations and Foreign Wealth and Power?

While the biodiversity in Papua New Guinea (PNG) is grand and unique, of world relevance and globally fascinating, the science-based conservation of this biodiversity and its track record is not; far from it. This is a typical state of affairs in conservation and tropical nations, and it's certainly true for PNG.

Still, PNG is part of the Megadiverse countries (see also public knowledge about the associated Cancun Initiative, and the Like-Minded Nations which PNG is not part of though (see public information for the Cancun initiative, etc. https://www.wikiwand.com/en/Megadiverse_countries)).

Measured by western metrics, the amount of protected land in PNG is very low, e.g. < 1%, and with just 3 RAMSAR sites (Mongabay, 2022; https://www.ramsar.org/); the RAMSAR sites are a EU centered and limited approach; they carry no relevant budget or protection levels anyways, hardly a meaningful administration or concept (nations of the highest density of RAMSAR sites are found in places like Holland and Switzerland, none of them have any large and true wilderness left while vast and quite lowly-populated nations and locations like Russia, Alaska (US), Antarctica or Canada have some wilderness left but are virtually free of any RAMSAR sites, and the ones that do exist carry almost no enforcement or wider recognition in the respective nations). Some biological reserves are outlined, and some are pondered—specifically for marine protected areas (MPAs; which are not 'no take' zones)—but many other key areas are left unresolved, and widely unprotected, water tables go down one-by-one, major rivers get spoiled, and precious ocean areas get misused; many bad examples exist about it, unrooting local people (see already Cousteau & Richards, 1999 for examples within).

And then there are works of this kind, e.g. for Papua New Guinea's Hindenburg Wall and its rapid biodiversity assessment (https://www.researchgate.net/publication/274961216_A_rapid_biodiversity_assessment_of_Papua_New_Guinea's_Hindenburg_Wall_region). Those type of assessments state things like "*The Hindenburg Wall, along with the Muller Range and Nakanai Mountains, is a part of a proposed UNESCO World Heritage Site called The Sublime Karst of Papua New Guinea (Hamilton-Smith 2006).*"

Sounds impressive indeed, but then, what does a UNESCO World Heritage Site really mean, and what do the landowners say, what research design and taxonomy standard really was used, and where do things stand now, 15 years later? Are such efforts really and truly meaningful if conservation and sustainability are the goal? According to D'Eramo (2014) likely not, the UNESCOcide just makes money for some, but ignores wider conservation needs for all.

And so, as per Beehler and Laman (2020) there are no real national parks in PNG neither (but see McAdam National Park, and then the Varirata National Park, near Port Moresby, mentioned in https://www.britannica.com/place/Papua-New-Guinea/Daily-life-and-social-customs) because of the public land tenure ownership in PNG. Resources in Melanesia—and also PNG—are more community-based and managed that way (Baraka, 2001; Nakaorobi, 1983); intense fighting about land is known to occur. Western-style national parks are not compatible with the PNG system of landownership. A typical example for New Guinea is provided by Beehler and Laman (2020) in Western Guinea's Lorentz National Park (NP): It's hardly legally valid and accepted; people that live there do not really know about it and the applicable laws. The Lorenz NP got actually fully redrawn around the Freeport mine to keep operating. As shown already by Buckley (2020) and many others, such national parks can easily just be lip service and are green-washing destructive practices in wilderness areas and elsewhere. Those are 'paper parks.'

Instead, well-guided and advised community-based landscape projects, such as YUS (see subsequent chapter; details in Beehler & Laman, 2020; Montgomery & Bishop, 2006) are likely the best, and probably the only good and true way to go. Local landowners must be on board and vested. And even then, there are still problems with such land conservation though (see YUS chapter in this book for a wider review).

Whereas, the parsimonious 'bean counting' of species and diversity per land unit using convenient taxonomies—now aided with DNA—serves little; hardly for the proponents themselves, constantly seeking for funding flashes to stay afloat in their own offices and their staff. That's the reality of biodiversity work these days in the absence of solid underlying national sustainable business models. One can there hardly be cynical that the PNG biodiversity reference book of Beehler and Laman (2020) is—in some part—funded by "Porgera Joint Venture", actually a major mining corporation (but that does not even mention in the funding and operation the term 'Mine' in its title.[2] It's clear to anybody who just looks at it closer that commercial mining and industry like to use wildlife photography, experts, science and biodiversity conservation as their strategic and PR tools and to their advantage, as they can. It's textbook marketing 101. Many examples exist, e.g. Richards (2018) for such a science, and the wide lack of specific mining references and direct mentioning of conservation problems for PNG in works like Diamond (2011), Martin (2005) and Laman and Stoles (2012) must not be so surprising for readers.

On such a PNG reality, many species of conservation concern are to occur, as per Beehler and Laman (2020, p. 348). It's quite a tragic state of affairs on the birds etc. due to the governance, data quality and data coverage (Davis et al., 2018; see for instance http://www.earthsendangered.com/search-regions3.asp?mp=&search=1&sgroup=allgroups&ID=283[3]). As a matter of fact, conservation status-wise, virtually all large 'tasty' mammals are under pressure now in PNG, many are extinct or at least getting there. Similar can be said for most other large species, certain birds, and if they have meat or value. Precedence exists: By the end of the Pleistocene twenty species of Australian kangaroos and wallabies became extinct (Beehler & Laman, 2020; Flannery, 2002). And according to these authors, (p. 221 in Beehler & Laman, 2020) already three species of Long-beaked Echidna, Mrs. Scott's Tree

[2] It's clearly a deflecting and misleading project title, and hide-and-play with words when using 'Porgera Joint Venture' while it actually says in the subtitle of the project site that '*The Porgera Joint Venture owns the Porgera Gold Mine*' as per http://www.porgerajv.com/. The public does notice.

[3] There are several sources and categories for endangered species in PNG; none fully agree or have deeper data even. This source here is a public web portal and can serve as an entry point, just as the IUCN Red Species List for PNG can. Based on personal experience, there are many inconsistencies and problems with the official sources and their meanings.

Kangaroo (Tenkile) and the Golden-mantled Tree Kangaroo are threatened and close to extinction; see also Martin (2005).

And then, for endemic places like PNG invasive species are a big threat and those topics are on the rise. Exotic species like common myna, house and tree sparrow, trout in mountain streams, Rusa deer, crab-eating macaque, house mouse, black rat, Norway rat, house cat, house shrew and axis deer are all found in PNG now. And on lakes and river systems, many more exotic species are found, e.g. floating weed, often helped by official agents and even international programs (Coates, 1984; Thomas & Room, 1986 for carp, Beehler & Laman, 2020 for a devastating assessment statement).

As a matter of fact, PNG as we know it, is probably strongly based now on 'exotic' species, namely dog and pigs and betel nut, which actually all have a relatively short history in PNG (Beehler & Laman, 2020; Diamond, 2011).

So what has the 'modern' aspect of biodiversity and conservation research really bought us, other than ongoing ignorance, inefficient conservation, decay and destruction on most fronts? "*More research needed*" and staying in the western mindset is not an answer in such cases and for PNG. Considering major progress in technological and industrial schemes, who then really gained, and by how much, and on what costs (see Taber & Payne, 2003 for North America)?

Using biodiversity and its study, PNG—the way of PNG and how its people operate (Narokobi, 1983)—simply shows us the mirror on how poor our modern western and dominating global society really is and stands.

> *"The reason why these animals survived is that no one has ruined their homes..."*
>
> Montgomery and Bishop (2006, p. 22)

References

Alcorn, J. B. (1993). Papua New Guinea needs assessment. A USAID-funded consortium of world wildlife fund. The Nature Conservancy, and World Resources Institute Washington, D.C. with the Government of Papua New Guinea Department of Environment and Conservation, Boroko, Papua New Guinea. https://www.sprep.org/attachments/VirLib/PNG/png-conservation-needs-assessment-alcorn-1993.pdf. Accessed June 2, 2022.

Baraka, R. (2001). The Meanings of Independence in Post-Colonial Papua New Guinea. A Plan B Research Paper Submitted to the Faculty of the Center For Pacific Islands Studies in the Candidacy for the Degree of Masters of Arts. School Of Hawaii, Asia and Pacific Studies. The University Of Hawai'i At Manoa.

Barclay, K., & Cartwright, I. (2008). *Capturing the wealth from tuna: Case studies from the Pacific* (p. 268). ANU Press.

Beehler, B., & Alonso, L. E. (2001). Southern New Ireland, Papua New Guinea: A biodiversity assessment. *RAP Bulletin of Biological Assessment, 21*, 1–101.

Beehler, B., & Laman, T. (2020). *New Guinea: Nature and culture of earth's Grandest Island*. Princeton University Press.

Blaha, M., Patoka, J., Kozák, P., & Kouba, A. (2016). Unrecognized diversity in New Guinean crayfish species (Decapoda, Parastacidae): The evidence from molecular data. *Integrative Zoology, 11*, 457–468.

Boyden, C. R., Brown, B. E., Drucker, R. P., & Tuft, S. I. (1975). *Ok Tedi environmental study.* Report of the 1974 Cambridge Expedition to the Western District, Papua New Guinea.

Buckley, M. (2020). A first-hand narrative account on Tibet's paper parks: How China's greenwashing in Tibet flies under the radar. In *Hindu Kush-Himalaya watersheds downhill: Landscape ecology and conservation perspectives* (pp. 197–212). Springer.

Burgin, S., (1980). A review of crocodile farming in Papua New Guinea. *Science in New Guinea 7*(2): 73–88.

Coates, D. (1984). *The occurrence, spread and potential effects of common carp, Cyprinus carpio L., in the Sepik River*. Report No. 84-13, Department of Primary Industry, Fisheries Research and Surveys Branch, Port Moresby, Papua New Guinea.

Cocker, M. (2006). *A tiger in the sand: Selected writings on nature.* J. Cape Publisher.

Cousteau, J. M., & Richards, M. (1999). *Cousteau's Papua New Guinea journey*. Henry Abraham's Books.

Czech, B. (2020). *Shoveling fuel for a runaway train.* California University Press.

Daly, H., & Farley, J. (2010). *Ecological economics: Principles and applications* (2nd ed.). Island Press.

Davis, R. A., Dutson, G., & Szabo, J. K. (2018). Conservation status of threatened and endemic birds of New Britain, Papua New Guinea. *Bird Conservation International, 28*(3), 439–450.

D'Eramo, M. (2014). UNESCOcide. *New Left Review, 88*, 47–53.

Demeulenaere, E., Yamin-Pasternak, S., Rubinstein, D. H., Lauren Lovecraft, A., & Ickert-Bond, S. M. (2021). Indigenous spirituality surrounding Serianthes trees in Micronesia: Traditional practice, conservation, and resistance. *Social Compass SAGE Journal.*

Diamond, J. M. (1973). Distributional ecology of New Guinea birds. *Science, 179*, 759–769.

Diamond, J. (2011). *Guns, Germs and Steel: The Fate of Human Societies.* Random House.

Diamond, J. M. (1976). Island biogeography and conservation: Strategy and limitations. *Science, 193*(4257), 1027–1029.

Dinnen, S. (2017). *Internal security in Papua New Guinea: Trends and prospects.* Seven Snapshots of a Nation; Lowy Institute.

Dixon, A. M., Forster, P. M., Heron, S. F., Stoner, A. M. K., & Beger, M. (2022). Future loss of local-scale thermal refugia in coral reef ecosystems. *PLOS Climate, 1*(2), e0000004. https://doi.org/10.1371/journal.pclm.0000004

Earthworks, Deep Sea Mining Campaign et al. (2015). World's first deep sea mining proposal ignores consequences of its impacts on oceans (September 28, 2015). https://earthworks.org/releases/worlds_first_deep_sea_mining_proposal_ignores_consequences_of_its_impacts_o/. Accessed July 14, 2022.

Faith, D. P., Nix, H. A., Margules, C. R., Hutchinson, M. F., Walker, P. A., West, J., Stein, J. L., Kesteven, J. L., Allison, J. L., & Natera, G. (2000). The BioRap biodiversity assessment and planning study for Papua New Guinea. *Pacific Conservation Biology, 6*, 279–288.

Flannery, T. (1990). *Mammals of New Guinea.* The Australian Museum.

Flannery, T. (2002). *The future eaters: An ecological history of the Australasian lands and people.* Grove Press.

Flannery, T. (2000). *Throwim Way Leg: Tree-kangaroos, Possums, and Penis Gourds--on the Track of Unknown Mammals in Wildest New Guinea.* Grove Press.

Freeman, B. G., Class, A., Mandeville, J., Tomassi, S., & Beehler, B. M. (2013). Ornithological survey of the mountains of the Huon Peninsula, Papua New Guinea. *Bulletin of the British Ornithologists' Club, 133*(1), 4–18.

Gentile, O. (2009). *Life list: A woman's quest for the world's most amazing birds.* Bloomsbury Publishing.

Gillison, G. (1993). *Between culture and fantasy: A New Guinea highlands mythology.* University of Chicago Press.

Gillison, D. (2002). *New Guinea ceremonies.* Harry N Abrams Inc.

Glavin, T. (2008). *The last great sea: A voyage through the human and natural history of the North Pacific Ocean.* Greystone Books.

Gotelli, N. J., Chao, A., Colwell, R. K., Hwang, W. H., & Graves, G. R. (2012). Specimen-based modeling, stopping rules, and the extinction of the ivory-billed woodpecker. *Conservation Biology, 26*(1), 47–56.
Graham, C. H., Ferrier, S., Huettmann, F., Moritz, C., & Peterson, A. T. (2004). New developments in museum-based informatics and applications in biodiversity analysis. *Trends in Ecology and Evolution, 19*, 497–503.
Gulyaeva, M., Huettmann F., Shestopalov, A., Okamatsu, M., Matsuno, K., Chu, D.-H., Sakoda, Y., Glushchenko, A., Milton, E., & Bortz, E. (2020). Data mining and model-predicting a global disease reservoir for low-pathogenic Avian Influenza (AI) in the wider pacific rim using big data sets. Sci Rep 10, 1681. https://doi.org/10.1038/s41598-020-73664-2
Hays, T.E. (2003). *New Guinea Bibliography*, October, 2003. https://nanopdf.com/queue/hays-2003-ngbibliography_pdf?queue_id=-1&x=1656200487&z=MjYwMDozODc6NDo4MDM6Ojcy. Accessed June 1, 2022.
Hamilton, R., Green, A., & Almany, J. (Eds.). (2009). *Rapid ecological assessment: Northern Bismarck Sea, Papua New Guinea*. Technical report of survey conducted August 13 to September 7, 2006. TNC Pacific Island Countries Report No. 1/09.
Heinemann, J. A., Abate, T., Hilbeck, A., & Murray, D. (2009). Agriculture at a crossroads: The synthesis report of the International Assessment of Agricultural Knowledge, Science and Technology for Development.
Herre, A. W. (1936). *Fishes of the crane expedition* (Vol. 21, pp. 22–432) (Cocos Island to the Solomons; pp. 433–460 Sepik River). Field Museum Press of America.
Hiepko, P. (1987). The collections of the Botanical Museum Berlin-Dahlem (B) and their history. *Englera, 7*, 219–252.
Holthuis, L. B. (1949). Zoological results of the Dutch New Guinea Expedition 1939. No. 3. Dccapoda Macrura with a revision of the New Guinea Parastacidae. *Nova Guinea, 5*, 289-32.
Holthuis, L. B. (1978). Zoological results of the British Speleological Expedition to Papua New Guinea 1975. 7. Cavernicolous shrimps (Crustacea Decapoda. Natantia) from New Ireland and the Philippines. *Zoologische Mededelingen, 53*(19), 209–224.
Huettmann, F. (Ed.). (2015). *Central American biodiversity: Conservation, ecology, and a sustainable future*. Springer.
Huettmann, F. (2020). Investigating Matschie's tree kangaroos with 'Modern' methods: Digital workflows, big data project infrastructure, and mandated approaches for a holistic conservation governance. In L. P. Dabeck, P. Valentine, J. Blessington, & K. Schwartzet (Eds.), *Tree kangaroos: Science and conservation* (pp. 379–391). Academic Press.
Humphries, G., Magness D. R. & Huettmann, F. (2018). *Machine learning for ecology and sustainable natural resource management*. Springer.
Jackson, J. (2013). *Dive atlas of the world: An illustrated reference for the best sites*. IMM Lifestyle Books.
Jackson, J. B. C., Kirby, M. X., Berger, W. H., Bjorndal, K. A., Botsford, L. W., Bourque, B. J., Bradbury, R. H., Cooke, R., Erlandson, J., Estes, J. A., Hughes, T. P., Kidwell, S., Lange, C. B., Lenihan, H. S., Pandolfi, J. M., Peterson, C. H., Steneck, R. S., Tegner, M. J., & Warner, R. R. (2001). Historical overfishing and the recent collapse of coastal ecosystems. *Science, 293*, 629–637.
Karger, D. N., Kessler, M., Lehnert, M., & Jetz, W. (2021). Limited protection and ongoing loss of tropical cloud forest biodiversity and ecosystems worldwide. *Nature Ecology & Evolution, 5*(6), 854–862.
Kirsch, S. (2014). *Mining capitalism*. University of California Press.
Krishna, P. A., Chand, R., Huettmann, F., & Ghimire, T. R. (2022). Rabies elimination: Is it feasible without considering wildlife? *Journal of Tropical Medicine* (in press).
Laman, T., & Scholes, E. (2012). *Birds of paradise: Revealing the world's most extraordinary birds*. National Geographic Books.
Leatherwood, S. (1991). Unpublished report on the marine mammals of PNG, including 2 species of river dolphins and the dugong. South Pacific Regional Environment Program, New Caledonia.
Liditker, W. Z., & Ziegler, A. E. (1968). Report on a collection of mammals from eastern New Guinea, including species keys to 14 genera. *University of California Publications in Zoology, 87*, 1–64.

Liu, J., Dou, Y., Batistella, M., Challies, E., Conno, T., Friis, C., Millington, J. D. A., Parish, E., Romulo, C. L., Bicudo Silva, R. F., Triezenberg, H., Yang, H., Zhao, Z., Zimmerer, K. S., Huettmann, F., Treglia, M. L., Basher, Z., Chung, M. G., Herzberger, A., Lenschow, A., et al. (2018). Spillover systems in a telecoupled Anthropocene: Typology, methods, and governance for global sustainability. *Environmental Sustainability, 33*, 58–69. https://doi.org/10.1016/j.cosust.2018.04.009

Long, K. E., Schneider, L., Connor, S. E., Shulmeister, N., Finn, J., Roberts, G. L., & Haberle, S. G. (2021). Human impacts and Anthropocene environmental change at Lake Kutubu, a Ramsar wetland in Papua New Guinea. *Proceedings of the National Academy of Sciences, 118*(40), e2022216118.

Lowe, C. (2004). Making the monkey: How the *Togean macaque* went from "new form" to "endemic species" in Indonesians' conservation biology. *Cultural Anthropology, 19*(4), 491–516.

Ludlam, S. (2021). *Full circle: Power*. Black Inc.

MacArthur, R. H., & Wilson, E. O. (1963). An equilibrium theory of insular zoogeography. *Evolution*, 373–387.

Mack, A. (2014). *Searching for PekPek: Cassowaries and conservation in the New Guinea rainforest*. Cassowary Conservation & Publishing.

Martin, R. (Ed.). (2005). *Tree-kangaroos of Australia and New Guinea*. CSIRO Publishing.

Marsh, H., Grayson, J., Grech, A., Hagihara, R., & Sobtzick, S. (2015). Re-evaluation of the sustainability of a marine mammal harvest by indigenous people using several lines of evidence. *Biological Conservation, 192*, 324–330.

Mavea, W. B., Belonovich, O., Nagombi, E., Boslogo, T., Gwynn, L., Russ, R., & Bird, J. P. (2021). A survey of cetaceans in New Britain, Papua New Guinea. *Australian Mammalogy*. https://doi.org/10.1071/am21005

Mayr, E., & Diamond, J. M. (2001). *The birds of northern Melanesia: Speciation, ecology & biogeography*. Oxford University Press on Demand.

Mongabay. (2022). Papua New Guinea. https://rainforests.mongabay.com/deforestation/forest-information-archive/Papua_New_Guinea.htm. Accessed March 29, 2022.

Montgomery, S. Y., & Bishop, N. (2006). *Quest for the tree kangaroo: An expedition to the cloud forest of New Guinea*. Houghton Mifflin Company.

Myint-U, T. (2006). *The river of lost footsteps: Histories of Burma*. Farrar, Straus and Giroux Publishers.

Narokobi, B. (1975). *Foundations for nationhood*. Papua New Guinea Press and Bookshop.

Narokobi, B. (1983). The Melanesian way. Institute of Papua New Guinea Studies, Boroko, Papua New Guinea.

New South Wales State Library. (2022). Birds of paradise. https://www2.sl.nsw.gov.au/archive/discover_collections/society_art/png/birds/index.html. Accessed April 28, 2022.

Orchid Society of Papua New Guinea Inc. (2006). *Spatulata orchids Papua New Guinea*. Moore Printing Ltd. ISBN 9980-86-070-7

Paijmans, K. (Ed.). (1976). *New Guinea vegetation*. National University Press.

Rauzon, M. J. (2016). Isles of Amnesia: The history, geography, and restoration of America's forgotten Pacific islands.

Robbins et al. (2016). Isles of Amnesia: The history, geography, and restoration of America's forgotten Pacific islands.

Richards, S. J. (Ed.). (2018). *Rapid biological assessments of Wau Creek, Uro Creek and Lake Kutubu: Documenting biodiversity values to promote forest conservation in the Kikori River basin, Papua New Guinea*. ExxonMobil PNG Limited.

Richards, S. J., & Whitmore, N. (Eds.). (2015). A rapid biodiversity assessment of Papua New Guinea's Hindenburg Wall region. Wildlife Conservation Society Papua New Guinea Program, Goroka, PNG.

Robbins, R. G., Saunders, J. C., & Pullen, R. (1976). Lands of the Ramu-Madang area, Papua New Guinea. 7. Vegetation and Ecology Land Research Series 37. CSIRO, Australia. http://hdl.handle.net/102.100.100/305517?index=1

Rhodin, A. G. J., Spring, S., & Prichard, P. E. H. (1980). Glossary of turtle vernacular names used in the New Guinea region. *Polynesian Society, 89*(1), 105–117.

Savery, L. C., Wise, S. S., Falank, C., Wise, J., Gianios, C., Jr., Thompson, W. D., Perkins, C., Zheng, T., Zhu, C., & Wise Sr, J. P. (2014). Global assessment of oceanic lead pollution using sperm whales (*Physeter macrocephalus*) as an indicator species. *Marine Pollution Bulletin, 79*(1–2), 236–244.

Snetsinger, P., & Pratt, H. D. (2003). *Birding on borrowed time*. American Birding Association.

Steadman, D. W. (2006). *Extinction and biogeography of tropical Pacific birds*. University of Chicago Press.

Steiner, M., & Huettmann, F. (2021). Justification for a taxonomic conservation update of the rodent genus Tamiasciurus: Addressing marginalization and mis-prioritization of research efforts and conservation laissez-faire for a sustainability outlook. *The European Zoological Journal, 88*(1), 86–116. https://doi.org/10.1080/24750263.2020.1857852

Stresemann, E. (1923). Ornithologische Ausbeute im Stromgebiet des Sepik. Ein Beitrag zur Kenntnis der Vogelwelt Neuguineas. *Archiv fuer Naturgeschichte ser. A.A., 89*(7)–*96*(8), 92.

Sullivan, N. (2015). Introduction to the prospects for sustainable development in Papua New Guinea. https://www.academia.edu/4492757/Introduction_to_The_Prospects_for_Sustainable_Development_in_Papua_New_Guinea. Accessed July 3, 2022.

Supuma, M. (2018). *Endemic birds in Papua New Guinea's montane forests: Human use and conservation* (Unpublished doctoral dissertation). James Cook University, Australia.

Taber, R. D., & Payne, N. F. (2003). *Wildlife, conservation, and human welfare: a United States and Canadian perspective*. Malabar, Florida: Krieger Publishing Company.

The Guardian. (2017). What David Attenborough should have said at the end of Blue Planet, December 12, 2017. https://www.theguardian.com/commentisfree/2017/dec/12/david-attenborough-blue-planet-bbc. Accessed August 14, 2022.

The Guardian (2020). Major climate changes inevitable and irreversible – IPCC's starkest warning yet. https://www.theguardian.com/science/2021/aug/09/humans-have-caused-unprecedented-and-irreversible-change-toclimate-scientists-warn. Accessed 1st July 2022.

Thibault, J. C., & Cibois, A. (2017). *Birds of Eastern Polynesia: A biogeography atlas*. Lynx Publishers.

Thomas, P. A., & Room, P. M. (1986). Successful control of the floating weed *Salvinia molula* in Papua New Guinea: A useful biological invasion neutralises a disastrous one. *Environmental Conservation, 13*, 242–248.

Taylor, A., Zotz, G., Weigelt, P., Cai, L., Karger, D. N., König, C., & Kreft, H. (2022). Vascular epiphytes contribute disproportionately to global centres of plant diversity. *Global Ecology and Biogeography, 31*(1), 62–74.

Vasilita, C., Moser, M., & Krogmann, L. (2022). Mission possible: An optimised protocol for the unbarcodable Ceraphronoidea (Hymenoptera). *Biodiversity Data Journal, 10*, e84860. https://doi.org/10.3897/BDJ.10.e84860

Webb, C. O., Boucher, T., Sheppard, S., & Summers, M. (2005). Vegetation of the Adelbert range, Madang Province, PNG. http://camwebb.info/files/pubs/webb2005_adelbert_veg.pdf. Accessed January 31, 2022.

Webb, V.-L. (1995). Photographs of Papua New Guinea: American expeditions 1928–29. *Pacific Arts, 11*(12), 72–81. http://www.jstor.org/stable/23409736

Webb, V. L. (1996). Framing time: Photographs of New Guinea from the crane Pacific expedition, 1928–29. Columbia University.

West, P., & Kale, E. (2015). The fate of crater mountain: Forest conservation in the eastern highlands of Papua New Guinea. In J. Bell, P. West, & C. Filer (Eds.), *Tropical forests of Oceania: Anthropological perspectives. Asia-Pacific Environment Monograph* (Vol. 10, pp. 155–178).

West, P. (2006). *Conservation is Our Government Now: The Politics of Ecology in Papua New Guinea*. Durham, NC: Duke University Press.

Williams, C. K., Guthery, F. S., Applegate, R. D., & Peterson, M. J. (2004). The northern bobwhite decline: scaling our management for the twenty-first century. *Wildlife Society Bulletin, 32*(3), 861–869.

Wildlife Conservation Society. (2009). Study on Papua New Guinea's long-beaked echidna reveals elusive habits. *ScienceDaily*, June 23, 2009. www.sciencedaily.com/releases/2009/06/090609220725.htm. Accessed June 24, 2022.

Wikramanayake, E. D., Dinerstein, E., & Loucks, C. J. (2002). *Terrestrial ecoregions of the Indo-Pacific: A conservation assessment* (Vol. 3). Island Press.

Wilson, E. O., & MacArthur, R. H. (1967). *The theory of island biogeography* (Vol. 1). Princeton University Press.

Zandri, E., Prentice, C., et al. (2009). The experience of UNEP GEF and partners in flyway conservation, UNEP GEF portfolio outlook and evolution. Biodiversity Issue Paper BD/001. UNEP, Nairobi, Kenya.

Zoeckler, C. (2012). Status, threat, and protection of Arctic waterbirds. In F. Huettmann (Ed.), *Protection of the three poles* (pp. 203–216). Springer.

Zöller, H. (1891). Deutsch-Neuguinea und meine Ersteigung des Finisterre-Gebirges: Eine Schilderung des ersten erfolgreichen Vordringens zu den Hochgebirgen Inner-Neuguineas, der Natur des Landes, der Sitten der Eingeborenen und des gegenwartigen Standes der deutschen Kolonisationsthatigkeit in Kaiser-Wilhelms-Land, Bismarck- und Salomoarchipel, nebst einem Wortverzeichnis von 46 Papua-Sprachen. Stuttgart: Union Deutsche Verlagsgesellschaft; 1891, xxxii, 546 pp. + Plates + Foldout Figure + Map.

Chapter 7
Papua New Guinea as a Global Sustainability Leader: Confronting Per Capita Metrics with GIS Layers and Social Engineering

The sin of PNG has been the failure of either the national government or provincial governments to carry out formal resource planning as a way of defining future conservation action balanced against smart and well-located economic development
Closing remarks by Beehler and Latam (2020, p. 353)

...those responsible for defining the new state's independence realized the Australian vision for PNG's future was not entirely appropriate. Rather, PNG's first generation of national leaders sought to develop PNG by emphasizing what they believed was most suitable for PNG' s overwhelmingly rural village-based societies. Yet the need for continued economic support from Australia constrained the degree of freedom these leaders actually had in charting a new course for the nation.
Baraka (2001, p. 7/8)

Abstract Using best-available online data, and when compared to its colonial powers, Papua New Guinea (PNG) has a tiny human footprint, even more so when compared globally. Being a rural island society, usually with a nomadic component and public ownerships, that's how most of PNG lives and operates. Private wealth accumulation tends to be very small, hardly needed. And PNG has done that way fine for over 47,000 years. The per capita consumption in PNG ranks as one of the lowest in the world resulting in a carbon sink for global society. Seen from that angle, PNG does the global community a favor and provides a huge service, but is not much recognized or rewarded for it. It's even widely called a 'failed state' while staying in the tribal Wantokc governance for its natural resources. Blended with 'modernity,' here it's shown how PNG can be a global policy lead and how a society is to look like and to remain if global well-being, world peace, is actually wanted.

Keywords Papua New Guinea (PNG) · Sustainable governance · World peace

7.1 Introduction

The current resource consumption footprint of a Papua New Guinea (PNG) citizen—per capita—ranks among the world's lowest. In PNG people have their own way and

F. Huettmann, *Globalization and Papua New Guinea: Ancient Wilderness, Paradise, Introduced Terror and Hell*, https://doi.org/10.1007/978-3-031-20262-9_7

Fig. 7.1 Map of human footprint (dark = low, light = high) for Papua New Guinea, a nation with a Gross Domestic Product (GDP) of app $23.59 billion (GIS map source Huettmann, 2020) Note the different data sources for PNG sectors within, e.g. the Australian claimed\supported areas in the south

its locally celebrated, as part of being Melanesian (Narokobi, 1983) and the PNG statehood (Narokobi, 1975). This is not widely known, hardly promoted anywhere (but see Golub, 2014 for a critique). However, it's quite a typical situation for many tropical nations, e.g. in equatorial areas where there is almost no need for winter heating and thus PNG emits virtually nothing on that topic, e.g. when compared to Arctic or high elevation and latitude nations such as Canada or England! The human population is low, has little consumption and few private goods. Publically owned land sits at the core of people's attention, its shapes their identity (Baraka, 2001) (Figs. 7.1 and 7.2).

PNG also has just few private cars, little roads, few big cities, and it features just a few public transport options (= buses) with no railway. Subways are not heard of in PNG. People of PNG also eat primarily a vegetable-based local garden diet ("*greens*"). Industrial mass production of meat is very little; and all commercial meat and food (= sugar) production was actually imposed from the outside, usually through Australia, and mostly done by the western world for their own use and profit, e.g. chicken and strawberry. Also, PNG Airlines has a very tiny fleet, domestically and internationally. Despite the very benign PNG impact, PNG citizens are affected unduly by economic growth and CO_2-inflicted climate change: this economic growth is mostly just due to foreign-driven industrialization, globalization, international fishing, mining and LNG plants; it's the prime cause for PNG's microplastics, sea level rise and ocean acidification. It also creates impacts reaching from global change and temperature rise over to melting glaciers and associated change in water tables, coastal erosion, pandemics and global warfare as well as subsequent immigration

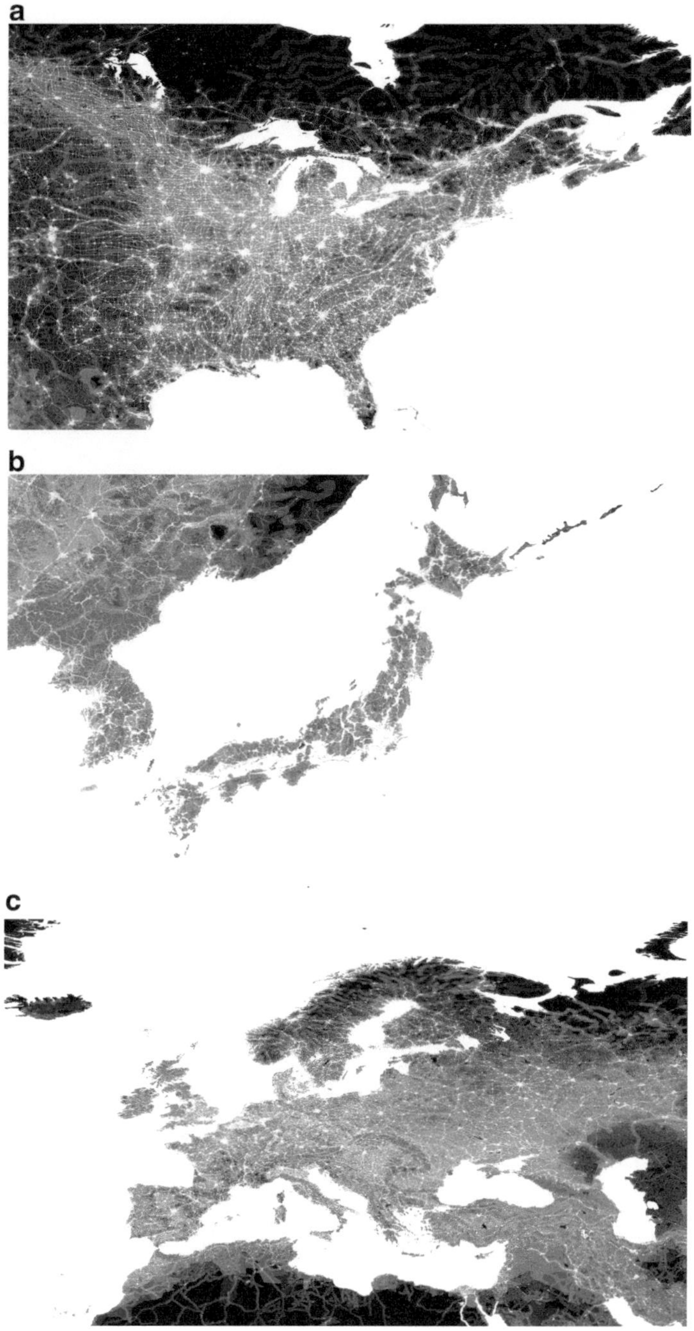

Fig. 7.2 For comparison, a map of human footprint (dark = low, light = high) zoom in for **a** Eastern US, **b** Japan and S. Korea, and **c** EU (see respective GDP values of app. $20.94 trillion, $5.06 trillion and $17.90 trillion, respectively)

Fig. 7.3 A typical garden and village site in Papua New Guinea: the engine of global sustainability; and perhaps the future of the world's wilderness?

chaos and human tragedy and suffering. The socio-economic impacts sit at the core of this discussion. Many of those activities act in synergy, with a lag effect and mostly caused 'by the western nations' and 'for the western nations' while PNG pays most of the true costs, clean-up and otherwise, e.g. long-term effects affecting future generations. It's more than a double whammy for PNG. Many of those impacts came from across the border (e.g. Lasslett, 2012), and globally (Stiglitz, 2003), fully approved by academic advisors, now many are also applied in PNG itself, but against the will of many of the PNG people, e.g. in the Fly River pollution (Kirsch, 2014) or Sepik river issues (Cousteau & Richards, 1999). For Ramu River, as one example of many, the author heard of, and saw, similar environmental issues (Examples of a natural live in PNG are shown in Figs. 7.3, 7.4, 7.5, 7.6, 7.7, 7.8, 7.9, 7.10, 7.11 and 7.12; Table 7.1).

7.2 Papua New Guinea Compared with the Colonial Outside World

Compared to the western lifestyle, the citizens of PNG live essentially without harming much the global environment, that is, true socially, economically and environmentally. When it comes to the carbon footprint specifically, PNG is one of the lowest consumers in the world releasing man-made CO_2; and instead the large forests and wetlands of PNG sequester carbon (e.g. the Guardian, 2015 for a topical review and citations within); it acts as a global carbon sink and as a temperature

Fig. 7.4 Pigs and coconuts in the garden: A genius but well-proven approach to sustainability overall (much better than any neoliberal industrialization effort ever can be)

regulator for land, sea and atmosphere. This is known but not much awarded by the U.N. (example of PNG consumption patterns shwon in Fig. 7.13).

PNG provides essential ecological services to the world; all done for free and based on a huge expertise accumulated for millennia (many details shown in Flannery, 1998)! Same can be said for consumption of nuclear power and waste production, plastics, electricity, industrialized foods and water or industrial contamination. In PNG, the +700 languages have no words for such terms. The creation, consumption and release of such contamination products in PNG are once more miniscule when compared to most other nations in the world. And having one of the highest rainfalls in the world, traditionally PNG's rural community has virtually no water-use footprint neither. All of this occurs while the world, certainly adjacent Australia, is struggling and fighting for fresh water and for good quality drinking water and man-made CO_2 release (e.g. Mercer et al., 2007). Water is not really the core issue in PNG, not to be (it actually now is a concern and that's just relatively recent artefact where the outside world takes advantage of cheap mining, low costs of environmental violations, inexperience, absurd governance structures, strange plumbing schemes, and remoteness affecting the initially pristine water cycles and handlings in PNG; some global resource examples provided in Stiglitz, 2003).

And that tiny environmental footprint of ancient PNG—more than 47,000 years old—is not a small feat whatsoever! It's a gift to the world widely unrecognized. Many nations wish to have a low carbon footprint by now, if they just could!

Fig. 7.5 A 'castle' in Papua New Guinea, fabricated from palm trees and some tropical timber, and earthquake-proof

With such a tiny human footprint, the PNG culture does indeed offer a sustainability model for the world. PNG runs a modified natural 'garden' to feed its people on a finite space. PNG is doing a fine-tuned and sophisticated gardening already for millennia. It does that without industrial fertilizers and industrialized debt-causing or seed-destroying DNA-bred GMO plants. It does no relevant harm to nature or the ecosystem or the nation, or for the world for that matter. It's for a small and rel. appropriate human population on the PNG landmass. The PNG society lifestyle is as benign as it gets, and it is quite sustainable. PNG is a cradle of human farming for many species, e.g. yams.

This IS a modern life and culture and must be seen that way (compare with Golub, 2014 for 'grassroots' in PNG), and in the days of 2023. It's what most of the western and industrial world seeks (e.g. German 'Coconut cult' https://en.wikipedia.org/wiki/August_Engelhardt) but what they do not achieve (see O'Coonor et al., 2020; compare with Dublin & Tamaka, 2015 for Japanese Sayotoma and Ainu approaches practiced on Pacific islands just north of PNG).

So where is the reward for PNG, and how recognized? Where is the PNG-owned copyright for such a global remedy and management? There is not a single mentioning and recognition of such an OUTSTANDING contribution in the U.N., in the U.N. Aichi targets and SDGs, or in any of the COPs on Climate Change, nor in the IPCC.org or with majpr NGOs for the matter of fact. It simply goes by the wayside. And even major experts on the PNG and environmental subject remain quite silent

Fig. 7.6 Water taxi, part of the coastal fabric and a life line in Papua New Guinea (Note the generic lack of live vests)

about it, e.g. Martin (2005), Diamond (2011a), Laman and Scholes (2012), Beehler and Laman (2020).

Why is that?

Here sits exactly the crux: low impact metrics usually come with modern-style (created) poverty and thus those metrics are not so interesting—hardly achievable or wanted—for western nations, whatever they say and claim (win–win situation, e.g. 'have a small car, a small house and combat climate change' as promoted in "Inconvenient Truth" by Al. Gore, etc., https://en.wikipedia.org/wiki/An_Inconvenient_Truth).

Overall, and despite its reputation (Gosarevski et al., 2019; Smith, 1974), PNG is very peaceful for the world overall; already the bartering and trading of goods all over PNG requires a functioning system of human interaction for survival (see Golbu, 2014 for Yame; another typical example shown in the oil of the tigasso tree (*Campnosperma*), that was known, harvested and traded for its amazing properties throughout ancient times in the Sepik region and beyond among different tribes and ecosystems (it was later picked up as a global commodity by international business people for the organic health market, etc.). '*PNG ingenuity*' indeed is 'a thing' (Narokobi, 1983). And it exists for everybody who just looks at PNG.

Intense resource conflicts for oil and gas in PNG versus the world—as a type of international warfare—are very few. There is no global warfare that PNG really engages in, or is occupied with. PNG has no real international enemies and virtually no drones to kill people, nor a relevant number of airplanes, or warplanes or

Fig. 7.7 Papua New Guinean village and garden foods are very healthy indeed

bombs for that matter. PNG can hardly defend its coastline and EEZ (see Radio New Zealand 2018 reporting on 18 missing PNG fisheries observers). The actual harm that truly comes from PNG is miniscule, certainly when compared on a per capita basis with nations like Australia, NZ, Switzerland, Canada, the U.S., or surely the EU—as an acclaimed center of civilization and colonial power; but just think of Iceland, or Finland and their heating costs and its wider lifestyle subsidized by the EU (milk, energy, diesel and coffee, sugar, broccoli, oranges and bananas). Finland now gets support from the NATO to remain a western-style democracy (vs. a Russian-style influence). Or even worse, think of Norway—as a nation and a royal kingdom—a stench whaling supporter but all just kept afloat by the NATO and oil consumption (now using mining-intense electro cars). Norway would otherwise likely be consumed by its neighbors, namely Russia, or Sweden and Denmark (see the track record of Norwegian history and Royal courts in Scandinavia, specifically Denmark and its crown). Arguably, ancient Viking raids onto existing and neighboring kingdoms are not sustainable but added to some Norwegian wealth while the Vikings exploited civilzations and used slaves extensively and in a rather cruel manner; Brink (2021). Instead, PNG is not given any credit for being among the lead on global metrics of peace, low consumption and no relevant impact for a local island that left its neighbors mostly alone; the exchange with Micronesia and Polynesia was through bartering, at best (but see modern aspects in Figs. 7.14, 7.15 and 7.16). Historically, PNG took not a strong role outside of PNG.

Fig. 7.8 Want some lemon tea? It's grown in the garden

So instead, and going with the colonial narrative, one may read that PNG is 'violent' and has one of the highest crimes in the world (e.g. Salak, 2001, or West, 2006). Its capitol Port Moresby is flagged accordingly and its rural areas are high in crime rates like rape, domestic violence and gang violence, rascal'ism (Table 7.2).

And yes, compared to the west the industrial-driven education level in PNG is pretty low, and so is its income. PNG lacks the western education designed for industrial processes, including computing. But most PNG citizens know how to operate a garden and the forests; grassroots! PNG can hardly afford to operate an international airline nor a relevant industrial workforce, army or reliable police force (e.g. Gosarevski et al., 2019). Just like many tropical nations (e.g. see Huettmann, 2015 for Central America as a similar example), PNG has no car brand or car manufacturer, no relevant software company and no hardware producer. PNG makes no military weapons, did not contribute to a genocide or produces airplanes nor is it involved in the weapon trade (other than smuggling perhaps driven by Australia or Asia, Schloenhardt, 2006). PNG languages have virtually no words or concepts for it. PNG features the 'bow and arrow,' the ancient ax, and gangs can produce shotguns from water pipes (Salak, 2001). Killings are frequently done by machete. While embedded in the local bartering and markets, PNG as a nation has a relatively small (financial) market and a low amount of registered taxpayers even (Baraka, 2001;

Fig. 7.9 Watercress maintained in a riverbed

Chan, 2016). It all sits in the garden instead; not metal coins or plastic money printed in Australia drives the local economy but betel nuts and bartering for garden products do. PNG should be a feast for ecological macroeconomists and sustainability but has received little of such study recognition (Table 7.3).

PNG does have a few tuna canneries—foreign-owned—and those are not only an environmental nightmare, but also create large social problems (see Sulivan et al., 2003; Beehler & Laman, 2020 for examples). Same can be said for mining, certainly with a long list of very serious impacts (see Cousteau & Richards, 1999; Kirsch, 2014, PNG Mining Watch https://mine.onepng.com/, etc. Note the detailed absence of such problems reported in Beehler & Laman, 2020 as supported by a Gold Mine Join Venture but celebrating PNG's beauty; see chapters in this book).

The mentioned education level deserves a second look though: '*Learning' for what*? Who benefits and what for? PNG citizens are fully trained in PNG life, what else is needed really? Trained workforce tends to serve industry, often located abroad, which not only makes rich people richer but also exploits the environment, and often the worker and the family itself. Man-made climate change is primarily caused by industrialization and globalization; not as a PNG product! Remember, PNG is running 'gardens' which is widely organized in family groups instead. Wantok rules and PNG have virtually no feudal system. So what again is a modern education good for? Considering that religious missions have such a stronghold in PNG education, what did they really achieve last 200 years, for PNG and its citizen? They have no good track record.

Fig. 7.10 Harvesting some watercress

Needless to say, PNG has no top ivy league school and the relevant goal often consists of getting into an Australian or New Zealand school, if ever. How many graduates of PNG are in the UK or the U.S. for that matter?

And then, no wonder, the PNG education consists by now of an Australian remake (see O'Donoghue, 2009; Rena, 2010 for a syllabus review), while the schools are still widely in the hands of the church and Australian syllabi, also based and funded outside of PNG, e.g. through AussiAID, helped by New Zealand, and the U.S. (http://www.pngembassy.org/religion.html). PNG has little of its own on those matters. This is social engineering in its pure form, and it has no good outcome. Why would it?

7.3 Reality View on PNG Metrics, Crime and Socioeconomics and Human Footprint

When just looking at Tables 7.1 and 7.2, compared to Figures 7.1, 7.2, 7.3 and 7.4 it becomes clear that it is not the PNG citizen who drives the problems, it's primarily the industry, their governance and efforts helped from abroad all operating under a wider framework (*sensu* Ostrom, 1990; Ostrom et al., 1993). The U.N. mandate via

Fig. 7.11 Some forest beetles, those ones are eatable

Australia proves to be quite disastrous for PNG and its wilderness lifestyle indeed. This is a common feature, as seen in the Pacific Rim, Central American side (see Huettmann, 2015 for details and reviews), the connected Arctic of the Pacific, Alaska and its disease problems affecting industry, people, landscapes and its wider set up (Gulyaeva et al., 2020 for poultry industry and wildlife influenza hotspots and spread in Asia and Pacific Rim; Huettmann & Hueffer, 2021 for examples of rabies, humans and its species).

While many people will not admit it, the western governance system comes with massive problems, catering a one-sided industrialization the most; wildlife and habitat come last (Taber & Payne, 2003 for problems with one of the largest budgets on wildlife management worldwide). Sustainability is virtually impossible to achieve within that. This is based on the western system of governance and its law, which is widely failing (details exposed for instance in The Secret Barrister, 2018 with thorough detail). Needless to say that 'modern' PNG is to follow exactly those types of legal systems and concepts now, compared with the older tribal system and community governance that worked for millennia and had sustainability at the root.

Fig. 7.12 Stairs of an ancient forest trail in rainforest Papua New Guinea

7.4 Education What for and for Whom?

See it that way, why pursuing industrial efforts and educations so costly when all it achieves is just destruction, certainly destruction of the ancient PNG life, fabric and environment wholesale? The more English is taught, the more the many local languages will disappear; effects are shown, e.g. Kulik (2019) for lost culture, lifestyles and sustainability. English is a language of money, not well evolved and suited for PNG and the western money world is in a massive crisis anyways, certainly on the sustainability aspect. So now what?

The presented numbers and maps and pixels beg the question: modernity what for, education what for, and education for whom? (Rooney & Papoutsaki, 2004) Arguably, educated urbanized people are pretty wasteful for nature and its resources as they have higher demands and thus higher consumption and subsequent footprints. Computers make that point well and none pay back to nature, whatever they say and claim otherwise (Kuehr & Wiliams, 2007) (Figs. 7.17, 7.18, 7.19, 7.20 and 7.21).

Table 7.1 A typical selection of per-capita metrics of PNG citizens

Metric	Value (=How often occurring)	Source	Comment
CO_2 emissions	Among the lowest on earth	Various sources	Exact and agreeing numbers are hard to come by, but in virtually all sources PNG ranks at the bottom of the list
Starting of international or global wars	1	Public record	Coconut war A strange brief conflict involving French and US and asking PNG to fight Vanuatu at Espiritu Santo
Participation in a global war	3	Veenendaal (2021)	PNG joined fights in WW1 and WW2, but was essentially dragged into it to a Japanese invasion and being part of different colonial empires at that time Coconut war
Heating costs for winter	Nil	FH pers. obs.	
Cars per capita	<1	FH pers. obs.	This number changes a bit in urban areas

Table 7.2 A selection of per-capita socioeconomic metrics of PNG citizens

Metric	Meaning	Source	Comment
Infant mortality	Conditions at birth	Various	Usually birth is given in the bush, not all in hospitals; exact metrics are difficult to obtain
Life expectancy	A generic index of healthcare and pension support	Various	Exact metrics are difficult to obtain; like with the above numbers those are usually best for local areas
Literacy rate	Western schooling	Various	Fails to express Gardening or Conservation and most Sustainability efforts
Cancer rate	Difficult interpretation and depending on cancer type	NA	A complex disease not 'just' related to industrial factors Betel nut chewing is cancerogenus also
Domestic violence	A complex PNG problem in society	NA	PNG is notorious for such high metrics with women and children being affected the most; but exact numbers are virtually not collected
Break-ins	Poverty-related	NA	PNG urban areas are notorious for crime rates, details are poorly kept

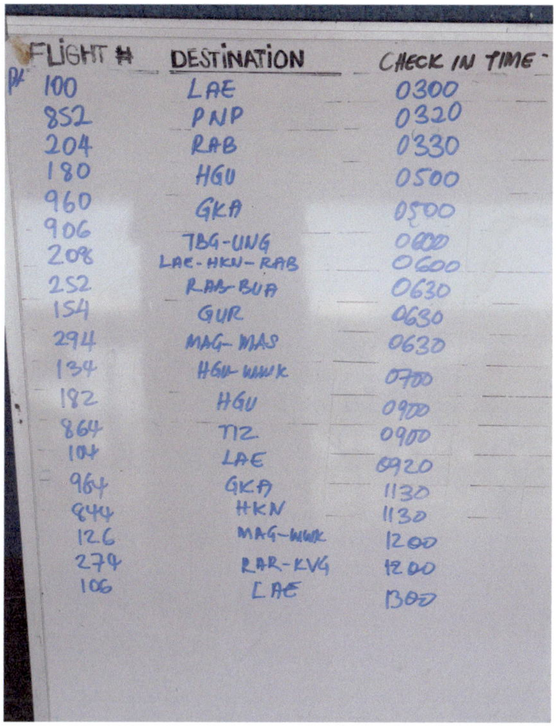

Fig. 7.13 Flight schedule format and details in a domestic Papua New Guinea flight; compared to other nations and airports that is a very low impact

7.5 Papua New Guinea as the Global Sustainability Leader and Role Model

PNG and its citizens remain a global showcase for low environmental impact. Living a PNG live will result in less global environmental suffering; certainly the atmosphere and the water will benefit. It's a live with a local emphasize embedded in global actions, and with resolving local issues, including its challenges for sure. Live is never easy and the ying-and-yang remains; the good and bad forces of live and the universe. Having a different governance emphasize, PNG is actually quite free of any feudal structure and can provide happiness and balance for the seeker.

There are very absurd but ongoing tendencies though to engage and entangle PNG into the wider capitalistic, neoliberal and global marketplace that failed elsewhere (Gosarevski et al., 2019). PNG gets involved in maintaining an airline, in stock markets, a treasury, plastic money, and even worse, a supermarket economy and army support abroad. The recent import of 'clean' drinking water from Australia, a dry nation already running short of water while PNG has one of the highest rainfalls in the world, is a classic scheme in that (see Figs. 7.17, 7.18 and 7.19 on water and practical ingenuity in PNG). Using neoliberalism in education makes it even worse (Carter, 2017).

Table 7.3 Islands and a selection of their conservation issues compared with Papua New Guinea

Island name	Conservation issue	Citation	Comment
Japan	Human footprint, various species extinctions, e.g. wolves and fisheries	Brazil (2022)	A major civilization with one of the world's leading GDPs, but still in debt and in a vast ecological crisis, e.g. overfishing and nuclear energy; global disputes about whaling and fisheries (including thuna)
Guam	Human footprint, various species extinctions, habitat loss invasive species	Rogers (2011)	A small island nation Invasive species problems like Brown Tree Snake
Iceland	Human footprint, various species extinctions, Climate change	Thordarson and Höskuldsson (2014) Huettmann et al. (2016)	An important western contributor but based on a rel. small human population still with a vast per capita human footprint, including tourism, whaling, fisheries and bauxite processing
Greenland	Human footprint, Climate change, Indigenous people conflict (via Denmark)	https://www.atlanticseabirds.info/greenland	A large landmass but ice-covered and dominated by Denmark. Much mining efforts, vast problems related to contamination and climate change melting ice sheets
Borneo, Malaysia	Human footprint, Habitat loss, e.g. oil palms, Indigenous people conflict	Cushman et al. (2017)	Massive loss of rainforest habitat
Hawaii	Human footprint, Habitat loss Indigenous people conflict		A U.S. state with a very large GDP and many environmental problems, e.g. due to industrialization: habitat modification, invasive species and climate change

Fig. 7.14 Domestic industrial chicken, quite a new food in Papua New Guinea that otherwise as a nation could feed itself as one of the Megadiversity countries. In such places, it should be impossible to starve, or to be poor for that matter

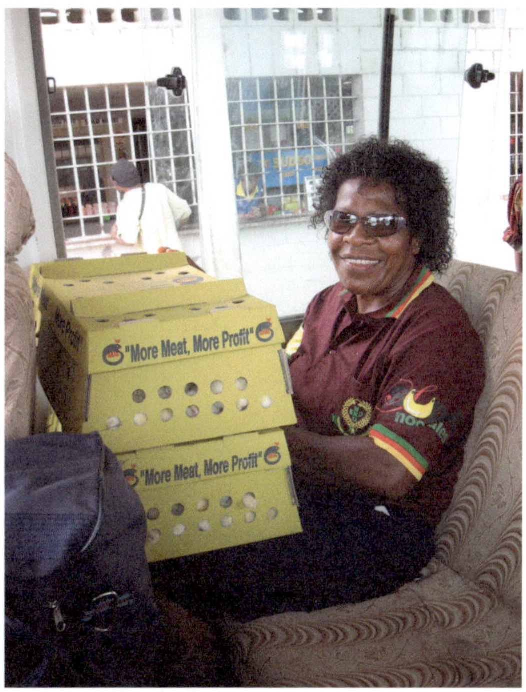

Fig. 7.15 Chicken transport in a public bus, reality to food security and zoonotic diseases in Papua New Guinea

Fig. 7.16 A typical mass chicken transport into the remote villages. Packaged chickens on a crate at an airport. Another reality of livestock, food and disease handling in tropical nations, western style

Fig. 7.17 Plumbing, pipes and running water, a reality need for anybody on earth

Fig. 7.18 Natural fence in Papua New Guina, sustainable gardening, e.g. to keep the pigs at bay

Fig. 7.19 Papua New Guinea ingenuity

Fig. 7.20 Clouds and the atmosphere: a safeguard for Papua New Guinea's environment and life style but widely ignored and highly mismanaged

Fig. 7.21 Beauty in the eye of the beholder: Papua New Guinea and its lifestyle must not be destroyed

As a token participation, PNG is also part of UN Blue Helmet peace-keeping efforts abroad (2 soldiers listed; see also background information here https://news.un.org/en/story/2010/09/353842-papua-new-guinea-moves-closer-contributing-first-blue-helmets-un-missions), and PNG is part of the Antarctic Treaty System (ATS). PNG famously spoke up at COPs against the industrialized world on climate change within U.N. events (see Kevin Conrad in Bali's COP; The Guardian, 2015), but how much CO_2 was truly reduced for PNG and its citizens?

And so for climate change and PNG: What comes, what causes and what gives? PNG does 'make a weather' overall (see Fig. 7.20 on clouds as an essential icon and for its rainfall).

The issue here is less to make PNG a sole global leader rather than to acknowledge, rethink, reflect and adjust to a culture and approach similar to PNG and where PNG can showcase its achievements for the world to learn from. Why not?

References

Baraka, R. (2001). The meanings of independence in post-colonial Papua New Guinea. A Plan B Research Paper Submitted to the Faculty of the Center for Pacific Islands Studies in the Candidacy for the Degree of Masters of Arts. School of Hawaii, Asia and Pacifci Studies. The University of Hawai'i at Manoa.

Beehler, B., & Laman, T. (2020). *New Guinea: Nature and Culture of Earth's Grandest Island.* Princeton University Press.

Brazil, M. (2022). *Japan: The natural history of an Asian archipelago* (Vol. 72). Princeton University Press.

Brink, S. (2021). *Thraldom: A history of slavery in the Viking age.* Oxford University Press.

Carter, L. (2017). Neoliberalism and STEM education: Some Australian policy discourse. *Canadian Journal of Science, Mathematics and Technology Education, 17*(4), 247–257.

Chan, J. (2016). *Playing the game: Life and politics in Papua New Guinea.* University of Queensland Press.

Cousteau, J. M., & Richards, M. (1999). *Cousteau's Papua New Guinea journey.* Henry Abraham's Books.

Cushman, S. A., Macdonald, E. A., Landguth, E. L., Malhi, Y., & Macdonald, D. W. (2017). Multiple-scale prediction of forest loss risk across Borneo. Landscape Ecology *32*(8), 1581–1598.

Daly, H. E., & Farley, J. (2010). *Ecological economics: Principles and applications* (2nd ed.). Island Press.

Diamond, J. (2011a). *Guns, germs and steel: The fate of human societies.* Random House.

Diamond, J. (2011b). *Collapse: How societies choose to fail or succeed* (Revised ed.). Penguin Books.

Dublin, D., & Tanaka, N. (2015). Comparison between Satoyama and the Ainu way of life. *Kawistara Jurnal Sosial Dan Humaniora, Universitas Gadjah Mada, Indonesia, 5*, 221–328.

Flannery, T. (1998). *Throwim' way leg: Tree-kangaroos, possums, and penis gourds.* Grove Press.

Flannery, T. (2002). *The future eaters: An ecological history of the Australasian lands and people.* Grove Press.

Gosarevski, S., Hughes, H., & Windybank, S. (2019). Is Papua New Guinea viable? *Political Economic Bulletin 19.* https://www.semanticscholar.org/paper/Is-Papua-New-Guinea-viable-Gosarevski-Hughes/6d71c70951c12fa82dea578a36b66923997f3f92. Accessed June 1, 2022.

Gulyaeva, M., Huettmann, F., Shestopalov, A., Okamatsu, M., Matsuno, K., Chu, D. -H., Sakoda, Y., Glushchenko, A., Milton, E., & Bortz, E. (2020). Data mining and model-predicting a global

disease reservoir for low-pathogenic Avian Influenza (AI) in the wider pacific rim using big data sets. *Scientific Reports, 10*, 1681. https://doi.org/10.1038/s41598-020-73664-2
Golub, A. (2014). *Leviathans at the gold mine: Creating indigenous and corporate actors in Papua New Guinea*. Duke University Press.
Huettmann, F. (Ed.). (2015). *Central American biodiversity: Conservation, ecology, and a sustainable future* (p. 805). Springer.
Huettmann, F., Riehl, T., & Meissner, K. (2016). Paradise lost already? A naturalist interpretation of the pelagic avian and marine mammal detection database of the IceAGE cruise off Iceland and Faroe Islands in fall 2011. *Environment, Systems and Decisions*. https://doi.org/10.1007/s10669-015-9583-0
Huettmann, F. (2020). Chapter 24—Investigating Matschie's tree kangaroos with 'modern' methods: Digital workflows, big data project infrastructure, and mandated approaches for a holistic conservation governance. In D. L. P. Valentine, J. Blessington, & K. Schwartzet (Eds.), *Tree kangaroos: Science and conservation*. Academic (pp. 379–391).
Huettmann, F., & Hueffer, K. (2021). The ecological niche of reported rabies cases in Canada is similar to Alaska. *Zoonoses Public Health, 68*, 677–683. https://doi.org/10.1111/zph.12835
Kirsch, S. (2014). *Mining Capitalism*. University of California Press.
Kuehr, R., & Williams, E. (Eds.) (2007). *Computers and the environment: Understanding and managing their impacts* (Vol. 14). Springer Science & Business Media.
Kulik, D. (2019). *A death in the rainforest: How a language and a way of life came to an end in Papua New Guinea*. Algonquin Books. 9781616209049
Laman, T., & Scholes, E. (2012). *Birds of paradise: Revealing the world's most extraordinary birds*. National Geographic Books.
Lasslett, K. (2012). State crime by proxy: Australia and the Bougainville conflict. *British Journal of Criminology, 52*, 705–723.
Martin, R. (Ed.). (2005). *Tree-kangaroos of Australia and New Guinea*. CSIRO Publishing.
Mercer, D., Christesen, L., & Buxton, M. (2007). Squandering the future—Climate change, policy failure and the water crisis in Australia. *Futures, 39*(2–3), 272–287.
Narokobi, B. (1975). *Foundations for Nationhood*. Papua New Guinea Press and Bookshop.
Narokobi, B. (1983). The *Melanesian way*. Institute of Papua New Guinea Studies.
O'Connor, B., Bojinski, S., Röösli, C., & Schaepman, M. E. (2020). Monitoring global changes in biodiversity and climate essential as ecological crisis intensifies. *Ecological Informatics, 55*, 101033.
O'Donoghue, T. (2009). Colonialism, education and social change in the British Empire: The cases of Australia, Papua New Guinea and Ireland. *Paedagogica Historica, 45*(6), 787–800.
Ostrom, E. (1990). *Governing the commons: The evolution of institutions for collective action*. Cambridge University Press.
Ostrom, E., Schroeder, L., & Wynne, S. (1993). *Institutional incentives and sustainable development: Infrastructure policies in perspective*. Westview Press.
Papoutsaki, E., & Rooney, R. (2004). Who is research for: An observation from Papua New Guinea. In *Conference paper, UNECO Forum Colloquium on Higher Education, Research and Knowledge*, Paris.
Rena, R. (2011). Challenges for quality primary education in Papua New Guinea—A case study. *Education Research International, 3*, 11–15.
Rogers, R. F. (2011). *Destiny's landfall: A history of Guam*. University of Hawaii Press.
Salak, K. (2001). *Four corners: One woman solo's journey into the heart of Papua New Guinea*. Counterpoint Publisher.
Schloenhardt, A. (2006). Drugs, sex and guns: Organised crime in the South Pacific. In *Regionalising international criminal law in the pacific* (pp. 159–184).
Smith, H. (1974). Internal conflict in an independent Papua New Guinea: Problems of Australian involvement. *Australian Journal of International Affairs, 28*(2), 160–167.
Stiglitz, J. (2003). *Globalization and its discontent*. Norton Paperback.

Sullivan, N. et al. (2003). *Tinpis Maror: A social impact study of the proposed RD Tuna Cannery at Vidar Wharf, Madang*. https://www.academia.edu/4492602/Tinpis_Maror_A_social_impact_study_of_the_proposed_RD_Tuna_Cannery_at_Vidar_Wharf_Madang?email_work_card=title. Accessed July 22, 2022.

Taber, R. D., & Payne, N. F. (2003). *Wildlife, conservation, and human welfare: A United States and Canadian perspective*. Krieger Publishing Company.

The Guardian. (2015). The incredible plan to make money grow on trees, 24th November 2015. https://www.theguardian.com/world/2015/nov/24/redd-papua-new-guinea-money-grow-on-trees. Accessed May 28, 2022.

The Secret Barrister. (2018). *Stories of the law and how it's broken*. Picador Publishers.

Thordarson, T., & Höskuldsson, Á. (2014). *Iceland*. Dunedin Academic Press Ltd.

Veenendaal, W. (2021). How instability creates stability: The survival of democracy in Vanuatu. *Third World Quarterly, 42*(6), 1330–1346.

West, P. (2006). *Conservation is our government now. In Conservation Is Our Government Now*. Duke University Press.

Chapter 8
Fundamental Humanitarian Sustainability Issues in Papua New Guinea: SDGs and Global Top-Downs to Fail Us Further on Biodiversity and Wilderness Conservation with an Indigenous Vision

Abstract Papua New Guinea (PNG) as a culture in that region has been around for over 47,000 years, and it has not harmed really the global community. However, when it comes to relevant modern metrics of progress and society, PNG looks poor. Here some of those metrics are discussed and put into context of earth and human history, and get shown how poor those 'modern' metrics perform. The 'modern' concept of the UN Sustainable Development Goals (SDGs) is confronted with reality and get exposed as a wide failure to global mankind. PNG, with over 80% of the people living in bush communities and in the 'Wantok' tribal system is part of 'Indigenization' and it serves as a sustainability and governance role inspiration for any society really helping global mankind to survive.

Keywords Papua New Guinea (PNG) · International development aid · Humanitarian aid · Urbanization · Health metrics

8.1 Introduction

Compared with western nations, many people will find that Papua New Guinea (PNG) is characterized by rather odd-looking, often very tragic, humanitarian metrics; specifically when investigating the capitol of Port Moresby, many of the rural and mining hubs, or some remote areas in PNG, e.g. in the former Australian claim area. As a colonial nation, slavery is not new to PNG neither and it still exists in its modern-odd forms, e.g. Imbun (2006) for employment practices. However, PMG just abandoned its death penalty, as part of being a 'Christian' nation (https://www.theguardian.com/world/2022/jan/21/papua-new-guinea-repeals-death-penalty-30-years-after-reintroduction), a feat that other nations have not achieved, e.g. the U.S. but where many missionary actions are funded from.

Certainly on the human welfare side many socioeconomic metrics—the civil strife and as 'valued by the west' (Gosarevski, 2019)—can look pretty grim in PNG, e.g. regarding domestic abuse, crime rate, alcoholism and drugs, literacy rates, health and life expectancy (Table 8.1). It's widely accepted and published, e.g. Matthiessen (1987), West (2006), Salak (2001). However, many aspects of those problems and

F. Huettmann, *Globalization and Papua New Guinea: Ancient Wilderness, Paradise, Introduced Terror and Hell*, https://doi.org/10.1007/978-3-031-20262-9_8

Table 8.1 Performance of Papua New Guinea in regards to some socio-economic metrics when compared globally (source WWW; these metrics are relatively robust over the last 3 years and across sources. The actual source matters less and that's why those public metrics serve well the argument presented here)

Metric	PNG	Globally	Comment
Average live expectancy	~ 64 years	~ 73 years	Note any gender differences, and the archeological record for Melanesia showing people's age well into the 70s
Average literacy rate	App 66%	App 86%	Literacy here means the western alphabet, which is an artefact and not always globally shared even
Crime rate (Homicide)	9.75 deaths per 100 k	6.60 deaths per 100 k	PNG stands better than Nicaragua for instance
GDP (2020)	23$billion	84.71$trillion	Very incomplete metric for nations like PNG where female labour plays a main role, and other currencies are used like betel nut or bartering
GDP per capita	$2636	$10,925	See above, also note the lack of registered citizens making a good computation of the base widely impossible
Estimated proportion of population living in modern slavery[a]	10.28/1000	NA	Humanity metrics, arguably unacceptably high
Government Response Rating[a]	C	NA	Hinting towards governance failure

[a]*Source* https://www.globalslaveryindex.org/2018/data/country-data/papua-new-guinea/

sufferings are man-made and western-made, if not directly caused by Australia and its mining industry and with international aid services let loose onto PNG (examples of effort shown in Lasslett, 2012; Kirsch, 2014; Chan, 2016). A watchdog is needed to expose and halt the bad practices.

A good example for this is the refugee crisis rolling out in PNG territory, and that one is not new, nor a PNG problem only. For instance in 1988, thousands of Irian Jayan people crossed into the PNG province of East Awin, West Sepik. They escaped from Indonesia soldiers and then staying in PNG camps for their safety, and subsequently the UN had to engage and trying to resolve. The artificial border between PNG and Indonesia, essentially a straight line across same habitats and

Fig. 8.1 Village life in Papua New Guinea, cradle of a sustainable society and life for over 47,000 years; is this harmony as good as it can get?

tribes, creates refugees in the own land, so to speak. It's a border and line that was not approved by the local tribes and simply imposed from remote Europe onto them.

Many natural disasters are also adding to the tragedies. On January 21, 1951, a violent eruption in Higaturu, Popondetta killed over 2900 people; one of the greatest natural disasters in the recorded history of PNG. International efforts between Japan, Australia rebuild schools, roads and infrastructure with Prince Charles from the UK attending briefly the celebrations and re-openings. Those are always good PR stunts but hardly address the issues longer term. PNG remains widely in its state as before, and that is wanted that way.

And then there are also widely mis-labeled 'rural problems' with young 'rascals' getting pushed out of the villages (e.g. described for impacts in the capital city by Salak, 2001). Instead, these are colonial problems with the village structures and PNG fabrics breaking down as caused by western colonial nation cultures, so-called development aid, globalization and industrialized education. It's a feature seen anywhere where industry hits a landscape, and a nation (see Diamond, 2011a, 2011b; Hayan, 1990 for a PNG example). Those PNG metrics are not really a 'PNG cause' because tribal society actually has a record of long-proven family and social values and a certain safety net all embedded in nature (Baraka, 2001). At least it's a system that worked well for over 47,000 years (Figs. 8.1, 8.2, 8.3 and 8.4 for such and similar PNG looks), so why not now, last 50 years?

Fig. 8.2 Fieldworkers around a fire; a wrongly understood camp romance? Not as far as I can judge

Already the occurrence of big cities, of administrative units and of the actual 'nation-state of PNG' in a federal Westminster governance scheme and its international engagement are essentially artifacts that most western nations have not resolved well for themselves (see an example with the UK and its Brexit, or Northern Ireland and the independence of Scotland); the latter nations tend to run huge debts and spurious welfare metrics (see U.S. metrics as one of the richest nations in the world on depression, infant mortalities or lack of a 40h compensated work-week with a decent social welfare scheme, Twenge, 2020; Wilcox et al., 1995; see Grauwe et al., 2017 for the ongoing EU debt crisis). Clearly, the current PNG society is part of a design and mélange created from the outside (see Sullivan et al., 2011 for the EUROTHON zone; a dubious economic PNG-EU partnership centered around PNG thuna 'for the EU').

While PNG's metrics in Table 8.1 appear relatively clear on the surface, those metrics deserve to be looked at more closer. These are not really PNG homebrews. There are clear discrepancies between UN metrics, the Australian Aid perspectives (https://www.dfat.gov.au/geo/papua-new-guinea/development-assistance) and The World Bank views (https://www.worldbank.org/en/country/png) when compared with local realities for PNG. It reads like a *deja vu* from Cammack (2002), but then it unfolds in a PNG landscape with a 47,000 years old civilization in the year 2022. Even more so, the SDGs for the world and for PNG are not aligning well here, and have not done so for a long time (see Rich, 1994 globally, and Gosarevski et al.,

Fig. 8.3 Grass dance at an airstrip, a handed-down tradition and festivity in Papua New Guinea meeting cultures and linking with the universe and part of the wider cosmology! It's one of many rituals and ceremonies found in PNG (see Gillison, 2002 for other examples)

2019 for modern PNG). The SDGs are mismatching and make things worse for most people.

8.2 Human Metrics in PNG: A Quick Selection of Artefacts?

A few more detailed examples are provided on those metrics next:

National income: PNG has one of the lowest national incomes in the world; it thus is ranked 'poor.' This has many implications for the nation and the region. However, what is missing here is the notion of 'subsistence lifestyle' and which fuels the local economy and its markets for livelihood. Easily over 50% of the real GDP and its workforce are not accounted for because people are not registered well, while the industrial model does not include gardening and female family work/labor in a system where the schools are ineffective for that. Considering that in PNG pigs are valued as a bank and betel nut is a currency, equally or more valued than money, the national income is not so good of a metric, nor is the GDP and nor should such

Fig. 8.4 A Sunday afternoon stroll on the rugby field downtown/village

purely western measures really be pushed upon PNG. See Narokobi (1983 for the Melanesian way).

That's because most western nations are essentially bankrupt, certainly not sustainable, and the GDP is a relatively new construct in that scheme and world order which is known to fail for its meaning, certainly for sustainability and for nations such as PNG. 'Cash crops' are not so suitable for PNG really (Baraka, 2001) and are nothing to go by. Already the reasoning and justifications of poverty, and in PNG, remain dubious. CSIRO for instance states for the South Fly district in PNG "*...It is one of the poorest places in the world, due to the minimal opportunities for livelihoods caused by isolation from markets and regulations which limit cross-border trade.*" (https://research.csiro.au/laad/home/our-projects/). Considering that PNG has vast biodiversity and marine resources, one in the best of the world, and it's relative closeness to Australia—quite a rich nation and global middle power—how can such a site be the poorest? Poverty is created here (Bougrine, 2016) and the Fly river region was a tightly run protectorate of Australia for almost a century with active missionary; now who is to blame? How many years does Australia need for its protectorate to turn it into an independent and sustainable nation that is not failing, and not poor?

Literacy rate: PNG's literacy rate is rather low, officially. But in reality, it is difficult to measure in a meaningful way for PNG, as a nation with a rural economy where reading and writing matters less but over 700 languages are spoken! As PNG features those 700 languages most citizens speak at least 4 and can operate in PNG overall. Written language is less relevant for a garden society; that is certainly the case for Pisin English. Expressing literacy in Moto, Pisin English or English creates a certain artifact which is of little value in a 'Wantok' system overall where local issues and bartering dominate.

Domestic violence: Domestic violence is fueled by poverty, and usually by lack of a healthy social structure, including education and values. Modern poverty and violence, in most parts, is created (e.g. Bougrine, 2016). Measured by starvation or happiness, PNG as a megadiversity nation—and with pigs and fisheries—was never poor; natural regrowing resources existed throughout the history of PNG. Conceptually, tropical nations are not, and cannot really be, poor as they are living in such an effluent ecosystem, *'kissed' by the sun*' it's one of the wealthiest areas in the world that way; that must easily be true for the geological riches, and for forestry, gardening and for fisheries. PNG remains a garden provisioning for virtually all human needs; PNG was never any other the last 47,000 years.

While this question remains a complex topic for many people to comprehend (Diamond 2011a, b), the PNG society living in tribal networks has little modern-style poverty; at least it has a stable fabric to go by. Having a dishwasher, or lacking one, is a no real issue in rural villages, same applies to laundry machines, hair dryer or central heating, or even cars and GPS navigation. People of PNG are mostly linked-in, and associated with, tribal village communities and their lives, living in sophisticated adobe-style housings embedded in nature. If it rains, it will dry; move on. In villages, there is a social network either way. The concept of humans economically and socially expelled from the villages and pushed into cities, and subsequent gang crime (rascal'ism), is more of a modern colonial phenomenon. Port Moresby is its dominant outlet for that. People living in tribal and village bands instead have certain rules of treatments and taboos to be sustainable. Group crime was not much allowed, while the 'original' society was 'raw' and usually male-dominated. That can include female and child cruelty in the initial human sense. However, whoever was in such tribal villages know that these have been pretty happy societies and full of issues of humanity; much more than any industrial society I know of.

Gang violence: As stated, the rascals formed in cities due to expelled youngsters from the villages and their communities. It's widely recognized and described as a problem for long times (e.g. Harris, 1990), but not resolved whatsoever.

Arguably, PNG has a modern court and policing problem. Australia frequently comes in trying to fix and resolve the problems—'Aussi' style (Lasslett, 2012; Gorovetsky et al., 2019)—but it widely fails long term (Smith, 1974; the Australian missions always keep coming back to PNG and POM, simply because problems are not solved longtime). Whatever they plan, do and build from the outside is not lasting nor is it sustainable in PNG. Neither the 'modern' actions nor the vision really work, people and the environment pay the price (Henton & Flowers, 2007; Kirsch, 2014 for examples).

And in reality, PNG is there not so much different than the global discussions elsewhere, namely U.S., Brazil, South Africa or Russia. The global economic framework runs *havoc* (Stiglitz, 1993). PNG indeed suffers from an instable and unsuitable legal system and its funding and missing reality feedbacks, anything far from the Australian role model imposed onto PNG (Chan, 2016 for first-hand account). PNG nation-making was a rushed job and one that creates problems along the way while the PNG actully forms (Gorevsky et al., 2019) (Fig. 8.5).

8.3 Complex Lives, Complex Habitats, Urbanization and Being Embedded in the Ring of Terror from Abroad

Like with many other so-called third-world nations, the list of socioeconomic metrics where PNG ranks internationally low is long, and it is a rather complex one indeed. Is PNG just a misfit? But then, ask yourself why PNG is part of the Australian Arc of Instability and Terror (Henton & Flower, 2007; May, 2016) and what fueled it and

Fig. 8.5 This landscape view shows the north coast of Papua New Guinea's Huon Peninsula, down from the mountains to the ocean in the far distance. For most parts it has been virtually occupied by humans for over 47,000 years but remained in a relatively 'untouched' benign state.

Fig. 8.6 Ancient and sustainable flow of things; where does it lead to?

where does much of the expertise come from, using what underlying governance model? (Fig. 8.6).

Urbanization and its business model are clearly on the rise, anywhere (e.g. Davis, 2004); but that speaks strongly against wilderness and PNG's rural lifestyle. Urbanization cannot provide for proper wilderness conservation, regardless of what is claimed and hoped (Sanderson et al., 2018). It's best to further strengthen the rural lifestyle instead, and trying to avoid moving into more slums and decayed suburbs. Going with a blended indigenous approach might be the answer.

References

Baraka, R. (2001). The meanings of independence in post-colonial Papua New Guinea. A Plan B Research Paper Submitted to the Faculty of the Center for Pacific Islands Studies in the Candidacy for the Degree of Masters of Arts. School of Hawaii, Asia and Pacifci Studies. The University of Hawai'i at Manoa.

Bougrine, H. (2016). *The creation of wealth and poverty: Means and ways*. Routledge.

Cammack, P. (2002). Attacking the global poor. *New Left Review, 13*, 125–141.

Chan, J. (2016). *Playing the Game: Life and Politics in Papua New Guinea*. University of Queensland Press.

Davis, M. (2004). Planet of slums. *New Left Review, 26*, 5–11.

Diamond, J. (2011a). *Guns, germs and steel: The fate of human societies.* Random House.
Diamond, J. (2011b). *Collapse: How societies choose to fail or succeed* (Revised ed.). Penguin Books.
Gillison, D. (2002). *New Guinea ceremonies.* New York: Harry N Abrams Inc.
Gosarevski, S., Hughes, H., & Windybank, S. (2019). Is Papua New Guinea viable? *Political Economic Bulletin, 19.* https://www.semanticscholar.org/paper/Is-Papua-New-Guinea-viable-Gosarevski-Hughes/6d71c70951c12fa82dea578a36b66923997f3f92. Accessed June 1, 2022.
De Grauwe, P., Ji, Y., & Steinbach, A. (2017). The EU debt crisis: Testing and revisiting conventional legal doctrine. *International Review of Law and Economics, 51*, 29–37.
Harris, B. (1990). The rascal gangs of Port Moresby. *Criminology Australia, 1*(3), 3–4.
Hayan, D. M. (1990). *Road through the rain forest: Living anthropology in Highland Papua New Guina.* Waveland Press.
Henton, D., & Flower, A. (2007). Mount Kare gold rush: Papua New Guinea, 1988–1994. Published by Mt Kare Gold Rush, Brisbane Australia. ISBN 10: 0646482815, ISBN 13: 9780646482811.
Imbun, B. Y. (2006). Local laborers in Papua New Guinea mining: Attracted or compelled to work? *The Contemporary Pacific*, 315–333.
Kirsch, S. (2014). Mining Capitalism. University of California Press.
Lasslett, K. (2012). State crime by proxy: Australia and the Bougainville conflict. *British Journal of Criminology, 52*, 705–723.
Matthiessen, P. (1987). Under the Mountain Wall: A Chronicle of Two Seasons in Stone Age New Guinea. Penguin Classics.
May, R. J. (2006). Papua New Guinea: Disorderly democracy or dysfunctional state? In *Australia's arc of instability* (pp. 151–170). Springer.
Rich, B. (1994). *Mortgaging the Earth: The World Bank. Environmental Impoverishment and the Crisis of Development.* Earthscan Publishers.
Salak, K. (2001). *Four corners: One woman solo's journey into the heart of Papua New Guinea.* Counterpoint Publisher.
Sanderson, E. W., Walston, J., & Robinson, J. G. (2018). From Bottleneck to Breakthrough: Urbanization and the future of biodiversity conservation. *Bioscience, 68*(6), 412–426. https://doi.org/10.1093/biosci/biy039. Epub 2018 Apr 22. PMID: 29867252; PMCID: PMC5972570.
Smith, H. (1974). Internal conflict in an independent Papua New Guinea: Problems of Australian involvement. *Australian Journal of International Affairs, 28*, 160–167.
Stiglitz, J. (2003). *Globalization and its Discontent.* Norton Paperback.
Sullivan, N., Warkia, R. K., & Lewis, A. (2011). A Social, Economic and Environmental Impact Assessment on the EU-Pacific Economic Partnership Agreement For Eurothon. https://www.academia.edu/4492625/A_Social_Economic_and_Environmental_Impact_Assessment_on_the_EU_Pacific_Economic_Partnership_Agreement_For_Eurothon. Accessed July 2, 2022.
Twenge, J. M. (2020). Increases in depression, self-harm, and suicide among US adolescents after 2012 and links to technology use: Possible mechanisms. *Psychiatric Research and Clinical Practice, 2*(1), 19–25.
West, P. (2006). *Conservation is our government now. In Conservation Is Our Government Now.* Duke University Press.
Wilcox, A., Skjaerven, R., Buekens, P., & Kiely, J. (1995). Birth weight and perinatal mortality: A comparison of the United States and Norway. *Jama, 273*(9), 709–711.

Chapter 9
In Bed with a Big Bad Neighbor for Life: The Middle Power of Australia as a Domestic, Cultural, Political, Material and Environmental Sustainability Problem for Papua New Guinea and Beyond?

The Australian Government has proposed an Enhanced Cooperation Package to place 230 Australian police and 64 legal
justice and economic advisors in Papua New Guinea to restore the rule of law and to make economic growth possible.
(*Gosarevski et al.*, 2019)
"Steering corrupt cash into Australia from PNG: a how-to guide"
Front news article in The Sydney Morning Herald (2015)

Abstract The 'Sahul' is essentially the ancient area of Australia and Papua New Guinea (PNG) combined, sitting on the Australian 'craton' and connected by the Torres Strait and its islands and people. Australia is the closest and biggest neighbor of PNG; it's also the most influential one. In an arc-like fashion Australia is surrounded by several smaller, poorer and subsequentially weaker nations like Solomon Islands, PNG, Fiji and Vanatu, including political and wider terror conflict zones like East Timor and Bougainville. It's actually of global relevance. These nations and areas have a unique biodiversity and cultural diversity, and often receive direct support and (widely conflicting) advise from Australia for many decades while Australia and the neocolonial set up itself shows poor performance on environmental and many social and economic metrics for a relevant sustainable vision and leadership. Here, I show how sustainability is compromised by this geopolitical set up dominated and lead by Australia, helped by the UK, EU and U.S. and their promotion of self-serving economic growth while robust home-grown solutions exist for millennia but are widely ignored and often destroyed even.

Keywords Papua New Guinea (PNG) · Australia · Australian Aid · Ring of terror · Bougainville · British Dominion · Globalization

F. Huettmann, *Globalization and Papua New Guinea: Ancient Wilderness, Paradise, Introduced Terror and Hell*, https://doi.org/10.1007/978-3-031-20262-9_9

9.1 Introduction

The history and legacy of former '*New Holland,*' now called Australia[1]—as a cultural continent, and neighboring PNG—is well known and documented. Australia is a large but relatively little-populated continent offering a home for parts of European's willing immigrant and (prison) populations who then settled modern industrial Australia—but just after the aboriginees were already present there for over 60,000 years and essentially got chased away and removed (e.g. Barta, 2008; Diamond, 2011a, 2011b; Flannery, 2002). All of this happened with direction and approval of the powers in charge, certainly supported by 'the crown' (Britain; Twormey, 2006). Prison-life and labor was used by 'the crown' once more to extend the British Empire (a common scheme used world-wide; see Belanger, 2018 for mining example). Virgin forest landscapes got cut for farming, and the ancient habitat schemes are widely lost with serious implications to this very day (Lines, 1999, see also Ludlam, 2021 for forestry, mining, indigenous lands and fires, as well as Tang et al., 2021 even for subsequent ocean impacts, blooms; for global context see Huettmann & Young, 2022). It's a regime of abuse and exploitation that changed little over time but now using modern tools like computers, airplanes and policy science.

And while extinctions occurred in the region during the aboriginee time (Beehler & Laman, 2020; Flannery, 2002, and discussion and citations within), those actions of the last 60,000 years in the Sahul equaled by no means the modern global overkill of the last 300 years of 'enlightenment' (Flannery, 2002; Lines, 1999). Much of the Sahul also had virtually no relevant global footprint, or when compared with spending the last 50 years record budgets in research, arts, PR and science-based management but with increased destructions ongoing (Flannery, 2002; Sanderson et al., 2018). During that time of 'enlightenment,' science ministries and entire institutions and international agencies as well as NGOs were established. Some had the specific aim of sustainability and conservation of biodiversity. They advised the respective government apparatus on all relevant aspects, from space exploration to mobile phone communication, computer education, policy, infrastructure and the environment done for best-possible outcome (see for instance Wildlife and Forestry Departments with Aldo Leopold, Meine, 1988; Taber & Payne, 2003). Instead it turned into '*eating the future*' and eliminating the wealth for future generations (e.g. Flannery, 2002; Lines, 1999, Ludlam, 2021). Old-growth forest wilderness and much of the biodiversity simply went away! This western enterprise became more complicated than people thought...

[1] One may usually extend this area and Australian views to include New Zealand. That's because both nations tend to act as a southern block of civilization and NZ often fully supports Australian projects, visions and it participates. But here I will focus on Australia the most. It's understood that both nations align and operate in the region and in Papua New Guinea together having similar goals and aims. It makes for a (eco-political) justification for the wider Sahul region, which includes PNG, as provided by Flannery (2002).

9.2 Australia as a Middle Power

Australia and its governance and democracy clearly struggle (Evans et al., 2016; Ludlam, 2021). It was labeled a middle power (Abbondanza, 2021); it's not so strong and dominant on the global arena but it has a lot of influence in its region and 'the arc' of island nations around it (Ayson, 2007; May, 2006; White & Wainwright, 2004). Also, it must be recognized by a global audience that just now—after a tremendous push by a global public for decades—the Australian government apologized officially to the aboriginees for how they treated them in virtually all relevant aspects of live (= land-grab, schooling, society, housing, economy and integration, including history writing; Barta, 2008, but see for still ongoing controversies Jennett, 2011, Ludlam, 2021, The Guardian, 2022). The public will ask why that is not a genocide, while internal racism continues (Jennett, 2011)? Still, Australia still remains closely aligned with Britain and its policies—culturally, financially and strategically as well as environmentally (Twomey, 2006). While Australia somewhat voted out the Union Jack flag (Fisher & Sonn, 2002 for change of identity), a truly independent and new Australian nation - a true New Holland - and sustainable culture still awaits its coming and development; a long process when wanting all relevant stakeholders on board. At least the environmental, climate and aboriginee flag is not really part of Australia, yet and hardly recognized there for relevance (details in The Guardian, 2022; Ludlam, 2021). And various policy and cultural failures can easily be seen in Australia itself, reaching from historic overcutting (Lines, 1999), Tasmania's old-growth forests and Australia forest management and clear cuts (Ludlam, 2021) and the fire regimes (aka the 'Australian Landscape BBQ'; see for instance Altangerel & Kull, 2013; Scherl, 2005 for protected area management) to the Great Barrier Reef, endless mining and land debates, water mismanagement, climate change denial and the recent bushfires (more details in Ludlum, 2021, see Textbox 9.2). Socio-economic factors to add the most. But further, and probably least known is that Australia also set *havoc* in nations around itself, namely Solomon Islands, Fiji, Papua New Guinea, Vanatu, East Timor, etc. (May, 2006 for PNG within Australia's Arc of Instability). And that is nothing new neither but ongoing (see The Guardian, 2022 for nations of the Australian Arc complaining about insufficient climate actions by Australia). For the Pacific Theatre - an area of concern to the U.S. and globally - Australia acts out and creates instabilities.

After WW1 and WW2 that new nation of Australia—essentially just another continental construct 'down under' shaped by remote royal courts in Europe (Holland and UK) and their advisors and beneficiaries—was given a renewed chance of its lifetime on the international stage: to be the assigned protectorate power for Papua New Guinea through the U.N. And so during WW1 PNG was simply handed over now as a global trust resource to Australia. And after WW2 for Australia this got intensified and Australia became the schooling and training ground for the future leaders of PNG (example shown in Chan, 2016) leading it to a so-called 'independence.' Australia was given that designing role by the wider international community due to a UN mandate (Baraka, 2001) and it was to fully act upon it. As we know now, obviously that happened with direct self-interest (e.g. details in Baraka, 2001 for saving Australian taxpayer's expenses; for more recent reasonings of mining see Guvarevski et al., 2019; Henton & Flower, 2007 and for many examples).

And so, a former European penal colony and last-rescue reserve for immigrants Australia was to take care of what was referred to as so-called '*stone age*' cannibals and primitive people, savages (e.g. Hoffmann, 2015; Levis-Strauss, 1966; Matthiessen, 1987; Mead, 1932, 1967). Can that work out well? The global legacy remains doubtful on that and it asks for a correction, fairness and balanced oversight with a vision free of self-interest.

Over times, Australia has fully taken on, and claimed and then maintained a well-described power position in PNG (Gosareveski et al., 2019). It was often based on political, policing, financial and mining projects (see subsequent chapters), most of which resulted into an acknowledged Australian 'ring of terror' (e.g. Bougainville, Fiji, Torres Islands and others; Henton & Flowers, 2007; Reilly, 2004,). The Manus island migrant problem clearly showed us no other.

Textbox 9.1: A Rather Short Primer on Australian Foreign Policy

Australia is essentially a Dominion and thus part of the British Commonwealth. Similar to Canada, etc. it fully reports to the late Queen of England, who signs most Australian operations. Australia has a role to play for the Pacific Theatre that is it serves a lot for western interests.

Geographically speaking, Australia is couched between northern Antarctic waters and SE Asia. PNG is its closest neighbor, followed by other nations in the Arc, namely East Timor, Solomon Islands, Vanatu, and Fiji. In recent times, Australia has always used those nations as a certain political influence zone, a geograpgical shield and part of a political united block on the international front. It became a foreign battle ground for domestic Australian security and immigration problems.

More is to be said on this, but Australia is seen as a 'western', and thus white, a beacon of culture and civilization in the Pacific. The immigration policies set by Australia show no other. While China was excluded for long time, this situation now has flipped and China gets catered by Australia. Similar can be said for Australia in regards to India. Both nations now also play a bigger role in PNG, Fiji, etc.

What is widely missing here is a good vision of Australia on environmental issues, specifically for the economic growth conflict, human consumption, equality and integration, as well as man-made climate change, and how Australia sees itself within the international community, certainly with the Australian Arc? Australia is more than just global mining, right?

Textbox 9.2: Some domestic Australian environmental problems spilling over the roof: Great Barrier Reef, Tasmanian Old-Growth Forests and Wildfires

Like for most industrial nations, it's fair to say that Australia has a lot of environmental problems. This is caused by the underlying wealth model and structure, but Australia has a clear and unique profile and much better paths can easily be taken. For instance, Australia struggles with climate change. Due to its mining lock-in how Australia is funded it does not accept much of the dominant climate policy, and consequently it has impacts on the landscape, ocean and socio-economics. The recent bush fires are good examples, leaving wide and dramatic impacts behind, on land, at sea and in the atmosphere and also globally (Ludlam, 2021).

The landscape and forestry history of Australia is well described in books like Lines (1999) and Flannery (2002). Already the devastating old-growth forest cutting regime in Tasmania reaches easily world recognition (Ludlam, 2021). It shows some serious governance and direction problems allowing for those overcuts and the greenwashing coming with it.

Another fascinating story of the management specifics of unique Australian assets is provided for the Darling river (now in a massive water crisis, e.g. Garcia, 2019, like many other watersheds in 'down under').

The topic of invasive species in Australia is by now also of global legacy (Hoffmann & Broadhurst, 2016) affecting virtually all of the immigration and trade policies as a life-line for Australia

Australia also suffers from large shorebird declines (Clemens et al., 2016), another international trust asset and as a classic indicator for the status of habitats, wetlands and the nation itself (see Nebel et al., 2008 for water extraction impacts). But one of the most spectacular environmental failure stories likely comes from the Great Barrier Reef. This reef is among the most stunning ones in the world and matters internationally. It certainly links ecologically with PNG. But it gets affected by climate change and certainly by tourism and development, e.g. an approved coal mining plan from India (The Guardian, 2014). The widely overfished Australian Orange Roughy makes a point in time for the natural resource management style in Australian waters; widely overfished and mis-managed.

There actually is a long list of Australian species of serious conservation concern (https://tsx.org.au/tsx/#/) and subsequent failures (Flannery, 2002).

It remains clear that Australia itself has not its eco-assets in good order. It gets widely critiqued by its neighbors even for climate change denial and lack of sufficient efforts in the region (The Guardian, 2022). Presumably that indicates a lack of good and appropriate management and expertise on those issues. It's also easy to conclude that any advice and aid that comes from such a set up cannot result anywhere into sustainability, indigenous or wilderness conservation. The problem cannot be divorced from the solution. And that's precisely what is seen and experienced when Australia engages abroad, and when it engages as a monopoly, a true middle power, such as witnessed for over 100 years with the League of Nations and the UN in PNG and beyond.

To this very day, Australia remains a political power house for the Southeast Pacific championing larger western perspectives, e.g. for the U.S., UK and Europe. The associated role of the IMF and also the G8 should be considered here for sure. This then represents the western civilization, law and order with a block of western money and culture, e.g. linked with London ('The City'), U.S., Singapore, Taiwan (and till recently, Hongkong). But to what good outcome?

9.3 The Middle Power of Australia Leashes Out in Mean Terms

China plays a big and relevant role for Australia either way. For long time, Australia actually treated Chinese immigrants badly, as a minority to be 'Australianized', but that's a policy that now has switched (Jupp, 1995). It's simply because China now dominates a lot of Asia, including Australia and Australian foreign policy and economy (mines), e.g. how Australia approaches its position on Tibet, and Chinese fisheries, Solomon Islands, or mining ethics. Compared with c. 1 billion Chinese the Australian market and population must appear like a dwarf. The GDP of China is far bigger than that of Australia and PNG combined (or that of most nations in the world). The Australian foreign policy does not confront and challenges China much, and thus, PNG is now essentially divided, catering larger sections directly to China. In this context, India should be recognized also, as can be seen for the Great Barrier Reef and its coal mine from India (The Guardian, 2014). And so, Australia and its mining industry also then caters to Indian interests. This makes Australia a formidable neighbor for the much smaller PNG representing Asian interests. And for those who rejected this to be true for PNG, it certainly is obvious for Fiji and Solomon, as part of the Australian arc of influence and policy (May, 2006). It now all comes as a package (see Kirsch, 2014 for 'Mining Melanesia' strategy with direct Australian and Asian interests and promotion).

The foreign policy for any nation usually comes out of interior domestic policies, supposed to be driven by the people, democracy (Stiglitz, 2003). However, embassies stand in the way of it for implementation and often those ones act in their own interest because they are soft-funded 'independent' entities looking for income, and a stable and powerful position. The role of international agreements increases when nations are domestically weak, and thus those international tools set themselves up for a top-down scheme on local issues. For Australia, international policies transferred as part of the British Commonwealth (Twomey, 2006) and its dominions. They have the financial headquarters often in London's "city" (Norfield, 2016) or related but widely unregulated stock markets, including Dubai or Qatar. Like New Zealand 'inherited' remote Pacific islands and thus it has to cater them (The Guardian, 2021a, 2021b for islands not wanting or not able to afford independence), Australia overlooks the case here and was 'given' PNG, and it always has been. It started with the League of Nations and continues with the UN mandate. Without an act of war, Australia directly takes its claims for PNG essentially from the commonwealth membership and such a culture, as determined by the late queen and its agencies and funders. Bougainville and similar cases, usually linked to massive mining projects, show that very clearly (Lasslett, 2012).

But the other stronghold for Australia comes more indirect, from the U.S. In the 'Pacific Theatre,' the U.S. is afraid to lose its power, namely Taiwan is close to get swallowed by China and thus, one capitalistic ally of the scheme is already 'down.' Beyond Guam (and Hawaii) other U.S. strongholds are found in South Korea and Okinawa, Japan. Clearly, PNG plays a major role in this region and consideration, arguably, it's neocolonial. While Australia is an explicit U.S. ally, it's part of the UK block and does not directly host U.S. military bases (see Lockyer et al., 2021 for Australian-supported strategic ports and submarines with PNG as a central agenda item). Already in WW2, the link between Australia and U.S. became clear and was played out—in part—in PNG waters and territory (Cousteau & Richards, 1999 for documentation and legacies). All the western forces were quite united in their fight against Japan (and earlier against the allied Germans, owning till then a land industry with exploitive labor in northern PNG).

Another relevant aspect comes from Australia's geography. Australia as a continent is large, but in reality, most people live in Sydney and Melbourne regions, and all the way up the coast to Brisbane and then on the other side of southwest Australia around Perth. Southern coastal regions dominate. Of course, there are settlements and 'stations' all over Australia, and many other places in the outback desert and in the bushlands. A few smaller settlements are found in Tasmania, but largely the Australian powerhouse is on the east coast, and some cities there are closer and better connected to PNG, e.g. Brisbane, Darwin and Cairns, also Sydney. Australia as an entity is surprisingly small though because the vast desert and dry zones are not much inhabited. Compared to the U.S. or EU, it certainly has no big 'market' or economy (in terms of GDP already the state of California competes with the Australian nation overall). Australia has app. 22 million citizens (over c. 3 times less than Germany); and even the notion of 'Big Australia' (a policy to grandize Australia, its standing, population, subsequent taxbase and eventual power role to 36 million Australians in 2050; https://en.wikipedia.org/wiki/Big_Australia) will not change that much any time soon. With that, just a few municipal centers in Australia call the shots. It's a small 'circus' (as per Ludlam, 2021). Brisbane—very close to PNG—is a natural player in this scenario. It has a large international airport catering PNG and many other cities, nations and their airlines, e.g. in Asia, even linking with Latin America and certainly with the EU. Just like Sydney, Brisbane, as an investment hub and the municipality, plays a role for PNG [see for instance The Sydney Morning Herald, 2015 (video document shown here https://www.youtube.com/watch?v=u-ScQAp-uYs) and The Guardian, 2021b linking business via Brisbane with PNG and Singapore etc.]. Darwin is another city of such a role.

And then there are sectorial policies that come to play. For instance, the Australian farming areas need water, which Australia really is short of (Mercer et al., 2007);

Table 9.1 Selected listing of Australian development projects and aid. The role and engagement in those nations is widely debated and critiqued

Aid project	Citation	Comment
Tibet	There is no direct Tibet aid or support from Australia, but see for instance Australia Tibet Council (https://www.atc.org.au/)	The Autonomous Region of Tibet is part of China and Australia has not interfered much, likely due to mining dependencies with China
PNG	See citations in this chapter	Details in text
Solomon Island	https://www.dfat.gov.au/geo/solomon-islands/development-assistance/development-assistance-in-solomon-islands	Australia as a deep history affecting the Solomon Islands
East Timor-Leste	https://www.dfat.gov.au/geo/timor-leste/development-assistance/development-partnership-with-timor-leste	Australia leads military support and is the largest bilateral donor for East Timor-Leste; it affects directly political realities there
Fiji	https://www.dfat.gov.au/geo/fiji/development-assistance/development-assistance-in-fiji Gounder (2001)	Part of the wider Australian Arc topic, role of China becomes a major item while India plays a role already for many years

colonial water rights assured it til recently Forestry as well as climate change-driven fires have made Australia a less wild place, somewhat in crisis, e.g. due to that water shortage (Ludlam, 2021 for a reality perspective). Still, Australia sells and exports water products to PNG (a nation with the highest rainfall in the world). Water now comes also in different shapes and forms, and companies like Coca Cola operate in PNG from Australia (pers.com.). Still, Australia provides a lot of international development aid and resource assistance to other nations (Table 9.1).

9.4 Australia and Its wider Environmental Problem

But even "just" on the environmental side, Australia remains a conservation bully at home (Australian Conservation Foundation, 2022; Ludlam, 2021) as well abroad (Henton & Flower, 2007; Lasslet, 2012 for PNG and Solomon Islands, see Ludlam, 2021 for East Timor) and it spoils the surroundings. The Chevron-led oil development at Moro, and followed by Exxon LNG and a pipeline, got all reviewed by Australian National University staff and was made fit for Australian terms/laws (but not to the terms of PNG's people, hardly the PNG nation). Another classic ecology example is found with the now occurring algae blooms, e.g. caused by Australian wildfires (Tang et al., 2021); often attributed to a lack of a relevant climate change policy and poor forestry practices). Another one is repeatedly seen with the Great Barrier Reef where Australia allows for massive tourist operations, and foreign coal mines to operate and thus impacts this unique ecosystem legacy (A typical example of an Australian dependency is shown in Fig. 9.1).

Fig. 9.1 Air Nuigini directly links to Brisbane and Australia, as its frequent flight schedule shows. It's the designed hub for travel and business, all else follows from that

The actual human problems, including 'mixing' and 'half-breeds' as an inherent and difficult part of the PNG-Australian relationship get rarely discussed, certainly not resolved (e.g. Wilson, 2019). Racism in Australia is known to occur (Jennet, 2011). Being between 'two cultures' remains the problem of the day, globally.

9.5 A Solution to the Australian Problem?

To overcome problems in the Australian vicinity and its 'arc' (Papua New Guinea, Solomon Islands, Fiji, Torres Islands) can hardly be done without minimizing immigration into Australia—a topic of great self-interest and linked with PNG (Details shown in The Sydney Morning Herald, 2015). To achieve it, that 'assistance' is tried by Australia with helping the PNG court system and sending judges and other legal helpers to PNG, or in emergency policing campaigns sending over a variety of peace officers to stabilize crime hotspots like in Port Moresby or rural areas like Ramu when there are village fights, riots or other conflicts (e.g. an unrest or a mass-outbreak of prisoners). The net effect is to have less immigrants to Australia. That's part of so-called humanitarian missions (Gosirevsky et al., 2019). Now, this might be helpful sometimes and stabilizing. But just think about how you would feel if another nation comes over and does the legal policy and policing for you, providing 'safety on the streets', all the time?

The absurdity of this Australian argument for PNG is seen in the case that the number of immigrants from PNG into Australia are really small. It's simply not the case that PNG people will flood Australia. As a matter of fact, the U.S. for instance has already a very open walking-in policy for any PNG citizen (visa upon arrival). PNG immigration to the west, and having a concern about it, simply is no real issue anyways. PNG immigration to Germany is a minor topic, or for the EU. So why is Australia that much concerned about is own borders? The answer likely is: self-interest, developing a portfolio on the international arena and a generic fear against anything that is foreign and not Australian, avoidance to be overrun from the arc and Australian's self-created terror, and from Asia, loss of control and money. And there is hardly more to such an 'aid'.

Accordingly, the Australian aid comes in many shapes and forms (https://png.embassy.gov.au/pmsb/cooperation.html). Australia remains the location of choice for any training and education for PNG such as police forces, judges, security guards or the health sector and teachers of course. It has been critiqued accordingly (e.g. Shek, 2009 and others for a long list of discussion teams).

Along the same lines, Australia trains future doctors and academicians from PNG, e.g. at the James Cook University and in the wider Australian governance and hospital system (the northern regions are more prone to this due to their affiliation with PNG and Torres Strait). The latter is often linked with Cambridge and Oxford, U.K. to maintain and raise their own prestige, putting all effort once more onto remote pseudo-elite schools in Europe; it's Cecile Rhodes re-visited.

This sounds all great initially, and looks great to the superficial on looker. It helps some people but also makes rich people richer catering a foreign elite once more on the cost of the majority; the trends and metrics show it clearly. So what are the real world impacts of such a policy and for the PNG public and the PNG nation, helped by Australia? Is that helping peace and harmony, and how?

And in the meantime, as common in international aid, Australia is happy to provide aid, in exchange for certain favors, to PNG and other nations (Gosarevski et al., 2019 for role of Solomon Islands for PNG; O'Neill, 1972 for a historical view to start up PNG).

Clearly, the idea of a trickle-down economy failed everybody early on (e.g. Arndt, 1983) and boosting Economic Growth makes things worse; it's certainly not sustainable (Czech & Daly, 2004; Rich, 1994; Stiglitz, 2003). In PNG, economic growth got once more boosted at all costs, and PNG is set up and to deliver accordingly (as shown and critiqued in Gosarevski et al., 2019). Even, and specifically, with Australia being app. the 4th biggest gold producer in the world. The history and legacy for the Australia-PNG relationship shows it for a long time. Australia lead a gold rush in Wau in the 1920s with quite devastating consequences for the social fabric in PNG (despite the narrative stating the opposite); many other mines and aid projects are just the logical consequence (see subsequent chapters). The actual argument to pacify the rural areas and stop them from fighting turned around; see Wilson (2019) for some descriptions.

So what's new, what was learned and what was improved? Just see Bougainville and beyond (see subsequent chapters). It's the Australian culture and approach, the mind set, with a poor track record, in Australia itself (Ludlam, 2021), in PNG and beyond. And that must be greatly exposed, acknowledged and improved.

References

Abbondanza, G. (2021). Australia the 'good international citizen'? *The Limits of a Traditional Middle Power, Australian Journal of International Affairs, 75*, 178–196. https://doi.org/10.1080/10357718.2020.1831436

Altangerel, K., & Kull, C. A. (2013). The prescribed burning debate in Australia: Conflicts and compatibilities. *Journal of Environmental Planning and Management, 56*(1), 103–120.

Arndt, H. W. (1983). The "trickle-down" myth. *Economic Development and Cultural Change, 32*(1), 1–10.

Australian Conservation Foundation. (2022). Aggravating extinction: How the Australian government is greenlighting destruction of the habitat our threatened species need to survive. Canberra, Australia. https://assets.nationbuilder.com/auscon/pages/20116/attachments/original/1647489840/Aggravating_extinction.pdf?1647489840

Ayson, R. (2007). The 'arc of instability' and Australia's strategic policy. *Australian Journal of International Affairs, 61*(2), 215–231. https://doi.org/10.1080/10357710701358360

Baraka, R. (2001). The meanings of independence in Post-Colonial Papua New Guinea. A Plan B Research Paper Submitted to the Faculty of the Center For Pacific Islands Studies in the Candidacy for the Degree of Masters of Arts. School of Hawaii, Asia and Pacific Studies. The University of Hawai'i at Manoa.

Barta, T. (2008). Sorry, and not sorry, in Australia: How the apology to the stolen generations buried a history of genocide. *Journal of Genocide Research, 10*(2), 201–214. https://doi.org/10.1080/14623520802065438

Beehler, B., & Lamann, T. (2020). *New Guinea: Nature and Culture of Earth's Grandest Island*. Princeton University Press.

Belanger. (2018). *Extraction empire: Undermining the systems, states, and scales of Canada's Global Resource Empire, 2017–1217*. MIT Press.

Cousteau, J. M., & Richards M. (1999). *Cousteau's Papua New Guinea Journey*. Henry Abraham's Books.

Chan, J. (2016). *Playing the Game: Life and Politics in Papua New Guinea*. University of Queensland Press.

Clemens, R., Rogers, D. I., Hansen, B. D., Gosbell, K., Minton, C. D., Straw, P., & Fuller, R. A. (2016). Continental-scale decreases in shorebird populations in Australia. *Emu-Austral Ornithology, 116*(2), 119–135.

Czech, B., & Daly, H. E. (2004). In my opinion: The steady state economy—what it is, entails, and connotes. *Wildlife Society Bulletin, 32*(2), 598–605.

Diamond, J. (2011). *Guns, germs and steel: The fate of human societies*. Random House.

Diamond, J. (2011). *Collapse: How societies choose to fail or succeed* (revised). Penguin Books.

Evans, M., Stoker, G., & Halupka, M. (2016). A decade of democratic decline: How australians understand and imagine their democracy. In *From Abbott to Turnbull: A New Direction? Australian Commonwealth Administration 2013–2016* (pp. 23–41). Echo Books.

Fisher, A. T., & Sonn, C. C. (2002). Psychological sense of community in Australia and the challenges of change. *Journal of Community Psychology, 30*(6), 597–609.

Flannery, T. (2002). *The future eaters: An ecological history of the Australasian lands and people*. Grove Press.

Garcia, E. (2019). Rivers in crisis: Water theft and corruption in the Darling River system. *Green Left Weekly, 1207*, 10–11.

Gosarevski, S., Hughes, H., & Windybank, S. (2019). Is Papua New Guinea Viable? *Political Economic Bulletin,19*. https://www.semanticscholar.org/paper/Is-Papua-New-Guinea-viable-Gosarevski-Hughes/6d71c70951c12fa82dea578a36b66923997f3f92. Accessed June 1, 2022.

Gounder, R. (2001). Aid-growth nexus: Empirical evidence from Fiji. *Applied Economics, 33*(8), 1009–1019.

Henton, D., & Flower, A. (2007). Mount Kare Gold Rush: Papua New Guinea, 1988–1994. Published by Mt Kare Gold Rush, Brisbane Australia. ISBN 10: 0646482815, ISBN 13: 9780646482811.

Hoffmann, C. (2015). *Savage Harvest: A Tale of Cannibals, Colonialism, and Michael Rockefeller's Tragic Quest*. William Morrow Paperbacks; Reprint edition.
Hoffmann, B. D., & Broadhurst, L. M. (2016). The economic cost of managing invasive species in Australia. *NeoBiota, 31*, 1–18.
Huettmann, F., & Young, B. D. (2022). The so-called modern 'sustainable forestry' destroys wilderness, old-growth forest landscapes and ecological services worldwide: A short first-hand review and global narrative on the use of 'growth-and-yield' as a destructive and even impossible goal. In M. Kumar, S. Dhyani, & N. Kalra (Eds.), *Forest dynamics and conservation*. Springer, Singapore. https://doi.org/10.1007/978-981-19-0071-6_3
Jennett, C. (2011). Internal colonialism in Australia. In *First world, first nations: Internal colonialism and indigenous self-determination in Northern Europe and Australia* (pp. 106–129). Sussex Academic Press.
Jupp, J. (1995). From 'white Australia'to 'part of Asia': Recent shifts in Australian immigration policy towards the region. *International Migration Review, 29*(1), 207–228.
Kirsch, S. (2014). *Mining Capitalism*. University of California Press.
Lasslett, K. (2012). State crime by proxy: Australia and the Bougainville conflict. *British Journal of Criminology, 52*, 705–723.
Levi-Strauss, C. (1966). *The savage mind*. University of Chicago Press.
Lines, W. (1999). Taming the Great South Land: A history of the conquest of nature in Australia. University of Georgia Press.
Lockyer, A., Burke, J., Lim, Y. H., & Smith, F. (2021). *Manus Island and the Lombrum Naval Base: Five options for Australia's geostrategic gateway*. Royal Australian Navy.
Ludlam, S. (2021). *Full circle: Power, hope and the return of nature*. Black Inc.
Matthiessen, P. (1987). Under the Mountain Wall: A Chronicle of Two Seasons in Stone Age New Guinea. Penguin Classics.
May, R. J. (2006). Papua New Guinea: Disorderly democracy or dysfunctional state? In Australia's arc of instability (pp. 151–170). Springer.
Mead, M. (1932). An investigation of the thought of primitive children, with special reference to animism. *The Journal of the Royal Anthropological Institute of Great Britain and Ireland, 62*, 173–190.
Mead, M. (1967). The Rights of Primitive Peoples: Papua-New Guinea: A Crucial Instance. *Foreign Affairs, 45*(2), 304–318.
Meine, C. (1988). *Aldo leopold: His life and work*. Univ of Wisconsin Press.
Mercer, D., Christesen, L., & Buxton, M. (2007). Squandering the future—Climate change, policy failure and the water crisis in Australia. *Futures, 39*(2–3), 272–287.
Nebel, S., Porter, J. L., & Kingsford, R. T. (2008). Long-term trends of shorebird populations in eastern Australia and impacts of freshwater extraction. *Biological Conservation, 141*(4), 971–980.
Nelson, H. (1982). *Taim bilong masta, The Australian involvement in Papua New Guinea* Australian Broadcasting Commission, Sydney. ISBN 9780642975669, 0642975663. Google Books https://www.google.com/books/edition/Taim_Bilong_Masta/y2MhAAAAMAAJ?hl=en
Norfield, T. (2016). *The city: London and the global power of finance*. Verso Books.
O'Neill, R. (1972). Australia's future defence relations with Papua New Guinea. *Australian Outlook, 26*, 193–203.
Reilly. (2004). State functioning and state failure in the South Pacific. *Australian Journal of International Affairs, 58*(4), 479–493. https://doi.org/10.1080/1035771042000304742
Rich, B. (1994). *Mortgaging the Earth: The World Bank, Environmental Impoverishment and the Crisis of Development*. Earthscan Publishers.
Sanderson, E. W., Walston, J., & Robinson, J. G. (2018). From bottleneck to breakthrough: Urbanization and the future of biodiversity conservation. *Bioscience, 68*(6), 412–426. https://doi.org/10.1093/biosci/biy039. Epub 2018 Apr 22. PMID: 29867252; PMCID: PMC5972570.
Scherl, T. (2005). *At the crossroads: A comparison of current social, scientific and political influences on fire management in Australia and the USA*. World Forest Institute.

Shek, E. (2009). Australian aid to Papua New Guinea: Where is this aid really going? *Cross-Sections, 5*, 66–69.

Stiglitz, J. (2003). *Globalization and its Discontent*. Norton Paperback.

Taber, R. D., & Payne, N. F. (2003). *Wildlife, conservation, and human welfare: A United States and Canadian perspective*. Krieger Publishing Company.

Tang, W., Llort, J., Weis, J., et al. (2021). Widespread phytoplankton blooms triggered by 2019–2020 Australian wildfires. *Nature, 597*, 370–375. https://doi.org/10.1038/s41586-021-03805-8

The Guardian. (2014). The Great Barrier Reef and the coal mine that could kill it (1st August 2014). https://www.theguardian.com/environment/2014/aug/01/-sp-great-barrier-reef-and-coal-mine-could-kill-it. Accessed July 12, 2021.

The Guardian. (2021a). 'Is it to satisfy you or to satisfy us?' Why New Zealand's Pacific colony doesn't want independence. 29th October 2021a. https://www.theguardian.com/world/2021a/oct/30/is-it-to-satisfy-you-or-to-satisfy-us-why-new-zealands-pacific-colony-doesnt-want-independence. Accessed July 12, 2020.

The Guardian. (2021b). 'Too good to be true': The deal with an Isis-linked Australian family that betrayed PNG's most marginalized. 12th Feb 2021b. https://www.theguardian.com/world/2021b/feb/13/too-good-to-be-true-the-deal-with-an-isis-linked-australian-family-that-betrayed-pngs-most-marginalised. Accessed June 2022.

The Guardian. (2022). 'Far from adequate': Former Pacific leaders group urges Australia to increase 43% emissions cut. 7th July 2022. https://www.theguardian.com/world/2022/jul/08/far-from-adequate-former-pacific-leaders-group-urges-australia-to-increase-43-emissions-cut. Accessed May 15, 2022.

The Sydney Morning Herald. (2015). Steering corrupt cash into Australia from PNG: a how-to guide. 24th June 2015. https://www.smh.com.au/national/steering-corrupt-cash-into-australia-from-png-a-howto-guide-20150623-ghv1sx.html. Accessed July 1, 2022

Twomey, A. (2006). *The chameleon crown*. Federation Press.

White, H., & Wainwright, E. (2004). *Strengthening our neighbour: Australia and the future of PNG*. Australian Strategic Policy Institute.

Wilson, M. (2019). *A true child of Papua New Guinea: Memoir of a life in two worlds*. MacFarland Publishers.

Chapter 10
Betel Nut, Coconuts/Copra, Chocolate, Strawberries, Coffee, Apples, Spam and Fish in Papua New Guinea: From Ancient Farming, Highly Bred Species and Sustainability Concepts over Diseases and DNA into Global Market Repercussions and Wholesale (Environmental) Bankruptcy

Thanks to the work of women only a very few Papua New Guineans go hungry
Gosarevski et al. (2015, p. 144)

Abstract Papua New Guinea (PNG) is a tropical nation and blessed with biodiversity as a cradle of farming for many of the world's species deeply enjoyed by humans. It holds one of the first known and documented human agriculture and sustainable products and lifestyles. However, in the modern global agenda and with a sole focus on 'cash crops' PNG has a difficult time to compete and to integrate itself in the wider global market forces of globalization and its production scheme. Using examples of betel nut, coconuts/copra, chocolate, strawberries, coffee, apples, spam, fish and others, I show how PNG copes, what the typical and specific problems are of tropical nations and with the highly commercialized and wrongly copyrighted food commodity market neoliberal in nature harming Mother Earth and food security in a hyped-up boom-and-bust cycle spiraling downwards on finite soil and earth.

Keywords Papua New Guinea (PNG) · Natural resources · Tropical resources · Ancient farming · Food security

10.1 Introduction

Papua New Guinea (PNG) is seen as one of the world's first/oldest centers of agriculture, c. 9000 years old taro fields have been found (Bourke, 2009; see also Beehler & Laman, 2020). Further, it is likely one of the first areas in the world for the cultivation of globally relevant food items such as bananas, sugarcane, and greater yams. The Eastern Highlands of PNG sustained themselves through ancient farming, located in elevated valleys of app 1500–1800 m above sea level, and consequently, they have among the highest human densities in PNG with an associated deep culture and sustainable society.

F. Huettmann, *Globalization and Papua New Guinea: Ancient Wilderness, Paradise, Introduced Terror and Hell*, https://doi.org/10.1007/978-3-031-20262-9_10

In subsequent times, PNG was able to cultivate sweet potatoes and pigs for their great benefit (Diamond 2011a, 2011b) and in a more or less sustainable management (as much as that is possible in wilderness areas). Coastal areas and their marine resource extraction and management play a huge role also, and they connected with the highlands through a landscape-scale bartering network (e.g. Dalzell et al., 1996; Cousteau & Richards, 1999; Golub, 2014). One would therefore assume that PNG has a head start in sustainable food production showing everybody a template to global mankind on food security (Diamond, 2011a). But far from it, PNG is instead sent into a receiving status by the Western World, being essentially told on what and how to produce.

As instructed by its powerful international advisors, 'modern' PNG tries at all costs to join the global commodity market for economic growth (examples provided in Gosarevski et al., 2019; see Rich 1994 for The World Bank policies); PNG then extends its 'portfolio' accordingly with products it can produce (WTO 2000 for a typical overview, for natural resources and its 'supply shocks' see an example of PNG's mother-of-pearl export market as per Simard et al., 2022). PNG has already introduced exotic species to achieve better, e.g. freshwater fish (see Beehler & Laman, 2020 for 21 examples changing widely endemic ecosystems such as the Sepik). It's absurd for a nation as rich, deep and sophisticated as PNG to obtain help on food security, or having to learn from abroad. It was meant as an effort 'to stay afloat' for local PNG bush communities but usually such actions are very problematic and destroyed endemic species, ecosystems and services and sustainable practices putting people into shanty towns (Srijinda, 2019 for the PNG experience). Impacts of such policies can be seen in the Mt. Hagen area with the very fertile Whagi valley of app. 80 km in length heavily catering now (western) cash crop production (see Baraka, 2001 for impacts of cash crops on bartering and traditional PNG lifestyle).

This is a typical strategy pushed onto tropical nations though to reach and cater global and western markets, designed to follow and to make money for the investors, e.g. as advised by The World Bank (see Paris Declaration, Washington Consensus, and Accra Agenda https://www.oecd.org/dac/effectiveness/parisdeclarationandaccraagendaforaction.htm; The Accra Agenda for Action 2008, U.N. 2008 and its neoliberal policies (see Rich, 1994; still ongoing almost unabated). It was meant to be a win–win for everybody, but PNG lost! This is internationally done for many tropical nations in a template fashion, applied globally for decades, [1] and fails widely. The policies by the Food and Agriculture Organization (FAO) show us not other (e.g. International Assessment of Agricultural Knowledge, Science, 2009; Jarosz, 2009; Dutilleul, 2009). However, it also does not work so well for themselves at all, but it does for their investors (see an example in the 'Quinoa hack' Berson, 2014; or Gupta et al., 2019 for cauliflower). And those are often subsidized schemes to start them

[1] This is usually achieved though agronomists. Similar to foresters, or resource economists, those tend to be people that advise farmers on what to cultivate and how. As those experts tend to come from the western world and are trained there, they often are urbanuzed and have a much higher salary than the farmers themselves, which makes it for an odd trust and advisee relationship. Cynical people speak of Agronomy, spelled as 'AgronoMe'. However, the recent global trend to promote for GMOs is extremely problematic while local cultivates are getting lost on a global rate.

off (they actually can hardly work without subsidies due to their inherently unsustainability). For the locals then, it's a loss-loss and worse (=for the resource itself); consequently, the triple whammy hits home again leading to unnecessary decay and destruction. In the tropics and beyond, smal-scale farmers are pushed out and find themselves in a dramatic crisis. Typical tropical products of this portfolio list consist of items demanded by the western world consumers for their convenient lifestyle such as coffee, chocolate, coconuts/copra, betel nut,strawberries, pineapples and fish. Now such basic commodities are often sold and promoted as a niche market product such as a VIP option or 'organic' (see also Sullivan, 2005 for Vanilla as a PNG example). However, none of those resource concepts work out so well and with profits expected for the lower ranks that actually do the bulk of the work; marketing took over instead (Figs. 10.1, 10.2, 10.3, 10.4, 10.5, 10.6, 10.7, 10.8, 10.9, 10.10, 10.11, 10.12, 10.13 and 10.14).

On a so-called open market it hardly works internationally with profit for the sellers neither, unless they are supported and 'insured' (which is where the money sits to make this sad state of affairs seem profitable; another political deal but devaluing the initial labor and small farmer). A typical support for the seller comes through the fuel prize, e.g. relevant for operating tractors, heating and for bringing the harvest to the market (those can be markets abroad and thus involve airplane fuel or bunk oil for ships etc; having a refinery would be ideal. This is part of the classic research

Fig. 10.1 Container ships off Papua New Guinea's coast: shipping away all the fine local raw goods for a global economy as a classic example and indicator of colonialism

Fig. 10.2 The core of an economy: a local marketplace with common products like betel nuts and 'greens' Papua New Guinea runs on those concepts of bartering for millennia. It's almost impossible to account for that in modern economic metrics such as the GDP or taxation

of Odum (1993, also summarized by Madison, 1997 now also heavily discussed for climate change questions in regards to the human carbon footprint.

Another support scheme for or against farming is also political, and it is played via tariffs, or invited labor with support visas (Srijinda, 2019; in the U.S. and in the EU for instance farming is supported by cheap labor, seasonal farm-hands with temporary or no visa and no relevant welfare costs paid). And it does not end there because product certification presents another barrier to enter the market. Whereas, PNG products are often not well certified due to the buy-in cost of that scheme and to feature the label (Beehler & Laman, 2020, p. 341). If PNG would produce for international corporations like Walmart, it would make more money but have additional dramatic impacts on the production resource and the need to produce 'in bulk'. However, PNG is not in that league really to meet all requirements joining the supply chain, and thus dwindles in secondary markets that pay little while PNG is a key farming nation with top world biodiversity and genetic material How can such a reality be defended.

On the international market, many of such market admission 'games' can be applied, e.g. sabotaging supply chains and blackmailing entire brands, and it affects the local farmers and the landscape overall (an example of such impacts can be seen

Fig. 10.3 Betel nut, on a bilum bag for the market; betel nuts are used like coins (see the white lime powder, traded up the coast from reefs)

in the news of 2018 that "Strawberries in Australia carried needles" https://en.wikipedia.org/wiki/2018_Australian_strawberry_contamination; PNG sells strawberries too). It would affect any producer and seller of that product, the regional markets and associated with Australia, even if it's not their fault. The media is of big relevance in such 'product promotions' and to shape brand reputation; consider to include here the social media also).

Overall, other than physically accumulating materials, being highly wasteful and transferring wilderness sites and societies to industrial production sites, and educating their cheap labor force accordingly, the triple-down economy—including farming—has not worked well. Most money stays in the outside of (rural) PNG and the actual PNG resources, and their workers pay the price (typical template and examples provided in ecotourism, e.g. Schellhorn, 2010). The locals pay likely for a long time to come while the resource needs time to recover from the exploitation and trauma, such as seen in the soil degradation, erosion, diseases, invasive species, clean up costs, landscape maintenance and rewildering processes once old-growth forest rainforest is cut down, wilderness is destroyed and soil erosion started as an inherent part of the modern industrial farming concept (see Madison, 1997 for details and for reality see International Assessment of Agricultural Knowledge, Science, 2009).

While pointed out for decades, e.g. Ladurie (1976, as a classic work for farming, taxation, harvest time series and local climate change impacts) the failure of this

Fig. 10.4 Betel nuts offered at a local market; a common sight in Papua New Guinea. One may treat it like a 'bank' and where value is traded. Such currency literally grows on trees in the gardens

scheme might still be surprising to many. But already the profit margins in the 'modern' global supermarkets are actually marginal to start out with (Eisenhauer, 2001), and it forces waste and highly bred crops, usually run by subsidy schemes that cost the consumer more than it makes (e.g. Friedmann, 1993; see International Assessment of Agricultural Knowledge, Science 2009). The local small farmer cannot really compete anymore with his/her own product in such a market. This easily applies to entire landscapes, nations, workforces and their joined trade schemes (e.g. EU zone or NAFTA; Stiglitz, 2003). It's only viable when it gets subsidized from elsewhere, e.g. politically, with military support, through oil and energy, or 'crazy' investment money like hedge funds and national debt trading using politically designed bottlenecks to actual make money at all. That's what much the western world food production runs on and replies to. It's quite mindless and not sustainable, and also running PNG and such nations and their small-scale farmers and associated landscapes easily to the ground (for details see UNEP's Agriculture at the Crossroads Report https://wedocs.unep.org/handle/20.500.11822/8590; Czech, 2000 for scheme and underlying theory; Okello, 2012 for real-world examples).

This is easy to show—as a concept to be true—and when just looking at the energy flow- energy input and energy output—and the costs in such a farming scheme; it's part of laws of thermodynamics and as many classic studies were done

Fig. 10.5 Betel nut bags; each in excess of $50. This little van on the public road easily holds $5,000 of currency and becomes a strategic target for robbery and worse

by Eugene Odum 1960s onwards (see Madison, 1997). An agricultural field can only produce as much as what was invested into it, specifically on soils that have been overused and exploited for decades, if not for centuries (as done in many feudal and colonial regimes). The tropics with their laterite soil provide a classic example for this (Brockett, 1994 as a classic citation on the topic with Central America as the example). Only a loose swidden farming and forestry concept with a nomadic lifestyle in low intensity and few humans can keep it running sustainable over the years; adding coastal products from the ocean reduces pressure. But that is not what industrial food production is about for cash crops.

PNG benefits from geological activity and volcanoes with rich soils, but there is a reason why deserts are on the rise and soil erodes, globally. With human population increase the soil has to provide more to feed the world (but can't, regardless how rich it initially is). It then just becomes a fertilizer bottleneck game, e.g. phosphorus (Vaccari, 2009; with Morocco as a global monopoly, and the nearby Pacific nation state of Nauru now almost exploited for good [public information provided here https://en.wikipedia.org/wiki/Phosphate_mining_in_Banaba_and_Nauru]; all just as the guano fertilizer trade did before and ran afoul a century earlier, e.g. Rosenthal, 2012). To obtain such fertilizers for farming, it heavily relies on fossil

Fig. 10.6 Cacao plantation. I have seen in Papua New Guinea just small operations, namely that has to do with the topography because wide flat areas with the appropriate soil in private ownership are not easily available and not used for cacao, e.g. when compared to West Africa

fuel in the first place to be extracted, shipped and refined cheaply. And where does the fossil fuel now really come from, and which military keeps it open and running?

While it is well known that the global commodity market is very risky and exploitive (Friedman, 1993) with a predictable scheme of failure (see Huettmann, 2015, Fig. 2.11 for patterns and a generic tropical example), nations like PNG still get pushed into it, and one will find shaky and academic advisors and speculative investors to still go that route, all based on non-sustainable practices and subsequent education and finances. Austerity adds to that trend, used as a fear tool. A wide disconnect with the local farmers is frequently found as a consequence. The trusting locals pay the price for it; it's done on their cost. Political advisors are easily found who recommend no other; it's easy for them if they are vested. The Ramu company (see subsequent chapter; it was planned during a game of Golf) is a point in time of such a scheme. In such a setup, PNG and its people have no real chance, considering loans and polices being linked to it to be viable. The consumer pays, so does the landscape and the nation. PNG shows such schemes clearly.

But despite the lack of sustainability one sees here an underlying technological scheme—with strong involvement of the former colonial powers—that gets pushed and that is to overcome the global food gap, or any food uncertainty really. In reality

Fig. 10.7 Cacao fruit

though, this tech-hope, growth number and food certainty is virtually untrue—and it's stressing, even killing people. Many examples and bad outcomes of these techno schemes can be found, e.g. the increasing divide between a few rich and many poor, malnourishment, type II diabetes (=a wrong and off diet boosted by cheap supermarket food) and obesity across all ages (=a trend that is found specifically in low income sections of the society). Technology gets used here in full, e.g. precision farming, but not to make food quality and quantity better, instead it just helps saving costs for some. In PNG, its gets observed that with the advent of cash crops lifestyle diseases are on the rise (Baraka, 2001). In the meantime, small farmers are driven to the edge (see suicide rates in farming now almost worldwide, e.g. Shiva, 2004). In PNG, people then try to make a living in the cities instead, Port Moresby population increases for a reason.

Typical examples for such promoters of techno-farming can be found in the agro-sciences—as the engine of DNA promotion for the life sciences. 'Scientific' journals like NATURE and SCIENCE come peppered with microbiology editors and their reviewers since the 1960s, e.g. provided and trained by Ivy League schools with banking schemes, and who promote all but the same thing: narrow DNA and food industrial growth concepts for higher industrialization levels to feed the world, new tech-generation methods, latest species breeds, precision farming with GPS and

Fig. 10.8 Want some sugar fruit? Those are frequently offered to travelers on the trail and very delicious and juicy; it's kid's food

Remote Sensing, to get more cloned and privatized food done making rich people richer (and leaving the poor in the dark), and done with *ex situ* seeds (seed banks accessed by an elite taken from the public good; e.g. Seed Vault in Svalbard, Norway, Eastwood et al., 2015; see DNA genome projects (‚e.g. by Rockefeller University see YouTube video https://www.youtube.com/watch?v=5PwnQ9IIVr4). Universities in a constant run for funding are not shy of joining and reselling it as 'modern' and 'visionary' setting up future generations with such minds. The latter are widely divorced from wilderness, or farm work realities. The latest cry now is high-precision farming, involving satellites in space and Artificial Intelligence to determine and tell farmers—in-time- where exactly to put the seeds, plants and how much fertilizer so that no investment goes to waste. Farmers become tools, micromanaged by remote tech. It's meant to be efficient but as far from ecology, from PNG (Baraka, 2001) and Mother Earth ("*Demeter*")—or from humanity—as it can get. This profile is obvious for everybody who just looks at it and the global trend (Elkington, 1998). The global poverty gap and the conservation crisis show us no other. In PNG, the Special Agricultural and Business Leases (SABLS) are clearly going that way, helping privatization of a public good promoted from abroad, and thus are very controversial (ATBC, 2011).

While wilderness and natural resources are all on the decline, on the finite landmass the DNA breeding and high tech farming dreams are still on the rise and

Fig. 10.9 Palm tree plantations, e.g. for coconut

prominent in the leading political and governance circles.[2]And the public goes with it as they are not much informed better (e.g. International Assessment of Agricultural Knowledge, Science, 2009). Change for no change; the modernity in its decline while the vast poor mass of people pays for it. It gets pushed even if it goes against fcats and common sense; it and remains suicidal for the landscape, for the people and managers, and for almost everybody else. It's often a lock-in. PNG is on the receiving end once more. This is clearly shown when just looking at the PNG currency, the Kina and its exchange rate (Garnaut & Baxter, 1984). According to classic economic principles the national currency is to reflect the value of the land, its products (Daly & Farly, 2010; see Brockett, 1998), but it does not anymore. Other factors come to play and then overrule the local farming value and the value of labor, products and the soil itself.

[2] For the curious mind, there is a very interesting shism in 'conservative' farming. Considering much of the farming community hails itself as 'conservative' and connected to the ground and nature, promoting hunting, subsistence life and the control of wolves and coyotes, etc. When it comes to their political lobby, they are widely on the other side though, very far progressive aligning with latest commercialization, high-tech tractor machinery, seeds, DNA clones and international fertilizer trade to make ends meet. Very few of those 'conservative' political farming lobbies are truly on the organic or vegetarian side, or truly 'green' (which might carry more the label of a 'hippy commune'). While this is not found directly in PNG it's certainly found in Australia's farmers, as well as with the western agronomists involved in PNG advice. It also affects the farming products for western markets.

Fig. 10.10 Tabacco with paper, makes for a smoke

10.2 A Selected List of PNG Farming Commodities

And there are many clear evidences that show us no other. In the following, I will present some details about a selection of the tropical products that the western nation is so eager to get for their markets (but not all are truly needed—and in that amount and composition—when they would be less convenience-oriented and more subsistence-minded and, local. It would also help to overcome the encroaching civilization diseases, as stated in Baraka, 2001 for PNG):

Betel nut (*Areca catechu*): For Papua New Guinea, this is a 'bread-and-butter' species, similar to a currency. It is more than 'just' a nut. In PNG, a betel nut is actually money ("*bua*"). Green coins that get traded with your close friends, and to appease enemies (I have also seen it used as a bribe and appeasement for the rascals and criminal gangs for not being robbed; widely used by bus drivers crossing dangerous road systems infested with gangs. Either way betel nuts and its palm tree and products keep many parts of the nation going). Those betel nuts are grown 'on trees' and used all over tropical Asia, but PNG remains a hotspot now. Whereas, Australia makes the public use of betel nut more or less illegal. That's mostly because betel nut gets

Fig. 10.11 Cuscus, an animal on the list of bushmeat

chewed and then it creates a foamy red liquid in the mouth that gets spit out on the pavement… Australia is not much compatible with PNG.

The use of betel nut is unhealthy creating cancer and similar diseases of the throat and stomach. It should be stated that betel nut, like several other utilized species in PNG (pigs, dogs, Casuarina trees etc.) arrived just recently in PNG c. 1200 years ago.

A specific place for growing betel nut is the Sepik river area, the coastal road connecting to Lae and Madang. Different varieties of betel nut are found and grown. Betel nut trees can carry diseases which has real-world currency and economic impacts, and for local friends, enemies and peace in the region of PNG.

Coconuts (*Cocos nucifera*)/copra: This is one of the classic export products from Papua New Guinea. It is exotic and it matches nicely the stereotype of the Pacific (e.g. beaches with coconut palms) and what it can deliver; the (northern/wealthy) world market just wants it. And coconuts can be shipped around the world. There was an infamous coconut cult in Germany based in PNG (Bernard, 2017, see also public information on the topic: https://en.wikipedia.org/wiki/August_Engelhardt). Coconut is a pen-ultimate tropical fruit, and thus most people rarely ask where the coconut comes from, how harvested and how shipped. For the German colonial times, it was again Hamburg, as a colonial and hanseatic central european port and for PNG products, where many coconut got shipped to. Coconut are nowadays highly bred and carry many subsequent problems (some details in Huettmann, 2015, Fig. 21). And there is a side product, copra; a dried coconut product. This product comes from the coconut palm and shells producing sisal (ropes) with a globally competitive market.

Fig. 10.12 Watercress; always makes for a fresh salad

This is a product widely used for ropes in shipping, fishing, farming and industrial activities, and it has been traded around the globe for several centuries. As a result, the coconut, copra and sisal markets are fully established and saturated, and it's not a lucrative business much.

Chocolate (*Theobroma cacao*): Arguably, chocolate is the pen-ultimate southern product for the urbanized consumer from the North. The 'moor' shows it no other, branding a classic tropical product (widely politically incorrect by now; Eatough, 2003; Moss & Badenoch, 2009)! However, growing chocolate is far from benign or friendly. Chocolate products not only got accused using child labor (like the case in West Africa where the majority of the production is located; Hosseinzadeh-Bandbafha and Kiehbadroudinezhad (2022). The modern version of chocolate—sweetened—is quite far from being a freak health product, this also true for organic chocolate (which likely is more labor intense even and invokes union issues). Chocolate has a heavy metal contamination aspect to it and the high sugar level added into modern chocolate can easily take on—and take out—any tooth. Dentists will like it (and gave such sweets to kids with great teeth…I remember that as child myself). Chocolate means labor and profit for the farm and industry owner. How is it settled well in PNG? The chocolate plantations I saw in PNG tend to come from small to mid-scale farmers and are often located at the coast (see Figs. 10.6 and 10.7).

Strawberries (*Fragaria sp*): It's a certain wilderness forest floor species and also found in PNG mountain forests. This is a relatively new industrial product from PNG. It's happily consumed all over the world but in PNG it is designed for the Australian and Asian markets (not local PNG). Many 'western' consumers might

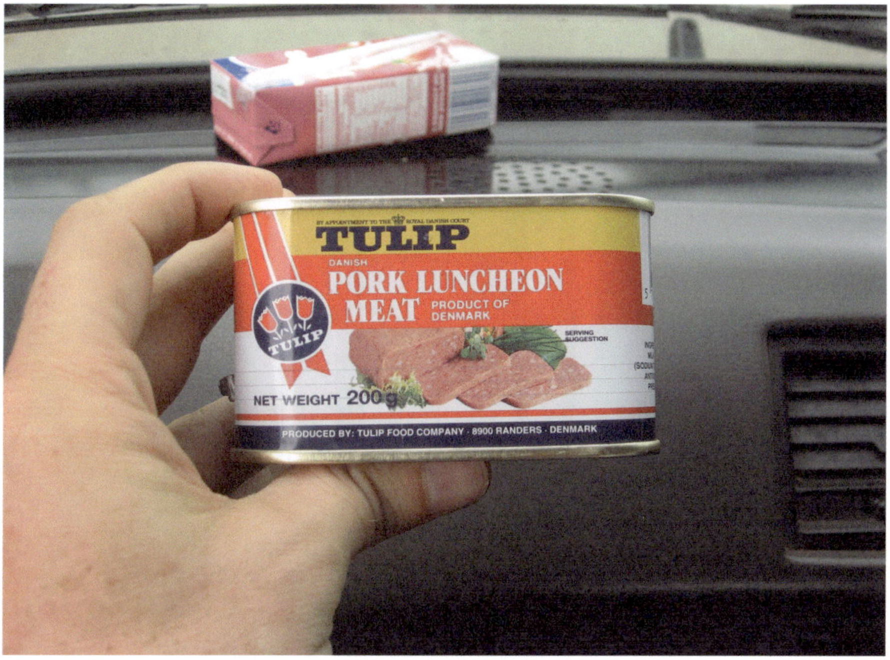

Fig. 10.13 Want some spam? A product offered from Denmark, subsidized overall by the EU

believe strawberries come straight out of somebody's local garden or romantic greenhouse. That can probably somewhat be true, but it's industrial style now. And so while many people might think strawberries come just from your next door neighbor or grandmother-style relative, it actually is a highly industrial, designed luxury product. Somebody has to do the year-round soil and groundwork, fertilize and to pick and clean them. It a ground/earth product that is full of 'dirt' (sand; as the fruit's skin and its leaves are not easily cleanable). Of course in the modern urban centers a strawberry, e.g. dazed into chocolate, must not be missing; if they come from PNG it makes it even more exotic and special, likely perceived as helping the little PNG nation? And strawberries are also supposed to be health food, e.g. for pre-pregnancy.

Now, PNG is initially not so famous for its strawberries, but Mount Hagen became famous for it. That's because it has a good humid mountain climate, soil and a labor base to grow strawberries. Those are meant for export, mostly Australia and Asia, but can be produced with local PNG labor; done with a fraction of the cost compared to where they are consumed. Adding fertilizers, pesticides and insecticides will make it profitable in the tropics; like with many tropical farming products. What a great business model then; a win–win for the easy-believing, mislead and ill-informed urban consumer: treat yourself with tropical strawberries. In my own assessment, it's probably best to stay clear of such industrial strawberry production as it has many flaws and involves international transport and product waste that hardly is

Fig. 10.14 Swine diseases are not a good news for Papua New Guinea and its rural economies

sustainable. If given a choice I would collect a few strawberries myself from the wild forest instead (as I did in PNG and elsewhere, delicious).

Coffee (*Coffea arabica and robusta*): Although originally from Ethopia, the '*Arabica*' and '*Robusta*' coffee bean is famous worldwide. It got exported and grown in many tropical places essentially like an invasive species: Costa Rica, Nicaragua and Columbia. Coffee grows on forest soil and replaces it. And who does not want to have coffee from the tropics, ideally, shade-grown (a scheme that is far from eco-friendly)? Now, PNG coffee is a relatively new addition to market of coffees available and offered to the consumer, while the coffee trade is nothing for beginners and amateurs indeed. It's done and fine-tuned over centuries of colonial business. The crux—and money—in coffee actually sits in the roasting, and then also, in the marketing and product distribution and delivery. In Hamburg/Germany—as a colonial nation for PNG—where a traditional port and associated roasting place is found, including for Papua New Guinea, this business is tightly controlled by just a few traditional families, many of them accused of being too close to politics and during WW2 (see Wierling, 2017 for deep German colonial history and during WW1 and WW2). Families can here easily not only determine the national market, but also the PR and entire supply chain of coffee for Central Europe, including the ships, employees and the growing areas and coffee buyers and villages as laborers! And think of another big coffee market in Seattle, located much closer and more direct to PNG.

The PNG coffee is famous for its remoteness and because it is harvested by local tribes and farmers in PNG. It gets collected into villages, traded through CO-OPS and companies via coffee buyers, and from remote airstrips maintained by machetes flown out to market.

But be aware, coffee is not a local or endemic product of PNG; far from it. PNG just serves the global market using its resources like cheap labor and soil and water that otherwise could be used for other more relevant aspects like food production and a sustainable society, let's say.

The PNG coffee has two variants, *robusta* is traditionally grown in the Sepik area, *arabica* more in the wider Goroka area.

Apples (*Malus pumila*): Papua New Guinea has one of the wettest, and best, tropical environments to grow virtually any food product. If done well, cherries, prunes, pears and apples could easily be among those. Apples comes originally from Central-Asia. However, I see in PNG apples imported from New Zealand for c. 1$ per apple. How can that ever make sense? And who pays for it, considering apples are to be shipped long-distance and to be kept fresh along the journey? What a carbon footprint, work force and salary would that be? I like to state that I see similar schemes elsewhere in the tropics, e.g. Nicaragua selling apples on the market street from Oregon, perhaps produced there with cheap Latino immigrant labor (see Huettmann, 2015 for an example).

Spam (meat): This product is another unique case and thus the reason why it is included here. PNG received pigs in the sixteenth century from the colonial powers, likely earlier, and that event changed the entire dynamics and ecosystems in PNG (Flannery, 1998, 2002; Diamond, 2011a. Pigs have a lot of 'lard' and spam. In the highlands they were a real game changer to the nation and culture of PNG (Flannery, 2002; Diamond, 2011b; Beehler & Laman, 2020). For people in PNG there is a need and demand for 'fat.' But the spam that I refer to here is not produced in PNG; it actually is imported. The spam cans I saw came from…Denmark/EU. How exotic of a product can it get? How on earth can one buy spam in PNG, produced and somewhat supported by the EU; shipped around half the world to feed the local community in PNG's remote bush on a product that got already introduced there over 400 years ago and was well maintained by local PNG citizens themselves? One needs to be a marketing wizard to come up with that idea and to make it viable! I saw a can of spam in PNG supermarkets for c. $3. That's where globalization and its sustainability easily breaks down; it can hardly be exemplified more than with the spam can from Denmark (the colonial links that PNG has with Denmark are marginal, if even that).

Textbox 1: Yes, PNG society and culture eats and wants pig meat for centuries; but industrial pigs are a different beast

Pigs, wild boar, are a currency in Papua New Guinea and highly valued (Diamond, 2011a, Flannery, 2002). But while PNG was populated for over 47,000 years the pigs likely just arrived 2500 years ago, with certainty 400 years ago brought by the colonialists. Pigs changed the ecosystem, namely the ground vegetation and bird life, e.g. ground nesters (Beehler & Latam, 2020). Wild boars and their introduction turn much of the world's soil around, and worse (Barrios-Garcia & Ballari, 2012). In PNG, local wars are fought around those pigs, they are used as a currency and bride price, and the value is in the range of several thousand dollars but come with some payback. Clearly, wild boars are a certain sustainable lifestyle and 'organic' food in the bush. It's a highly prized food item and settles virtually all conflicts in PNG society (see Flannery, 1998 for likely saving his life).

Modern global society is aware of pigs and meat production; it widely centered around it (Diamond, 2011a, 2011b). Industry and society has started to mass-produce such food items early on; certainly the meat aspect of it. That way meat becomes very convenient, stable and available for many people.

However, such a meat supply comes with a huge cost. Virtually all nations that have so much meat available 'in time,' start showing diseases in parallel (Clark et al., 2019). Baraka (2001) shows now 'lifestyle disease' occurrence in PNG. It also results into a loss of appreciation of wild pigs and their habitats: why bother when good meat is readily available in the super market, whatever the source and supplier really?

Similar has happened with domesticated chicken and ducks, e.g. McKenna (2017) using antibiotics. There is a wider discussion whether that wild-domesticated poultry interface has contributed to the rise of zoonotic diseases, a global pandemic and its hotspots (e.g. Gulyaeva et al., 2020 for Avian Influenza).

Like elsewhere in the industrialized world, in PNG, at a few locations pig and chicken are mass produced now. It's a new trend in the otherwise widely dispersed bush production of such meat supply.

Still, already without so much industrial labor, and in a different governance framework, PNG had its food production well worked out on a landscape scale, sustainably and without global footprint. Why discarding that experience and skill?

Fish: 'Fish' is a big term and includes many diverse species; salt and freshwater; often those are confused and mislabeled (Khaksar et al., 2015; see Pazartzi et al., 2019 for PNG sharks marketed in Greece/EU)**.** Now, the fish story in PNG is as bizarre and as absurd as many other tropical fish stories under capitalism, 'cash crops.' Great examples of this failing scheme have been provided by now for Indonesia and Philippines (Pauly, 1979; Pauly et al., 1998), or even Australia (e.g. orange roughy, Bax et al., 2005). A very high number of fish stocks under management and sale are not sustainable. As a tropical ocean nation, there are many fish species that PNG offers, saltwater as well as freshwater and exotic invasive ones (e.g. carp, trout; Beehler & Laman, 2020) and it's difficult to show they are well managed, or reasonably administered or sustainable even. Where are the data to justify those practices? There is virtually no sustainable yield in those tropical coral reef area multispecies fisheries, as per legacy and track record, e.g. Jackson et al. 2013 (See chapters of this book).

When it comes to saltwater fish, albacore and thuna might well lead the pack (see Sea Around Us Project https://www.seaaroundus.org/data/#/eez/598?chart=catch-chart&dimensiontaxon&measure=onnage&limit=10. Thuna gets fished for instance by Japanese and Taiwanese vessels (Cousteau & Richards, 1999), and then ends up in canneries (Sullivan et al., 2003, 2011). The EU tried to setup a EU-PNG Tuna catch zone; EUROTHON to get its quotas. Thuna is a predatory species, the fallacy of a sustainable tuna fishing is well documented worldwide for over a decade but still ongoing to the extreme now (e.g. Safina, 2010, see mislabling as a common practice The Guardian 2021), also for problems in PNG (e.g. Parris, 2010).

Tilapia is a freshwater farmed species; with fish farms run worldwide. Coming initially from Africa, it is widely subsidized by The World Bank and others for (tropical) protein production now done worldwide. However, it comes with 'issues' (Huettmam, 2015 for Central America). Key questions are centered around water quality, diseases, invasives and fish food for instance. In PNG, other aspects come to play like fish theft of the actual fish product, lake drainage sabotage, and even more so, cooling problems of the harvest when it is to be slaughtered and to come to market. The latter is not easily done in remote tropical areas and when the electric system and the roads are unreliable. Spoilage of the product leads to bankruptcy of the entire operation.

Other products: PNG is asked to try consistently to join, explore and offer its natural wealth to the wider international community and markets. Would it not be nice to have a PNG product on your table, and eat or consume it, and for a very low price? Such products are the classic colonial ones like vanilla, rubber or tea, which PNG grows in the Western Highlands (Beehler & Laman, 2020) but is also known for many impacts, e.g. Sulivan (2005). Tea came to PNG in the 1960s, Western Highlands, and much land was modified accordingly. Shrimp farming is a big business in Asia anywhere, now also in PNG.

One may add now 'anything' that sells and makes money, such as aquarium fish (see Militz et al., 2018). But the real value to mankind coming from PNG, such as its ancient taro, yam, sugarcane and banana variants and local food items are not industrially produced yet; perhaps those food items and species are finally recognized for their value soon. But the production limit sits in the land area and soil available. PNG is not really designed to feed the world in bulk.

10.3 International Trading and Instability

Those commodities shown here - and many more exist - are traded internationally. But already every 'world war' (and at least seven of those big war types are currently raging in 2022) can create a major disruption in the markets. In addition, currency crashes and national currency adjustments and inflations add to the gamble.

As PNG is part of the British Commonwealth, the city of London plays a central role for lenders and investors (Norfield, 2016). As per marketing textbook, products and their markets are to be developed to be successful. How can PNG do that, and how is it going for PNG? Well, according to Gosarevski et al. (2019) and Srijinda (2019) it has been not a great success; certainly not for PNG and its people or the resources in an otherwise public land tenure. Setting up private land and products for industrial export production remains an exotic concept to PNG and its people (Baraka, 2001, Foster, 2002). It's an old problem with cash crops where undeveloped land in a public land tenureship gets grabbed, it destroys the soil and wilderness (Baraka, 2001). The PNG conflict about 'Special Agricultureal business leases' (SABLS) shows exactly that (ATBC, 2011; Laurence, 2011; Australian Network News (ABC), 2014; Beehler & Latam, 2020).

So what is the real advantage for PNG to engage in such a global commodity market, when it functioned well and better by itself for over 47,000 years? The only relevant model that PNG followed was the garden, a certain 'grass roots' effort evolved to perfection with an entire tribal and semi-nomadic sustainable income culture around it (see Narokobi, 1983). Who can beat that (see also comparison with Mennonite farming Loewen, 2021), and who wants to destroy it, and who actually can destroy it, considering that tribal society and its big men remain to dominate in PNG? As shown with vanilla (Sullivan, 2005), many of those western farming and business ideas come and go leaving destruction behind.

To me, the poinsettia flower remains the icon of a sustainable garden world that can be followed, within adjustments. It certainly looks nice, leaves no major impacts and provides a happy nation, not? Flower power to the rescue once more.

References

ATBC. (2011). *The urgent need to curtail 'special agricultural and business leases' in Papua New Guinea.* Resolution, Association for Tropical Biology and Conservation.

Australian Network News (ABC). (2014, February 11). PNG's land scandal inquiry names an Australian-led company. https://www.abc.net.au/news/2014-02-12/an-png-land-scandal-inquiry-names-an-australian-led-company/5255528?nw=0&r=Gallery. Accessed May 1, 2022.

Barrios-Garcia, M. N., & Ballari, S. A. (2012). Impact of wild boar (Sus scrofa) in its introduced and native range: A review. *Biological Invasions, 14*(11), 2283–2300.

Baraka, R. (2001). The meanings of independence in Post-Colonial Papua New Guinea. In *A plan B research paper submitted to the faculty of the center for Pacific Islands studies in the candidacy for the degree of Masters of Arts.* School of Hawaii, Asia and Pacific Studies. The University of Hawai'i at Manoa.

Bax, N. J., Tilzey, R., Lyle, J., Wayte, S. E., Kloser, R., & Smith, A. D. M. (2005). Providing management advice for deep-sea fisheries: Lessons learned from Australia's orange roughy fisheries. In *FAO fisheries proceedings* (Vol. 3, pp. 259–272).

Beehler, B. M., & Laman, T. (2020). *New Guinea.* Princeton University Press.

Bernard, Z. (2017, November 30). The curious case of August Engelhardt, leader of a coconut-obsessed cult. Gastro Obscura. https://www.atlasobscura.com/articles/august-engelhardt-coconut-cult. Accessed 25th June 2022.

Berson, J. (2014). The quinoa hack: A critique. *New Left Review, 85*, 117–132.

Brockett, C. D. (1998). *Land, power and poverty: Agrarian transformation and political conflict in Central America (Thematic studies in Latin America)* (2nd ed.). Westview.

Bourke, R. M. (2009). History of agriculture in Papua New Guinea. In *Food and agriculture in Papua New Guinea* (pp. 10–26).

Cousteau, J.M., & Richards, M. (1999). *Cousteau's Papua New Guinea Journey*. Henry Abraham's Books.

Clark, M. A., Springmann, M., Hill, J., & Tilman, D. (2019). Multiple health and environmental impacts of foods. *Proceedings of the National Academy of Sciences, 116*(46), 23357–23362.

Czech, B. (2000). *Shoveling fuel for a runaway train*. University of California Press.

Czech B., Angermeier, P., Daly, H., Pister, P., & Hughes, R. (2000). Fish conservation, sustainable fisheries, and economic growth: No more fish stories. *Fisheries 29*(8), 36–37. https://steadystate.org/wp-content/uploads/Czech_et_al_Fish_Stories.pdf

Daly, H., & Farley, J. (2010). *Ecological economics: Principles and applications* (2nd ed.). Island Press.

Dalzell, P., Adams, T. J. H., & Polunin, N. V. C. (1996). Coastal fisheries in the Pacific Islands. *Oceanography and Marine Biology: An Annual Review, 34*, 395–531.

Dutilleul, F. C. (2009). Failure of negociations on agricultural trade at the WTO, of the FAO Summit and of the Copenhagen climate change Conference in November and December 2009. In *Analysis and "Roadmap" proposition of the Lascaux Program 2009* (pp. Ffhal–00716810f).

Diamond, J. (2011a). *Guns, Germs and Steel: The Fate of Human Societies*. Random House.

Diamond, J. (2011b). *Collapse: How Societies Choose to Fail or Succeed* (Revised Ed.). Penguin Books.

Eastwood, R. J., Cody, S., Westengen, O.T., & Bothmer, R. (2015). Conservation roles of the millennium seed bank and the Svalbard global seed vault. In *Crop wild relatives and climate change* (pp. 173–186).

Eatough, G. (2003). Columbus, cannibals, and chocolate. *The Classical Review, 53*(1), 233–234.

Eisenhauer, E. (2001). In poor health: Supermarket redlining and urban nutrition. *GeoJournal, 53*(2), 125–133.

Elkington, J. (1998). Partnerships from cannibals with forks: The triple bottom line of 21st-century business. *Environmental Quality Management, 8*(1), 37–51.

Flannery, T. (1998). *Throwim way leg: Tree-Kangaroos*. Atlantic Monthly Press.

Flannery, T. (2002). *The future eaters: An ecological history of the Australasian lands and people*. Grove Press.

Friedmann, H. (1993). The political economy of food: A global crisis. *New Left Review, 197*, 29–57.

Foster, R. J. (2002). *Materializing the nation: Commodities, consumption, and media in Papua New Guinea*. Indiana University Press.

Garnaut, R, & Baxter, P. (1984). *Exchange rate and macro-economic policy in independent Papua New Guinea*. Report to the Government of Papua New Guinea, Reprinted as Pacific Research Monograph No. 10, National Centre for Development Studies, The Australian National University, Canberra, Australia.

Golub, A. (2014). *Leviathans at the gold mine: Creating indigenous and corporate actors in Papua New Guinea*. Duke University Press.

Gosarevski, S., Hughes, H., & Windybank, S. (2019). Is Papua New Guinea Viable? Political Economic Bulletin 19. https://www.semanticscholar.org/paper/Is-Papua-New-Guinea-viable-Gosarevski-Hughes/6d71c70951c12fa82dea578a36b66923997f3f92. Accessed 1st June 2022

Gulyaeva, M., Huettmann, F., Shestopalov, A., Okamatsu, M., Matsuno, K., Chu, D.-H., Sakoda, Y., Glushchenko, A., Milton, E., & Bortz, E. (2020). Data mining and model-predicting a global disease reservoir for low-pathogenic Avian Influenza (AI) in the wider pacific rim using big data sets. *Science and Reports, 10*, 1681. https://doi.org/10.1038/s41598-020-73664-2

Gupta, G., Kumar, V., Paruthi, M., & Mendiratta, P. (2019). The cauliflower dilemma! *International Journal of Indian Culture and Business Management, 18*(3), 291–297.

Hosseinzadeh-Bandbafha, H., & Kiehbadroudinezhad, M. (2022). Environmental impacts of chocolate production and consumption. In *Trends in sustainable chocolate production* (pp. 229–258).

Huettmann, F. (2015). If leaders like Alexander Von Humboldt, Ronald Reagan, and Xi Jinping would have been environmental justice radicals using public internet and drones: A short political ecology of Central America with regard to global agenda setting in 2014. In F. Huettmann (Ed.), *Central American biodiversity: Conservation, ecology, and a sustainable future* (pp. 25–76). Springer.

International Assessment of Agricultural Knowledge, Science. (2009). *Agriculture at a crossroads—Global report*. https://wedocs.unep.org/20.500.11822/8590

Jarosz, L. (2009). The political economy of global governance and the world food crisis: the case of the FAO. *Review (Fernand Braudel Center)*, 37–60.

Khaksar, R., Carlson, T., Schaffner, D. W., Ghorashi, M., Best, D., Jandhyala, S., Traverso, J., & Amini, S. (2015). Unmasking seafood mislabeling in US markets: DNA barcoding as a unique technology for food authentication and quality control. *Food Control, 56*, 71–76.

Ladurie, E. L. R. (1976). *The peasants of Languedoc* (Vol. 635, No. 6). University of Illinois Press.

Laurence, B. (2011). "Special Agricultural and Business Leases" imperil forests in Papua New Guinea. *Pacific Conservation Biology, 17*, 297–298. https://doi.org/10.1071/PC110297

Loewen, R. (2021). Mennonite farmers: A global history of place and sustainability. A comparative global history of Mennonites from the ground up. In *Young center books in Anabaptist and Pietist studies*. https://jhupbooks.press.jhu.edu/title/mennonite-farmers?utm_source=newsletter&utm_medium=email&utm_content=Mennonite%20Farmers&utm_campaign=f21_new-nov-books_hnov_promo

Madison, M. G. (1997). 'Potatoes Made of Oil': Eugene and Howard Odum and the origins and limits of American agroecology. In *Environment and history 3:2 ecological visionaries/ecologised visions* (pp. 209–238). White Horse Press.

McKenna, M. (2017). *Big chicken: The incredible story of how antibiotics created modern agriculture and changed the way the world eats*. National Geographic.

Militz, T. A., Kinch, J., & Southgate, P. C. (2018). Aquarium trade supply-chain losses of marine invertebrates originating from Papua New Guinea. *Environmental Management, 61*(4), 661–670. https://doi.org/10.1007/s00267-018-1006-9

Moss, S., Badenoch, A. (2009). *Chocolate: a global history*. Reaktion Books.

Narokobi, B. (1983). *The Melanesian way*. Institute of Papua New Guinea Studies.

Nelson, H. (1982). Taim Bilong Masta, the Australian involvement in Papua New Guinea. Australian Broadcasting Commission, Sydney. ISBN 9780642975669, 0642975663. Google Books https://www.google.com/books/edition/Taim_Bilong_Masta/y2MhAAAAMAAJ?hl=en

Norfield, T. (2016). *The city: London and the global power of finance*. Verso Books.

Odum, E. (1993). *Ecology and our endangered life support systems*. Sinauer Publishers.

Okello, J. J. (2012). Smallholder agriculture: The dilemma. In *FAO/UoNairobi regional workshop on smallholder commercialization held at Norfolk Hotel, Nairobi on June* (pp. 6–7).

Parris, H. (2010). Tuna dreams and tuna realities: Defining the term "maximising economic returns from the tuna fisheries" in six Pacific Island states. *Marine Policy, 34*(1), 105–113.

Pazartzi, T., Siaperopoulou, S., Gubili, C., Maradidou, S., Loukovitis, D., Chatzispyrou, A., Griffiths, A. M., Minos, G., & Imsiridou, A. (2019). High levels of mislabeling in shark meat–Investigating patterns of species utilization with DNA barcoding in Greek retailers. *Food Control, 98*, 179–186.

Pauly, D. (1979). *Theory and management of tropical multispecies stocks: A review, with emphasis on the Southeast Asian demersal fisheries* (No. 1, 35p.). ICLARM Studies and Reviews. International Center for Living Aquatic Resources Management, Manila. ISSN 0115-4389.

Pauly, D., Christensen, V., Dalsgaard, J., Froese, R., & Torres, F., Jr. (1998). Fishing down marine food webs. *Science, 279*, 860–863.

Rich, B. (1994). *Mortgaging the Earth: The World Bank, Environmental Impoverishment and the Crisis of Development*. Earthscan Publishers.

Rosenthal, G. (2012). Life and labor in a seabird colony: Hawaiian guano workers, 1857–70. *Environmental History, 17*(4), 744–782.

Safina, C. (2010). *Song for the blue ocean: encounters along the world's coasts and beneath the seas*. Holt Paperbacks.
Schellhorn, M. (2010). Development for whom? Social justice and the business of ecotourism. *Journal of Sustainable Tourism, 18*(1), 115–135.
Shiva, V. (2004). The suicide economy of corporate globalization. Z Space, pp. 1–3.
Simard, N. S., Militz, T. A., Kinch, J., & Southgate, P. C. (2022). Shocks within a Pacific island fishery: A historic study of events impacting the mother-of-pearl fishery in Papua New Guinea. *Marine Policy, 143*, 105173.
Srijinda, P. (2019). Papua New Guinea: Problems of economic integration at the beginning of the twenty-first century. *Actual Economy: Local Solutions for Global Challenges,* 223–227.
Stiglitz, J. (2003). *Globalization and its Discontent*. Norton Paperback.
Sullivan, N. (2005). *Vanillia's Aftertaste: the social impact of the vanilla boom and bust on the Abelam people of the East Sepik Province, PNG.* https://www.academia.edu/4492623/Vanillias_Aftertaste_the_social_mpact_of_the_vanilla_boom_and_bust_on_the_Abelam_people_of_the_East_Sepik_Province_PNG. Accessed July 3, 2022.
Sullivan, N., et al. (2003). *Tinpis Maror: A social impact study of the proposed RD Tuna Cannery at Vidar Wharf, Madang*. https://www.academia.edu/4492602/Tinpis_Maror_A_social_impact_study_of_the_proposed_RD_Tuna_Cannery_at_Vidar_Wharf_Madang?email_work_card=title. Accessed July 22, 2022.
Sullivan, N., Warkia, R. K., Lewis, A. (2011). A *social, economic and environmental impact assessment on the EU-Pacific economic partnership agreement for Eurothon.* https://www.academia.edu/4492625/A_Social_Economic_and_Environmental_Impact_Assessment_on_the_EU_Pacific_Economic_Partnership_Agreement_For_Eurothon. Accessed July 2, 2022.
The Accra Agenda for Action (AAA). (2008). https://www.oecd.org/dac/effectiveness/45827311.pdf. Accessed May 13, 2022.
The Guardian. (2021, June 22). *Lab analysis of Subway tuna sandwiches fails to identify tuna DNA.* https://www.theguardian.com/us-news/2021/jun/22/subway-sandwiches-tuna-lab-test
United Nations. (2008). *UNCTAD XII Accra Accord and the Accra Declaration.* New York and Geneva. https://unctad.org/system/files/official-document/iaos20082_en.pdf. Accessed May 13, 2022.
Vaccari, D. A. (2009). Phosphorus: A looming crisis. *Scientific American, 300*, 54–59.
Wierling, D. (2017). German history as global history: The case of coffee. In *German history in global and transnational perspective* (pp. 69–90). Palgrave Pivot.
World Trade Organization (WTO). (2000). *Trade Policy Review: Papua New Guinea 1999.*

Chapter 11
Papua New Guinea: Australia's Failed Dream, Some Failed States and the Widely Acknowledged Outer Fascism

If the Carmichael coal mine is a global story, and the Great Barrier Reef a global asset, then the issue should not be left to Australia alone to decide. The citizens of the world deserve a say on whether their children should have the opportunity to see the wonder that is the reef.
*Tim Flannery in The Guardian (*2014*)*
The Bismarck Sea is the jewel in Papua New Guinea's crown, with abundant natural resources including fish, reefs and mineral deposits. There are many opportunities for mining, agriculture, fisheries and tourism.
*CSIRO Australia; the governmental research branch (*https://research.csiro.au/eap/adaptive-livelihoods-development/*)*

Abstract As the closest and biggest neighbor of Papua New Guinea (PNG), Australia steers many details of governance and civil strife in PNG and the region. Helped by colonial powers, the United Nations and globalization, Australia actually designed most of PNG's nation state reality, including social, and environmental ones. In Australia's Arc of Instability (= several poorer and subsequentially weaker nations like Solomon Islands, PNG, Fiji and Vanuatu), it is the Australian governance itself that created much conflict and terror. It was envisioned that PNG can step out of 47,000 years of human history within less than 30 years and just be 'modern' from now on; to save costs for Australia. Here it is shown how such concepts compromised sustainability, namely through the Australian self-promotion of economic growth and other neoliberal metrics with a wider dysfunct patch-worked framework.

Keywords Papua New Guinea (PNG) wilderness · Australian protectorate · Outer fascism · Racism · Foreign affairs · International politics · Globalization · Modernity

11.1 Introduction

For some, Papua New Guinea (PNG)—the megadiversity nation—lists officially as a '*failed nation*' (e.g. May, 2006; Reilly, 2004; see details in Gosarevski et al., 2019). If biodiversity is to be well governed, PNG is apparently not 'it' because

F. Huettmann, *Globalization and Papua New Guinea: Ancient Wilderness, Paradise, Introduced Terror and Hell*, https://doi.org/10.1007/978-3-031-20262-9_11

the wider framework for doing so is missing and operates counterproductive. PNG could hardly unleash its own governance potential, e.g. for sustainability. As a former Australian protectorate from 1902 onward (Hawksley, 2006)—it was to turn into a happy Australian PR showcase, and Australia really had a lot of time to achieve a good outcome for PNG and the region. The biodiversity and wilderness laid bare for over a century for a good sustainable national management…and this resource could actually be used for the PNG nation and nation-building (see Bougainville and its mining projects and as an inherent part of it, with major colonial powers involved and earning also, Baraka, 2001; Golub 2014; for former colonial powers see German role here: http://www.bougainville-copper.eu/comment-kommentar02-3.html).

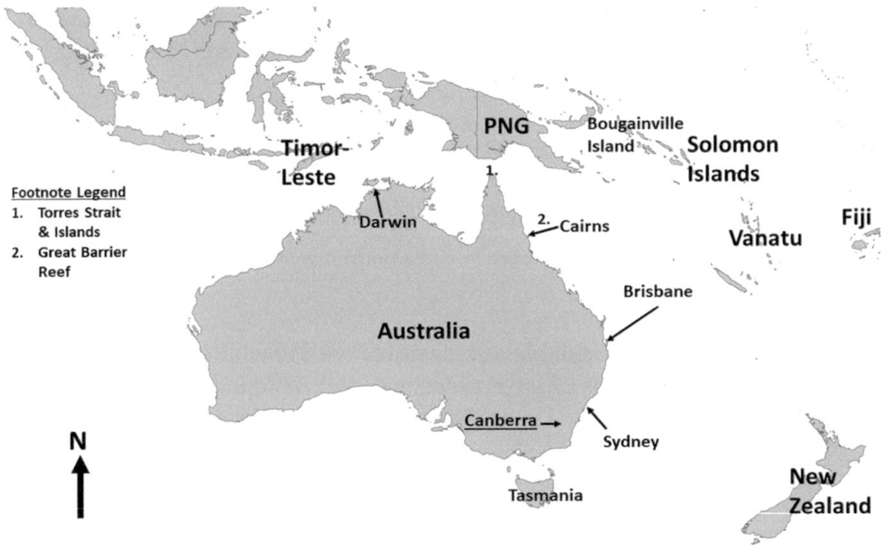

It further mattered because—as shown by Baraka (2001)—PNG got pushed into independence because Australia was not willing to further pay the costs for the UN mandated Protectorate any further, e.g. covered to c. 70% by its tax payers.

Baraka (2001) outlined that Australia viewed its responsibilities to the territory of PNG as a "*civilizing*" project. It's widely communicated that way, certainly in the public Australian narrative (e.g. Nelson 1982). The primary task of colonial administration was making contact and pacifying the people, and introducing them to "Western ways" (Baraka, 2001; this attitude still can be found today; see Gosarevski et al., 2019 for Australian efforts in PNG). The first major Australian colonial patrols reached the Southern Highlands in the mid-1930s, and those efforts were ongoing directly till the 1970s (afterward it switched into a political oversight and development aid style; Baraka, 2001). Roads were built in the southern highlands of PNG by Australians 1960s onward, and it became a dominating force for infrastructure and domination. It fits well in what Hyndman (1998) described that Melanesians resisted the ecocide and genocide (Fig. 11.1).

Fig. 11.1 PNG Day, the celebration of an independent new nation that is not independent nor really new

With the early provision of an Australian education and further training to the PNG leadership (see for instance Michael Somare and specifically PNG Prime Minister Julius Chan, 2016), it was to turn it into a western role-model of success. It was to be simple, a fast process of transition, and to show the world that Australia became a global advisor and power, one that is up to the task on the global cutting edge. A power on the international stage that can be trusted and that can handle successfully complex international problems, turning wicked colonial leftovers from the stone age into modernity; all just within a few years and approved by the U.N. And PNG was just to be one showcase among others in the region, e.g. Timor-Leste that received major aid donations and help in building up its military for independence from Indonesia (a policy that now somewhat was let go or switched; The Guardian, 2019; see Thomson, 2022 for a review of the ethical failures of Australia in Timor and acting as an internationally recognized bully). Modern PNG essentially then is an Australian 'homebrew'—the implementation of a UN mandate through Australian tools—for the international community, all achieved by foreign policy and aid investment, using public money to do so, but driven by just a few 'leaders'; a clique. PNG is to be controlled (Saffu, 1993). That's not modern and not a deviation from a colonial mindset driving PNG-western relations for over 300 year; decolonization was not achieved then (see also Shibuya, 2004).

But like elsewhere in the tropics of this world, and when such protectorate principles are simply applied top-down (e.g. Erler, 1985), they carry an ideology and using commercial tools (Stiglitz, 2003), the reality showed the failure of that effort, how nation-building abroad and how Australian Aid projects and their education inside and outside of Australia have not achieved (see Kolova, 2015 for Bougainville aid by Australia, as an essential problem for the PNG nation; see Fowke, 2006 for more details). PNG is actually not to be controlled much (Saffu, 1993), and it can easily fail as a nation (Gosarevski et al., 2019; May, 1998, Mietzner and Farelly 2013). And the current performance metrics and the legacy are clear on this, e.g. wilderness loss, forest loss, cultural loss and decay, civil strife issues and corruption (Gosarevski et al. 2019; Transparency International, 2005; see The Guardian, 2020, 2022b for real-world examples). It just exemplifies what 'outer fascism' is (Henton & Flower, 2007, p. 290 Solomon Islands; Lasslett, 2012), how it is thought of (example shown here: Woodbury, 2014), how it looks like in 'modern times,' that forests and their biodiversity processes suffer from that, and that it can actually be found in PNG with an Australian footprint.

Australia is not really a world power, it's 'just' a middle power (Abbondanza, 2021), if even that (Ludlam, 2021) and it fully struggles with itself, with its environment and the surrounding neighbors—all of them in deep conflict also but supposed to follow an Australian lead. The struggles can easily be seen in the environmental issues there because those are all deeply linked and connected with economic and social ones. Already on the climate change aspect of it, Australia's neighbors are far from happy with the Australian performance, or seen well cared for (e.g. The Guardian, 2022c). While Australia essentially does not accept human causes of climate change, and does not much to change it, the neighbors are starting to drown due to sea level rise, while there is a refugee crisis with an Australian footprint also.

This raises some relevant questions, for instance, how does the PNG lead and vision, Australia, actually fare itself with its own domestic policies (see also other chapters of this book)? Already Table 11.1 exposes Australia as a widely imperfect nation itself, certainly for the environment; so how on earth should it lead others then, and why? It is here where the UN mandate misleads, but this is also not so surprising when knowing the UN struggles and failures itself (see for instance Mace et al., 2010; Ziegler, 2013 for missed biodiversity targets; Crowley, 2007 for failures on man-made climate change; many more can be named, e.g. Ruanda or Irian Jaya).

Simply looking at some socio-economic and environmental metrics, it becomes quickly clear that Australia does not really have a great success story to report on, or to export to others, e.g. Abbondanza (2021). Australia's legacy is not a shining example on environmental aspects and wilderness biodiversity conservation. Already from the ecological side, Paul Ehrlich from the U.S. has frequently exposed this issue clearly (Fraenkel, 2016; see also https://newsroom.unsw.edu.au/news/science-tech/pioneering-ecologist-paul-ehrlich-present-unsw-sydney-sixth-mass-extinction). And according to the Australian Conservation Foundation (2022) mining accounted for app. 72% of the total habitat destruction approved under the national environment law; this trend is increasing.

Such crisis state is also true for many domestic problems but reaches out wider from there, e.g. use of PNG's Manus Island, for refugees and migrants to

Table 11.1 A selection of Australian performance metrics on the environment affecting the wider region also

Item	Impact	Citation	Comment
Great Barrier Reef (GBR) decay	Coral reef bleaching, physical decay, ocean acidification	Veron (2008)	Most coral reefs in the world might be affected. Papua New Guinea has equal or higher diversity than GBR, the GBR being a global asset
Tasmania old-growth forest decline	Unsustainable cutting of old-growth forest	Ludlam (2021)	PNG has one of the largest intact tracts of tropical old-growth forests in the world left
Foreigner integration and handling	Integration issues with outsiders	Miller (2000), Aquino (2017), The Boston Review (2022)	While now a global problem, the Australian approach has been critiqued for many years
Aboriginee handling	Integration, nation-building, indigenous people	see citations in the book	A global example of western failure
Aboriginal history and legacy	Integration and self-understanding of the Australian nation	e.g. The Guardian (2022a)	Still unsatisfactory efforts critiqued by the world's indigenous communities
Climate change denial	Global warming	Ludlam (2021)	A major impact unresolved and denied spilling across borders
Darling river drying	Water management	Garcia (2019), Kerr (2019)	A wider and generic problem of water treatment
Heatwaves and bushfires	Climate change	Ludlam (2021)	This handling of an environmental tragedy reached a global attention
Overfishing	Orange roughy		Part of a global scheme
Forest loss and modification	Landscape change	Lines (1999)	Part of the colonial legacy
Endangered species loss	Conservation	Australian Conservation Foundation (2022), Kearney et al. (2022)	Consequence of a neoliberal policy applied internationally affecting the world's (endemic) biodiversity and wilderness
Journalism	A free society, inquisitive journalism	Ludlam (2021), Tanter (2020)	As seen with Wikileaks, Australia does not support well its own journalists and citizens abroad etc.

Australia (Fraenkel, 2016, Australia uses PNG islands and hotels for unwanted international immigrants[1]), for the Arc of Terror and for Australian areas abroad, including its Exclusive Economic Zone (EEZ) and remote islands (Macquarie Island, Antarctic territory; contaminated with invasive species, etc.; Bergstrom et al., 2009). And so the problems are certainly true for Australian Aid (Heinecke et al., 2008), and Australian handlings and involvement in climate change, handling of indigenous people, warfare such as Iraq and Afghanistan, many Australian mining operations, and for much of the environment (see Ludlam, 2021). It was New Zealand's media—not Australia—that exposed the 18 missing PNG fisheries observers (Radio New Zealand, 2018; presumably that points toward something serious on that topic). On those essential instances—the environment and beyond—the great Australian nation is not really a showcase to be exposed, and rolled out, to others, certainly not for PNG nor internationally. Other than the vast space and the 'down-under' mystique, Australia and its governance construct leaves much to be desired for the world's sustainability (e.g. Hamilton, 1998; Whiley et al., 2019; see Ludlam, 2021 for generic review; Lien, 1999; Flannery, 2002 for track record examples). Heim (2017) provides a wider Pacific-wide perspective on this showing island dynamics and reasonings.

Arguably, Australia is not alone in this poor performance on relevant issues including the environment and sustainability, but those issues come embedded in its chosen and wider Western Governance model; it's the source of its very origin. One will agree that capitalism plays a role in it, and so does neoliberalism and economic growth policies (Daly & Farley, 2010) pursued and viciously defended by Australia and its foreign interventions all over the world (including Africa). Considering the large western society crisis ongoing (Buell, 2004; Czech, 2002; see Cockburn, 2013 for U.S.), Australia cannot, and unlikely will, fix those problems all by itself (Flannery, 2002; The Guardian, 2022c). One may easily conclude then that such a strong Australian engagement abroad is not recommended much, likely outright dangerous to many people. It certainly is to PNG (see Busilacchi et al., 2018 as an example for Torres Island). Just throwing industrial money and governance at problems, and bound by a return of investment, is not helping, it's not aiding. Metrics speak here for themselves for decades.

Textbox: Use of PNG's Island of Manus to resolve Australia's immigration problem

While accepted practice now, Australia started early to detain immigrants to Australia on non-Australian ground (Fraenkel, 2016). It's meant to be a Processing Center (public information provided here: https://en.wikipedia.org/wiki/Manus_Regional_Processing_Centre). This would make the entire proceedings much easier as people can simply not be allowed to enter Australian legislation, and thus, it avoids elongated court cases and return procedures. It was meant to be cheaper and run smooth when all done abroad. Anyways, it did not run smooth and Australia, as a continent of immigrants itself, got a black eye (see a

[1] "*The roots of this outrage stretch back through twenty years of cruel and increasingly preposterous anti-immigrant policy in Australia*" (https://bostonreview.net/articles/the-humanitarian-disgrace-of-australias-immigration-regime/).

public news headline; https://bostonreview.net/articles/the-humanitarian-disgrace-of-australias-immigration-regime/). Eventually, after several court cases, Australia had to abandon the entire procedure and its camp. Refugees went into cheap chain hotels instead or get 'processed' elsewhere.

The initial idea that it is to be an elegant way of handling the global refugee crisis, looking good and professional about it in the public eye, and contract it all out abroad, went bust once more.

11.2 Australia's Environmental Problem Export

But while some people perceive Australia as the beacon of civilization in the region—if not even in the world (Greasley et al., 2017)—many might perceive Australia rather as a brute but failed governance scheme from terror-like colonial times, now widely overruled by Asia, sidelining the UN + its values, and intervened and fine-tuned by the U.K. and U.S. for their purposes with the EU watching and somewhat engaging from the remote in times of globalization (but having no relevant power to truly engage). Australia handles an accused genocide at home (The Guardian, 2022a). Australia has obviously a very hard time with climate change, and with its virgin forest maintenance, wilderness management, fire regimes, ocean resources, sustainable mining and a good water management (Table 11.1; Ludlam, 2021). So what comes recommended abroad?

Of course, much of Australia itself does not see it that way (Ludlam, 2021). And so, with large environmental and policy decay ongoing, climate change in wide denial, Australia continues in this race to the bottom, toying with refugees, but along this path tries in parallel to gain influence and get more power in wider and international circles and to impress there, also making money. The recent APEC summit in Port Moresby, PNG but actually run and organized out of Australia (https://www.apec.org/groups/committee-on-trade-and-investment/market-access-group/ntm/papua-new-guinea) is a good example. Or the immediate and strong support of virtually any international warfare by the western world, including the war on terror, speaks to that argument. Australia is always available for tasks of international security as needed, usually in full support of the UK and its royal court, as well as in line with the U.S. But see the outcome!

Modernity in Australia looks great in the heavily designed PR image and online governmental webportals perhaps, but is not achieved much on the ground or in the 'arc,' certainly not for PNG and its wilderness and 'bush' people. Limits are set by the environment, geography, natural resources and its social problems. Australia is finite and remote, and it is widely indeed used up. What good can come from it then, e.g. for PNG? (Fig. 11.2).

Part of a global question - water management - remains key in hot but dry nations like Australia and it shows all what is wrong with the underlying business plan and governance (see Garcia, 2019 for overuse, corruption, theft, etc.). While PNG has one of the highest rainfalls in the world, the water administration problem in Australia shows us no other, e.g. Darling River (Garcia, 2019; Kerr, 2019), as linked with personal favoritism, urbanization, farming, mining and overfishing and climate change. As remote as Australia is it can only be part of the western power clique when it connects via money and the WWW, and physically, by airplanes, cargo and ships (the Australian fleet of long distance planes helped significantly to shape modern Australia and its politics). These items by themselves are an inherent part of global contamination, invasive species and climate change affecting 'the arc' as most travel to PNG is through air and from Australia (steered and designed that way).

All of those mentioned Australian issues touch on internal domestic Australia problems embedded in the wider global fabric and geography. It gets more serious though when those ones get exported, e.g. as a culture and action items to neighbors like PNG (Gosarevski et al., 2019). Australia already exports economic growth schemes all over the southern Pacific and has done so proudly without deeper reflection for over a century (public record seen here: https://en.wikipedia.org/wiki/Foreign_relations_of_Australia). Australia is also very quite on exposing its own economic growth problems but blames other factors (see shorebird example in Studds et al., 2017 for promoted outside factors).

In the wider Australian region, the promotion of economic growth (= the overconsumption of goods and habitats, e.g. from abroad) is usually done through advisors, policing and aid (Gosarevski et al., 2019; Table 11.2). The nation of Australia

Fig. 11.2 Care Center for Manam Islanders on mainland Papua New Guinea; note the German colonial name; also co-funded by the EU and Australia. Manam island is a coastal site affected by vulcanoe outbreaks and with some subsequent local disputes!

Table 11.2 Selection of Australian spill-overs into PNG

Project topic	Category	Citation	Comment
Policing	Military	Gosarevski et al. (2019)	Highly strategic, long-term efforts failed
Education of leaders	Political	Gosarevski et al. (2019)	Highly political
Education	Humanitarian	Gosarevski et al. (2019)	Biased due to western views
Court support	Civil	Gosarevski et al. (2019)	Not a long-term solution

and its money is PNG's largest source of Official Development Assistance (https://www.dfat.gov.au/geo/papua-new-guinea). But what is less recognized is that those mind sets and efforts come with a certain self-interest and a certain ignorance of others, long-term. International aid tends to benefit the aid-donor nation the most (e.g. app. 130% return in German and EU-based aid projects; compare also with Erler, 1985). Arguably, the world will not turn due to Australian efforts any time soon; nor will PNG then. And there sits the long-term dilemma.

Australia operates in self-interest indeed and that has aspects of racism in its very policy and daily actions (e.g. The Guardian, 2022c; see also Forrest & Dunn, 2007; Forrest et al., 2016; https://time.com/6176970/australia-election-china/); it is founded upon it (Twomey, 2006), cannot control it well internally (Jennett, 2011) and relays it to the region with a certain wider approval in the 'Pacific Theater' (Adamson et al., 2018). It so has indeed featured an outer racism (Henton & Flower, 2007, p. 292 Solomon Islands; Lasslett, 2012). Australia actively puts other nations and people into misery. Those details exist and are factual, certainly for Australia's arc (May, 2006). A 'ring of terror' shows us no other (Henton & Flower, 2007).

11.3 How Does that Link Directly to Papua New Guinea?

Papua New Guinea (PNG) is the closest real neighbor of Australia. The large Australian culture does spill over and is fully institutionalized (Gosarevski et al., 2019). Other efforts from Australia can easily be found in PNG e.g. media (AusAID, 2004) or direct impact of certain municipalities like Brisbane, Darwin or Sydney (see also Lutton, 1981 for the PNG national university library—a cultural strife, deep source and pride item in any nation—moved from Port Moresby to Perth). According to the Australian Embassy online, at any time, there are over 10,000 citizens from PNG in Australia and Australians in PNG. Modern society in PNG and the wider south Pacific region is affected by Australia and how it runs 'business'; the culture spills over (see, for instance, Blair, 2013 for drug smuggling) and is virtually unavoidable. This has been critiqued for long time already, e.g. former long-term presidents Michael Somare (for the famous 'shoe incident' caused by Australia see here: https://en.wikipedia.org/wiki/Australia%E2%80%93Papua_New_Guinea_relations) and by prime minister Julius Chan (2016).

Already from a forestry perspective—a natural resource of global importance—one can easily see the *laissez-faire* attitude by Australia; it comes as a wider template:

(a) For over 50 years Australia has no relevant policy in place for betterment of the forestry sector in PNG. It's clearly a by-stander and sees the decay.
(b) In PNG over 400 tree species can be used for market (out of 600 up to 3000 species or so). The timber market is rife with mis-identification and mis-labeled species products though, knowingly. In the meantime, Australia oversees, studies, documents and knows most of it—e.g. through the Sydney Museum and James Cook University—but does virtually nothing effective on the matter. That's then Bandura (2007)'s paradigm of selective moral dis-engagement in pure form. It's certainly not a pro-active or an effective management, or a good showcase.

And Australia—the western management of it—is far from being a considerate, multi-national and multi-perspective nation that sees and treats PNG as an acclaimed equal partner, "family" or who wants to learn and apply PNG values back home (see statement below). Such Australian approaches tend to be very destructive for PNG and the region instead. PNG values are frowned upon, and only western values and approaches seem to count. Despite the international trend, we see very little indigenization and decolonization of the Australian PNG-Agenda (see here for details https://en.wikipedia.org/wiki/Foreign_relations_of_Australia#Oceania; compare with Demeulenaere et al., 2021; Narokobi, 1983 for generic Melanesian tree and associated land management values) or for the ongoing Australian Arc of Influence (May, 2006; Wight and Wainwright 2004).

Based on those realities, or official selectve PR messages like below, who has a good faith and trust in Australian efforts abroad, and with PNG?

> *The Australia-PNG partnership is special: we are neighbours, friends and family. Australia is committed to working with the PNG government to continue to deepen our comprehensive bilateral relationship and tackle shared regional challenges.*
>
> Australian prime minister Anthony Albanese, tweet to congratulate Papua New Guinea to the new elections 2022.

References

Abbondanza, G. (2021). Australia the 'good international citizen'? The limits of a traditional middle power. *Australian Journal of International Affairs, 75*, 178–196. https://doi.org/10.1080/10357718.2020.1831436

Adamson, F., Goh, E., Theatre, F., & Xu, R. (2018). Shaping Australia's role in Indo-Pacific security in the next decade. Women in International Security series, speech by the Secretary of the Department of Foreign Affairs and Trade, Australian National University, Canberra.

Aquino, K. (2017). *Racism and resistance among the Filipino diaspora: Everyday anti-racism in Australia.* Routledge.

AusAID. (2004). PNG Media for Development Initiative Draft Design Document, April 2004.

Australian Conservation Foundation. (2022). Aggravating extinction: How the Australian government is greenlighting destruction of the habitat our threatened species need to survive. Canberra, Australia https://assets.nationbuilder.com/auscon/pages/20116/attachments/original/1647489840/Aggravating_extinction.pdf?1647489840. Accessed August 1, 2022

Bandura, A. (2007). Impeding ecological sustainability through selective moral disengagement. *International Journal of Innovation and Sustainable Development, 2*(1), 8–35.
Blair, S. (2013). Refocusing the Australia-PNG relationship. *Australian Defence Force Journal* (190), 104–111.
Baraka, R. (2001). The meanings of independence in post-colonial Papua New Guinea. A plan B research paper submitted to the Faculty of the Center for Pacific Islands Studies in the Candidacy for the Degree of Masters of Arts. School of Hawaii, Asia And Pacific Studies, The University of Hawai'i at Manoa.
Bergstrom, D. M., Lucieer, A., Kiefer, K., Wasley, J., Belbin, L., Pedersen, T. K., & Chown, S. L. (2009). Indirect effects of invasive species removal devastate World Heritage Island. *Journal of Applied Ecology, 46*(1), 73–81.
Buell, F. (2004). *From apocalypse to way of life: Environmental crisis in the American century*. Routledge.
Busilacchi, S., Butler, J. R., Van Putten, I., Maru, Y., & Posu, J. (2018). Asymmetrical development across transboundary regions: The case of the Torres Strait Treaty Region (Australia and Papua New Guinea). *Sustainability, 10*(11), 4200.
Chan, J. (2016). *Playing the Game: Life and Politics in Papua New Guinea*. University of Queensland Press.
Cockburn, A. (2013). *A Colossal Wreck: A road trip through political scandal*. Verso Publishers.
Crowley, K. (2007). Is Australia faking it? The Kyoto protocol and the greenhouse policy challenge. *Global Environmental Politics, 7*(4), 118–139.
Czech, B. (2002). *Shoveling fuel for a runaway train: Errant economists, shameful spenders, and a plan to stop them all*. University of California Press.
Daly, H., & Farley, J. (2010). *Ecological economics: Principles and applications* (2nd ed.). Island Press.
Demeulenaere, E. S., Yamin-Pasternak, D. H., Rubinstein, A., Lovecraft, L., & Ickert-Bond, S. M. (2021). Indigenous spirituality surrounding *Serianthes* trees in Micronesia: Traditional practice, conservation, and resistance. *Social Compass SAGE Journal*.
Erler, B. (1985). *Tödliche Hilfe: Bericht von meiner letzten Dienstreise in Sachen Entwicklungshilfe*. Dreisam-Verlag Publisher.
Flannery, T. (2002). *The Future Eaters: An Ecological History of the Australasian Lands and People*. Grove Press.
Forrest, J., & Dunn, K. (2007). Constructing racism in Sydney, Australia's largest EthniCity. *Urban Studies, 44*, 699–721.
Forrest, J., Lean, G., & Dunn, K. (2016). Challenging racism through schools: Teacher attitudes to cultural diversity and multicultural education in Sydney, Australia. *Race Ethnicity and Education, 19*, 618–638.
Fowke, J. (2006). Getting it wrong in Papua New Guinea. *Quadrant, 50*(12), 28–31.
Fraenkel, J. (2016). Australia's detention centres on Manus Island and Nauru: An end of constructive pacific engagement? *The Journal of Pacific History, 51*(3), 278–285.
Garcia, E. (2019). Rivers in crisis: Water theft and corruption in the Darling River system. *Green Left Weekly, 1207*, 10–11.
Gosarevski, S., Hughes, H., & Windybank, S. (2019). Is Papua New Guinea viable? *Political Economic Bulletin, 19*. https://www.semanticscholar.org/paper/Is-Papua-New-Guinea-viable-Gosarevski-Hughes/6d71c70951c12fa82dea578a36b66923997f3f92. Accessed June 1, 2022.
Golub, A. (2014). *Leviathans at the gold mine: Creating indigenous and corporate actors in Papua New Guinea*. Duke University Press.
Greasley, D., McLaughlin, E., Hanley, N., & Oxley, L. (2017). Australia: A land of missed opportunities? *Environment and Development Economics, 22*(6), 674–698.
Hamilton, C. (1998). Economic growth and social decline: How our measures of prosperity are taking us down the wrong path. *AQ-Australian Quarterly, 70*(3), 22–30.
Hawksley, C. (2006). Papua New Guinea at thirty: Late decolonisation and the political economy of nation-building. *Third World Quarterly, 27*(1), 161–173.

Heim, O. (2017). Island logic and the decolonization of the Pacific. *Interventions, 19*(7), 914–929.
Heinecke, D., Dollery, B., & Fleming, E. (2008). The Samaritan's dilemma: The effectiveness of Australian foreign aid to Papua New Guinea. *Australian Journal of International Affairs, 62*, 53–71. https://doi.org/10.1080/10357710701842686
Henton, D., & Flower, A. (2007). *Mount Kare Gold Rush: Papua New Guinea, 1988–1994*. Mt Kare Gold Rush. ISBN 10: 0646482815; ISBN 13: 9780646482811.
Hyndman, D. (1998). Melanesian resistance to ecocide and ethnocide: Transnational mining projects and the fourth world on the Island of New Guinea. In H. J. Bodley (Ed.), *Tribal peoples and development and issues: A global overview* (pp. 281–298). Mayfield Publishing Company.
Jennett, C. (2011). Internal colonialism in Australia. In *First world, first nations: Internal colonialism and indigenous self-determination in Northern Europe and Australia* (pp. 106–129). Sussex Academic Press.
Kearney, S. G., Carwardine, J., Reside, A. E., Adams, V. M., Nelson, R., Coggan, A., Spindler, R., & Watson, J. E. (2022). Saving species beyond the protected area fence: Threats must be managed across multiple land tenure types to secure Australia's endangered species. *Conservation Science and Practice, 4*(3), e617. https://doi.org/10.1111/csp2.617
Kerr, S. (2019). Aboriginal rights must be respected if the Murray-Darling is to survive. *Native Title Newsletter, 1*, 12–14.
Kolova, S. (2015). Critique of Australian aid to Bougainville. *Contemporary PNG Studies, 23*, 83–92.
Lasslett, K. (2012). State crime by proxy: Australia and the Bougainville conflict. *British Journal of Criminology, 52*, 705–723.
Lines, W. (1999). *Taming the Great South Land: A history of the conquest of nature in Australia*. University of Georgia Press
Ludlam, S. (2021). *Full Circle: Power, Hope and the Return of Nature*. Black Inc.
Lutton, N. (1981). From Port Moresby to Perth. *Archives & Manuscripts*, 30–38.
Mace, G. M., Cramer, W., Diaz, S., Faith, D. P., Larigauderie, A., Le Prestre, P., & Mooney, H. A. (2010). Biodiversity targets after 2010. *Current Opinion in Environmental Sustainability, 2*(1–2), 3–8.
May, R. J. (2006). Papua New Guinea: Disorderly democracy or dysfunctional state? In *Australia's arc of instability* (pp. 151–170). Springer.
May, R. (1998). From promise to crisis: a political economy of Papua New Guinea. *Governance and reform in the South Pacific* 54–73.
Miller, J. M. (2000). Language use, identity, and social interaction: Migrant students in Australia. *Research on language and social interaction, 33*(1), 69–100.
Mietzner, M., & Farrelly, N. (2013). Mutinies, coups and military interventionism: Papua New Guinea and South-East Asia in comparison. *Australian Journal of International Affairs, 37*, 342–356. https://doi.org/10.1080/10357718.2013.788128
Narokobi, B. (1983). *The Melanesian way*. Institute of Papua New Guinea Studies, Boroko, Papua New Guinea.
Radio New Zealand. (2018). PNG parliament told about fisheries observers who disappear, February 16, 2018. https://www.tunapacific.org/2018/02/16/png-parliament-told-about-fisheries-observers-who-disappear/. Accessed June 1, 2022.
Reilly, B. (2004). State functioning and state failure in the South Pacific. *Australian Journal of International Affairs, 58*(4), 479–493. https://doi.org/10.1080/1035771042000304742
Saffu, Y. (1993). Papua New Guinea in 1993: Toward a more controlled society? *Asian Survey, A Survey of Asia in 1993: Part II (Feb., 1994), 34*, 133–138. https://doi.org/10.2307/264511
Shibuya, E. Y. (2004). *Australia-Papua New Guinea Relations: New Pacific way or neocolonialism?* Asia-Pacific Center for Security Studies Honolulu HI.
Stiglitz, J. (2003). *Globalization and its discontent*. Norton Paperback.
Studds, C. E., Bruce, E. K., Nicholas, J. M., Wilson, H. B., Rogers, D. I., Clemens, R. S., Gosbell, K., et al. (2017). Rapid population decline in migratory shorebirds relying on Yellow Sea tidal mudflats as stopover sites. *Nature Communications, 8*(1), 1–7.
Tanter. (2020). Chap. 2: WikiLeaks, Australia and Empire. https://rest.neptune-prod.its.unimelb.edu.au/server/api/core/bitstreams/d097b1f9-86d5-51d2-ac8e-a4d584cca95f/content. Accessed August 10, 2022.

The Boston Review. (2022). The humanitarian disgrace of Australia's immigration regime, January 18, 2022. https://bostonreview.net/articles/the-humanitarian-disgrace-of-australias-immigration-regime/. Accessed May 20, 2022.

The Guardian. (2014). The Great Barrier Reef and the coal mine that could kill it, August 1, 2014. https://www.theguardian.com/environment/2014/aug/01/-sp-great-barrier-reef-and-coal-mine-could-kill-it. Accessed July 12, 2021.

The Guardian. (2019). Australia's history with East Timor isn't pretty but it must be told truthfully, November 12, 2019. https://www.theguardian.com/australia-news/postcolonial-blog/2019/nov/13/australias-history-with-east-timor-isnt-pretty-but-it-must-be-told-truthfully, August 8, 2022.

The Guardian. (2020). Papua New Guinea police arrest former PM Peter O'Neill over alleged corruption, May 23, 2020. https://www.theguardian.com/world/2020/may/24/papua-new-guinea-police-arrest-former-pm-peter-oneill-over-alleged-corruption. Accessed June 29, 2022.

The Guardian. (2022a). Not calling a massacre a massacre is ridiculous': A model of truth-telling at Myall Creek, May 2, 2022a. https://www.theguardian.com/australia-news/2022a/may/03/not-calling-a-massacre-a-massacre-is-ridiculous-a-model-of-truth-telling-at-myall-creek. Accessed August 1, 2022.

The Guardian. (2022b). PNG prime minister denies connection to suitcase full of cash found as voting starts, July 5, 2022b. https://www.theguardian.com/world/2022b/jul/06/png-prime-minister-denies-connection-to-suitcase-full-of-cash-found-as-voting-starts. Accessed July 1, 2022b.

The Guardian. (2022c). 'Far from adequate': Former Pacific leaders group urges Australia to increase 43% emissions cut, July 7, 2022. https://www.theguardian.com/world/2022/jul/08/far-from-adequate-former-pacific-leaders-group-urges-australia-to-increase-43-emissions-cut. Accessed May 15, 2022.

Thomson, D. (2022). How Australia's ethical failures with Timor-Leste should inform a future shaping strategy for the Indo-Pacific. *The Forge*. https://theforge.defence.gov.au/publications/how-australias-ethical-failures-timor-leste-should-inform-future-shaping-strategy-indo-pacific. Accessed August 8, 2022.

Transparency International. (2005). Global Corruption Report. Transparency International, Berlin. https://www.transparency.org/en/publications/global-corruption-report-2005-corruption-in-construction-and-post-conflict. Accessed July 13, 2022.

Twomey, A. (2006). *The Chameleon Crown*. Federation Press.

Veron, J. E. N. (2008). *A reef in time: The Great Barrier Reef from beginning to end*. Harvard University Press.

Whiley, H., Willis, E., Smith, J., & Ross, K. (2019). Environmental health in Australia: overlooked and underrated. *Journal of Public Health, 41*(3), 470–475.

White, H., & Wainwright, E. (2004). *Strengthening our neighbour: Australia and the future of PNG*. Australian Strategic Policy Institute.

Woodbury, J. (2014). Bougainville: The promised referendum and its implications for Australia. *Australian Defence Force Journal, 194*, 5–13.

Ziegler, J. (2013). *Betting on famine: Why the world still goes hungry*. The New Press.

Chapter 12
Destruction by Mining and Drilling on a Global Scale, with Papua New Guinea Fully Embedded: A Meta-analysis Contribution About the Resource Curse, Global Bankruptcy, Destruction, Terror and Death Favors Mother Earth Concepts

This is why we're here; unobtainium, because this little gray rock sells for 20 million a kilo. That's the only reason. It's what pays for the whole party.
Reference in Carlson (2012) *as per quote in AVATAR movie*
Mining is responsible for app. 72% of approved habitat destruction overall, and 75% of approved habitat destruction where habitat loss is a medium- or high-impact threat.
Australian Conservation Foundation (2022)
Ramu Nickel (NiCo) has inspired a flood of public concern. […] The PNG National Fisheries Authority initially stated "the Ramu Nickel mine project is an unsustainable project, socially, economically and environmentally and cannot be allowed to proceed… mining tailings dumped into Basamuk Bay will gradually create food losses to Papua New Guinea's rich and renewable fisheries resources of the Bismark Sea." […] Scientists from James Cook University, Flinders University, UPNG and the Australian Institute of Marine Science found that the company's plan was "so inadequate that no realistic assessment of the mine's impact can be made", and that the environmental impacts "are significantly greater than the company has indicated, including the contamination of local reef systems and parts of Astrolabe Bay with mine waste."
Sullivan (2015)

Abstract The 'resource curse' is a globally applicable pattern suggesting decay and bankruptcy after major resources were discovered and developed. Papua New Guinea (PNG) was seen by western explorers and colonial powers as a territory that holds much promise for gold and other resources. While it took a while it proved correctly after the Australian exploration and development of such natural resources 1890s onwards, the resource curse came to act on all stakeholders involved in full force. PNG and Australia alike suffer to this very day from massive mining development with various problems while the actual wealth has not really entered PNG society, education, health sector or public infrastructure. Rather the PNG nation is on the edge

F. Huettmann, *Globalization and Papua New Guinea: Ancient Wilderness, Paradise, Introduced Terror and Hell*, https://doi.org/10.1007/978-3-031-20262-9_12

of falling apart due to those mines, e.g. Bougainville. Here, an overview is presented of those major mining projects in PNG (e.g. Ok Tedi, Mt Kare, Lihir, Bougainville, Porgera, Frieda), some of their problems and contaminations with wider implications for PNG, the nation construct, the environment, the people and the economy overall, as promoted by Australia and China. The recent oil and gas developments are also included in this perspective while man-made climate change is widely ignored

Keywords Papua New Guinea (PNG) · Curse of mining · Gold mines · Rare earths

12.1 Introduction

Despite 'Green Mining' being promoted these days (Herrington, 2021), it is known for a long time that mining runs a 'curse,' virtually anywhere (Belanger, 2019). This curse means that despite some wealth the countries, companies and cultures that engage in mining get bankrupt and run a crisis mode for decades, if not centuries ever after. This easily applies when looking at such operations from a global and historical perspective, but also local. One cannot divorce the impacts from mining; they come instead as an ecological whole (see Hyndman, 1998 for ethical impacts in Melanesia). Removing earth on a large scale has nothing sustainable (see Dove et al., 1974). Virtually no culture—no humans—are prepared and can handle the fast wealth, demands, impact and all what comes with it from such a natural resource extraction. Once mining or oil production hits a place, ancient human cultures change, usually go down and have a fight (e.g. Ross, 2001 showing that civil wars are more likely to occur in producing countries; see also in Hyndman, 1998). Typical examples would be found in Iran-Iraq, former Soviet States, or Scotland and UK; see Chand & Levantis, 2000; Eichstaedt 2016; for PNG see for instance Henton & Flower, 2007; Kirsch, 2014).

Table 12.1 Short but quite typical selection of crooks and scandals known from mining projects of the world

Project	Location and Nation	Issue	Citation
Alaska Gold Rush claims	Alaska	Disputed mine ownerships, gold strike claims and land transfer from Felix Pedro and others	Public record https://en.wikipedia.org/wiki/Felix_Pedro Cravez (2017)
Bre-X, M. De Guzman	Malaysia	Tainted gold dust in sample, Canada investment One of the biggest scandals in mining	Freeman (1997)
Rio Tinto	Australia	Harassment, Bullying, Racism	Reuters (2022)
Ok Tedi	Australia and PNG	Misuse of funds	NME (2021)

Table 12.2 Basic list of mines in Papua New Guinea[a]

Name of mine	Mining detail	Company details	Comments
Mount Kare	Gold	Various ownerships over time	Created much civil unrest
Porgera	Gold and silver	Barrick Gold and others	
Misima Island	Gold (placer)	Placer Dome	Now closed
Lihir (New Ireland)	Gold		Among largest in the world
Panguna (Bougainville Island)	Copper	Now owned by PNG state	Created essentially a civil war in PNG

[a]Those mines are part of the wider 'Mining Melanesia' effort, Kirsch (2014); more details found in the text of this chapter

Table 12.3 Basic list of oil, gas and other resource extraction projects in Papua New Guinea

Name of mine	Extraction detail	Company details of major operator	Comments
Chevron Iagifu near Lake Kutubu close to Mendi	Oil drilling	Chevron	See also Cousteau and Richards (1999)
Eddie Creek	Gold and silver 1926 onwards		A very old mine set near Wau
Lihir	Gold mine	Merged with Newcrest	Controversial island site with ocean tailings
Ok Tedi	Gold	Government owned	Fly river pollution; see text of this chapter
Panguna Mine	Copper	Bougainville Copper Ltd	This mine created a civil war-like state for PNG
Porgera	Gold	Barrick Gold	https://www.canadaland.com/podcast/mining-7-barrick-and-the-cruelty-of-gold/
Ramu	Nickel	China Metallurgical Group Corporation	Famous for a dam break and subsequent contamination
Solwara 1	Seafloor Massive Sulfide	Nautilus Minerals	First tried seafloor mining, went bankrupt in Canada
Woodlark	Gold	Geopacific Ltd.	

Table 12.4 Selected list of famous failures of mine sites in Papua New Guinea

PNG mining event	Impact	Citation	Comment
Mt Kare	In-fighting, direct Australian involvement	Henton and Flower (2007)	A smaller mine project gone bad
Ok Tedi	Generic contamination, then tailing dam broke	Boyden et al. (1975) Kyle et al. (1986) Kirsch (2014)	Classics on mining failures
Crater Mountain	In-fighting	Mack (2014) West (2006)	Another classic for infighting in the community even before the production starts
Morobe Goldfields	Miners killed locals 1920s	https://en.wikipedia.org/wiki/Wau,_Papua_New_Guinea	A wider landscape of gold mining and its long-term impacts
Nickel Mine Madang China	Tailing dam broke	Beehler and Laman (2020)	An example for a Chinese mine failure and impacts, and how dealt with
Frieda River mine	Environmental and human right violations	The Guardian (2021)	Widely found narratives of "*The largest*" mine, etc.

Arguably, mining has always attracted crooks, and public trust is lost. This is known to everybody who just looks at it (see Table 12.1 for a selection). Typical examples of that curse can be seen in Latin America, with the Spanish Empire and its chaos and fast fall (ongoing and never recovering; see for instance civil war, Franco regime and recent resignation due to WWF scandal by Juan Carlos; BBC, 2012), or see the state of Congo (past and present; Eichstaedt 2016), with Russia, Mongolia (e.g. Ganguli 2020), Tibet region (e.g. Buckley, 2020) and Australia itself (Australian Conservation Foundation, 2022; Ludlam, 2021). And PNG is here no exception but rather it confirms the rule, e.g. mines at Ok Tedi (copper and gold), Ramu (nickel), Porgera Mine (gold), Mt Kare (gold), Lihir (gold) and Panguna Bougainville (copper and gold) (see Tables 12.2, 12.3 and 12.4). All of them are widely set up, managed and funded with an Australian involvement and with related international entities—usually with a South African-UK link. Now China—partly India—being much in the driver seat for an even more global market (The Guardian, 2021 for a typical set up of China via Australia; see for rare earths Zhanheng, 2011. See The Guardian, 2014 for India's coal mining in the Great Barrier reef region with Australian governance approval). Noteworthy are the direct involvements of former colonial powers for PNG such as Germany/EU in Ok Tedi (Kirsch, 2014) and in Bougainville (http://www.bougainville-copper.eu/comment-kommentar02-3.html). To no big surprise, those mining companies and their owners and funders happen to be those same powers that consume and need much of those rare metals and similar mining products, e.g. used for their batteries, cars and mobile phones, for export products of that industry as well as to maintain party ideologies afloat, e.g. socialism in Germany linked with the Volkswagen VW company which subsequently needs mining and oil to fuel the industry (VW actually has non-democratic Saudi Arabia as a large shareholder and investor). It drives industrialization and such a lifestyle. In the German case, together with France it's the stronghold of the EU itself with a dominating budget power over the 27 member nations and beyond. This is global power, and of course it spills into PNG and its mines and inherent fabric! The origin of this power and demand, the handling-expertise and privilege of PNG access usually links with colonial efforts and head starts. Typical examples for industrial headstarts are described for rubber (Revkin, 1994 for Amazonia; also linked to the car industry and warfare including Nazism, the US and WW2 and the rise of global capitalism).

12.2 Papua New Guinea: "Just Mine It"?

Due to its dynamic and uprising geology, PNG brings economic relevant resources of gold, copper, nickel, oil and natural gas products into easy human reach and offers it cheaply (as per Beehler & Laman, 2020) but that widely undervalues what PNG really has to offer. It certainly affects biodiversity wilderness. In PNG, it all connects: the earth, the live and the universe in one site.

To stay with the 'curse,' the actual gain of mining within PNG is rather little when compared to the money that actually is made elsewhere, e.g. in expectation of a presumed profit. The stockmarkets and investments are hyped up from it (e.g. Kowalewski & Śpiewanowski, 2020). This link of PNG with its resources and subsequent wealth was promoted by Michael Somare, etc., for a long time (e.g. Gosarevski et al., 2019; See also Kraal, 2019 for taxation, inefficiency, assessment and quite poor performance overall). Conflicts follow from there in a typical pattern; see the PNG cases in MacIntyre (2007) and Banks (2008). Already, the needed labor market will be affected by mining (see Baraka, 2001; Imbun, 2006 for forced labor questions, slavery, genderized abuse). Judged by most metrics, it's dubious to state that Australia brought civilization to PNG (e.g. as expressed in Nelson 1980 and 2016 that in rural PNG killings were on the order of the day and what Australia tried to settle and to improve on, including outlawing cannibalism). And then, for environmental, social and economic bankruptcies, the PNG mining enterprise is not short of those, a famous and recent case is the sea floor mining project—Solwara 1—by Nautilus Minerals Inc. (Mining Watch Canada, 2019) headquartered in Toronto/Ontario, Canada. It declared insolvency just a few years after its existence but still created many bad impacts (Mining Watch Canada, 2019) with clean up costs carried by te public (the PNG citizen and new generation to come). Further, the BHP Billiton company exited from the large mine of Ok Tedi (https://www.bhp.com/news/media-centre/releases/2002/02/bhp-billiton-withdraws-from-ok-tedi-copper-mine-and-establishes-development-fund-for-benefit-of-papua-new-guinea), and the Bougainville Panguna mine with Rio Tinto shut down, eventually after a

Fig. 12.1 The deeper spirits are watching

civil war and much other damage (https://news.mongabay.com/2017/04/rio-tinto-walks-away-from-environmental-responsibility-for-bougainvilles-panguna-mine/) (Figs. 12.1, 12.2, 12.3, 12.4 and 12.5).

And it's not just mining, and it's also drilling for oil and gas that runs such a curse; the associated warfare related to religion (Islam) and fueled by an ultra-conservative Saudi Arabia shows us no other (as found in New Guinea which is dominated in the Indonesian part by a Muslim governance, as described in Beehler & Laman, 2020). For a similar type of template-like conflict hotspot, one may easily look at Texas in the U.S.—as one of the global headquarters and think tanks of the oil and gas lobby—and just see its resource curse with refineries, refinery contamination, associated lawsuits and society. Certainly the very own Texas energy grid fails. And may one go that far and relate it to Texas's recent winter energy chaos, the death penalty (one of the highest in the nation and western world overall), new anti-abortion laws and the many political conflicts? And again, PNG is fully a part of the same reality and the same set of countries; it's widely documented in the international public (see The Guardian, 2018 for an example on such works exploiting PNG's economy knowingly); another example is provided by The Guardian (2019) as well as with investigating journalists, politicians and authors (Ludlam 2021).

If it would be true for what was promised to PNG's people, the mines, oil and gas, and their income should easily pay for all relevant costs in PNG, such as schools,

Fig. 12.2 Gas stations are increasingly popping up in PNG reliance on fuel and its industry grows accordingly while nature its people pay again the bills

medicine, pensions and more (M. Somare promised that such a model would be so wealthy it would allow PNG even to also support its neighboring nations; Gosarevski et al., 2019. The latter virtually did not happen, e.g. see Manu Island refugees, Timor-Leste or Irian Jaya). PNG is still in the receiving mode and ranks at the global bottom of the list for most industrial and society performance metrics!

So just see reality, the industry and such PNG governmental branches simply do not pay those relevant bills, and hardly engage (see U.N. Blue Helmet missions without relevant PNG involvement; https://news.un.org/en/story/2010/09/353842-papua-new-guinea-moves-closer-contributing-first-blue-helmets-un-missions). Australia did not set it up that way, an underlying sustainable business model—sharing with a mutual partnership—was not emphasized, and eco-expertise is not what modern Australia pursues or has available (see previous chapters). Instead, the social services are paid by entities aligned with western money, and roads and bridges are poorly funded, if at all (see Hayan, 1990; Beehler & Laman, 2020). PNG's education and health system are in a bad shape. The PNG economy and social system is quite in shambles, so is the generic wealth of the society and its progress in western terms (Gosarevski et al., 2019). What taxes are paid to PNG's people and their direct needs?

It's the typical catch-22 associated directly with mining and the associated resource exploitation set up by 'the west'. That's why it's referred to as a 'curse,' as per historical legacy.

Fig. 12.3 Industrial hubs and around airports carry their usual set of contamination, waste and pollution with them; plastics all over (ocean included) (an image from the remote coastline of eastern PNG)

Fig. 12.4 Beauty and pristine surroundings in the eye of the beholder. Image by Porgera Joint Venture, as per their website http://www.porgerajv.com/Company/Image-Gallerys

Fig. 12.5 Nature's beauty and the pristine sits in the eye of the beholder: Ornamental plants along a trail and watercourse in the wilderness

The infamous Rhodes scholars—the ones that are paid by mining money etc. and support such an economy and who are supposed to lead the Commonwealth, academia and the world (Ndlovu-Gatsheni, 2018)—arguably seem to be at the end here, and their arguments appear rather hollow by now. What does such a system—globally—really lead and contribute to? The UK and the Rhodes scholars themselves are widely absent though from space travel or modernity and its governance for all, by design! And what a new education should this really be, for what and for whom? STEM education (https://ncses.nsf.gov/interest-areas/stem) and its cyberinfrastructure training (Ainsworth et al., 2005; Council, 2007) has not reached PNG yet, nor does it perform well in the first place, is not sustainable at all, or can be achieved in any realistic time frame for most of PNG any time soon!

Classic alluvial mining instead—and potentially done by any public member—remains a widespread and relevant mining 'technique' and income for PNG's reality (see Cousteau and Richards 1999, p. 80 for documentation and photos); they are widely destructive to the river system and to ownership of land and resources. Other methods to mine in PNG consist of placer mines (~wild west style), open pit mines and some underground mines (often with tragic safety standards, and then the recent and ill attempts of sea floor mining. As typically done in mining, all of them are hyped up for money and investment, with great support from the Asian Development Bank, The World Bank (e.g. Fujita 2013 for human rights) and others (see Rich 1994 for generic overview). In modern PNG, the hungry horde is reality.

Instead of industrial styles, for thousands of years the indigenous people of Papua New Guinea have traded stone implements and ochre, and they used clay to make pottery (Source: https://www.sl.nsw.gov.au/stories/papua-new-guinea-forty-years-independence/exploration-gold). Mining was a marginal activity and resource (Burke, 1994); it's virtually not mentioned in treatises about PNG culture, e.g. Flannery (2002), Diamond (2011a, 2011b). Gold was first discovered in Papua New Guinea in 1852 as accidental traces in pottery from Redscar Bay on the Papuan Peninsula (see State Library of New South Wales, Australia above. The next paragraph sections below come from the same source).

From their earliest sightings of New Guinea, Europeans assumed they would find gold. It would finally set their efforts and discovery into value; a commercial justification to explore and dominate PNG after all. Already the initial geography of the landscape, e.g. rugged mountains, denseness of vegetation and the vast unknown interior, seemed to convince explorers that there was treasure to find; must be. A small amount of gold was actually found later on the Gira river and on the Yoda trail near Kokoda in 1890. But it took until the early 1900s for the more successful gold discoveries. The first Gold Rush in PNG occurred then in 1889 on Tagula Island. In 1896, a gold strike was made in the Oro Province, and 1909 in Lakekamu, the Wau Bulolo. Arthur Darling shared with 'Shark Eye' Park in 1921 the Koranga Creek in Wau Valley. The Bulolo region, located at tributaries of the Markham River adjacent to Wau, became a core center of gold dredging in PNG (see also details in Cousteau and Richards 1999; see Mining Data Base details for Wau: https://www.mindat.org/loc-26128.html). It's the sole reason for the road system til today (Beer & Church, 2019), and a large bridge in the area eventually and connecting with Lae and its airport (Sinclair, 1978). But presumably, most people in the west have never heard of Bulolo, and thus, it shows the irrelevance of the undertaking to most people. Beyond dusty museums showcasing PNG specimen and narratives, does anyone really care, other than for the money?

Initial industrial-style mining in PNG consisted mostly of gold-related resource extraction, and it picked up in the 1970s. Eventually, Mr Park got out of PNG gold mining as a wealthy man. And some of the gold mining left behind now operates as a South African subsidiary Harmony Gold (https://www.harmony.co.za/business/png). The story is ongoing, in PNG, in Melanesia and worldwide (see Burke, 1994 for context and outlook). But what was really gained and learned?

As mentioned and described in Cousteau and Richards (1999, p. 80), then there are also many independent miners ('free agents') that try their luck all over the landscape. They are essentially squatters, and it creates many conflicts widely unresolved; a form of piracy triggered by poverty and markets, used up by the larger mines who actually buy up such gold happily; promotes the business, controls markets better and offers scapegoats! The Da Costa brothers for instance, initially from France, report a legal loophole that then allowed them to plan a new mine near Wau 'leveling' an entire mountain, being fully aware of the bad implications for the region and its people.

And then there also were the three Leahy brothers—Michael, Patrick and James—who found gold in Goroka, highlands of PNG in 1930s (Cousteau and Richards 1999 p. 80; see Wilson, 2019 for first-hand account and legacy). This left a massive impact in the area, and it even shifted—in part—and stabilized some village fights and conflicts (Cousteau and Richards 1999 p. 88; "*daily killings could be found*' as per Nelson, 2016). One of the Leahy brothers, Patrick Leahy, went on with the business and became a millionaire, and much racial mixing occurred between Australians and the locals (including the Australian brothers). Clearly, that's a big impact and social engineering on a landscape scale. But who asked for it?

Like almost anywhere in the world, the mining in PNG comes with massive impacts (further details provided below). But to overcome those public resistances, it usually gets presented then as a template phrase, such as this one:

> … *one of the world's most prospective yet under-explored terrains for* …[insert the name of the metal to be pursued] …

Who then would resist such a mining effort to go ahead?

As the generic and widely heralded narrative goes—well promoted globally—the gold miners from Australia put law and order to PNG society, as tribal warfare and killings were part of the day (e.g. Nelson, 2016; see also Matthiessen 1987). But that's like putting a fire out with fuel from a hose. Considering that PNG has lived for over 47,000 years rel. sustainably, Australians are not needed for peace keeping or for PNG to function; rather vice versa. That's because by now, it's obvious that PNG and its wilderness are worse off than at the start in the 1970s. Australia and other nations just it for their own gain the most. The colonial narrative is not honest but selective and self-serving. The metrics of forest cover, overfished fish stocks and civil strife in PNG indicate us nothing else. And Australia keeps running its own type of terror, in Australia (Madley, 2008) and outside (e.g. Henton & Flower, 2007; Lasslett 2012 for Australia's Arc of Terror); so what can Australia really contribute to PNG and the mining industry, other than extracting the resource away from PNG with money made abroad leaving PNG and its environment quite empty-handed?

To further exemplify the approach, the first big gold nugget from Mount Kare (Henton & Flower, 2007) is presented in shiny nice details by Beehler and Laman (2020), who got funded with support by "Porgera Joint Venture" (officially called Porgera Joint Venture Gold Mine as per http://www.porgerajv.com/). But in reality, there is nothing shiny and nothing nice or artistic about this gold piece that resulted in the tragedy for so many people, for an entire culture and tribes. Many wished Mt Kare gold find and exploration would not have happened.

And for PNG, it does not even end there. Australia comes with partners; a global system and web deeply encroaching into PNG. South Africa, Cape Town investors backed up by the City of London, is for instance a strong player in PNG. By now, the Barrick Gold Corporation and Zijin Mining Group each own 50% of Barrick (Niugini) Ltd. Porgera Gold Mine as the second largest mine in Papua New Guinea, and it is regarded as one of the world's top ten producing gold mines ever The style and track record of Barrick Gold can be reflected in the Karida massacre (The Guardian, 2019; see also The Guardian, 2021).

Overall, PNG is just a tragic part of the wider international mining world, including 'Mining Melanesia' and the British Commonwealth, headquartered in London city, with other relevant trading places in Singapore, Dubai and Cape Town. PNG is part of the wider empire of mining, a concept well described by Belanger (2019) for Canada, Commonwealth and beyond.

And mining is just one resource extraction of many. Starting with colonial efforts and then mining, the international industry lost virtually any limits and taboos in PNG. It's not humble. For oil extraction for instance, the story hardly differs. PNG was widely untouched for oil and gas but just recently embarked on the boom. Projects of discussion are found with Exon Mobile and then also Chevron at Lake Kutubu (see Cousteau & Richardson, 1999 as well as Richards, 2018 for such funded biodiversity assessment studies). Chevron led the oil development at Moro (background details: https://www.thefreelibrary.com/Drilling+Papua+New+Guinea%3a+Chevron+comes+to+Lake+Kutubu.-a018322243), which then was followed by Exxon LNG and pipeline, all reviewed by Australian National University staff involvement, and approved. For associated massacres and human tragedy from such efforts, see The Guardian (2019). Like with oil and gas production elsewhere, it raises major questions of legal justice (see Sawyer, 2022 and oil and gas in Kikori river basin).

Textbox 1: OK Tedi is not 'ok'!

Many mines are explored and exploited in PNG; mining plays a big role for the nation of PNG. A typical case for mining is found with the Ok Tedi mine, as one of the biggest projects in the South Pacific and for 'Mining Melanesia' (Kirsch, 2014; see also Hyndman, 1998). Ok Tedi consists essentially of a mountain of copper topped by a gold cap (Cousteau & Richards 1999, p. 210). It's nothing but the biggest gold reserve outside of South Africa. The detailed story of Ok Tedi is greatly explained by Kirsch (2014); the changes with the people and their villages are shown in Jorgensen (2006).

It was envisioned as a thirty-year project, with 5 years gold mining, another 5 years of copper and gold extraction, and then 20 years of copper. It was meant to bring true wealth to PNG; but did not.

Companies that engaged were a consortium from PNG, Australia, America and Germany. For a while, the Ok Tedi mine actually had one of the busiest airports in the world! But after 3 years, the project already ran into a hick up when the Ok Ma river tailing dam broke and polluted the Fly river, etc. It's nothing but the largest river in PNG. To contaminate such an essential lifeline for PNG is quite a feat! The PNG agency in charge was afraid and approved more work in order not to lose the future money of PNG.

Then, a mine barge flipped and dropped cyanide. Then, other accidents happened killing Fly river live further (details found in Cousteau and Richards 1999). Ok Tedi turned into a mess (Beehler & Laman, 2020 p. 74/5 for more mining details).

The Ok Tedi mine and its bad money still hounds PNG and Australia to this very day, see recent events and scandal reported in NME (2022) affecting some of the leading Australia entertainment outfits. It's goes global.

What is to be concluded from the Ok Tedi mine experience; was it worthwhile? It was a boom-and-bust, and it left a mark, and a bad example in PNG, and beyond. Ok Tedi is just another global mine project gone awful with much damage.

Textbox 2: How a trickle-down science fails just like a trickle-down economy: Examples for biodiversity wilderness inference in PNG

When looking at PNG research efforts, one easily recognizes the gaps. Studies are often only done once a mining or oil and gas industry hits and spends 1% or so of its budget, usually with a legal requirement to do so or to look good in public.

This sets up research without a good objective and on 'free' footing. Science is not in the driver seat but became just a pitty token exercise. Instead, it is all embedded with the natural resource extraction industry and scholars, e.g. from abroad. Already the study site, and often even the research question is highly biased, usually parsimonious and thus biased and ignorant of relevant wider views.

A good example was already mentioned in this boom by Boyden et al. (1975), linking research efforts with a mine site question, all done from a globally prestigious university. Many of those examples can be found in PNG, for instance, Richards (2018) by Exxon Mobile as an example for Wau Creek, Uro Creek and Lake Kutubu. It's nothing unusual that such works then enter the national colonial biodiversity literature (essentially as a subsidized industry product with all its constraints; Woxvold et al., 2019). It's how science operates and produces western knowledge. There is nothing indigenous in such work, nor de-colonized. Instead, it's just the plain old colonial approach rolling out in PNG further, 'window dressing' at its finest.

Much of the deeper research in PNG essentially co-occurs linked with industry hotspots and thus misses a wider more balanced and widespread inference, but as needed for an objective and unbiased science effort, and quite frankly, for progress. That way, looking at science locations and knowledge about PNG is similar to the reported beer consumption peak as an indicator when a gold rush occurs with a people influx (Henton & Flowers, 2007 for Kare Gold Mine). It must be seen as an amazing detail that the western so-called high level science education fully approves of it and self-clones, examples from Bourke (1975) onwards. It becomes just a biased and self-fulfilling prophecy because other sites have been entirely left out of the equation and consideration. But for PNG, that's most of what is to be inferred on still.

It just shows, what those so-called science-based sustainable approaches are, industry first (to get the money) and then left-over nature afterward to be studied (examples shown in Taber & Payne, 2003). Such efforts just cannot achieve, as all metrics show us now.

And the list in Tables 12.1, 12.2 and 12.3 is far from complete, nor is it really done with a research design or precise standards of meta-analysis (Stegenga, 2011). But by now it's not really needed to establish the 'curse' with hard-core science. It's not rocket science and 'the evidence sits in the pudding'; the signal is that strong. Simply considering that these projects are large and that they carry a very big footprint, they cannot be ignored for impacts in an ancient somewhat pristine PNG society and fabric, nor the trend overall. That message should be clear to anybody (e.g. Hayan, 1990 for road impacts, Mack 2014 for mining exploration). The industrial lifestyle itself is not so beneficial to people, their psychological sanity or physical health (e.g. Drucker, 2017, Amiri, 2021). In mining projects, people do get injured or can die, despite the slogans of an accident-free industry, or approved environmental impacts, or a claimed obtained public license. Typical examples are found in the social realm and in natural resource sectors, see Willions et al., (2022; Willions and LeBillion, 2017).

Already in quick summary here, the track record and the gains of mining when it comes to long-term sustainability of the PNG nation, its culture and the environment remain very poor and failing (Burton 2001, Gosarevski et al., 2019). At best, it reminds of the movie Avatar (https://en.wikipedia.org/wiki/Avatar_(2009_film)) and it has been globally well exposed by Tim Flannery (1998), Jacques Cousteau family (e.g. Cousteau & Richards, 1999), David Attenbourogh (e.g. https://www.reuters.com/article/us-climate-change-australia-attenborough/david-attenborough-criticizes-australia-on-climate-coal-policy-idUSKBN1W90R7 etc.; for seabed mining see here: https://www.ecowatch.com/david-attenborough-deep-sea-mining-2645477565.html) and Ludlam (2021), besides many others.

12.3 The Generic Pattern of Mining: Hype, Transitional Ownership and Liability Questions, Money Elsewhere and the Locals Pay the Bills While the Environment and Social Fabric Gets Knowingly Destroyed, Long-Term

Mining comes usually with a huge promise: Everybody can get rich; great infrastructures are to come, and our children will do even better. The suffering now would be worth the outcome later (aka '*no pressure, no diamonds*'). Once the exploration was successful, speculation starts and plans get unleashed how the resource can get accessed and exploited, and wealth would be made available to everybody with all its infrastructure. Often those are pipedreams, but the party begins anyways... For instance, the Mt Kare Goldrush was supposed to be kept secret but leaked and created a vast and immediate gold rush internationally from Australia to Asia (Cousteau & Richards, 1999, p. 211; Henton & Flower, 2007). Before this became the news, many people already knew that 'something was cooking': The local beer sales exploded at the local hubs. Suddenly, a huge demand existed for 'beer in the bush' and thus something happens locally! (Cousteau & Richards, 1999, p. 210): Boom, destruction got on the way!

The money in mining is perhaps less in the operation of the mine itself—which usually comes flawed with many problems (see examples in Ok Tedi, or Newcrest https://ramumine.wordpress.com/2020/03/12/png-production-issues-plague-newcrest/)—but the real gain sits in the speculation and growth parts. Hyping up a mine is good for business and stock markets either way. Even a bad reputation can help some people and groups like insurance, investments companies, facilitators and strategic mining decisions by an international corporation that has many mining projects to pick from for their best way to go. It can be a gambit. There is a big market for debt trading. Those actors can blackmail entire nations and stakeholders and then switch projects around within their wide portfolio as they see fit; surfing subsidies and waiving clean-up costs for own profit; playing the game!

Clearly, mining is 'dirty'; a massive amount of soil gets removed and it creates much dust and waste. Most mine sites suffer from the hype and assumed promise of a vast wealth. But as dynamite and chemicals like cyanide are used by mines, questions about water contamination loom (see Cousteau & Richards, 1999 for the Wau region; or Boyden et al., 1975 for an old study subject in PNG). Ok Tedi details are studied for a long time (Boyden et al., 1975; Kyle et al., 1986) which is located in the headwaters of the Fly river with toxic residues from a tailing dam, as polluted with lead, arsenic, mercury and cadmium. This matters because fish harvest in PNG supplies entire families for millennia and the actual catching is usually done by kids and women (Beehler & Laman, 2020) who then suffer from the food chain early on. It's a human tragedy unfolding, I witnessed that first hand in Sepik river area (see also @savethesepik), Ramu river area and in Madang's coastal area.

And those 'technical' problems are not limited to mines with a western ownership. The Chinese Nickel mine had similar problems and contaminated vast stretches of areas (Beehler & Laman, 2020), etc. It's a generic mining and industry problem (see Pebble Mine in Alaska, e.g. Hauser 2007), thus 'the curse'.

A typical story simply goes: The tailing dam breaks and toxic effluents are spilled regardless

In many mines, the actual clean-up and impacts are paid by the local community (but not by the mining company who reap the profit and leave; example provided by Jorgensen, 2006). As a typical pattern, Cousteau and Richards (1999, p. 28) show that waste water from mining gets directly dumped into rivers and then reaching the ocean, its seagrass and coral reefs (see Harris-Charles & Pemberton, 2007 for copper mining, effluents and sperm whales as indicators). And in addition to the physical and chemical impact, the social impact is even stronger. For instance, in Milne Bay mining destroys the underlying fabric of PNG's society for the next two generations or more (Cousteau & Richards, 1999). The discussion about Lae is hardly any other, or exploration work in Crater Mountain (West, 2006; Mack, 2014). In other areas, some age-old fights and conflicts actually ended but just because the locals are now united over the environment and try to halt the gold mining jointly! Mining as the new and essential enemy to fight (see Hyndman, 1998)!

The resource curse is not just a curse, but it's a 'true'ism'; it's a plain governance failure and mostly driven by greed left unresolved. "*The promise of endless wealth*": who would not like that, or giving it a try? Many needy people will fall for it in a broken-down society. But in reality, mining etc has always brought in and attracted 'crooks' and such projects (see Table 12.1).

And there is another fallacy in mining, as expressed in the commonly known Jevons Paradox (see https://en.wikipedia.org/wiki/Jevons_paradox) in economics. A higher efficiency will not make the resource or mining business better, obtain a higher income, but instead it will exhaust it faster, drive down the value of the product as the market gets flooded, actually resulting into less profit, and the mining operation runs into a boom-and-bust spiral downwards. It's widely seen in the world.

It is stated for PNG that while its living resources are visible and very precious assets (Beehler & Laman, 2020). An even bigger value lies underground (Cousteau & Richards, 1999, p. 210). The mainstream media promotes it daily that way, and thus is pushing for its extraction. It sets PNG up for more mining and its wealth extraction going abroad. Clearly, the media turns into a promotion tool (Rooney et al., 2004), as they are funded that way and favor it that way. It's a message by design.

In reality, for most projects that the author knows of, mining—or oil and gas—has not provided the good promise of wealth to most people and nations; see California Goldrush, Yukon Rush, Alaska (see Ott, 2005), Inner Mongolia, etc. And many people who retire from mining are usually happy they are 'out'; the disputes and conflicts of such projects are accumulating quickly and going through the roof; those are almost not controllable anymore. Mining is a template of problems and arguments; now similar perhaps to a man-made soap opera. I know of personal acquaintances that had to retire early even; and they are happy to be 'out' indeed, mental and health problems aside.

Some typical time steps found in the mining business and its projects:

1. Locate a good exotic-sounding place
2. Invest and explore the ground/soil
3. Report a finding
4. Set up a basic infrastructure
5. Get serious with mining the ground; first business plan
6. Run the mine and with a major business plan
7. End of a mine and a new business plan
8. Declare bankruptcy, company break-up and departure
9. Public pays the clean-up costs and Aftermaths, including broken trust and decay of society fabric

Once everybody realizes what mining really entails, the curse of mining becomes an understandable fact. It's real, unavoidable and hard to resolve, if ever. One of the better options is to leave it in the ground, untouched (see for Bolivia's Mother Earth Rights; see also UN's clear message: https://news.un.org/en/story/2021/04/1090262). That's what Mother Earth teaches us. Now, not withhold the fact that the 9 steps described above can be fully interfered with by nation states and international aid and donors. But that hardly makes it better.

12.4 Obtaining the Public/Social License and Buy-In Widely Failed in PNG

There is a process in mining and similar projects where resource extraction companies are trying to obtain 'the public buy-in and support' ('social license'; *sensu* Imbun, 2007): Get the public to agree moving ahead with the mine, then coined a 'Partnership' (=often a PR term that actually has little partners nor are they all on an equal footing, nor it's done voluntarily by either members of that relationship). Nowadays, such an approach is almost a 'must'; everything is to look good and perfect—perceived harmony—in the public eye. It's a facade and window-dressing often. Indeed, that would be great for the industry to act in a true partnership because then there is no resistance in the operation by the generic public, and a happy and smooth mining and resource extraction can occur. A win-win, no critical voices, no critical thinking. It would be an ideal case for the operator and its supporters working on a public trust resource going ahead in full steam... And anybody who interferes there would just be seen as a 'traitor' and gets mobbed out. Voices are silenced.

Has that happened? Of course. Examples of various sorts are shown in Mack (2014) and elsewhere (e.g. Kirsch, 2014; The Guardian, 2021); see Macintyre (2007) for overview of the PNG experience. Table 12.5 shows already a short list of tribes and areas in PNG that were described for big impacts.

How is that public opinion shaped? It's a classic script applied in extractive industries (see for instance with Alaska Exxon Valdez oil spill the buying out the newspapers first thing; Ott 2006). It's achieved through at least three processes usually:

(a) media campaigns, which can be soft and done through public meetings, or social media adds, or,
(b) buying out entire media branches by interest groups (see Ott, 2005 for an example), or

Table 12.5 Selection of tribes and areas affected by mining in Papua New Guinea

Tribe and area	Mine	Citation and comment
Pawaia and Gimi	Crater Mountain	West (2006), Mack (2014)
Ipili	Pogana	Golub (2014)
Hewa	Various mines	Pushing indirectly developmental landscapes and man-made climate change on remote landscapes

(c) engaging in full-scale mobbing, including firing of inconvenient voices and people with the 'wrong tone. ("tone policing") It has been witnessed that mining demonstrators, usually referred to as left-wingers, get chased out of the town, out of the landscape and certainly out of any public institutions and jobs, including government agencies and universities.

As a matter of fact, a typical case is presented by Ashley Cooper on climate change (https://www.imagesfromawarmingplanet.net/ashley-cooper) where even the police in Canada aligns with the industry and makes taking photos—any reporting—difficult. It's done in a cautionary fashion. See for more of such examples with Belanger (2019). But the people will not shut up (Hyndman, 1998) and large mines cannot work in secrecy, e.g. MiningWatch.com (2004), see mining elaborations in Ludlam (2021).

A more sophisticated way of massaging the public license gets done through science. It's also a frequently found scheme. While universities are not to get engaged with funders for objectivity, there are many ways how science is used, can support mining and is fully embedded in its curse. As a matter of fact, science in not objective and frequently does support mining, e.g. by doing nothing and having no position ("*bystanding*"; Baldura, 2007) or by seeking funding, by paying student funds, or through any funding and non-disclosure agreement, e.g. when data are not shared. A more direct way is the use of Environmental Impact studies and as done by contractors and NGOs, e.g. trained through the university. A wider 'army' of NGOs and contractors exist to do so.

While western universities and academia have and try to follow ethical procedures, those might be fully missing in tropical nations, or in many other places and in some high profile cases. Consider that most mines nowadays claim things like "*one of the world's largest resource for mineral x*" and thus we cannot stop the project....).

In the case of PNG, most of those concepts have been tried; but surprisingly not often with success. On the one hand, it is precisely the tribal structure that can reject industrial intrusion wholesale. On the other hand, tribal leaderships can, and have been, bought out leading into tribal warfare and wholesale destruction of the ancient society and the modern society. Examples can be found for instance in Mack (2014). Tribes can be split in their environmental opinion, often based on some earlier conflicts in deep time. Such conflicts can spill over and then even be tracked in Port Moresby, e.g. NME (2022 for Ok Tedi and Australian entertainment industry). PNG has many of those examples (see Table 12.4), so why do we need more?

12.5 A Reality Perspective Where PNG Stands After a Century of Mining: Mother Earth to the Rescue

Judged by most metrics, PNG is considered today a poor nation and one that subsequently needs help/aid; some people say PNG is a failed nation as part of an instable arc surrounding, and created by, Australia (Reilly, 2004, May 2006), e.g. many of

PNG's institutions are in a low estate, sometimes broke, corruption is reported, performance is low and crime is very high (Gosarevski et al., 2019; The Guardian, 2020, 2022 for a public state of the affairs). The road department (Ministry of Transportation) is a good example not really being able to provide a road service and maintenance throughout the nation; existing road projects loom accordingly. In PNG, much transport of goods remains 'on foot' and where there is no road (see Hayano, 1990; Mack, 2014 for road and development politics in PNG with real-world examples and first-hand experiences). Bush planes still play major role, e.g. certainly for gold mines and for hospital flights; many are operated by private services and missionaries (see Sinclair, 1978 for legacy). The university of PNG, as a flagship entity of modern PNG, simply is not listed among the top 100 universities; or even close to it. Nor is the university budget of international relevance allowing a fluent and modern exchange of research, scientists, students, or ideas for that matter. It shows what the state of the art is for a nation. The wealth—and subsequent stabilization—promised through decades long mining—largest mines in the world—and associated economic growth has simply not happened; the environmental decay has though (Czech, 2019 for underlying concepts). The decline of PNG is part of the resource curse ongoing. Whereas, if the claimed concept of a trickle-down economy would be true (Aghion & Bolton, 1997; Arndt, 1983), it should be so easy to achieve in poor nations such as PNG, where a little bit of wealth would bring us pretty far and could buy 'a lot'. But clearly, it does not happen. And there is no new mine site in view that actually will change that, certainly not sea floor mining (an activity that went bankrupt early on, is run from abroad and which carries no support from most of the community). In reality, for PNG and many places elsewhere, mining means usually a decay of life quality, of wilderness, with some rich nations and their highly trained employees reaping the profit out of the country.

In addition, PNG now has a vast poverty gap and a large percentage of poor people. Such a high amount of ghetto poverty and such a society type and destruction never really existed in the last 47,000 years of PNG; it's man-made and made by the west and its colonial approach and associated business model that was not adjusted for fairness, empathy, betterment and sustainability. It's still black and white.

PNG is a classic and prime example for mining NOT adding to better schools, better infrastructure or a better environment or society and international status for that matter. It widly disproves the industrial narrative (see Giljum et al. 2022 for typical additional mining impacts like deforestation). Most western metrics for PNG are either stagnant or they drop (Gosarevski et al., 2019); specifically, AFTER mining hit PNG in the 1970s (Jorgensen, 2006 for an example), or sea floor mining started with promised wealth and same slogans. It's a classic experiment showing mining impacts and failures. And so, how much longer do we all have to endure this? And leading companies based in South Africa, PNG, Congo, Russia, Canada etc., show us no other. Gold mining and its business are widely in the hands of nations like South Africa and Australia, now also dominated by China, while the actual gold value is not that high (likely an outcome and strategy of the global and Chinese supply and approach).

Like with most nations, and as governed with a western structure, PNG in its modern form has no suitable and fair distribution of wealth. Poverty remains a major problem, e.g. in cities, and it's on the rise. PNG and the shanty towns, as found around coconut plantations and tuna canneries for instance, are locations that really create problems, including many diseases (it's the shanty towns that provide social dynamite, as can be seen with the Rasta movement and Bob Marley as a generic pattern in the tropics, way beyond Jamaica).

There are simple concepts in investment, and with a natural resource economy: Let an interest account run in parallel of the similar amount of the resource under management and see which fares better: mining or banking?. If one would have invested a basic checking account with a simple interest rate, it would have fared much better (see example in Ludlam 2021) than real-world mining. It's typical rent paying. How come there is a difference (see Czech, 2019 for additional details)? So where did that money from mining really go, and how spent? Consider this done for the many mines in PNG, some of them with a boosted world class resource size, and from that one gets an idea of the waste in mining and that it actually is not working out and as claimed. In mining, forestry or fishery, coins do not double well (see examples on 'coins' provided in Ludlam, 2021). That's a fact.

And finally, if those mines of global size and relevance, more than 'just' PNG and Australia as the operator should have a say about such resources; it's a global commodity then. That's a statement by Tim Flannery (The Guardian, 2014) but which is widely ignored in those discussions (see Steiner 2011 for a public license in PNG).

12.6 Modern Life and Sustainability Do not Need Intense Mining: Where to Go Now with Mining and the Aftermath?

Mining has clearly not brought or kept what it promised. PNG is a classic example for it. Also, despite the argumentation mining is not much needed for a modern life or a modern society or for anything sustainable and what involves future generations. The rural bush life in PNG shows it no other, Wantok rules for over 47,000 years quite fine and is still the concept to go by for PNG.

So what is the alternative: more mining; 'Mining Melanesia', the seafloor and beyond? Unlikely. Leave it in the ground, *'keep it under the earth'*—as already promoted by the Universal Declaration of Mother Earth Rights (https://www.garn.org/universal-declaration/; see also Steiner 2010 for IUCN suggestions—remains the better concept and for future generations. The nation of Bolivia has promoted that for a long time. To add Citizens' Advisory Councils for Mining (Steiner 2011) is a good idea indeed. That's part of the classic 'Steady State Economy' and principles followed from Ecological Economics (Daly & Farley, 2010). Something better has to be found and it can be done (Ludlam, 2021), and it exists for most parts already, as shown in PNG's legacy.

Neither New York and London, nor Singapore, Tokyo, Dubai, Qatar and Brisbane are the standards to strive for; certainly not for PNG. A solid and sustainable village life—in the bush—will suffice, e.g. when compared to western metrics of depression, cancer rates, divorce rates and endless search for the meaning of life in highly urbanized dwellings and appartment blocks.

To many people, by now, PNG gets just reduced to a mining and resource outfit that has lost touch with its people and is widely removed from reality. And Beehler and Laman (2020, p. 220) stated:

> *The gold, gas and oil will be gone in decades*

To focus instead on sustainable practices, local food and fixed growth and with such a society and happiness should be worth the try and effort instead.

References

Aghion, P., & Bolton, P. (1997). A theory of trickle-down growth and development. *The Review of Economic Studies, 64*(2), 151–172.

Ainsworth, S., Honey, M., Johnson, W. L., Koedinger, K. R., Muramatsu, B., Pea, R. D., Recker, M., & Weimar S (2005). Cyberinfrastructure for education and learning for the future: A vision and research agenda.

Amiri, S. (2021). Prevalence of depression disorder in industrial workers: A meta-analysis. *International Journal of Occupational Safety and Ergonomics,* pp. 1–12.

Arndt, H. W. (1983). The "trickle-down" myth. *Economic Development and Cultural Change, 32*(1), 1–10.

Australian Conservation Foundation. (2022). *Aggravating extinction: How the Australian government is greenlighting destruction of the habitat our threatened species need to survive.* Canberra, Australia. https://assets.nationbuilder.com/auscon/pages/20116/attachments/original/1647489840/Aggravating_extinction.pdf?1647489840

Baraka, R. (2001). The meanings of independence in post-colonial Papua New Guinea. A plan B research paper submitted to the Faculty of the Center for Pacific Islands Studies in the Candidacy for the Degree of Masters of Arts. School of Hawaii, Asia and Pacific Studies, The University of Hawai'i at Manoa.

Banks, G. (2008). Understanding 'resource' conflicts in Papua New Guinea. *Asia Pacific Viewpoint, 49*, 23–34.

BBC. (2012). Spanish WWF sacks King Juan Carlos over elephant hunt. 22 July 2012. https://www.bbc.com/news/world-europe-18942736. Accessed July 14, 2022.

Beehler, B., & Laman, T. (2020). *New Guinea: Nature and culture of earth's grandest Island.* Princeton University Press.

Beer, B., & Church, W. (2019). Roads to inequality: Infrastructure and historically grown regional differences in the Markham Valley Papua New Guinea. *Oceania, 89*(1), 2–19.

Bélanger, P. (2019). *Extraction Empire: Undermining the systems, states, and scales of Canada's Global Resource Empire, 2017–1217.* MIT Press.

Boyden, C. R., Brown, B. E., Drucker, R. P., & Tuft, S. I. (1975). *Ok Tedl environmental study.* Report of the 1974 Cambridge expedition to the Western District, Papua New Guinea

Buckley, M. (2020). A first-hand narrative account on Tibet's paper parks: How China's green-washing in Tibet flies under the radar. In: G. Regmi, & F. Huettmann (Eds.), *Hindu Kush-Himalaya Watersheds Downhill: Landscape ecology and conservation perspectives* (pp. 197–212).

Burke, G. (1994). Mining in the pacific: Past history, present practice, continuing troubles. *The Journal of Pacific History, 29*(1), 105–109.

Burton, J. (2001). Morobe Gold and Silver Project Socio-Economic Impact Study. 3 volumes. Morobe Consolidated Goldfields Limited and Australian National University.

Carlson, D. J. (2012). International Polar Year IPY 2007–2008: Where threads of the double helix and sputnik intertwine. In *Protection of the three poles* (pp. 35–50). Springer.
Chand, S., & Levantis, T. (2000). Dutch disease and the crime epidemic: An investigation of the mineral boom in Papua New Guinea. *Agricultural and Resource Economics, 44*(1), 129–146.
Clark, K. (2021). *Rare earth metals will create a new foreign dependency.* Academia Letters, pp. 1–5. https://www.academia.edu/49923564/Rare_Earth_Metals_Will_Create_a_New_Foreign_Dependency
Council, C. (2007). *Cyberinfrastructure vision for 21st century discovery.* National Science Foundation. http://www.nsf.gov/pubs/2007/nsf0728/nsf0728.pdf. Accessed July 14, 2022.
Cousteau, J. M., & Richards, M. (1999). *Cousteau's Papua New Guinea Journey.* Henry Abraham's Books.
Cravez, P. (2017). *The biggest damned hat: Tales from Alaska's territorial lawyers and judges.* University of Alaska Press.
Czech, B. (2019). The trophic theory of money: Principles, corollaries, and policy implications. *In Journal and Proceedings of the Royal Society of New South Wales, 152*(471/472), 66–81.
Daly, H., & Farley, J. (2010). *Ecological economics: Principles and applications* (2nd ed.). Island Press.
Diamond, J. (2011a). *Guns, germs and steel: The fate of human societies.* Random House.
Diamond, J. (2011b). *Collapse: How societies choose to fail or succeed.* (Revised ed). Penguin Books.
Giljum, S., Maus, V., Kuschnig, N., Luckeneder, S., Tost, M., Sonter, L. J., & Bebbington, A. J. (2022). A pantropical assessment of deforestation caused by industrial mining. *Proceedings of the National Academy of Sciences, 119*(38), e2118273119.
Dove, J., Miriung, T., & Togolo, M. (1974). Mining bitterness. In *Problem of choice*, edited by Peter Sack. ANU Press
Drucker, P. (2017). *The age of discontinuity: Guidelines to our changing society.* Routledge.
Eichstaedt, P. (2016). *Consuming the Congo: War and conflict minerals in the world's deadliest place diamonds.* Laurence Hills Publisher.
Engler, Y. (2022). *Media ignore tragedies exploitation in Canadian mines.* https://yvesengler.com/2022/05/11/media-ignore-tragedies-exploitation-in-canadian-mines/. Accessed May 14, 2022.
Flannery, T. (1998). *Throwim way leg: Tree-kangaroos.* Atlantic Monthly Press, New York.
Flannery, T. (2002). *The future eaters: An ecological History of the Australasian lands and people.* Grove Press
Freeman, A. (1997). The Canadian gold scam. *Multinational Monitor, 18*(5), 8–10.
Fritsche, U. W. (2017). *Energy and land use—Working paper for the global land outlook* (working paper). https://doi.org/10.13140/RG.2.2.24905.44648
Fujita, S. (Ed.). (2013). *The World Bank, Asian Development Bank and Human Rights.* Edward Elgar Publishing.
Ganguli, R. (2020). Water in Mongolia: Sources, uses and issues, with special emphasis on mining. In: G. Regmi, & F. Huettmann (Eds.) *Hindu Kush-Himalaya watersheds downhill: Landscape ecology and conservation perspectives* (pp. 813–821).
Golub, A. (2014). *Leviathans at the gold mine: Creating indigenous and corporate actors in Papua New Guinea.* Duke University Press.
Gosarevski, S., Hughes, H., & Windybank, S. (2019). *Is Papua New Guinea Viable? Political Economic Bulletin 19.* https://www.semanticscholar.org/paper/Is-Papua-New-Guinea-viable-Gosarevski-Hughes/6d71c70951c12fa82dea578a36b66923997f3f92. Accessed June 1, 2022.
Harris-Charles, E. L., & Pemberton, C. A. (2007). Copper mining and environmental costs in Dominica. *Farm and Business-The Journal of the Caribbean Agro-Economic Society, 7*(533-2016-38351), 103–121.
Hauser, W. J. (2007). *Potential impacts of the proposed Pebble Mine on fish habitat and fishery resources of Bristol Bay.* Eye on Pebble Mine http://eyeonpebblemine.org/downloads. Accessed June 1, 2022.

Hayan, D. M. (1990). *Road through the rain forest: Living anthropology in Highland Papua New Guinea*. Waveland Press. Inc.
Henton, D., & Flower, A. (2007). Mount Kare Gold Rush: Papua New Guinea, 1988–1994. Published by Mt Kare Gold Rush, Brisbane Australia. ISBN 10: 0646482815; ISBN 13: 9780646482811.
Herrington, R. (2021). Mining our green future. *Nature Reviews Materials, 6*, 456–458. https://doi.org/10.1038/s41578-021-00325-9
Hochschild, A. (1999). *King Leopold's ghost: A story of greed, terror, and heroism in Colonial Africa*. Houghton Mifflin.
Huettmann, F., & Young, B. (2022). The so-called modern 'Sustainable Forestry' destroys wilderness, old-growth forest landscapes and ecological services worldwide: A short first-hand review and global narrative on the use of 'Growth-and-Yield' as a destructive and even impossible goal. In *Forest dynamics and conservation* (pp. 53–82). Springer.
Hyndman, D. (1998). Melanesian resistance to ecocide and ethnocide: Transnational mining projects and the fourth world on the Island of New Guinea. In H. J. Bodley (Ed.), *Tribal peoples and development and issues: A global overview* (pp. 281–298). Mayfield Publishing Company.
Imbun, B. Y. (2006). Local laborers in Papua New Guinea mining: Attracted or compelled to work? *The Contemporary Pacific*, pp. 315–333.
Imbun, B. Y. (2007). Cannot manage without the 'significant other': Mining, corporate social responsibility and local communities in Papua New Guinea. *Journal of Business Ethics, 73*, 177–192.
Jorgensen, D. (2006). Hinterland history: The Ok Tedi mine and its cultural consequences in Telefolmin. *The Contemporary Pacific*, pp. 233–263.
Kirsch, S. (2014). *Mining capitalism*. University of California Press.
Kraal, D. (2019). Petroleum industry tax incentives and energy policy implications: A comparison between Australia, Malaysia, Indonesia and Papua New Guinea. *Energy Policy, 126*, 212–222.
Kowalewski, O., & Śpiewanowski, P. (2020). Stock market response to potash mine disasters. *Journal of Commodity Markets, 20*, 100124.
Kyle, I. H., Tinkerame, J., & Haei, P. (1986). Concentrations of zinc, copper, lead and cadmium in three species of fish from the Ok Tedi Region. *Science in New Guinea, 12*, 150–156.
Ludlam, S. (2021). *Full Circle: Power, Hope and the Return of Nature*. Black Inc.
Lasslett, K. (2012). State crime by proxy: Australia and the Bougainville conflict. *British Journal of Criminology* 52, 705–723.
Macintyre, M. (2007). Informed consent and mining projects: A view from Papua New Guinea. *Pacific Affairs, 80*(1), 49–65.
Mack, A. (2014). *Searching for PekPek: Cassowaries and Conservation in the New Guinea Rainforest*. Cassowary Conservation & Publishing.
Mining Watch Canada. (2017). https://miningwatch.ca/news/2017/3/28/village-houses-burnt-down-again-barrick-mine-papua-new-guinea-violence-against-local
Mining Watch Canada. (2019, March 13). *Nautilus seeks financial protection: Key players looked after at the expense of small shareholders and Papua New Guinea*. https://miningwatch.ca/news/2019/3/13/nautilus-seeks-financial-protection-key-players-looked-after-expense-small. Accessed January 1, 2021
Madley, B. (2008). From terror to genocide: Britain's Tasmanian Penal Colony and Australia's History Wars. *Journal of British Studies, 47*: 77–106. http://www.jstor.org/stable/25482686
May, R .J. (2006). Papua New Guinea: Disorderly democracy or dysfunctional state?. *In Australia's arc of instability*(pp. 151–170). Springer.
Moher, D., & Olkin, I. (1995). Meta-analysis of randomized controlled trials: A concern for standards. *JAMA, 274*, 1962–1964.
Nelson, H. (1982). *Taim Bilong Masta, the Australian involvement in Papua New Guinea*. Australian Broadcasting Commission, Sydney. ISBN 9780642975669, 0642975663. Google Books https://www.google.com/books/edition/Taim_Bilong_Masta/y2MhAAAAMAAJ?hl=en

Nelson, H. (2016). *Black, white and gold: Goldmining in Papua New Guinea 1878–1930*. ANU Press, Google Books https://www.google.com/books/edition/Black_White_and_Gold/8M0PDQAAQBAJ?hl=en&gbpv=1&printsec=frontcover

New South Wales State Library. (2015). *Papau New Guinea Gold*. https://www2.sl.nsw.gov.au/archive/discover_collections/society_art/png/gold/index.html. Accessed August 15, 2022

NME. (2021, January 22). *Sheppard band members' father and former manager arrested in Papua New Guinea for alleged fraud*. https://www.nme.com/en_au/news/music/sheppard-band-members-father-and-former-manager-arrested-in-papua-new-guinea-for-alleged-fraud-2862298. Accessed July 18, 2022.

Maganga, F., & Jacob, T. (2016). Defying the looming resource curse with indigenization? Insights from two coal mines in Tanzania. *African Review 43*(2), 139–160. file://C:/Users/fhuettmann/Downloads/454-3354-1-PB.pdf

MiningWatch.com. (2004, September 22). *Environmental Disaster at Misima Mines Continues...Misima Landowners in Port Moresby to Seek Justice*. Accessed July 13, 2022.

Ndlovu-Gatsheni, S. J. (2018). Rhodes must fall. In *Epistemic freedom in Africa: Deprovincialization and decolonization*. Taylor & Francis.

Ott, R. (2005). *Sound truth and corporate myths: The legacy of Exxon Valdez oil spill*. Dragonfly Sisters Press.

Papua New Guinea Mine Watch. (2011). *Repeal of the amendments to the environment act 15th October 2011*. http://ramumine.wordpress.com/2011/10/15/repeal-of-the-amendments-to-the-environment-act/. Accessed May 14, 2022.

Reilly. (2004). State functioning and state failure in the South Pacific. *Australian Journal of International Affairs, 58*(4), 479–493. https://doi.org/10.1080/1035771042000304742

Reuters. (2022, February 1). *Report on Rio Tinto finds 'disturbing' culture of sexual harassment, racism, bullying*. https://www.reuters.com/business/rio-tinto-report-finds-disturbing-culture-sexual-harassment-racism-bullying-2022-02-01/. Accessed July 20, 2022.

Revkin, A. (1994). *The burning season: The murder of Chico Mendes and the fight for the amazon rain forest*. Penguin Books.

Richards, S. J. (Ed.). (2018). *Rapid biological assessments of Wau Creek, Uro Creek and Lake Kutubu: Documenting biodiversity values to promote forest conservation in the Kikori River basin, Papua New Guinea*. ExxonMobil PNG Limited.

Rich, B. M. (1994). *Mortgaging the Earth: The World Bank, environmental impoverishment and the crisis of development*. Earthscan Publishers.

Rooney, D., Papoutsaki, E., Pamba, K. (2004). *A country failed by its media: A case study from Papua New Guinea*. Paper presented at the Annual Conference of the Asian Media Information and Communication Centre, Bangkok

Ross, M. (2001). Does oil hinder democracy? *World Politics, 53*(3), 326–361.

Ross, M. L. (2015). What have we learned about the resource curse? *Annual Review of Political Science, 18*, 239–259.

Sawyer, S. (2022). *The small matter of suing chevron*. Duke University Press.

Sinclair, J. (1978). *Wings of gold: how the aeroplane developed New Guinea*. Pacific Publications. ISBN 0-85807-033-2.

Stegenga, J. (2011). Is meta-analysis the platinum standard of evidence? *Studies in History and Philosophy of Science Part C: Studies in History and Philosophy of Biological and Biomedical Sciences, 42*(4), 497–507.

Steiner. R. (2011). Establishing the Papua New Guinea (PNG) Deep Sea Mining (DSM) Citizens' Advisory Council (CAC) Richard Steiner, Professor, Oasis Earth

Steiner, R. (2010). El Salvador—Gold, Guns, and Choice: The El Dorado gold mine, violence in Cabanas, CAFTA claims, and the national effort to ban mining. International Union for the Conservation of Nature (IUCN) Commission on Environmental Economic, and Social Policy (CEESP).

Taber, R. D., & Payne, N. F. (2003). *Wildlife, conservation, and human welfare: A United States and Canadian perspective*. Krieger Publishing Company.

The Guardian. (2014, August 1). *The Great Barrier Reef and the coal mine that could kill it*. https://www.theguardian.com/environment/2014/aug/01/-sp-great-barrier-reef-and-coal-mine-could-kill-it. Accessed July 12, 2021.

The Guardian. (2018, May 2). *Exxon Mobil's project in PNG is economic parasitism*. https://www.theguardian.com/commentisfree/2018/may/02/exxonmobils-project-in-png-is-economic-parasitism. Accessed May 13, 2022.

The Guardian. (2019, July 22). *The Karida massacre: fears of a new era of tribal violence in Papua New Guinea*. https://www.theguardian.com/world/2019/jul/23/the-karida-massacre-the-start-of-a-new-era-of-tribal-violence-in-papua-new-guinea. Accessed May 14, 2019.

The Guardian. (2020). *Papua New Guinea police accused of gun running and drug smuggling by own minister 10th May 2022*. https://www.theguardian.com/world/2020/sep/18/papua-new-guinea-police-accused-of-gun-running-and-drug-smuggling-by-own-minister. Accessed November 5, 2021.

The Guardian. (2021, December 9). *Villagers file human rights complaint over plan for giant PNG goldmine*. https://www.theguardian.com/world/2021/dec/10/villagers-file-human-rights-complaint-over-plan-for-giant-png-goldmine. Accessed May 1, 2022.

The Guardian. (2022, July 6). *PNG prime minister denies connection to suitcase full of cash found as voting starts*. https://www.theguardian.com/world/2022/jul/06/png-prime-minister-denies-connection-to-suitcase-full-of-cash-found-as-voting-starts. Accessed July 6, 2022.

West, P. (2006). Environmental conservation and mining: Between experience and expectation in the eastern highlands of Papua New Guinea. *The Contemporary Pacific, 2*, 295–313. https://core.ac.uk/download/pdf/5103144.pdf. Accessed July 1, 2022.

Williams, A., & Le Billion, P. (2017). *Corruption, natural_resources and development*. Edvar Elger Publisher. https://www.google.com/books/edition/Corruption_Natural_Resources_and_Development/16ffDQAAQBAJ?hl=en&gbpv=1&dq=Resource+Curse+defying&pg=PR1&printsec=frontcover

Williams, B. A., Grantham, H. S., Watson, J. E. M., et al. (2022). Reconsidering priorities for forest conservation when considering the threats of mining and armed conflict. *Ambio*. https://doi.org/10.1007/s13280-022-01724-0

Wilson, M. (2019). *A true child of Papua New Guinea: Memoir of a life in two worlds*. McFarland Publishers.

Woxvold, I. A., Diamond, J. M., Bishop, K. D., & Legra, L. (2019). Avifauna of the lake kutubu wildlife management area, Papua New Guinea. *Bulletin of the British Ornithologists' Club, 139*(3), 266–292.

Zhanheng, C. H. E. N. (2011). Global rare earth resources and scenarios of future rare earth industry. *Journal of Rare Earths, 29*(1), 1–6.

Chapter 13
Seafloor Mining Around Papua New Guinea: Just Another 'Big Deal' Gone Bust

Seabed minerals have a potentially critical role to play in the decarbonisation of the planet by providing a vital and reliable alternative source of critical minerals for among other things clean energy including battery technologies. We intend to continue to work towards the realization of this potential opportunity with our valued stakeholders and partners.
Peter Ruddock
Director of UK Seabed Resources Ltd and chief executive of Lockheed Martin UK (https://theminernetwork.org/deep-sea-mining-part-3-awash-in-sustainability-speak/. *Accessed 3rd May 2022*)

Abstract Our planet is mistakenly called 'The Earth' where it should instead be referred to as '**The Ocean**,' covered by saltwater. It makes easily for over 70% of the planet's coverage and includes the seafloor as the largest habitat on the planet. Instead of flying to the moon, why not using it, the 'Blue Economy'? This habitat is pursued by mining for the earth resources, and experimental mining operations have explored and started now this effort in earnest. Located close to several major global deep sea trenches, PNG offers a massive geological hotspot that provides urgently pursued minerals for exploration, on land and at sea. Seabed mining in PNG pursues for instance Sulfide (SMS) deposits with copper, gold, zinc and silver from the Bismarck Sea. Mining the ocean floor is new territory that mankind has no experience with and has not co-evolved. The project Solwara1 in PNG waters was pursued by an international mining company (Nautilus) for seabed mining headquartered in Canada, with global market consumption. Like typically found in PNG, many environmental impact statements showed the serious flaws of such project early on, because such mines will affect future ocean generations for decades to come, likely longer. While the Solwara 1 project went down in history bankrupt, it took PNG and the ocean already with it. Seabed mining is of deep global concern, a socio-economic drama on the grand scale, and an ongoing and expanding business failure that affects ocean ecosystems and ecological services worldwide. One can easily conclude that staying out of such type of an 'Blue Ocean' economy is wiser for future generations: "*leave it under water.*"

F. Huettmann, *Globalization and Papua New Guinea: Ancient Wilderness, Paradise, Introduced Terror and Hell*, https://doi.org/10.1007/978-3-031-20262-9_13

Keywords Papua New Guinea (PNG) · Seafloor mining · Deep sea · Blue Economy

13.1 Introduction

In the mining industry, much money is made with speculation. Such a mining is primarily just a belief system and the gamble with the deep unknown (e.g. the earth, 'the deep' underground, the unknown and the uncertainty). It's hyped up by investors and stock holders through their media, and it's all just made possible with political approval, support and various upfront subsidies (usually public money operating on public land resources). No wonder that many religious groups are not opposed to mining and stand firm with the miners and their families and the employers and funders. A gold rush comes as a wider package. For instance, the seafloor was already referred to as the 'Alice in Wonderland world of the extremes' (https://www.huffpost.com/entry/deep-sea-mining-new-threa_b_8334428 "Alice-in-Wonderland world of extremes").

And what cannot be done in one country, e.g. due to laws and attitudes, gets easily pushed through in another country then, as a gamble enabled by globalization and a one-sided trade-off game by international mining corporations. Seen from that perspective, those companies might perhaps act not much different from a sneaky thief. 'Mining Melanesia' (Kirsch, 2014) is nothing but that; trying to grab as much as is possible, each time. Oil explorations add to that puzzle, similar to a game of chess for self-interest, making global ocean issues highly problematic, certainly for small nations (see Thomson, 2022 for how adjacent Timor-Leste was treated by Australia and how those treatments shifted toward Australia as the international bully, simply due to offshore oil considerations, etc.).

'Blue Growth' is part of ocean explorations, using the ocean in many ways, and seabed mining is part of this. In seabed mining—a new global approach (The Guardian, 2022a)—it even gets more broad and complex for accessing the same mining area from many different coasts and nations in parallel; it actually allows for more options and loopholes when international waters are involved or when EEZs are not well known or mapped, and when such laws like UNCLOS are changing boundaries, ownerships and liabilities (UNCLOS https://www.iucn.org/our-work/oceans-and-coasts). Liabilities get swapped while ocean currents operate across boundaries.

The actual process of mining, the physical labor and the destruction and pulverization of 'the rock and ground layer' for extraction of the resource in a remote and very difficult to work area remains the core problem. Compared to the speculative parts, it is actually real work and less lucrative (example provided in Kirsch, 2014 and Papua New Guinea Mine Watch 2020c for 'production issues,' which tend to be 'messy' and very costly); unions frequently get involved to proceed and social-wash it. Mining simply remains a game and a gamble, politically and otherwise. It always has been. It's heavily driven by such minds (see Sinclair, 1978 for flying gambles but which shaped the mining industry in PNG and beyond) while wilderness and other great assets become the collateral (Kirsch, 2014 for PNG; Williams et al., 2022 for forests; see Belanger, 2019 for British Canada's mining empire and entire nation set up to cater it).

And so, other than mining remote planets like the Moon and Mars, it comes as no surprise that seafloor mining is the latest cry on such issues of applying luck and speculation on the cost of others and their common good with a subsidy from abroad. There is now, of course, a gold rush for sea bed mining (The Guardian, 2022a) done without any relevant consideration (The Guardian, 2022b), hardly with a proof that it works (e.g. Earthworks, Deep Sea Mining Campaign et al., 2015). The widely untouched deep ocean around PNG and its ports happily offer themselves for such endeavors. People's concern get ignored, e.g. Hyndman (1993).

13.2 Mine and Drill PNG's Seafloor for Real?

For instance, the mid-ridge Bismarck Sea has massive sulfur deposits, app 1,800 m deep. But PNG is also attractive because when compared to dominating Australia it does not really have many relevant policy restrictions regarding the environment, labor and governance. Essentially, PNG lays there for anybody to grab and to use it, just located in the backyard of major mining monsters and markets. The wrongly perceived slogan goes like "*Drill Baby Drill: Mining is good for the environment and brings you progress.*" Like with forestry and fisheries, it's instead a very raw and barbaric colonial and destructive model, whatever other sources claim: grab the resources and run, wealth accumulates outside of PNG and the clean-up costs are for others to deal with; a future generation's problem (see Pauly, 1979; Pauly et al., 1998; Huettmann & Young, 2022).

Sea bed mining involves dredging, drilling and transporting the product to a processing plant on shore. Tailings are produced in the extraction process, and real risks exist contaminating the sea when mining-related ships travel through choppy waters to the shore. Like with terrestrial mining, a big question centers around the mining waste, tailings (see previous chapter). In deep sea mining that usually involves locations of steep slopes, knowing them and having them accessible so that the mining process can be carried out. As stated in Sullivan (2015), the Lihir mine for instance is located on an extinct but geothermically active volcanic crater on Aniolam Island 900 km northeast from Port Moresby. Lihir's submarine tailings disposal process has been in operation now for ten years. It makes their monitoring of benthic organisms, subsurface plumes and tailings dispersion considerably more substantive than either Ramu Nickel or Nautilus Minerals. Such an undisturbed tailings placement depends upon the steep submarine slopes off Lihir Island though (Fig. 13.1).

Deep-sea mining has not much been used yet, and it was just anticipated or experimented with (see example at Clipperton in Jones et al., 2021; see for mapping examples with Saeedi and Brandt 2020). But the first larger mining effort of its kind in the world was done in PNG, around 1997, while the negative impacts of it are not really known (Earthworks, Deep Sea Mining Campaign et al. 2015). Deep-sea mining

Fig. 13.1 Who wants to destroy the resources under the sea surface? And who knows what truly is underneath? Why not "keep it underwater"?

has simply used PNG as a massive showcase and it remains widely experimental (see The Guardian, 2020)!

A real-world descriptive, fact-based and critically holistic look on the issue of (seafloor) mining in PNG is still widely missing though (see also Kirsch (2014); and for a more global view Belanger 2019). A textbook does not exist to go by. And if done, unless issues get sidelined, such a mining, etc. would not happen due to the large impacts. Arguably, corruption is involved (see for instance Kirsch, 2014; Mining Watch Canada 2019b; see vested interested by the International Seabed Association ISA, as described by The Guardian, 2020, 2021).

From my own inquiries, and using public available information, for seafloor mining I doubt a proper environmental impact assessment is really made in PNG, or is hardly possible due to lack of high resolution data, lack of knowledge and experience, e.g. taking a 360 degrees view, including port impacts, supply chain and marine contaminations and seafloor destruction, or global market and socio-economic impacts including climate change (see for instance Wise et al., (2011) for dimensions of contamination. See Rey et al., (2017) for methods on environmental telecoupling analysis of ports and people). The Deep Ocean Science still sits at its infancy! Seabed mining is just in the hands of a few companies and nations (The Guardian, 2020). Having seen some partial efforts, in the public eye environmental impact studies carry wide aspects of demagogy indeed (Belanger, 2019; Kirsch,

2014; MacIntyre, 2007). It invokes the entire arsenal what the western society has to offer: NGOs, science, institutions, parsimonious statistics, PR visualization, money, banks, development aid, media, social online media and their governance structure including contractors. But it has not delivered well in PNG, on land, at sea and for the atmosphere.

13.3 How Would Seafloor Mining Work, and Around PNG?

Seafloor mining requires special exploration; it needs specific methods and techniques, namely deep seismic shockwaves, sonar and specific submarines to explore and locate mining sites. Many are not developed, lack the experience or policies are difficult to enforce; impacts are not well known. And then, it also needs specific towered vessels with drilling support, supply and support vessels and helicopters on standby. Workforces are international. Having a coastguard in the EEZ is essential for such operations. Many of those details are not worked out yet or do not carry a valuable/sustainable business model even. Oversight and compliance by air or ship is another requirement. Further, the special ships need special ports to enter and for potential maintenance; sophisticated crews need to be exchanged frequently on a shift-basis, usually coming from abroad and in-time via airplane of course; hotels are needed, cargo airports and roads, as well as hospitals. The typical PNG village and port lack such technology and infrastructure where even a basic WWW connection or electricity can be tricky. And many PNG citizens lack that high-tech education and cannot really participate in the exercise of their own land and ocean; they get excluded from high-paying jobs of that industry loosing out. Landing a decent job or obtaining a salary there is next to impossible. This essentially leaves PNG outside of the seafloor mining dea on their own territory. And that's precisely how seafloor mining in PNG will work and starts to operate now: It's exclusive, destructive, not sustainable and exploitative. It's neocolonialism in pure form and thus lacks the public support.

13.4 A 'License to Play and to Kill' Granted by a 'modern' PNG Government, UN International Seabed Authority and International Corporations (While Citizens are Excluded and a 47,000 Old Sustainable Culture Gets Overtaken)

Mining usually requires a license to operate. Those licenses are usually provided in a bidding process and involve proposals, impact statements, reviews and local support (public license; see Banks, 2008 for process as well as McIntyre 2007 for experiences). In PNG, it involves the Mineral Resources Authority and its associated ministries while actually many coastal communities are against it (Beehler & Laman, 2020; see for a real world example with Indigenous People of the Bismarck-Solomon Solomon Sea 2008). There are some major unresolved questions about citizen involvement, democracy, framework, the common good, and how managed, and for whom (e.g. Hyndman 1998). It's a classic 'common good' scheme covered by the late Nobel Prize winner Ostrom et al. (2002) but not resolved well. As this involves the wider seafloor and global resources and actors, the actual license—approval—to operate is to be coming from many actors, not just a sole village or land-owners.

Key questions are usually centered around tailings, the mining dirt and contamination (see for generic details also with Pennington, 2009). This is usually placed on the seafloor and requires a slope. Often such geological slope details are not well known and present already a risk, together with ongoing seafloor geology activity, as it can frequently be seen in the 'Ring of Fire' that PNG is located on.

As a process, a UN International Seabed Authority (ISA https://www.isa.org.jm/about-isa) has formed trying to overlook and steer those efforts globally. The UN International Seabed Authority is based in Jamaica and consists of 167 member states (U.S. is not a member state, Papua New Guinea is) following UNCLOS. It gets its share for each concession approved and thus has an interest in getting 'many' concessions (The Guardian, 2020); a considerate decision-making process is not possible. Much debate occurred around seafloor mining either way, and a science-effort is under way to address those bathymetry and geological and logistic questions, e.g. Weiet al. (2011) for predictive model support options and baseline.

It was noticed that the seafloor mining project Solwara 1 in the New Ireland Bismarck Sea seems to be correlated with recent earthquakes near the site adding to impacts and wider fears in the region. PNG has suffered many tragic consequences of earth plate dynamics. As a matter of fact, that project was heavily reviewed and showed a large array of problems (Earthworks, Deep Sea Mining Campaign et al., 2015; Steiner, 2009). This is significant but it's meant to be a role model for much of the ocean seafloor, how the mineral resources are treated and extracted. It's of universal impact indeed, with PNG and its people and the marine environment being the test case.

And clearly, the mining effort and process is driven by investment and hype but less so by facts, science or even true profit and the wider common good (e.g. https://miningwatch.ca/publications/2019/7/17/why-rush-sea bed-mining-pacific-ocean). The curse of mining also runs through to the seafloor (Mining Watch Canada, 2019a for bankruptcy of the entire PNG enterprise based in Canada).

For PNG, there is a wider request to cancel licenses (Papua New Guinea Mine Watch, 2020b). Opposing groups to seabed mining are diverse and include parts of the catholic church, the coffee grower association and political figures and 'big men' (see MacIntyre, 2007 for the PNG experience overall).

Companies that were already involved in seabed mining are for instance Nautilus Minerals Niugini Ltd headquartered in Toronto, Canada (see public record at: https://en.wikipedia.org/wiki/Nautilus_Minerals) and OceanaGold Philippines Inc. (https://oceanagold.com/ also based in Canada; their Didipio mine project was publicly asked to be shut down 2019).

Clearly, human right abuses are common in mining, also widely known for PNG (e.g. The Guardian, 2021); seafloor mining projects are apparently no exception.

The problems with seafloor mining are easy to understand because there is no culture of seafloor mining among locals, e.g. when compared to the traditional Pacific ocean commons (Hyndman, 1993), or as it can easily be seen with Giant Clam harvesting as a common good (Cousteau & Richards, 1999; see Lewis et al., 1988 for adjacent Fiji). In the human history, humans virtually never took anything directly from ~30 m below the ocean surface. There is no public ownership scheme or sustainability developed last 47,000 years, or in the history of mankind really (see for Melanesia Hyndman, 1993). Technology confronts PNG and mankind just with another dilemma, one that is hardly needed that one can easily avoid. The 'Cargo Cult' for Seafloor Mining does not exist yet in PNG while there is such a deep culture of exploitation and abuse in the western world, now exported to PNG also! Seafloor mining is in essence nothing but an outflow and direct consequence of colonialism and royal court mentality, as practiced in Europe for millennia, now being rolled out at sea and the deep.

13.5 PNG's Seafloor Set up

PNG has essentially three geological plates (Bismarck, Australian Plate, Pacific Plate, e.g. Alcorn, 1993). Located at the Ring of Fire, those are naturally very active. Geological activity resulted into resources of gold, copper and silver. Due to the uplift over time from the deep earth, it makes those resources accessible, easier than anywhere else in the world. The resources to be mined consist for instance of massive sulfide systems, potential sources of high-grade copper, gold, zinc, silver and others. Traditionally, seafloor mining looks also for mangan, e.g. in the form of nodules to be grabbed and brought to the surface (Boschen et al. 2013). Anything that makes money there will be for the taker. It's the ancient 'hungry horde' in action, year 2022 onwards with money as the motive.

PNG is part of Melanesia, featured with many islands. PNG is surrounded by, and blessed with, several ocean sections. The world's deepest part is the extensive Mariana Trench region PNG is associated with. But if given a choice, deep sea mining does not really want to be in the deep sea because that's too expensive and too risky. Operation issues increase greatly by depth. However, in the world, geologically active and relevant areas are rarely directly found under the sea surface and in convenient locations. Instead, those are in remote and deep areas; many of them are barely explored, pristine and with live unknown to science even. This is where seafloor mining tends to happen, or at least where the global seafloor mining world has a plan to do so, with 'easy' sites singled out first. PNG is the 'low hanging fruit' and happens to be such a site, with all implications that come with it. The world looks at PNG once again. It is here where the mining gamble -seafloor version-rolls out this time. It comes as a template popping up globally. The conflict about seafloor mining and PNG is well known and partly exposed, e.g. David Attenborough spoke up against it (Papua New Guinea Mine Watch, 2020b), so did the French president Emmanuel Macron (2022c).

13.6 The 'Ocean Pump' and PNG's Sea Mammals, Shark Calling and Marine Ecology

Oceans are complex systems centered around life and death, gravity, nutrients and ecology, and are affected by the sun and the wider universe. While remote and widely unknown to many people, deep sea areas, shelf edges and trenches, play an important part for the detritus, and for pumping nutrients back up. It matters for global processes as those pumping details interfere with the ocean–atmosphere interface and must not be interrupted to keep ecological processes intact. A good stragegy is needed, e.g. Dunn et al. (2018) for a discussion.

Sea mammals should simply be able to swim around ships and avoid any of the industry, as one might easily think. But think again, seismic shock waves and booms—as typically used for many seafloor explorations and navigation—are known to affect sea mammals, and sea mammals are migratory needing a lot of space. Sea mammals easily get 'in the hair of industry' and create a nuisance for them. It looks bad, the false win-win narrative falls apart in the public eye. It's a typical example that we are out of space on a finite planet; many examples exist (see Table 13.1).

PNG has many sea mammals in its EEZ, some are rare and/or poorly studied, and many pass by during their migration (for species and details see other chapters of this book). Those impacts are little known and described, but 'deep' (often global, see West, 2006; Mack, 2014 for a vast and complex network of people and institutions involved in such style impact work) and thus they currently cannot be mitigated well, if at all. A synthesis is missing for PNG. That's because impacts come with a delay and are part of a wider ecosystem and thus they are often just detected outside of PNG. In the meantime, impact knowledge needs to form and to establish itself among all actors too. Traditional local knowledge is not well taken serious yet.

A typical example is that the PNG seafloor mining case is already known to correlate with beaked whales, many not even studied and well described (Papua

Table 13.1 A selection of sea mammal conflicts related to ocean activities

Species	Location	Stressor	Known impact	Citation
Polar bear	Alaska	Loss of summer sea ice	Starvation	Molnar et al. (2010)
Marine Mammals	Arctic region	Seismic shockwaves	Generic bad impacts and stress	Halliday et al. (2020)
Grey Whale	Sakhalin Island	Exploration and drilling	Behavior and movement changes	Racca and Bröker (2018)
Beaked Whales	Papua New Guinea	Seabed mining	Overlapping occurrences	The Guardian (2022a, b, c)
Sperm whales	Papua New Guinea	Effluents from mining, etc.	High heavy metal contamination, e.g. lead	Savery et al. (2014)

New Guinea Mine Watch, 2018). This knowledge now slowly comes through and enters the wider community exposing the sensitivity of the seabed mining matter

Shark calling—the ancient practice in PNG (Cousteau and Richards 1999 for descriptions)—is another victim of sea bed mining operations; some more details also outlined in The Guardian, 2022b).

Further, PNG is world famous for its mangroves, seagrass and coral reefs, the wider marine ecosystem which is still relatively pristine there (Cousteau and Richardosn 1999). Spoiling and interrupting it creates *havoc* on the entire ecosystem and affect humans and global well-being alike.

And what will more shipping and more traffic bring to such a system? Clearly, plastics, plastic leakage and microplastics are already widely found in the system—including at the deep seafloor—and those features will not decline any time soon. Who wants a high-grade mining product spoiled with plastics?

13.7 PNG as a Special Case for Seafloor Mining and Seeing Its Negative Impacts

PNG has one of the most untouched marine areas in the world (Cousteau & Richards, 1999; Beehler & Laman, 2020). It's unique for its community ocean management (Hyndman, 1993) and also has a very rich resource regarding biodiversity relevant on a global level. And then, it has a rather deep ocean trench system full of mystery (Rosenbaum 2011, Steiner 2009, 2011, 2016). Why spoiling it?

A wide public of PNG does not want any seafloor or much other mining (e.g. The Guardian, 2020, 2021). Many examples exist. For an ocean mine site near Philippines, Mining Watch Canada and the Institute for Policy Studies did not support this operation, as its 2018 report found "*not only of lack of compliance, but also of unacceptable impacts of OceanaGold's Didipio copper and gold mine.*" Can it be worse?

"*Don't you still want to mine it for large profit*"? Sounds exotic, an assumed win–win but a massive gamble with an odd profit margin, and really, it's all still very exotic after all. It's AVATAR in its purest form. De-growth done in the ocean might suit everybody better though (Childs, 2020). Based on all we know, one simply suggests not mining this area and leaving it for other generations to deal with. Just like the Antarctic, which as an area closed for mining still and that is not so far located from PNG and where PNG participates in the Antarctic Treaty System (ATS). It has species traveling back and forth from, and it has already a widely accepted mining embargo. Leave it widely untouched.

Mother Earth Rights matter, leaving it under the ground for future generations. Many people support those views as shown here, e.g. https://miningwatch.ca/news/2019/9/5/moratorium-deep-sea-mining-welcomed-more-courage-required-pm-marape-png-citizens-call. So why mining it and stirring up even more trouble from below, Mother Earth's belly?

References

Alcorn, J. B. (1993). *Papua New Guinea needs assessment. A USAID-funded Consortium of World Wildlife Fund*. The Nature Conservancy, and World Resources Institute Washington, D.C. with the Government of Papua New Guinea Department of Environment and Conservation, Boroko, Papua New Guinea. https://www.sprep.org/attachments/VirLib/PNG/png-conservation-needs-assessment-alcorn-1993.pdf. Accessed June 2, 2022.

Banks, G. (2008). Understanding 'resource' conflicts in Papua New Guinea. *Asia Pacific Viewpoint, 49*, 23–34.

Beehler, B., & Laman, T. (2020). *New Guinea: Nature and Culture of Earth's Grandest Island*. Princeton University Press.

Bélanger, P. (2019). *Extraction empire: Undermining the systems, states, and scales of Canada's Global Resource Empire, 2017–1217*. MIT Press.

Boschen, R. E., Rowden, A. A., Clark, M. R., & Gardner, J. P. (2013). Mining of deep-sea seafloor massive sulfides: A review of the deposits, their benthic communities, impacts from mining, regulatory frameworks and management strategies. *Ocean & Coastal Management, 84*, 54–67.

Cousteau, J. M., & Richards, M. (1999). *Cousteau's Papua New Guinea Journey*. Henry Abraham's Books.

Childs, J. (2020). Performing 'blue degrowth': Critiquing seabed mining in Papua New Guinea through creative practice. *Sustainability Science, 15*, 117–129. https://doi.org/10.1007/s11625-019-00752-2

Dunn, D. C., Van Dover, C. L., Etter, R. J., Smith, C. R., Levin, L. A., Morato, T., Colaço, A., Dale, A. C., Gebruk, A. V., Gjerde, K. M. , Halpin, P. N., & SEMPIA Workshop Participants. (2018). A strategy for the conservation of biodiversity on mid-ocean ridges from deep-sea mining. *Science Advances, 4*(7), eaar4313.

Earthworks, Deep Sea Mining Campaign et al. (2015, September 28). *World's first deep sea mining proposal ignores consequences of its impacts on oceans*. https://earthworks.org/releases/worlds_first_deep_sea_mining_proposal_ignores_consequences_of_its_impacts_o/. Accessed July 14, 2022.

Halliday, W. D., Pine, M. K., & Insley, S. J. (2020). Underwater noise and Arctic marine mammals: Review and policy recommendations. *Environmental Reviews, 28*(4), 438–448.

Huettmann, F., & Young, B. (2022). The so-called modern 'sustainable forestry' destroys wilderness, old-growth forest landscapes and ecological services worldwide: A short first-hand review and global narrative on the use of 'growth-and-yield' as a destructive and even impossible goal. In *Forest dynamics and conservation* (pp. 53–82). Springer.

Hyndman, D. (1993). Sea tenure and the management of living marine resources in Papua New Guinea. *Pacific Studies, 16*, 16–16.

Hyndman, D. (1998). Melanesian resistance to ecocide and ethnocide: Transnational mining projects and the fourth world on the Island of New Guinea. In H. J. Bodley (Ed.), *Tribal peoples and development and issues: A global overview* (pp. 281–298). Mayfield Publishing Company.

Indigenous People of the Bismarck-Solomon Solomon Sea. (2008). Karkum national sea bed mining forum statement. Accessed June 25th–27th.

Jones, D. O., Simon-Lledó, E., Amon, D. J., Bett, B. J., Caulle, C., Clément, L., Connelly, D. P., Dahlgren, T. G., Durden, J. M., Drazen, J. C., Felden, J., & Huvenne, V. A. (2021). Environment, ecology, and potential effectiveness of an area protected from deep-sea mining (Clarion Clipperton Zone, abyssal Pacific). *Progress in Oceanography, 197*, 102653.

Kirsch, S. (2014). *Mining capitalism*. University of California Press.

Lewis, A. D., Adams, T. J. H., & Ledua, E. (1988). Fiji's giant clam stocks—A review of their distribution, abundance, exploitation and management. *ACIAR Monograph Series, 98*, 66–72.

Macintyre, M. (2007). Informed consent and mining projects: A view from Papua New Guinea. *Pacific Affairs, 80*, 49–65.

Mack, A. (2014). *Searching for PekPek: Cassowaries and Conservation in the New Guinea Rainforest*. Cassowary Conservation & Publishing.

MiningWatch Canada and Institute for Policy Studies. (2018, December 10). *miningwatch canada and the institute for policy studies: Response to OceanaGold.* https://miningwatch.ca/sites/default/files/mwcandipsresponsetoogpi10dec180.pdf (Accessed June 1, 2022)
Mining Watch Canada. (2019a). https://miningwatch.ca/news/2019a/3/13/nautilus-seeks-financial-protection-key-players-looked-after-expense-small
Mining Watch Canada. (2019b, July 17). *Why the rush for seabed mining? Intrigue, collusion and interesting bedfellows.* https://miningwatch.ca/news/2019b/7/17/why-rush-seabed-mining-intrigue-collusion-and-interesting-bedfellows. Accessed July 26, 2022.
Molnár, P. K., Derocher, A. E., Thiemann, G. W., & Lewis, M. A. (2010). Predicting survival, reproduction and abundance of polar bears under climate change. *Biological Conservation, 143*(7), 1612–1622.
Ostrom, E. E., Dietz, T. E., Dolšak, N. E., Stern, P. C., Stonich, S. E., & Weber, E. U. (2002). *The drama of the commons.* National Academy Press.
Papua New Guinea Mine Watch. (2018, August 23). *Beaked whales may frequent a seabed spot marked for mining.* https://ramumine.wordpress.com/2020/03/16/david-attenborough-calls-for-ban-on-devastating-deep-sea-mining/. Accessed January 1, 2022.
Papua New Guinea Mine Watch. (2020a, March 20). *David Attenborough calls for ban on devastating deep sea mining.* https://ramumine.wordpress.com/2020/03/16/david-attenborough-calls-for-ban-on-devastating-deep-sea-mining/. Accessed January 1, 2022.
Papua New Guinea Mine Watch. (2020b, March 17). *Group wants deep sea mine permit licence cancelled.* https://ramumine.wordpress.com/2020/03/17/group-wants-deep-sea-mine-permit-licence-cancelled/. Accessed January 1, 2022.
Papua New Guinea Mine Watch. (2020c, March 12). *PNG production issues plague Newcrest.* https://ramumine.wordpress.com/2020/03/12/png-production-issues-plague-newcrest/. Accessed May 1, 2022.
Pauly, D. (1979). *Theory and management of tropical multispecies stocks: A review, with emphasis on the Southeast Asian demersal fisheries.* ICLARM Studies and Reviews No. 1, p. 35. International Center for Living Aquatic Resources Management, Manila. ISSN 0115-4389
Pauly, D., Christensen, V., Dalsgaard, J., Froese, R., & Torres, F., Jr. (1998). Fishing down marine food webs. *Science, 279*, 860–863.
Pennington, S. M. (2009). *Deep-sea mining in Papua New Guinea: Policy frontier*. Master's project, Duke University. Retrieved from https://hdl.handle.net/10161/956.
Racca, R., & Bröker, K. (2018). Monitoring and mitigation of sound exposure from seismic surveys for a feeding whale population. *First Break, 36*(11), 71–76.
Raya Rey, A. N., Pizarro, J. C., Anderson, C. B., & Huettmann, F. (2017). Even at the uttermost ends of the Earth: how seabirds telecouple the Beagle Channel with regional and global processes that affect environmental conservation and social-ecological sustainability. *Ecology and Society, 22*(4), 31 [online]. https://www.ecologyandsociety.org/vol22/iss4/art31/
Rosenbaum, H. (2011). *OUT OF OUR DEPTH: Mining the Ocean Floor in Papua New Guinea.* MiningWatch Canada, CELCoR, Oxfam Australia, The Packard Foundation. https://www.deepseaminingoutofourdepth.org
Saeedi, H., & Brandt, A. (2020). Biogeographic Atlas of the deep NW Pacific Fauna. *Advanced Books.* https://doi.org/10.3897/ab.e51315
Savery, L. C., Wise, S. S., Falank, C., Wise, J., Gianios, C., Jr., Thompson, W. D., Perkins, C., Zheng, T., Zhu, C., & Wise, J. P., Sr. (2014). Global assessment of oceanic lead pollution using sperm whales (Physeter macrocephalus) as an indicator species. *Marine Pollution Bulletin, 79*(1–2), 236–244.
Sinclair, J (1978) *Wings of gold: How the aeroplane developed New Guinea.* Pacific Publications. ISBN 0-85807-033-2
Steiner, R. (2009). *Independent review of the environmental impact statement for the proposed nautilus minerals solwara 1 seabed mining project, Papua New Guinea.* Bismarck-Solomon Indigenous Peoples Council. Available online at: http://www.deepseaminingoutofourdepth.org/wp-content/uploads/Steiner-Independent-review-DSM1.pdf. Accessed January 18, 2013.

Steiner, R. (2011). Establishing the papua New Guinea (PNG) Deep Sea Mining (DSM) Citizens' Advisory Council (CAC) Richard Steiner, Professor, Oasis Earth.

Steiner, R. (2016). Comments: ISA working draft exploitation regulations for mineral resources in the area. OASIS Earth Consulting. https://www.oasis-earth.com

Sullivan, N. (2015). Introduction to The Prospects for Sustainable Development in Papua New Guinea. https://www.academia.edu/4492757/Introduction_to_The_Prospects_for_Sustainable_Development_in_Papua_New_Guinea. Accessed June 15, 2022.

The Guardian. (2020, December 8). *Deep-sea 'gold rush': secretive plans to carve up the seabed decried*. J. Watts. https://www.theguardian.com/environment/2020/dec/09/secretive-gold-rush-for-deep-sea-mining-dominated-by-handful-of-firms

The Guardian. (2021, December 9). *Villagers file human rights complaint over plan for giant PNG goldmine*. https://www.theguardian.com/world/2021/dec/10/villagers-file-human-rights-complaint-over-plan-for-giant-pnggoldmine. Accessed May 28, 2021.

The Guardian. (2022a, April 29). *'Deep-sea gold rush' for rare metals could cause irreversible harm*. https://www.theguardian.com/environment/2022/apr/29/deep-sea-gold-rush-rare-metals-environmental-harm. Accessed July 15, 2022.

The Guardian. (2022a, May 26). *The race to protect the deep sea*. https://www.theguardian.com/australia-news/audio/2022/may/26/the-race-to-protect-the-deep-sea. Accessed May 30, 2022.

The Guardian. (2022b). *'Shark calling': locals claim ancient custom threatened by seabed mining*. https://www.theguardian.com/world/2021/sep/30/sharks-hiding-locals-claim-deep-sea-mining-off-papua-new-guinea-has-stirred-up-trouble. Accessed May 31, 2022.

The Guardian. (2022c, July 1). *Stop deep-sea mining, says Macron, in call for new laws to protect ecosystems*. https://www.theguardian.com/environment/2022/jul/01/stop-deep-sea-mining-says-macron-in-call-for-new-laws-to-protect-ecosystems. Accessed July 12, 2022.

Thomson, D. (2022). *How Australia's ethical failures with Timor-Leste should inform a future shaping strategy for the Indo-Pacific*. The Forge. https://theforge.defence.gov.au/publications/how-australias-ethical-failures-timor-leste-should-inform-future-shaping-strategy-indo-pacific. Accessed August 8, 2022.

Wei, C., et al. (15 co-authors). (2011). A global analysis of Marine Benthos Biomass using Random Forest. *Public Library of Science (PLOS), 5*, e15323.

West, P. (2006). Environmental conservation and mining: Between experience and expectation in the eastern highlands of Papua New Guinea. *The Contemporary Pacific, 2*, 295–313. https://core.ac.uk/download/pdf/5103144.pdf. Accessed July 1, 2022.

Williams, B. A., Grantham, H. S., Watson, J. E. M., et al. (2022). Reconsidering priorities for forest conservation when considering the threats of mining and armed conflict. *Ambio*. https://doi.org/10.1007/s13280-022-01724-0

Wise, J. P., Sr., Thompson, W. D., Wise, S. S., & La Certe, C. (2011). A global assessment of gold, titanium, strontium and barium pollution using sperm whales (*Physeter macrocephalus*) as an indicator species. *Journal of Ecosystem & Ecography, 1*, 1–8.

Chapter 14
"Come and Dive" in Papua New Guinea: Surfing, Reefs, Paper Park MPAs and Marine Conservation Issues in PNG Beyond Sharks, Whales, Crocodiles, Seagrass and 'Developing a Resource'

Abstract Coastal ecosystems worldwide are collapsing. In parallel, diving is a relatively recent but intrusive leisure activity of industrialization and its holiday society with a vast human footprint in the atmosphere, socially, on land and certainly under water. It was little pursued by humanity for millennia but never reached such an intensity and depths as found today, globally, e.g. as enabled by plastic and rubber gear, cumbustion engines, including sling shots made available to kids in supermarkets. The human pursuit of pristine wilderness—coral reefs and seagrass beds—remains ongoing, and Papua New Guinea (PNG) offers such waters and coasts. Here, the PNG diving, snorkeling and surfing situation gets put into a conceptual context of the modern ocean crisis, including ecotourism, neoliberalism, technical developments, charismatic megafauna, marine protected area (MPA) failures, sustainable development, climate change and sustainability governance.

Keywords Papua New Guinea (PNG) · Diving · Snorkeling · Surfing · Coral reefs · Marine protected areas (MPAs)

14.1 Introduction

Diving is an outdoor sport and a major industry, pursued worldwide (e.g. Cousteau, 1979; Jackson, 2013)! It's a relatively recent activity in the history of 'The Earth' (which should better be called 'The Ocean' because salt water covers it to over 70%; Cousteau, 1979). Throughout history, humans were mostly diving for food, without much gear nor really for fun. As practiced and promoted now, diving and surfing are leisure activities, as part of the wider industrial income scheme with holiday time. While widespread and bringing tourists 'under water' worldwide, snorkeling and diving flourishes specifically in tropical areas and coral reef regions, seeking specifically pristine waters, new depths and thus increasingly remote areas to explore (Cousteau, 1979; Jackson, 2013). It's similar now to mountain climbing and seeking the extremes and unexplored extending the human niche.

The 'Coral Triangle' region—which PNG is part of—is world famous for its marine beauty (e.g. Stone & Obura, 2013) but also it is affected by tourism and

F. Huettmann, *Globalization and Papua New Guinea: Ancient Wilderness, Paradise, Introduced Terror and Hell*, https://doi.org/10.1007/978-3-031-20262-9_14

heavily threatened by climate change and the ocean crisis overall requiring better management concepts (e.g. Pitman et al., 2020).

New Guinea presents us with a key market area for diving, e.g. for effluent clients from nearby Australia, Asia (Singapore, Hongkong, Taiwan, Japan, etc.) and essentially from all over the world, including EU. PNG is famous for its coral reefs and remote islands (e.g. Lonely Planet Guide; https://www.lonelyplanet.com/papua-new-guinea) while on the Indonesia side, the Gam Island and its surrounding famous karst islands (Raja Ampa) are among the world's richest reef systems in the world, and very popular for snorkeling and diving. New Guinea makes the deep earth's diversity accessible just under the near surface water (Beehler & Laman, 2020, p. 28 for details).

Diving got more intensified and widespread with better plastic and rubber gear such as flippers, snorkels and goggles over the last 50 years. It extended the human niche into ever more and deeper waters (Jackson, 2013; see Cousteau & Richards, 1999, p. 205 for use of surgical rubber in PNG to fasten slings to their spears to make it more powerful and get a wider reach but wiping out more fish). While it pushed out even more sensitive marine wilderness creatures, it brought in many invasive species also. And this trend certainly got a major boost with the world-famous Cousteau's Aqualung apparatus after WW2 (https://www.cousteau.org/legacy/technology/aqualung/); Jacques Cousteau made it happen and provided the tools, and thus, he came early to PNG to apply these techniques further in a unexplored marine wilderness (https://www.cousteau.org/legacy/; Cousteau, 1979; Cousteau & Richards, 1999). It opened up a new world and an ecological niche for humans to explore for fun and commercialization: They can stay longer under water, dive deeper, get closer to marine creatures and explore and exploit more! But in return, this left Mother Earth with a major human footprint, now even found deep under water at the most remote sites imaginable. The Australian Great Barrier Reef adjacent to PNG, and closely related, shows us no other, e.g. Rouphael and Inglis (2001). The human footprint is now widely recognized even within the diving industry and its promoters (Jackson, 2013; see Halpern et al., 2015 for Global Ocean Footprint overall) as it spoils their own experience they are affecting.

The marine life of PNG is still stunning in its own right, from clown fishes to groupers, many species of fusiliers, chama shells (operated for wear and jewelry; Beehler & Laman 2020; Cousteau and Richards, 1999) or the scorpion fish (with toxic needles, if not careful one can walk into those). See Blaha et al. (2016) for an example of genetic diversity not even being fully recognized yet based on crayfish sampling; much more can be expected in other taxa.

And there are many islands to explore in PNG, and Cousteau and Richards (1999, p. 271) showed for one of the outer islands of PNG—Wuvulu island—fascinating killer whale experiences with unique behavior patterns.

Diving easily convinces anybody who can afford to do it in person; others watch it on TV or read the colorful coffee table books (Cousteau, 1979; Cousteau & Richards, 1999). Diving is a global mainstream activity. Once this sport became popular, PNG was visited early on by recreational divers and promoted worldwide as 'the new thing' via media to most homes of the world, e.g. through the Jacques Cousteau family and their TV teams (supported in France, with subsidies and with U.S. media contracts worldwide; Cousteau, 1979. For specific PNG expeditions, see Cousteau & Richards,

Table 14.1 Selection of major globally relevant marine areas and dive sites elsewhere

Site	Location, nation	Outstanding feature	Conservation status
Maldives	Indian Ocean	Underwater reefs around islands	Climate change problems, sea level rise
Red Sea	Egypt/Israel	Coral reefs	Coral bleaching, overuse including ship traffic
Belize	Caribbean	Coral reef	Invasive species, overfishing, climate change impacts
Papua New Guinea	PNG	Coral reef and islands	Overfishing and contamination
Great Barrier Reef	Australia	Coral reef	Various, mining contamination, coral bleaching
Galapagos	Ecuador	Cold current	Invasive species, tourism

1999). PNG reefs are of world esteem. It markets itself as one of the last pristine, untouched reef areas in the world, equally or better than the Great Barrier Reef in adjacent Australia (e.g. Jackson, 2013; see Cousteau & Richards, 1999; Beehler & Laman 2020 for an environmental and wider perspective).

The coastline, coral reefs, beaches and remote sites and islands of PNG are simply a major experience for anybody to see and to explore with a conservation and humanity perspective. Rightly so, it finds itself in the travel office catalogues as one of the last wilderness places left in the world, exotic and it's all found right under the ocean surface, just waiting to be explored for anybody who can come there (e.g. https://walindiresort.com/; https://www.facebook.com/niuginidivepng/).

But while these diving sites are promoted as a 'fun place,' the ecological services that those reefs bring are vast. It's serious. Mankind cannot live without it. It's fragile, sophisticated and simply nothing to be played with really (Table 14.1). In the meantime, PNG is wild but actually lacks some relevant diving infrastructure e.g. high-powered hospitals and diving (de-compression) chambers nearby dive sites, emergency flights 'out' can be remote and long.

These sites are a major seascape wilderness. But in parallel, it also fragile and starts to get compromised increasingly. That's a global feature found with most diving sites, where resorts get promoted (Jackson, 2013). Those resorts tend to be set up by the wealthy, and where the wider public is essentially excluded, while a larger workforce stands ready to serve the paying tourists. There is no free lunch in that business unfolding on the wider public good of the reefs (Fig. 14.1).

PNG is affected by the Wallace Line, and the Weber Line (specifically for fish; Beehler & Laman, 2020 and citations within). Another reason why PNG is so fascinating for its marine life simply is its geography and related remoteness. For long time of the earth's history, it was secluded and thus was left alone; and before the '*aqualung,*' nobody really ventured into the deeper underwater ecosystems, ever. PNG and its waters are simply new and exotic to the western world and western eyes; the Europeans are not used to coral reefs and thus mishandle them. Even in colonial times, that aspect was widely untouched, e.g. due to the lack of the aqualung!

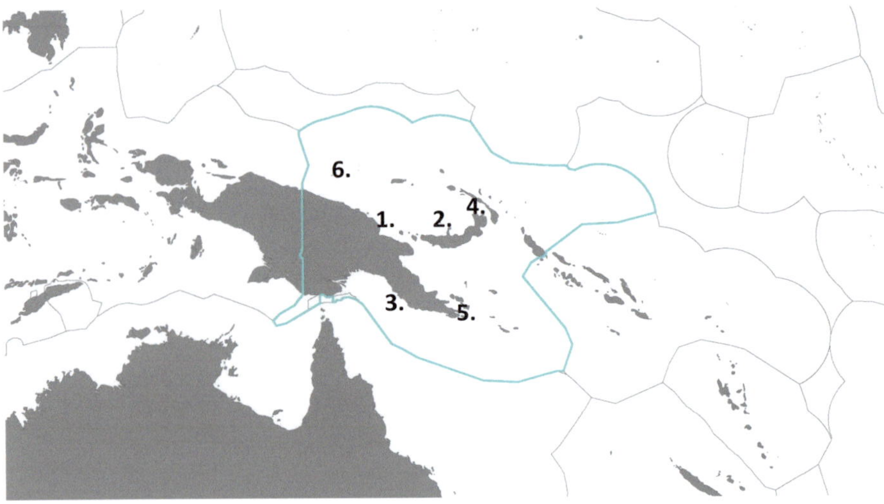

Fig. 14.1 Map of selected diving areas in Papua New Guinea 1. Madang, 2. New Britain, 3. Port Moresby area, 4. Kakopo, 5. Milne Bay region, 6. Wuvulu Island. Map also shows the EEZ of PNG (in blue)

Another reason for being let alone is PNG's fear factor ("cannibals"), and then, a third reason is the connection '*Ridge to Reef*': Tropical rainforests produce much nutrients, and together with the rainfall, it flows downhill and into the ocean, mixing, and that way feeding the currents and subsequently some reefs (e.g. Carlson et al. 2019). While this is a great support of the reefs, the downwash from the slope now spoils reefs and such habitats spoiling the ecology of the region. Other factors favoring biodiversity come to play also, e.g. ocean currents and the tropical location near the equator/sun.

But it is very important to state that it should be '*Ridge to Reef, and back.*' Because whatever happens on the watershed shows up further below in the ocean and vice versa (for examples see Cousteau & Richards, 1999). It's a main topic for any forestry activity in such areas, certainly in PNG (see Delevaux and Stamoulis 2020 for management).

14.2 What Do the 'Sea Around Us' Data Tell Us? A Nation-State Overfishes

A good source for any fisheries and EEZ data is the 'Sea Around Us' project (https://www.seaaroundus.org/). It shows for instance that the reported fisheries catch data on public record reported from PNG carry a reliability of ~1.1 (ranking from 4, the highest, to 1, the lowest). That does not instill much trust. But it's obvious from those records that tuna presents the largest harvest; skipjack and yellowfin tuna are seeked after as a global commodity. Further one can see very well when a 'sustainable fisheries' was started—statehood in the 1970s—and what it means: very high fishing rate, done in a boom-and-bust cycle. This can hardly follow maximum sustainable yield equations, or any good stable plan with a longer vision and stable approach and oversight. Arguably, the local artisan fisheries—as a major item in Melanesia—are not even included, nor are species like mother-of-pearl, giant clams or even herring-type fish. It simply makes the case that these data show us a vast underestimate of impact, albeit a very grim one.

14.3 Seagrass Anyone?

As one ocean habitat of many, the remaining seagrass beds are a main asset of PNG; they remain widely unassessed and but get now increasingly affected. Perhaps there are up to 13 different species in such beds of 'seagrass' (MacKenzie et al., 2021). It was also shown that those seagrass beds can be a singular plant clone, and thus perhaps are one of the largest plants in the world (Edgeloe et al., 2022). While the visibility in seagrass beds can be 'sanded,' they are the nursery grounds for fish and sea turtles for instance. An increased sedimentation has been observed in seabeds, e.g. due to watershed run-off and mining (details in Cousteau & Richards, 1999). Over-fertilization is an associated problem in such ancient seagrass beds. Much of the freshwater inflow comes from water sources downhill originating from mountain forests and feeding the adjacent reefs. With alluvial mining as a major mode of operation in PNG, this cannot be good news for seagrass beds, or beyond (see for instance Wise et al., 2011 for sperm whales being contaminated in such ocean regions). Those seagrass beds are vast and provide a massive ecological service to the world, certainly for the marine life! But those are not well studied, mapped or protected in PNG. It's a global heritage though and worth a diving trip indeed.

14.4 Come and See the Last Mangroves

Mangrove forests are on a massive decline worldwide (Sandilyan & Kathiresan, 2012); same applies now to PNG too. These coastal forests are simply cut down and get occupied by people and projects. Mangroves of New Guinea still exist in large coverage (Beehler & Laman, 2020). Mangroves are known to be large carbon sinks (Bouillon et al., 2008 for a discussion and relevance), as well as habitats, e.g. for birds, fish and turtles, as well as dugongs. Mangroves have vast ecological services. But while having been there for millennia, mangroves got in the way now, e.g. for development. It's commonly used too for fire wood. Like with regular forests in PNG, mangroves are not well mapped, inventoried and looked after.

14.5 Estuaries to Watch for

There are a few very large estuaries in PNG where rivers discharge in the ocean. The outflow creates a 'plume' which presents a consistent, dynamic and very sediment-rich nutrient flow into the wider ocean waters; a mixing regime offers vast resources for sustainable fisheries. It's those estuaries that are marine hotspots connecting with many other places in the Pacific, e.g. for migratory seabirds. And it's no wonder then that people live in such productive areas for a long time. And they live there well. Such estuary locations are found throughout PNG, e.g. the Fly river, the Sepik river or the Ramu river, as the three largest outflows in PNG. Those are classic entry points for people to get into interior PNG (see for instance Sauer, 1915 for German explorations of the Ramu river system in colonial times). However, with increasing river pollution, e.g. Fly river from the Ok Tedi mine (Kirsch, 2014), and the Sepik river also now covered with invasive species (Cousteau & Richards, 1999) those plumes and their food resources can become a hazard; they got spoiled. And added by ocean pollution and climate change things do not get any better for the estuaries and their food chains. While those estuaries are very muddy and come with sharks and crocodiles easily spoiling a diving trip, PNG's estuaries are special habitats to be protected, but lack most of the suitable efforts to do so. In the meantime, waterways that can be navigated will certainly have ship traffic, including tourist cruise boats on the Sepik or the Fly.

14.6 "Surf the Waves", as a New Participator in the Neoliberal Utilization Scheme of the PNG Oceans

Hawaii, Tahiti and California are traditional hotspots of the world surfing world; Australia and other sites followed suit (with few exceptions like Normandy/France and Portugal surfing is not so widespread on the Atlantic side but on the rise, like in Nova Scotia, Canada in winter). While it's a fun 'outdoor' activity, it actually has quite a commercial undercurrent and is a wider and intense business model. Surfing was likely a bit 'anti' but now got used for 'development,' virtually anywhere, e.g. from Alaska and Nova Scotia in winter to Nicaragua (https://www.surfline.com/travel/nicaragua-surfing-and-beaches/3617476), and Papua New Guinea (West, 2014). One can easily find now 'new' surf beaches in Wewak region, Sepik, and some surf dudes start to use them; all connected by Australian airlines (usually facilitated by including surf boards cheaply in the airline ticket) While it might have a rel. little environmental impact, it certainly has a social and carbon one. Surfing and those surf dudes are now part of the neoliberal ocean portfolio, and it's commercially serving a few investors and entrepreneurs, including hotel owners.

14.7 PNG as a Cheap Labor Operation for Diving and Surf Resorts and Others Work Places

And lastly, the pricing of a PNG diving trip is to be seen in the wider global context: PNG staff has little union power and a marginal salary in a nation that already ranks among the poorest in the world. The minimum salary per hour is app. 4 Kina, and per week it would be c. $160 US per month. That makes great business; a win–win…for some (for the tourists and for the operator abroad: but PNG tends to lose, including its economy). The large resorts are usually surrounded by villages, readily providing the labor for those resorts. It's the other side of the coin. The manpower needed for such resorts are linked to staff and kitchen personnel for instance, whereas the management and ownership remains outside of PNG and where the real money is made. It comes by design.

14.8 Sustainable PNG as the Treasure Trove

Some might see this (marine) biodiversity wilderness as a large treasure trove for the world's species richness, but it also is one for biomass and ecosystem services, and thus, a sustainable management of food and fisheries resources. And that's precisely what has happened in this region for over 47,000 years, e.g. PNG and Torres Islanders and the Great Barrier Reef (Table 14.2; Fig. 14.2).

This treasure trove gets exploited, as per the means available. Pirates were few, but 'beach combers' and escapees from the western colonial system are known there for a long time (for some statistics for northern PNG see Mayr & Diamond, 2001, p. 30). Dynamite fishing was commonly applied in the 1970s and earlier (Pauly et al., 1989, see Cousteau and Richards for details and amputees who tried to re-use war dynamite from WW2). The legacy of dynamite fishing left many victims, not only

Table 14.2 Components where PNG waters are still world leading

Marine feature	Justification	Citation	Ecological relevance
Seagrass	Large areas left and little disturbed or studied	e.g. Beehler and Laman (2020)	Global importance for ecological services
Mangroves	Large areas left		Global importance for ecological services
Islands	Vast number and diversity	e.g. Mayr and Diamond (2001)	Diversity, biogeography
Pristine waters	Relatively little industrialization	FH pers.com	Global importance for ecological services
Tuna	Rel. high abundance compared to other ocean areas	FH pers.com	Food chain

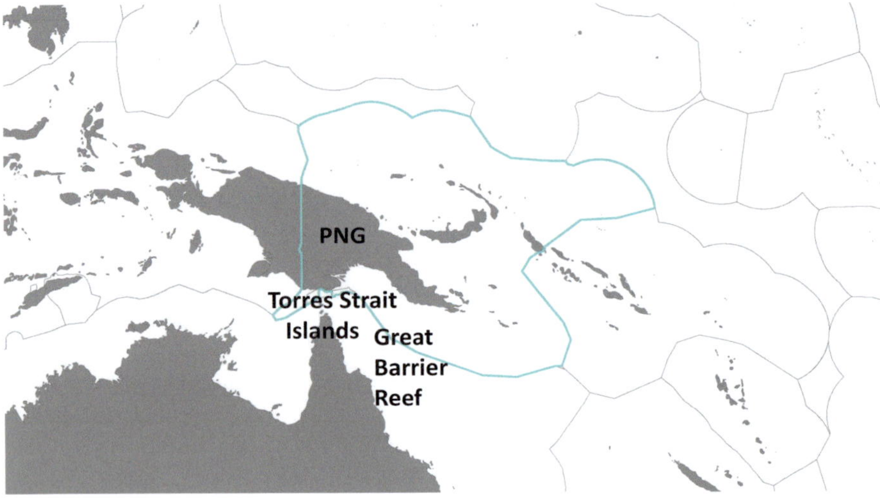

Fig. 14.2 Papua New Guinea in relation to Great Barrier Reef and Torres Strait Islands; all of those are major habitats of global relevance and with indigenous groups and civilizations co-evolved for millennia

reefs and ecosystems, but also for the wider impact and ecosystem. Many problems from this can be seen today.

Other 'modern' forms of exploitation and business concepts can widely be found beyond seafloor mining. For instance, sea cucumbers get caught in PNG and sold in Singapore, etc., with a tenfold gain (Cousteau & Richards, 1999; see Hair et al., 2019). What does such a business result into? As can be seen in the easily accessible traditional ocean resources (Hyndman, 1988; Hair et al., 2019), a shift from traditional village business models to new exploitive ones will make the natural resource and its sophisticated sustainable management expertise simply 'go away.' Shark-calling is among those, so is giant clam gardening, beyond many others.

14.9 Harvest a Crocodile, Eat a Dugong, Some River Dolphins and Others

PNG was famous in the 1960s and before for its (swamp) crocodile harvest (e.g. Bolton et al., 1982). And saltwater crocodiles (Webb et al., 2010) reached record trophy sizes. It was part pof the colonial exploit. But those harvest rates are now in the massive decline, if even that (Daltry et al., 2016). That is true for their abundance and their size; a clear symptom of their demise, just like elephants with big tusks are on the way out these days, or large land and sea turtles, or large mammals.

For Dugongs, those are not so easy to see anymore neither. One can sometimes find hunted penis bones on PNG markets (see previous chapter of this book). Whereas in Torres Islands—with a more direct Australian oversight—those next to PNG underly a more stringent regime and IUCN monitoring (Heinsohn et al., 2004) but remain very dubious for conservation effectiveness.

The status and narrative of river dolphins is another poor one anywhere in Asia (see Beasley & Brown, 2018 for Snubfin Dolphin in Australia and PNG). Those species are difficult to detect and to count, and thus to conserve. But it looks clear that those habitats are getting compromised and that their species are fading out also; like most river dolphins do these days. It's part of a wider trend found all over Asia and the world for such species (see Huettmann et al., 2020 for Ganges Dolphin & Huettmann, 2020 for China), with the richest countries having the worst records, e.g. China, Brazil. See also details on marine mammals with Mavea et al. (2021) and recent associated seabird species (Rayner et al. 2020).

14.10 And Where Are the Sharks Now?

Many divers are attracted by sharks, and vice versa. Like with so many other ecosystem components, without sharks, the tropical oceans are less relevant, hardly viable. Once sharks are gone, humans will be next. And the sharks are going...involuntary as they are overharvested. The global shark decline is for real though (Ferretti et al. 2010, Worms et al. 2013), and PNG is no exception (Beehler & Laman, 2020; The Guardian, 2022). With big ocean predators in stress, or decline, one sees a rise of slime in many oceans now.

With PNG being a shark fishing site, it's very noteworthy to look at the underlying structure how this all happened, who did it and who was responsible, and what does it mean for the foodchain structure overall!

PNG has a long tradition to catch sharks (shark calling; details in Cousteau & Richards, 1999; The Guardian, 2022 for seabed mining impacts). But clearly, the decline of sharks was not a PNG job, instead it came through the global and Asian demand of shark products, also fueled by a western market offering, e.g. Canada and Europe (Beehler & Laman, 2020). But then it got intensified with globalization and other overfished stocks while shark stocks are generally on the decline already (Butler et al., 2014; Vieira et al., 2017). In this regard, the mislabeling of sharks on international markets for fish products should be stated also (Pazartzi et al., 2019; it's quite common in fisheries and food markets anywhere though, including meat an on honey products).

14.11 "Come and Dive" to See the Deckchairs of the Titanic, in a Designated MPA and Outside...

While the oceans are in a global crisis, PNG remains a jewel to explore, including sunken ships and war planes shut down during WW2 ('wreck's; Cousteau and Richards 1999 for photos and details). But PNG also shows similar ocean problems than found in many other places of the world. Other than fish, the aforementioned Sea Around Us project for instance shows 93 vertebrate species for PNG; and that's arguably still an underestimate. Most of them remain unstudied, some undescribed.

Let's have it protected then! Well, for that ocean area a prioritization is called for, and then trying to locate hotspots to be protected…However, that widely used narrative fails on many grounds as it ignores the wider ecological connections required for such hotspots to function. Ocean hotspots cannot stand by themselves and in a vacuum. Many of the marine protected areas (MPAs) are 'paper parks,' as they have little budget and very poor enforcement records (see Buckley, 2020 for concept and terrestrial examples in Asia effecting rivers, estuaries and oceans). Secondly, many MPAs are not 'no-take' zones (protected) as people often believe, e.g. as known for most terrestrial Nationalparks. But rather allow a 'take', e.g. fishing (the commercial lobby would otherwise hardly agree with those efforts). Third, in PNG, mutual ownership drives the resource (Hyndman, 1988), and thus, an exclusive national park concept such as a MPA hardly gets a wider buy in. It fails almost anywhere else in New Guinea (Beehler & Laman, 2020). Fourth, even if a good outcome can be planned, it will take at least another 20 years for good progress outcome. For the fastly changing coral reefs, likely that is too late. The various ongoing MPA efforts in PNG—a world-relevant resource after all—must simply be seen in that light (Green et al., 2009; Hamilton et al., 2009; White et al., 2014). And the latest western trend of zoning will unlikely make it better; options remain few. PNG waters are on a same trajectory as outlined by Jackson et al. (2001, 2013) globally: a collapse. And with that ecological collapse comes a human collapse (Diamond, 2011).

In conclusion, diving in PNG remains a major experience of nature, of the marine life of Mother Earth, of PNG and of one's own soul. It's a life changer to most people who can afford it, while conservation management on an efficient level with a good vision is widely absent (Pittman et al., 2020) and currently it cannot achieve what it sets out to do. How to carry it out sustainably remains unknown and is not promoted hardly asked for by practitioners or a take-home message even; it's an abyss.

References

Beasley, I., & Brown, A. M. (2018). Australian Snubfin Dolphin: Orcaella heinsohni. In *Encyclopedia of Marine Mammals* (pp. 47–49). Academic Press.

Blaha, M., Patoka, J., Kozák, P., & Kouba, A. (2016). Unrecognized diversity in New Guinean crayfish species (Decapoda, Parastacidae): The evidence from molecular data. *Integrative Zoology, 11*, 457–468.

Bouillon, S., Borges, A. V., Castañeda-Moya, E., Diele, K., Dittmar, T., Duke, N. C., Kristensen, E., Lee, S. Y., Marchand, C., Middelburg, J. J., Rivera-Monroy, V. H., & Twilley, R. R. (2008). Mangrove production and carbon sinks: A revision of global budget estimates. *Global Biogeochemical Cycles, 22*(2).

Bolton, M., & Laufa, M. (1982). The crocodile project in Papua New Guinea. *Biological Conservation, 22*(3), (pp. 169–179).

Buckley, M. (2020). A first-hand narrative account on Tibet's paper parks: How China's greenwashing in Tibet flies under the radar. In *Hindu Kush-Himalaya Watersheds Downhill: Landscape Ecology and Conservation Perspectives* (pp. 197–212). Springer.

Butler, J. R. A., Skewes, T., Mitchell, D., Pontio, M., & Hills, T. (2014). Stakeholder perceptions of ecosystem service declines in Milne Bay, Papua New Guinea: Is human population a more critical driver than climate change? *Marine Policy, 46*, 1–13.

Beehler, B., & Laman, T. (2020). *New Guinea: Nature and Culture of Earth's Grandest Island*. Princeton University Press.

Carlson, R. R., Foo, S. A., & Asner, G. P. (2019). Land use impacts on coral reef health: A ridge-to-reef perspective. *Frontiers in Marine Science, 6*, 562.

Cousteau, J. (1979). *The Ocean World*. Abrams Publisher New York.

Cousteau, J. M., & Richards, M. (1999). *Cousteau's Papua New Guinea Journey*. Henry Abraham's Books.

Daltry, J. C., Langelet, E., Solmu, G. C., van der Ploeg, J., van Weerd M., & Whitaker, R. (2016). Successes and failures of crocodile harvesting strategies in the Asia Pacific Region. *Tropical Conservation: Perspectives on Local and Global Priorities*, 345–362.

Delevaux, J., & Stamoulis, K. (2020). *Assessment of ridge-to-reef management*. Seascape Solutions LLC.

Diamond, J. (2011b). *Collapse: How societies choose to fail or succeed*. Revised Edition. Penguin Books.

Edgeloe, J. M., Severn-Ellis, A. A., Bayer, P. E., Mehravi, S., Breed, F., Krauss, S. L., Batley, J., Kendrick, G. A., Sinclair, E. A. (2022). Extensive polyploid clonality was a successful strategy for seagrass to expand into a newly submerged environment. *Proceedings of the Royal Society* B.2892022053820220538

Ferretti, F., Worm, B., Britten, G. L., Heithaus, M. R., & Lotze, H. K. (2010). Patterns and ecosystem consequences of shark declines in the ocean. *Ecology Letters, 13*(8), 1055–1071.

Green, A., Smith, S. E., Lipsett-Moore, G., Groves, C., Peterson, N., Sheppard, S., Lokani, P., Hamilton, R., Almany, J., Aitsi, J., & Bualia, L. (2009). Designing a resilient network of marine protected areas for Kimbe Bay, Papua New Guinea. *Oryx, 43*(4), 488–498.

Hair, C., Foale, S., Kinch, J., Frijlink, S., Lindsay, D., & Southgate, P. C. (2019). Socioeconomic impacts of a sea cucumber fishery in Papua New Guinea: Is there an opportunity for mariculture? *Ocean & Coastal Management, 179*, 104826.

Halpern, B. S., Frazier, M., Potapenko, J., Casey, K. S., Koenig, K., Longo, C., & Walbridge, S. (2015). Spatial and temporal changes in cumulative human impacts on the world's ocean. *Nature Communications, 6*(1), 1–7.

Hamilton, R., Green, A., & Almany, J. (Eds.) (2009). Rapid ecological assessment: Northern Bismarck Sea, Papua New Guinea. Technical report of survey conducted August 13 to September 7, 2006. TNC Pacific Island Countries Report No. 1/09.

Huettmann, F. (2020). Chapter 31 Hydrodams: Death by over 500 cuts and blockages virtually build without Impact Assessment. In G. R. Regmi, & F. Huettmann (Eds.), *Hindu Kush-Himalaya watersheds downhill: Landscape ecology and conservation perspectives* (pp. 633–648). Springer Gland.

Huettmann F., Shrestha, M., Puri, G., & Regmi, G. R. (2020). Chapter 32 Hydrodams again: A closer look at expensive gharials, lost dolphins, dissected fish migration, and destructive energy for non-sustainable mining and societies abroad fueling a non-democratic governance for the designed '*homo electricus*' In G. R. Regmi, & F. Huettmann (Eds.), *Hindu Kush-Himalaya Watersheds Downhill: Landscape Ecology and Conservation Perspectives* (pp. 649–661). Springer Gland, Switzerland.

Heinsohn, R., Lacy, R. C., Lindenmayer, D. B., Marsh, H., Kwan, D., & Lawler, I. R. (2004, November). Unsustainable harvest of dugongs in Torres Strait and Cape York (Australia) waters: Two case studies using population viability analysis. *In Animal Conservation forum* (Vol. 7, No. 4, pp. 417–425). Cambridge University Press.

Hyndman, D. (1988). Melanesian resistance to ecocide and ethnocide: Transnational mining projects and the Fourth World on the island of New Guinea. *Tribal Peoples and Development Issues. A Global Overview Mayfield Publishing Company, Mountain view, CA*, (pp. 281–298).

Jackson, J. (2013). *Dive Atlas of the world: An illustrated reference for the best sites*. IMM Lifestyle Books.

Jackson, J. B. C., Kirby, M. X., Berger, W. H., Bjorndal, K. A., Botsford, L. W., Bourque, B. J., Bradbury, R. H., Cooke, R., Erlandson, J., Estes, J. A., Hughes, T. P., Kidwell, S., Lange, C. B., Lenihan, H. S., Pandolfi, J. M., Peterson, C. H., Steneck, R. S., Tegner, M. J., & Warner, R. R. (2001). Historical overfishing and the recent collapse of coastal ecosystems. *Science, 293*, 629–637.

Mayr, E., Diamond, J. M., & Diamond, J. (2001). *The birds of northern Melanesia: Speciation, ecology & biogeography*. Oxford University Press on Demand.

McKenzie, L. J., Yoshida, R. L., Aini, J. W., Andréfouet, S., Colin, P. L., Cullen-Unsworth, L. C., Hughes, A. T., Payri, C. E., Rota, M., Shaw, C., & Skelton, P. A. (2021). Seagrass ecosystems of the Pacific Island Countries and Territories: A global bright spot. *Marine Pollution Bulletin, 167*, 112308.

Mavea, W. B., Belonovich, O., Nagombi, E., Boslogo, T., Gwynn, L., Russ, R., & Bird, J. P. (2021). A survey of cetaceans in New Britain, Papua New Guinea. *Australian Mammalogy*.

Pazartzi, T., Siaperopoulou, S., Gubili, C., Maradidou, S., Loukovitis, D., Chatzispyrou, A., & Imsiridou, A. (2019). High levels of mislabeling in shark meat–Investigating patterns of species utilization with DNA barcoding in Greek retailers. *Food Control, 98*, 179–186.

Pauly, D., Silvestre, G., & Smith, I. R. (1989). On development, fisheries and dynamite: A brief review of tropical fisheries management. *Natural Resource Modeling, 3*, 307–329. https://doi.org/10.1111/j.1939-7445.1989.tb00084.x

Pittman S. J. , Yates, K. L., Bouchet, P. J., Alvarez-Berastegui, D., Andréfouët, S., Bell, S. S., Berkström, C., Boström, C., Brown, C. J., Connolly, R. M., Devillers, R., Eggleston, D., Gilby, B. L., Gullström, M., Halpern, B. S., Hidalgo, M., Holstein, D., Hovel, K., Huettmann, F., Jackson, E. L., James, W. R., Kellner, J. B., Kot, C. Y., Lecours, V., Lepczyk, C., Nagelkerken, I., Nelson, J., Olds, A. D., Santos, R. O., Scales, K. L., Schneider, D. C., Schilling, H. T., Simenstad, C., Suthers, I. M., Treml, E. A., Wedding, L. M., Yates, P., & Young, M. (2020). Seascape ecology: Identifying research priorities for an emerging ocean sustainability science. *Marine Ecology Progress Series, 663*, 1–29. https://doi.org/10.3354/meps13661

Rayner, M. J., Baird, K. A., Bird, J., Cranwell, S., RAINE, A. F., Maul, B., Kuri, J., Zhang, J., & Gaskin, C. P. (2020). Land and sea-based observations and first satellite tracking results support a New Ireland breeding site for the Critically Endangered Beck's Petrel Pseudobulweria beckii. *Bird Conservation International, 30*, 58–74.

Rouphael, A. B., & Inglis, G. J. (2001). "Take only photographs and leave only footprints"?: An experimental study of the impacts of underwater photographers on coral reef dive sites. *Biological Conservation, 100*(3), 281–287.

Sandilyan, S., & Kathiresan, K. (2012). Mangrove conservation: A global perspective. *Biodiversity and Conservation, 21*(14), 3523–3542.

Sauer, C. O. (1915). Exploration of the Kaiserin Augusta River in New Guinea, 1912–13. *Bulletin of the American Geographical Society, 47*(5), 342–345.

Stone, G. S. & Obura D. (2013). *Underwater Eden: Saving the last coral wilderness on earth.* The University of Chicago Press.

The Guardian. (2022). *'Shark calling': Locals claim ancient custom threatened by seabed mining.* https://www.theguardian.com/world/2021/sep/30/sharks-hiding-locals-claim-deep-seamining-off-papua-new-guinea-has-stirred-up-trouble. Accessed May 31, 2022.

Vieira, S., Kinch, J., White, W., & Yaman, L. (2017). Artisanal shark fishing in the Louisiade Archipelago, Papua New Guinea: Socio-economic characteristics and management options. *Ocean and Coastal Management, 137*, 43–56.

Webb, G. J., Manolis, S. C., & Brien, M. L. (2010). Saltwater crocodile crocodylus porosus. Crocodiles. In S. C. Manolis & C. Stevenson, *Status Survey and Conservation Action Plan*, 4rd edn. (pp. 99–113) Crocodile Specialist Group.

West, P. (2014). "Such a site for play, this edge": Surfing, tourism, and modernist fantasy in Papua New Guinea. *The Contemporary Pacific, 26*(2), 411–432.

White, A. T., Aliño, P. M., Cros, A., Fatan, N. A., Green, A. L., Teoh, S. J., Teoh, S. J., Laroya, L., Peterson, N., Tan, S., Tighe, S. & Venegas-Li, R. (2014). Marine protected areas in the coral triangle: Progress, issues, and options. *Coastal Management, 42*(2), 87–106

Wise Sr, J. P., Thompson, W. D., Wise, S. S., LaCerte, C., Wise, J., Gianios Jr, C., Zheng, T., Benedict, L., Mason, M. D., Payne, R., & Kerr, I. (2011). A global assessment of gold, titanium, strontium and barium pollution using sperm whales (Physeter macrocephalus) as an indicator species. Journal of Ecosystem and Ecography.

Worm, B., Davis, B., Kettemer, L., Ward-Paige, C. A., Chapman, D., Heithaus, M. R., Kessel, S. T., & Gruber, S. H. (2013). Global catches, exploitation rates, and rebuilding options for sharks. *Marine Policy, 40*, 194–204

Chapter 15
Timber and Forestry in Papua New Guinea: Still no Publicly Available Forest Inventory Data or Growth-and-Yield Tables Despite Google Earth, Microsoft Cloud, NGOs, International Development Aid, REDD, Satellites and Drones Applied Last Decade for a Science-Based Forestry

The forest is a source of my medicine My wealth My everything
Waiko (in Cousteau & Richards,1999, p. 212)
...the companies carry out their activities with no consideration whatsoever for their impacts on biodiversity, and they employ logging methods that cause considerable ecosystem degradation.
Beehler and Laman (2020, *p. 35).*
...lay people [...] seem to intuitively grasp the need for an ecological approach to forest management better than forestry professionals do.
Wohlleben (2015) in Kingsland (2018).

Abstract The largest intact virgin rainforest blocks in the Pacific are located in Papua New Guinea (PNG). They are recognized as world-relevant carbon sinks and species cradles in the hands of tribes with a complex public land tenure and society. Those areas are the livelihood of many complex indigenous tribes, who lived for millennia in a sustainable fashion far removed from western urbanization, industrialization and globalization. In the meantime, a ruthless forestry extracts timber from this region in a very high amount usually for the Asian market, in addition to an Australia-supported land scheme widely perceived as public theft; none of it is sustainable or considers PNG long term. International aid, development and NGOs for PNG have widely not resolved that issue well, hardly acknowledged the underlying policy and governance situation. Australia, the respective professions and the international public is fully informed and usually aware but remains passive. While science has not even described all tree species (e.g. Slik in PNAS 112:7472–7477, 2015), a certified and best-practice forestry with a transparent forest inventory and growth-and-yield tables for a 'sustainable cut' is virtually impossible—hardly done—in such a socio-economic science environment with climate change on the rise. Instead, indigenous management operated fine in such habitats for millennia. An outlook is presented to acknowledge this problem and for moving forward in the best-possible scenario with a grim outlook.

F. Huettmann, *Globalization and Papua New Guinea: Ancient Wilderness, Paradise, Introduced Terror and Hell*, https://doi.org/10.1007/978-3-031-20262-9_15

Keywords Papua New Guinea (PNG) · Sustainable forestry failure · Forest exploitation · Timber export · Wilderness · Overharvest · Corruption · Carbon sequestration

15.1 Introduction

Papua New Guinea (PNG) is covered by an estimated app. 65% with forests, of which app. 85% are prime forests (exact numbers are not available due to inventory problems but for details and context see Beehler & Laman, 2020; Shearman & Bryan, 2011; Global Forest Resources Assessment FRA 2015). PNG features the largest intact prime forest left in the Pacific (Beehler & Laman 2020. For overall New Guinea, over 20,000 plant species are found. In PNG, over 400 tree species are used for the timber market, out of c. 600 species, but up to 3000 species or so can likely be expected to be taxonomically detected in PNG forests still.

PNG forests are ancient, vast and unique; they are the livelihood of a deep human sustainable society. The forests feature for instance ancient fig species, Antarctica beaches, oaks and Dipterocarp species (the latter group has a lower diversity in PNG than elsewhere in SE Asia), and c. 150 rhododendrons. Noteworthy is also the Klipli pine of 45 m height, an Araucaria likely introduced centuries ago by human immigration from Asia, allowing for better wood constructions in the bush (see also the Casuarina tree that arrived in PNG c. 1200 years ago; Beehler & Laman 2020). Humans did indeed over time modify the forest tree composition in PNG for their gain, and this includes planting, and certainly fires (Flannery, 2002).

But for a long time, it's widely published and recognized in PNG—and also internationally—that the timber industry is among the chief corrupt industries in the business sectors and in the nation (e.g. Filer et al., 2009a, 2009b; Greenpeace, 2005; Laurence et al., 2011; Shearman et al., 2009, see also The Guardian, 2020). Statistically, the official annual deforestation rate is 0.44% (which is a lot and fast when occurring on a finite landmass and annually; see Hannah et al. 2020 for bad impacts). Most of the cutting occurred in low lands accessible by road, but more pressure is now put on steeper slopes (those habitats are of concern as the water runoff washes out soil and adds to sedimentation on seagrass beds, coral reefs, etc.). The amount of protected land is very low, e.g. <1%, with just 2 RAMSAR wetlands (Mongabay, 2022). Relevant aims, motives and actions become quickly clear from those metrics: In defiance of local people and customs simply get as much timber removed and brought to the market cheaply as possibly, e.g. with international dumping price and exchange trading via shell companies in SE Asia and China (details in Beehler & Laman, 2020; see also Cousteau & Richards 1999 for Japan and Australian forestry failures and major disputes with the local community). One of the many complications in PNG and Melanesia sits with the public and complex land tenure system. Land and natural resources in PNG are virtually not privately owned, and thus, it's 'social', and the community at large always is involved (example for PNG Forestry shown in Keenan et al., 2011) (Figs. 15.1, 15.2, 15.3, 15.4, 15.5, 15.6, 15.7, 15.8, 15.9, 15.10, 15.11, 15.12, 15.13, 15.14, 15.15 and 15.16).

Fig. 15.1 Fire wood, collected by a party of kids and females. A common sight in rural PNG

Fig. 15.2 A big tree cut down with an axe and manual saw: a rare sight in industrial forestry where chain saws and graders rule

Fig. 15.3 A large old-growth tree in Papua New Guinea; extremely precious these days for ecological services and associated culture of forest people

Fig. 15.4 Virgin ancient forests of Papua New Guinea; among the last ones left in the world certainly in the Pacific

Fig. 15.5 Fine people of the forest villages in Papua New Guinea; a typical family set up in 'the bush'

Fig. 15.6 A subsistence cut of tropical forest for gardening

Fig. 15.7 A palm tree grows fast and is widely used in subsistence forestry, but virtually not accounted for in any timber and forestry practice

Fig. 15.8 Trees on stilts, often found in flooded and wet areas, make for good 'timber' are widely used by the locals in Papua New Guinea; those kind of trees are usually not well accounted for in timber management, what are the growth-and-yield tables?

Fig. 15.9 Ferns growing on trees; just another 'ecosystem in an ecosystem' that is virtually not yet accounted for in meaningful forest management and in the forestry profession

Fig. 15.10 Staghorn ferns (Platycerium) also growing on trees; and again another essential 'ecosystem in an ecosystem' that also is virtually not yet accounted for in a relevant forest management

Fig. 15.11 Ancient fig trees in Lae city. While those are not in the forest, they are part of urban vegetation and make up for an impressive sight -widely found in Islands throughout the entire Pacific

15.2 Prime Forests of PNG: Something Very Different

Prime forests are something that many urbanized western people cannot relate well with and have no experience on; they do not really know how to operate and to manage, or to conserve them. The simplest understanding—profit-making—of the widely forest-covered landmass in PNG with precious virgin forests left just includes the cutting and selling timber from island rain forests quickly. This is done to deliver them for consumption abroad at the lowest cost. Shipping operations are an inherent part of such a forestry. Asia—China—is the core market for PNG (Beehler & Laman 2020), and this happens with an Australian oversight and research participation, e.g. taxonomic tree species, economic PNG assessments under 'aid' and institutional knowledge. Wiping out regional forest cover, and specifically on islands of PNG, can quickly result in erosion and destruction (Cousteau & Richards 1999, p. 207), wholesale habitat loss and species endangerment and subsequent destruction of the co-evolved human fabric, e.g. Baraka (2001). Forestry action was relatively low in the last 100 years but increased strongly after independence ('Nationhood'), and since the 1990s, it got into Asian predatory markets, usually to overcome the Asian financial crisis back then (as stated in Beehler & Laman, 2020). This pattern of cutting trees in PNG equals fisheries or mining intensities: the last 30 years after the so-called 'independence' were the worst in PNGs history!

Fig. 15.12 Mangrove forests, an overlooked forest type and which is even more far away from a sustainable forestry than the usual industrial cut blocks. This is partly due to mangroves having no real tree rings and are hard to access, inventory and manage

Fig. 15.13 The young generation of Papua New Guinea; how much forests will they have access to?

Fig. 15.14 An opening in the forest canopy: A glimpse of hope is important but how achieved? Good chances are very slim for Papua New Guinea and global wilderness

According to FAOSTAT (https://www.fao.org/forestry/95632/en/), PNG had in 1990 31.329.000 hectares of prime forest, but in 2017 it was 17.599.000 hectares (over 40% lost in just less than just 30 years, regardless of data quality [which apparently comes from modeled extrapolations in the absence of real ground inventories]; see also Sherman et al., 1972; Shearman & Bryan, 2015; Yen et al., 2005 for an average tropical deforestation rate in Asia—even in national parks, for the region see for instance Bhagwat et al. (2017) and Alamgir et al. (2020).

Prime forests are the most biodiverse forests in the world and have large intrinsic value (e.g. Simard, 2021; Wohlleben, 2016). Throughout the world, forests are known for their medical resources, e.g. (Katewa, 2009 or see Gray, 2011 for the boreal forest, Cousteau & Richardson, 2009 for PNG). And other than timber, alone the epiphytes (e.g. orchids and ferns) present app 1/4th of the species in such woods, but foresters do not account of them really nor are they part of their models nor are foresters trained that well (New Guinea overall features over 2800 orchid species; Beehler & Laman, 2020). An assault onto forests 'of the other kind' comes through modern poaching and trading. For instance, there is now a rise in exotic species trade with orchids as well as Birds of Paradise (Beehler & Laman, 2020); bush meat remains a high protein contribution and is common in PNG where over 80% are rural populations (Mack & West, 2018 for an example referring to 'Ten thousand tons' taken).

Despite decade-long colonial explorations and international research in PNG for over 300 years, forestry in New Guinea has received little or no environmental impact work whatsoever (Beehler & Laman, 2020, p. 339). Forests in the British Kingdom and Dominion are generally in a very poor shape and received unsustainable management, from UK itself to Jamaica over Canada, Kenya, Australia and PNG. This got ever worse for palm oil plantations, as a relatively new landscape feature and forestry

Fig. 15.15 **a** Landcover map for forestry schemes with the data source and legend explained in Table15.1 showing land categories. **b** Same map but categories 1–5 indicating forest type cover. In the absence of valid, publically available and spatially explicit forest inventory data for the year 2022, this data set represents one of the best-available forest information for Papua New Guinea. However, it falls short on a good resolution, and meaningful accuracy details, species information, growth-and-yield, DBH, tree height and canopy details needed for forest management and sustainable harvest quotas. Where here is the world's forestry profession? For more details see also Table 15.3

activity in PNG. Judged by the lack of good results elsewhere, and despite statements by Beehler & Laman (2020), the new Roundtable for Sustainable Palm Oil (rspo.org,) does not instill much trust for PNG on those issues at all. On finite land—PNG as an acknowledged island—any plantation must come on the cost of wilderness (*sensu* Daly & Farley, 2010; see Huettmann & Young, 2002 for real-world impacts in tropical regions).

Because in PNG most forest areas are cut in the low elevation areas, it affects big trees and much diversity. Typical examples of such an overcutting—without any

Fig. 15.16 Vegetation map expressing tree cover as a gradient from 0 (white) to 100% (black). Like the previous figure, this GIS map presents one of the best forest management data sets for Papua New Guinea available while major details required for a meaningful forest management are widely missing. For more details, see Table 15.3

consideration and plans of sustainability—are found around Port Moresby, in Wau, in the Markham Valley and in the Vanimo (Sepik area; as also shown in Beehler & Laman, 2020, p. 14, 110, 331).

Other forest types than the low land forests in PNG are Hill Forests, Montane Forests, Subalpine Forests, Swamp Forests, Mangrove Forests and Season Monsoon Forests. All of them carry unique features in their own rights, and many are cut now, too.

To get at the timber and money, some experimental forestry was done also, trying out what type of forestry would work, while the required roads needed for forestry are missing nor are they funded or hardly viable. Instead, the use of mobile saw mills brought in the bush was done on an experimental basis throughout New Guinea (Beehler & Lama, 2020), but those had little success. PNG sees little use of those ones though.

And so with PNG's virgin forest of global relevance held in a public land tenure system, the forest industry stands in wide contrast to what is stated in public, by agencies, by development aid funders, much of the world's common forest science, and best-professional practices (Quaglia, 2018 for FAO of the UN) and what is needed by the local people, and by PNG (see details in Cousteau & Richards, 1999, p. 206–207) or the wider global community. Due to the diverse resources, forests are the home for many PNC citizens (Baraka, 2001).

Those few facts by themselves should all be very concerning already. But the soil of PNG is also ancient and rich due to Pleistocene volcanoes and its exposed sediments and minerals. Those resulting forest types in PNG are of global relevance for wilderness forests, habitat to wildlife and human cultures alike, for wider ecological services, for PNG gardens and for the world's carbon sequestration. Humans lived in, and with, such forests for over 47,000 years, more or less sustainably; and they were able to obtain precious resources from them, such as medication, treatments, gear, food and housing materials (Flannery, 2002; examples in Flannery, 1998; Gillison, 1993; Hayan, 1990). It's part of the 'garden'. What else do you need in life?

Such forests are the medicine of many PNG citizens, a lifeline. In 98% of the PNG history, these forests and their landscapes never got to such a level of decay as we experience nowadays, last 50 years in times of global governance, modernity and 'science-based management.' Such a decay was not experienced socially, not commercially nor morally. While a colonial destruction started c. 300 years ago, the timber extraction out of the rugged landscapes and valleys was limited till the advent of modern harvest tools, namely the chain saw and the combustion engine (often used for electric generators in remote sites as roads were missing and the trucks and cars had a bit less relevance, timber travels poorly in planes…). Many ecological niches got lost ever since, e.g. Huettmann (2015) for a global review.

There is a general and easily detectable trend in urbanized landscapes in regard to forestry: "The last remaining virgin landscapes get turned into farming land and thus cut; it's a one-way street of habitat transition to habitat loss". Typical and devastating examples for PNG are found with the SABLS (Special Agricultural and Business Leases; ATBC, 2011).

As a matter of fact, SABLS are now the 'largest forestry project in PNG' involving a certain land scam orchestrated through an Australian company (Australian Network News (ABC), 2014). For PNG people, this is devastating as they lose all they have, forests and their resources in a public land tenure system they grew up in and trusted Ancient lives are further destroyed.

Along same lines, Plantation Forestry is now also tried in New Guinea, replacing virgin forests and supported by the Australian government, namely for exotic species such as *Pinus caribaea, Araucaria cunninghamii, A. hunsteinii*, Teak as well as fast-growing *Acacia mangium* (Beehler & Laman, 2020). Considering that land is 'finite,' e.g. on an island, and that those species are all invasives ones and excluding virgin rainforest, what is sustainable or good about it (*sensu* Huettmann & Young, 2022)?

The author has many years of first-hand experience and basic field work done in tropical forests and did over nine field campaigns in PNGs forests, walking and being with the 'forest people' of PNG, living in their villages, walking across cut-out gardens and in homesteads (upon invitation). I was fortunate and able to see some PNG wilderness forest live first hand, as glimpses, in many of its aspects and forms. And similar to Wohlleben (2006) and others, thus, my view differs from official statements like FAO (https://www.fao.org/3/ax277e/AX277E.pdf) or CSIRO (https://research.csiro.au/eap/adaptive-livelihoods-development/) and what is officially promoted regarding good forestry, best forestry practices, sustainable certifications and sustainable forestry overall, how it's done or as 'the only option we have left.' It applies to mainland PNG, but as well as to the islands and coasts (where much forests and trees are located). The public statements on PNG forestry made are just template narratives because none of them truly hold up to good scrutiny, or offer us a valid sustainability, or certification scheme (Clark & Kozar, 2011) of sustainability. Already the policy enforcement of the timber industry is the problem, besides the underlying concepts of best-professional book keeping, forest inventory data, growth-and-yield table-based harvest in a mega-diverse species forest, and sustainable forestry overall where wilderness simply does not fit the models or gets replaced wholesale by the forestry ideology and its cuttings (see global details with Huettmann & Young, 2022). Despite what is claimed (e.g. as the self-declared 'world's best' for one of the world's largest forest-science operations; Rodriguez & Conje, 2022), already the concept and administration of the forestry process itself presents a problem, and how it operates and funds itself. The institutions are part of this professional set up and of the obvious problems.

The Barnett Commission found in PNG widespread corruption surrounding the issue of government licensing of forestry concessions (see also Baraka, 2001; Laurance et al., 2011, for context). The name of prime minister Michael Somare turned up in connection with one of these concessions in his home area, the Sepik River Development Corporation. The following statement is from (Filer & Sekhran, 1998) and speaks to facts mentioned:

> …Thomas Barnett, to lead an inquiry into the forest sector. After two years, and in 20 volumes, Barnett described a "forest industry out of control" - dominated by foreign investors in questionable "partnership" with PNG leaders - in which the quantity of logs exported was maximized with no regard for environmental damage and to the detriment of local processing capacity. He called for a slow down in timber harvesting, and advocated the reformulation of national policy, establishment of a nationally integrated forestry service, consultation procedures in allocation of permits, and formalisation of detailed requirements for sustained-yield forestry. (Filer & Sekhran, 1998)
>
> (8) (PDF) Loggers, Donors and Resource Owners (researchgate.net).

Table 15.1 shows some reality views for Papua New Guinea; a sustainable forestry is virtually not possible.

15.3 What's Simply Wrong with PNG Forestry: An Easy List

Papua New Guinea has a deep legacy and a few modern policies; that is specifically true for the modern nation construct of PNG and where by coincidence most cutting occurred in the long PNG history of 47,000 years (https://pngforests.files.wordpress.com/2013/05/a-history-of-the-forestry-sector-in-png.pdf).

In the following, I will discuss some key items of a sustainable forestry and such a professional planning metrics required for PNG and how they fare, and what is to be done for betterment.

1. **Forest Inventory**: As a practitioner having worked in PNG for over a decade, whatever is said and stated, PNG has no nation-wide forest habitat inventory coverage and map that is current, available of on a decent resolution and based on research design and field data (see https://pngfa.gov.pg/images/articledocs/National_Forest_Inventory/Proceedings_of_the_second_NFI_Research_Conference_compressed.pdf; compare with U.S. FIA plots https://www.fia.fs.fed.us/ as one of the world's leading forest inventory schemes and well known by U.S. AID and other sponsors of PNG forestry). See Huettmann (2020) for an applied

example using canopy-living animals—tree kangaroos—of global concern. The International Tropical Timber Organization (ITTO; www.itto.int) reports on some 'plots,' but this is hardly publicly available for review nor peer-reviewed (Alder, 1998) nor does it allow for meaningful inference from a research design (generalization). Quality assurances are difficult to obtain for any PNG forest inventory data, e.g. on a 95% confidence interval, and vast data gaps exist (e.g. Bryan et al., 2010, Brack 2011). The data gaps will likely not be filled next 10 years or more. In the wider international world of forest research and plot-based data and time series network, PNG plays virtually no role (Dornelas et al., 2018; ForestPlots.net et al., 2021). This must come as a big surprise as plant and forestry research in PNG was done for a long time, and with direct Australian, U.S. and colonial oversight for over five decades. Tiber got cut in PNG by colonial powers for over 300 years, widely without any acknowledged or available sustainability plans (a scheme that is widely taught in colonial forestry schools for over a century; details in Huettmann & Young, 2022).

Table 15.1 Principles of forest practice and how those apply to PNG (required forest practices can be taken from any textbook in forestry)

Forest practice	Relevance	Applied in PNG	Comment
Species taxonomy and proper identification	Each species differs and needs a specific approach, management	Not really	Latest taxonomies used or known by most practitioners in the field are not agreed on, hardly exist for the 400+ species
Forest inventory	Basis for determining the harvest and what can be 'taken'	Not really known	A basic data set needed for any sustainable resource use, anywhere; widely missing though
Growth-and-yield	Determines the actual cut, what is to be harvested sustainably	Not known	Not possible to apply if species are unknown, used interchangeable and lumped, and if not 1–2 generational tree length data are collected or modeled. Spatial data is essential for this, including soil types, but widely missing
Institutionalized national and sustainable forestry department	Oversight of sustainable forestry process	Exists, but PNG is labeled a failed state due to broken institutions	If harvest rates are not set well, and not enforced, overcutting will occur. Forest industry has a foot in the door with the ministries and provides the infrastructure even

This lack of forest inventory must come as a true surprise because a nation and its government should know what it owns, and manages, all coming from a publicly reviewed and approved concept. Further, PNG actually has received much western science attention for over 300 years—the infamous Birds of Paradises (BoP) painted by Rembrandt and others live in the rainforest after all. They need such habitats, but in earnest, the BoP habitat status remains widely undocumented (see in Laman & Scholes, 2012, Beehler & Laman, 2020). And so, while western nations understood and undertook their first national forest inventories already over many hundred years ago (in Central EU easily 17th century onwards, locally 11th century such as in Austria's salt mining areas) no modern forestry was set up or undertaken in PNG (Brack 2011) despite the vast cutting. PNG has still no valid, accessible and digital nation-wide forest inventory maps to this very day. There is virtually nothing available online to be used and with documentation and a trackable record. And a missing, not agreed on and binding reference for tree taxonomy, or a flora (Webb et al. 2005), do not make it better at all.

There are a few global forest inventory layers that one can use for PNG and one can clip them out in GIS. But those have no relevant metrics needed for a sustainable forestry and lack a good ground-truthing for PNG. They are usually not up to date and have coarse resolutions; coastline problems—ocean, tidal ranges and freshwater lakes and rivers—are frequently found (this matters a lot for mangrove forests for instance), as well as geographic projection and coordinate format problems and lacking ISO-compliant metadata in XML in the GIS delivery of such maps (see Huettmann, 2020). Needless to say, they often come from Remote Sensing work done by NGOs and private contractors that is merged across images, years, cloud cover and even sensors (e.g. Hu et al., 2016). One will have a very rough time to obtain those forest inventory works for an area that is wild terrain, often steep-slopped and rugged, and for PNG as a whole, let alone the myriad islands and tribes that live there.

Arguably, in such a regime one has a difficult time to actually assess what forests were affected, lost and where and how. The metrics that do exist come from spatial summaries (data fudging, cruising) but lack to be explicit in space, on a pixel-basis. Arguably, not everybody wants to show and expose it. But many evidences help to provide a more complete picture of loss and of poor practice. For instance, Ebony wood on Kiriwina Island (Trobriand Islands) is widely overused due to carvers and the market demand (Cousteau & Richards, 1999). In Sepik, one finds similar situations as carved wood carries a high market prize. And historically, parts of the highlands have likely been de-forested due to the fertile soil and subsequent gardening. The native oaks, beeches, laurels and pines of PNG are reduced in range and abundance. Instead, locals grew there bananas, yams, sugarcane, peanuts, breadfruit, coconuts, pandanus, wild almond and sago modifying the actual tree composition. Local fires easily add to the complexity (Tables 15.2, 15.3 and 15.4). Still, recent industrial efforts made it much worse and eliminated much of the entire forest landscape overall!

2. **Species concepts and mislabeling**: It's not so clear how many tree species PNG really has, likely around 600 (http://www.pngplants.org/PNGtrees/TreeDescriptions/). Over 400 species of those can be used for the forestry market. That in itself is a problem as virtually no species carry meaningful harvest strategies based on data and science (the century-old Central EU Forestry Model is based on just

Table 15.2 Selection of some milestones in PNG forestry (Following Overseas Development Institute (ODI) 2007, appendix; as well as FAO: https://www.fao.org/forestry/14878-0a11fcc9fb23d06b1bb586fc398ffbf7.pdf)

Event	Time	Impact	Comment
Indigenous management	47,000 years ago	Sustainable tree extraction, fire management	Low-scale forestry, use of (larger) fires, and individual tree removal
Colonial timber extraction	<1922	Patchy removal of accessible and precious tree species	Timber removal, often done near the coast and rivers
Timber ordinance	1936	Policy on tree removal and land issues	More organized removal of precious timber for colonial efforts
Forest service established	1938	A first governmental agency in charge of forests	See wide failure for PNG, its forests and people
WW2	Cutting for military purposes	Overcutting	Little consideration of sustainability but strategic efforts
1946	First formal forest policy	A wider more strict forest policy	PNG as a nation did not exist yet. See little impacts on a nation scale
1958	A new forest policy	Like above	Like above
1975	Independence	Start of a real national forestry	Process of big hope and according to the Melanesian way (Narokobi, 1983)
1979	A new forest policy	A national plan	Reality shows industrial aims only
1987–1989	Barnett inquiry	Assessment of the poor status quo	Results were devastating for PNG showing reality and failed agency
1978–1997	Reform	A better plan	This process is very long to be considered efficient
1990	World Bank produces Tropical Forestry Action Plan Review for PNG	A plan supported by The World Bank with funds	The performance 'sits in the pudding'; plan for a plan of a plan
2001	Moratorium on issue of new timber concessions included in conditions of new Structural adjustment programme. Signing of loan agreement for the FCP	A typical example of progress when there is none while people suffer regardless	Forests and land turn into financial discussions. Conservation progress is not much achieved, inefficiency rules and big words and smooth neoliberal compliance terms are used but that fail on the ground for decades

Table 15.3 List of publicly available Forest Inventory data for Papua New Guinea in a GIS format (pixels) with basic metadata to actually understand and them[a]

Forest layer	Format	Citation	Detail accuracy	Comment
Land cover	geoTIFF	Kobayashi et al. (2017)	500 m (15 s); Landcover categories	One of THE best forest information for PNG and the world, but all relevant forest management attributes are missing and/or coarse, even basic ones like tree species or DBH Data carry no ISO-compliant metadata
	https://globalmaps.github.io/glcnmo.html			
Vegetation cover	geoTIFF	NA, but provided by: Geospatial Information Authority of Japan, Chiba University and collaborating organizations See URL below for citation details	15 arcseconds; Percent Tree Cover	Like above A great looking—and partly useful—layer for PNG and the world. But virtually all relevant forest management attributes are missing and/or coarse, including plant species, diversity and trees Data carry no ISO-compliant metadata
	https://globalmaps.github.io/ptc.html			
PNG local sources	YUS	Tree Kangaroo Conservation Program (2018)	NA	No relevant spatial forest management data exist
	Cut block concessions	NA	Those inventory data and maps must exist in a basic setting, e.g. aerial photos. But official and documented sources are not available or findable to a global public	Typical state of affairs in many forest projects in PNG and beyond. It's not a public participatory process and easy to understand

(continued)

Table 15.3 (continued)

Forest layer	Format	Citation	Detail accuracy	Comment
	Tree Kangaroo project	Huettmann (2020)	No consistent data truly exist of relevance for forest management; just proxies and indirect data are used	Data sets exist for canopy cover, e.g. from drones and using Google Earth but those are private and not really publicly available with ISO-compliant metadata in a relevant and projected OpenSource GIS format

[a]*Note 1* Satellite images like Landsat and RADAR are not directly mentioned here because (a) they are not really up to date and in time, (b) they are not free, (c) they are not easy to obtain for most people, (d) they are virtually impossible to use by the lay public, e) they are patchy in space and time, and f) they are impossible to interpret also for most people (e.g. spectral bands, and merged and heavily processed imagery).
Note 2 In the absence of good data, Google Earth becomes the data of choice for people to use and for reference. But Google Earth is not precise, nor a GIS, nor a good Remote Sensing tool, nor can it be used for relevant analysis or forest inventory whatsoever. Arguably, most people use Google Earth and are thus dumbfolded by this commercial source and tool, including for its widely known biases. It's a reality picture.

app. 20 tree species and still struggles with those!). They can hardly be tracked, get merged and can get used interchangeably. In addition, the vast unknown amount of new tree species to be described for science, e.g. through ever-changing DNA methods, subspecies, competing labs and computations (Silk et al. 2015), makes a reliable forestry on a relevant species level impossible, certainly undefendable.

In addition, there then is a product label problem at the markets when the taxonomy is so unresolved and will likely remain unresolved for decades to come. Like done with fish species, meat or honey, one can easily label and mis-label the item as wanted and its subsequent products for the commercial trade, import and export. It will throw off all book keeping, inventory and accountability. Enforcements are virtually impossible to track and to account for.

While all of this is ongoing, Australia knows about it first hand and for a long time. For instance, the Sydney Museum is a global resource of taxonomy for PNG. Royal Botanical Gardens Sydney, NSW and the PNG Forest Research Institute in Lae—supported by EU projects and Japan + the Forest Industry—are usually fully involved with strategy and paying the bills. Australia is also a host of many other tree-related and

Table 15.4 List of the world's forestry data that are typically not publicly available, not easily publicly available and/or lack a description/metadata to be understood and used well (there is a consistent argument brought forward in this on privacy concerns and in context of indigenous people; but those are often incorrectly phrased and can be addressed, specifically when using forests as part of the wider public good)

Location	Forest practice detail	Details
China	Harvest	This includes most of the forest inventory data, growth-and-yield curves, taxonomic tree species, as well as cut blocks and forest cover and non-timber products
Brazil	Harvest	Like above
Congo	Harvest	Like above
Australia and New Zealand	Harvest	Like above; national policies somewhat differ
Costa Rica, Nicaragua	Harvest	Like above
EU	Harvest	No clear guidelines and standards exist for national harvest data and maps in the EU or for essential forest metrics, but which is frequently subsidized
Research publications, e.g. Laurance (2007), Shearman et al. (2012)	Forest cover assessments	Used for conservation purposes; not directly linked to Forest Management either way and not really made available for colleagues or the public
IPCC.org, global	Carbon sequestration, climate assessments	Available but not easy to understand, to obtain and to use; many versions and implementations and confusing
REDD/REDD+	Plot data, carbon sequestration maps	This is the essence of high-stake multi-carbon trading schemes, but the foundation is not so clear and easy to understand nor all input data

forestry databases (https://www.idigbio.org/wiki/images/e/e9/Guide_to_trees_of_Papua_New_Guinea.pdf), including student research projects and degrees. So what comes out of it, while virgin rainforests get cut down further for Asia and with Australian-run SABLs?

3. **Growth-and-Yield table concept**: For forestry to be sustainable—successful, one needs to know how much one can cut and what regrows. This is usually

achieved and computed with growth-and-yield equations (Pretzsch, 2009) and subsequent models (e.g. Peng, 2000 for a discussion about uneven but reality forest stands). It's a science discipline within forestry, and a very strategic one, but one that cannot apply well to tropical virgin and old-growth forests and in PNG. It can be done for tree stands, for individual trees and for pixels and soil types and habitats, let's say. However, those growth curves are to be based on long-term data for single species and what grows in the harvesting units; it's classic western science. 'Long-term data' here mean to cover at least a generation length of a tree species, ideally two generations to confirm the trend with annual or seasonal measurements. Well, and that assumes tree species are actually known and taxonomically valid (a very questionable assumption, as shown above). One may easily add that different growth-and-yield curves are needed for the different soil types and treatments. So then where is the PNG soil map (considering that the mining companies have scouted out and mapped in good detail the geology for over 100 years very closely)? Because in PNG one can easily find over 600 tree species in a forest wilderness plot, growth-and-yield tables are difficult to apply and to generalize when widely missing and not specific. How then can sustainable forestry ever be practiced—or defended, e.g. in REDD/REDD + (details described in detail in The Guardian, 2015)—without such an essential foundation?

Alder (1999) shows such a growth model but one doubts it gets applied by Chinese and Australian forest projects. I am not aware about it and how it's done there; information is not publicly available (Beehler & Laman, 2020).

Real-world examples of such gross forestry and growth-and-yield failures are easily found with the pandanus palm tree as a major forest species in PNG (Cousteau & Richard, 1999). This species is not accounted for, hardly considered a tree with secondary growth (palms do resemble grass instead). Same can be said for sago, no growth-and-yield tables exist or are followed. And then, New Guinea features a world-record mangrove forest range (Beehler & Laman, 2020), but where a growth-and-yield table is biologically virtually impossible to obtain (due to the growth-specifics of the species involved, e.g. hardly showing tree rings even). Still, mangroves are heavily used up and not surprisingly, dramatically on the global decay and decline (Goldberg et al., 2020). In PNG, there is no relevant management using growth-and-yield tables because those simply do not exist, nor in time or in space, nor is there a good policy in place to do so.

4. **Management of cut blocks, bidding and timber sales**: Another core scheme of a sustainable forestry sits in the process and computation of what actually can be cut, where and how, by whom, and how enforced and checked? These decisions are usually done with strategic models and then, set and approved by a forestry ministry that has a bidding process to follow and to monitor for compliance. Those are core schemes in virtually any natural resource management and require expertise, independent decision-making and integrity, as well as a stable budget with a legal policy—including a good court system and legal framework and process—to adhere to. It's a culture and agency needed. Looking at PNG metrics on those issues might offer many insights. As a matter of fact,

it's a wide flaw in many royal democracies in the world (see, e.g. Nepal Shah regime with Prajapati et al., 2020; see Marold & Westholm 2016 for Swedish forestry now widely an industrial take-over). Many of those regimes—certainly the European ones—lack relevant expertise and handling of wilderness, tropical multispecies and old-growth forests for centuries (global examples provided in Huettmann & Young, 2022). Arguably such a governance is widely divorced from forest and nature realities. One easily sees those problems in PNG where a major intact wilderness forest block still can be found.

In addition, there is a serious outside overruling of the sustainable timber cutting process, see Bun et al. (2004), EIA-Telapak (2005) for China, and Greenpeace (2005) for the UK.

PNG is obviously not in charge of its own resource and of the international commodity; nobody really is, other than a few private power cliques (as stated in Beehler & Laman, 2020; Filer et al., 2009a, 2009b, 2011; Laurence et al., 2011 for New Guinea). Consequently, resources get grabbed and so gets the land (see SABLS in the previous citation, with ATBC, 2011 and in Beehler and Laman 2020).

5. **Renaturalization**: 'Modern-style' western forestry has taken the conceptual path to cut and replant, regrow and rebuild forests. It's seen like architecture, wrongly though! It is achieved through tree plantations. It happens to be a pretty lucrative business as it involves DNA-cloned trees, young tree keeping and then site preparation (usually done with chemicals and fertilizer, also adding soil treatments), a second round of tree plantation treatment, pruning, etc. It's a large business model in itself, assumed to be in support of a sustainable forestry, economy and society, where in reality, it is not (Huettmann & Young, 2022, see also Katovai et al. 2021 for Solomon Islands), e.g. the carbon footprint is not 'natural, but industrial and thus an energy source fueled by refineries and eventually: oil (see the applicable classic works by Odum from the 1950s summarized by Madison 1997 as it applies to such a forestry also).

   Doing such concepts in the tropics, with occurring pests, fire and the climate-failing plantations remains a large work which is not ready really to be rolled out for a market and for sustainability yet. How done here in earnest?

   So what is to be done about it? Instead of planting and maintaining trees very expensively one should focus much more on the existing forest maintenance and not overcutting prime forest. Prime forests have the largest value of all forests (e.g. Ramachandra & Bharath, 2020), are the home of indigenous societies, and they should be enhanced and improved, and maintained. Prime effort is needed for the prime forest, not for the replanted clearcuts.

   What is missing is a wider ecosystem approach, e.g. tree rejuvenation needs an ecosystem with species such as Cassowaries or Birds of Paradise for seed dispersal (Mack, 2014; Beehler & Laman, 2020). Indigenous perspectives are key to that also. So here, we lack much research as well as a law and principles to actually achieve sustainable forestry; unless it's a lie. How done then?

6. **Sustainability Institutionalized**: Ideally, in a generic concept, all of those steps for a sustainable forestry are to be run by strong public government sectors, e.g. a Ministry of Forestry, with a democratic and transparent foundation (tax-based), enforcement and with full public consent and overall, being sustainable. It's not to be influenced by outside nations or corporations but it is to serve the people of PNG in a democratic process, as a good governance principle. This concept, assumption and public perception still dominates the underlying idea and world of forestry, but reality widely differs. While colonial forest departments were historically strongly linked with serving imperial courts and they employed many military veterans (e.g. in Prussia, Austria, Hungary, Poland; see Imperial Forestry Institute in Oxford, UK) a change has occurred (Rajan, 2006, Lawrence, 2009; see Rangarajan, 1994 for an example with India) where forestry is now put onto the free market and privatized. It's widely neoliberal and makes for a major discussion item (e.g. reviewed and critiqued in Huettmann & Young, 2022) because the world is not really ruled by governments and good principles any more (Stiglitz, 2003), nor are those sustainable in the first place.

Corporations are now widely in charge, acting internationally, strategically for their self-interest and being profit-based; humans and habitats are frequently ignored (examples for PNG found in Cousteau & Richards, 1999; Montagu, 2001; Richards, 2018). The internet has changed many of the transactions regardless and such companies become the drivers (Lipschutz, 2000 on why there is no international forestry law; see also Brara & Berros, 2022; Leinen & Bummel, 2018; for an Environmental World Parliament, and Huettmann, 2011 for a Ministry of Climate). Arguably, the traditional forestry system is hardly sustainable neither, nor is the traditional one with the national government (e.g. as per track record last 300 years seen in the loss of old-growth forests and wilderness; Huettmann & Young, 2020). What to institutionalize, and how and whether a federated system has worked, e.g. for old-growth forests is already pretty clear: It has not (examples in Bialowieza/Poland, in Samojlik et al., 2020, and Tongass/Alaska in Schoen, 2021). Here sits a wider cultural and management crux that is not well resolved and which is not in favor of old-growth forests, the tropics, forests (e.g. Huettmann & Young, 2022; Laurence, 1998) or PNG for that matter.

The development and failed outcome of a traditional, sustainable forestry project is very well described for PNG in Cousteau and Richards (1999) and stated in Beehler and Laman (2020), for a Japanese-Australian project, here summarized in broad terms:

The JANT (e.g. Cousteau and Richards, 1999, p. 206, Odi, 2007) project is located near Madang with the help of Japan, major shareholders are Honshu Paper Company of Japan, for wood chips of paper and fine tissue in Japan as well as New Guinea Timber (a Commonwealth shareholder, essentially helped by Australia). Harvesting started in 1974. It's a longer term forestry project; some see it as a certain success and proof of concept. Locals initially welcomed it until they saw that actually all wood was cut out. Cutting wood has given them space for gardening with incomes provided for medical facilities and schools, with national employment. But once the virgin forest was taken, it was replaced with c. 13 fast-growing species, as well as coffee and cocoa trees. The locals got very disappointed and Madan banned such projects from now on. A JANT management letter text in reply sent to school children protesting the forestry project removing trees for pulp and paper stated instead:

> *…This is not a botanical garden nor a zoo called 'Papua' ….*
>
> (Cousteau and Richards, 1999, p. 207)

Whereas Odi (2007) heralded JANT, and described it as.

> …the longevity of the JANT project has demonstrated that industrial-sized wood *production for export is possible in PNG, although it comes with a social cost….*

Well, these 'social costs' are clearly marginalized, but are rather high and cannot be worth the damage and human lives damaged?
Along similar lines, most NGOs should be described as an aggressive delivery platform for whoever wants to buy and fund a NGO. NGOs are not elected by the people nor do they underlie a democratic foundation or concept. NGOs are funding driven and in environmental terms thus they can operate like the so-called 'hired gun'. NGOs do what they are paid to do, often even with tax exemptions (which is utterly bizarre why that actually is legal). It gets even more absurd when knowing the many NGOs that operate in forestry and natural resource sectors but then come from many different nations abroad, e.g. Russian NGOs, UK NGOs or U.S. NGOs; often they do not agree all in the national legal framework they are operating in, many receive conceptually opposing type of money, but all operate in the same nation, on the same topic and location even. What outcome does that have? In the NGO world, the most money wins. In PNG, such a thing can easily be seen in Goroka (highlands) or Madang and Lae. By now, NGOs are an inherent part of the forestry governance landscape and should be seen as an odd and likely ill addition to the institutionalized decision-making process during globalization, plain funding driven. With NGOs, there are some wicked visa issues also, and data-sharing policies to be considered. Reality remains, NGOs want to survive and want to get funded; as per their set up, they hardly (can) care for anything else. Online media campaign show it clearly. It's clear to any onlooker that this can never be sustainable really; PNG forestry reality shows us no other.

7. **Smooth Flow and Scales of (International) Governance**: In federated governance, forestry operates at different scales: National, Regional and Local. In times of globalization, wider international perspectives are to be added, e.g. via Asia and Australia, or PNG as a Dominion in the Commonwealth. While Forest Policies address some of those scales and hierarchies, in reality, in governance, much is to be desired for a smooth operation. In reality, most forest services have a very difficult time on those aspects across agencies, scales and hierarchies. It hardly can be done right, certainly not in very complex times of Climate Change, Carbon Sequestration and loan policies, made more difficult through

pandemics. It is not clear how this reaches well to the people of PNG relying on 'the bush' (Cousteau and Richards 1999 for JANT example).

8. **Carbon Sequestration**: Papua New Guinea has been the early posterchild for industrial debt swaps, namely carbon sequestration and REDD and REDD+ (https://pngreddplus.org/; see Filer, 2010 for the Cabon Cargo Cult; Fox et al., 2010; Melick, 2010 for real-world assessments of PNG). Norway and former colony-holder for PNG, Germany via the EU, paid initial record sums in exchange of pollution rights (some details shown in The Guardian, 2015). In the meantime, after nine field seasons in PNG and walking in several extensive forest areas I still know nobody in PNG who really has received that money or benefitted from it on the ground. Nor is carbon sequestration a real solution when polluting goes still through the roof. It's an important fact to know that the highest carbon sequestrations are in peatlands, specifically, tropical old growth and peatlands in adjacent Indonesia (New Guinea) for instance (e.g. Uda et al., 2017). Another carbon sink is the ocean, and also coral reefs and seagrass and mangroves. PNG has many of those, and none of those are yet much accounted for in the wider forest carbon sequestration discussion and actions. Lastly, forest fires must be added here too, but are not truly part of the equation (and when already considering adjacent massive forest fires in Australia; a nation that has widely denied man-made climate change, as per Ludlam 2021). Where does that leave us and PNG and its forestry?
9. **Use of so-called Alternatives and Social Approaches for Sustainable Forestry**: There are virtually tens of other suggestions to manage a forest than the classic western imperial growth-and-yield approach discussed here. Arguably, the Chinese and Asian approach—as a major force in PNG forests—is far from any science-based sustainable cutting concepts (Beehler & Laman,2020). In recent times, resilience management got pushed, and community forests, agroforestry or eco-forestry (e.g. Scudder et al., 2018). However, simply judged by the forest loss and state of the forests on a finite landmass (Forest Trends, 2006, Filer et al., 2009a, 2009b), those have not lived up to their promise neither. It's difficult to believe when acting on a finite space like PNG that those actually can work, anywhere.
10. **Adding Indigenous Views**: Land tenure plays a crucial role and who is in charge, and how done (see Bell et al., 2015; Demeulenaere et al., 2021 for an application in Melanesia; Hyndman 1998 for a reality review in the region). Like elsewhere, in PNG forests are not just timber or money, they are the entire life (Cousteau & Richard, 1999). PNG is the prime example for community-driven forestry and how it fares (Beehler & Laman, 2020). PNG forests make for a typical example that the western law cannot handle those concepts, well, or sustainably. Forests are just not the trees to be sold for money (Saulei & Aruga, 1994; Mack & West, 2016).

Textbox 1: Good Forestry practices? Western Science, Global Reality and Commercial Practices

Any forester who was trained by respected universities and institutions—certified—will agree that forest practices should be followed and done carefully and with a 'plan' for sustainable practices, e.g. less is cut than what regrowths so that future generations can still enjoy the forests and the trees (many definitions of SFMs exist and should be evaluated for their emphasize, e.g. Huettmann & Young, 2022). Entire degrees and their underlying textbooks exist to teach that and to make it happen in the real world.

Now, cultures differ and the perceptions differ.

But a standing tree is essentially money for whoever owns it, so are forests. Cutting it down with some effort and cost tends to make money, and thus, standing trees are money in the bank. Economic models in parallel using bank interest exist, and insurances with fire risk models look at it closely. While many forests cannot compete with interest rates of hedge funds or the stock market, they still provide wealth, usually to the land owner who owns the trees. It's a public resource, like in Canada for instance, and in PNG, it's usually a complex village land tenureship (Beehler & Laman, 2020). Timber felling profits then should be shared with the community at large (concepts described in Cousteau & Richards, 1999), but which virtually is not possible. This concept applies wider if forests are of global relevance. In those cases, the global community should have a say, but has not.

Science can provide growth-and-yield tables, models and rules how to achieve that wider sustainability. Science-based resource management is to assure good practices and that those are assessed and met. However, thus far, science-based management of forests differs from the practices done in PNG for over 47,000. PNG forest resource use has a strong spiritual component, a concept that is totally absent in western forestry, and not even wanted there (see discussions in Huettmann & Young, 2020; Suzuki, 1993, Simmard, 2021; Wohlleben, 2016). Whereas, most people understand the spiritual value of timber and forests, but practitioners and their agencies usually do not.

In reality, the sciences are not even present in most wilderness areas of the world, or are in the driver seat of companies who operate there. Money pays the labor and it is in the driver seat, making profit as the main paradigm. It easily overrules any practices. PNG is a typical example for that; people pay the price.

Usually, forest land is sold or made available in a bidding process. But in a land ownership of PNG, this is hardly possible unless the community agrees, or gets overruled (see ATBC, 2011 for SABLS in PNG). And that means it gives away access rights and global ecological services of that forest land, including clean water and clean air. In that case, forestry may equal more of a land grab (Filer 2011), taking livelihood aspects from the village and destroying lives, families and societies, spitting them out as compromised people in urban areas.

In the meantime, such produced timber goes to markets. The western markets are to sell good products, or products to good standards. But those are dubious to define, to assess and to enforce, hardly meaningful for PNG (see work by Clark & Kozar, 2011 for details).

Now, where does that leave us?

It must be stated that sustainable forestry is widely not achieved for areas where it really matters, e.g. in wilderness areas, in old-growth forest areas and certainly not in PNG; far from it.

There is no need to claim or state much different. As outlined in Huettmann and Young (2022), these realities should be stated in higher levels of forestry and its lobby, e.g. UN Forest Certification boards, timber trading and its international societies and associations, including the World Trade Organization and FAO, or in public media.

The fact that those details are clear and very obvious, even widely published for a long time in PNG, makes the case that the western concept and others who are engaged there are in violation of what is preached, knowingly.

Textbox 2: Non-timber products, so-called?

In 'modern' forestry nations, timber remains the money maker and the number 1 goal of the operation. Non-timber products then get promoted to add a bit more value. It's seen like an alternative. Non-timber products and efforts get usually seen as a side product not making much money or contributing, much. However, already when seen from an ecological perspective, it's clear that we have mis-valued forests and simply reduced them to timber production entities (Wohlleben, 2016). Still, the non-timber product aspect is vastly on the rise, starting with niche markets becoming mainstream, e.g. mushrooms, orchids, epithytes or health products. But already when ecological services of clean water and air are added the entire timber business gets challenged, it easily flips for value.

Already trees as such can be major items for people and societies to live by (e.g. Demeulenaere et al., 2021; Nakarobi, 1983 for Melanesia). In wilderness areas where indigenous people still live and where forests are the 'living room'—a so-called frontier—the western approach to non-timber products widely flips the timber production concept on its head. Classic cases can be seen in PNG, e.g. Saulei and Aruga (1994). Non-timber-products are many, kind of endless, and often highly valued (an easy list shown below):

- orchids
- tree ferns
- moss
- plant medicine
- fruits and berries
- mushroom
- bee-keeping
- land access/trails
- shade
- clean air
- clean water
- spirituality
- carbon sequestration
- bushmeat
- wildlife habitat, e.g. Birds of Paradise

Arguably, non-timber products are the core of forests (e.g. Taylor et al. 2022), and for forestry. Within that, timber is just a small structural side product really. How to manage them individually remains tricky, but is hardly needed if forests are treated and maintained areas (pixels) and as land.

11. **Get a Handle on Greenwashing in Order to Get to Real Sustainability**: Forestry by itself has an appeal to be sustainable, while it frequently is not, e.g. Lines (1999), Wohlleben (2016), Huettmann and Young (2022). Sustainable forestry cannot be put on a neoliberal business mode; it hardly functions in a capitalistic framework. To overcome this problem, many approaches are used, playing with words, branding and marketing strategies. One of them is certifications and labels. Clark and Kozar (2011) have reviewed some major certification forest schemes and found that those carry little authenticity, hardly truth and data; they are not much based on reality. PNG forestry must not fall into the same trap and carry a label that is not justified. One would challenge any forestry management in PNG to be valid and sustainable, unless it comes from a small-scale local village-driven and approved approach with cultural roots (see, e.g. Cousteau & Richards, 1999) or something similar to YUS landscapes (subsequent chapters).

    Along the same lines, the United Nations Aichi Standards, Sustainable Development Goals (SDGs; https://www.unep.org/resources/report/sdgs-and-aichi-targets) as the major global driving scheme for natural resources fall into the same category of 'Greenwashing.' They are not sustainable but promote unrealistic, neoliberal and untrue concepts (e.g. Huettmann & Young, 2022) and must be changed to be honest and real, globally to benefit PNG and the tropics overall. The UN has a longstanding track record of failing many of its goals, that is certainly true for the social and environmental ones, and for biodiversity (see Mace et al., 2013 for an example; see Krishna et al., 2022 for disease goals and see Ziegler et al. (2011) for a review of the UN apparatus and approach).

12. **Add Climate Change**: It's virtually impossible to plan for, and achieve, sustainability into the future without adjusting for climate change. It's the core scheme. That is certainly true for forestry. For instance, some palm tree species from coastal areas now are found up to 1200 m (Beehler & Laman 2020, p. 346). However, I am not aware of any PNG forestry, growth-and-yield models that have taken climate change truly into account, yet, or that a science capacity even exist to do so.
13. **Having a Reality buffer (= slack)**: While the ideal of sustainable forestry is out there, it's hardly met and achieved anywhere yet; certainly not in wealthy nations who promote it to PNG. Hardly, the concept of a wilderness and old growth maintenance of maximum sustainable yield (MSI) has been found and applied well, e.g. for fisheries or in hunting (Taber & Payne, 2003). So why should it be, and can it ever be, met by PNG and its forestry? While expectations should not be set too high then, the real-world outlook for PNG's wilderness forests remains grim and worse. Overall then, being humble and less confident on the sustainable forestry scheme would be a realistic acknowledgment, start and a sound position to be in. It must be acknowledged that there are no win–win situations, hardly any precision in forestry thus far, nor is sustainability achieved, or has international aid achieved much on the matter, nor do corporations help here much, nor has the Australian role model provided a relevant gain, or delivered a good future for PNG and its forests and the people who live there. Rather vice versa; facts speak for themselves. Despite decades of research, Australian forests themselves are not holding up for a valid sustainability assessment (see for instance Lines, 1999, or recent bushfires affiliated with insufficient forest practices, e.g. Ludlam, 2021). In PNG, El Nino-related fires occurred in 1997 and 2014 (Beehler & Laman, 2020) and more fires can be expected in the future, e.g. due to climate change or human-made. There is a lot of fuzziness and thus a certain 'slack' is needed to achieve.

    The reality is that forestry remains to operate in a messy environment and that makes any of those problems mentioned worse. To be 'ueber precise' does not help (e.g. Williams et al., 2022). Forestry is not a precise science and never has been really. It's about people and their wealth, culture and sustainability the most. One should be honest about that. So what is a sustainable cut and how to go about it?

15.4 Why then Does PNG Still not Have a Sustainable Forestry Practice, Yet?

With such an overall assessment in mind, a good question remains why PNG has not achieved more yet, or why not grown out of the western and commercial global shackles of a non-sustainable forestry and promote its own?

Well, PNG can't, under such views and perspectives, consider the colonial and neocolonial framework and its (Australian/Asian) funding that PNG operates under (see Gosarevski et al., 2019; Rumley, 2006). Already mining itself - as an inherent part of the Australian business aid and plans- is known to come with many problems, one is deforestation (Giljum et al. 2022) as it can easily be observed in PNG.

Arguably, science-based forestry needs suitable science, and a science institution (see Karger et al. 2021 for global forest status); ideally in PNG, or somewhat guided and supported outside of PNG by PNG expertise, funded well and all done in a democratic fashion meeting the sustainability needs of the people of PNG.

None of those really exist yet though.

It's not an ideal in the sky, but a reality need. If not met, PNG and its people and lifestyle and a 47,000-year-old culture will go down, and colonial nations and Asian involvement are primarily to blame with Australia as the closest proxy. There are no serious experts who would disagree, e.g. Beehler and Laman (2020), Diamond (2011), Flannery (2002). Such handling of foretsry applies to any western-style culture really, e.g. Simard (2021), Wohlleben (2016).

"*No trees, no life,*" and it's that simple. One must wonder why David Attenborough—who reported a lot about Birds of Paradise and its tree habitats and PNG—has never said it (The Guardian, 2017)?

15.5 Going Indigenous—Decolonized—For a Meaningful Forest Sustainability in Papua New Guinea

Arguably, Papua New Guinea (PNG) and the market—capitalism—values forests wrongly, in a destructive non-sustainable fashion, too narrow and too little (see Cousteau & Richards, 1999, p. 209; Simard, 2021; Wohlleben, 2016). Instead, forests provide the roof for our live and for many generations. It's the pharmacy, the living room, the hunting ground, and the spiritual balance all in one; past and future (Suzuki, 1993). Destroying the forest destroys people living in those areas (Hyndman 1998). It destroys the economy at large, and that applies to PNG and beyond. Since deep time, PNG has lived in forests, used forests and kept it relatively healthy (e.g. Beehler & Laman, 2020). Arguably, PNG inhabitants promoted a garden, which is not prime forest but a certain benign human-modified landscape which operates within the forest framework. It's part of the swidden forest concept; a sustainable form of '*slash and burn*' agroforestry. However, PNG as a human society for over 47,000 years; it simply achieved more without the western man and his institutions than with the new era of 'modernity' in the last 200 years or so. Like almost anywhere in the world (Huettmann & Young, 2020), with the rise of the Enlightenment, James Cook's discoveries, PNG forests started to change and to decline and it got worse in the last 50 years with independence given and overseen by Australia! A crisis state is reached.

For a role model and precedence, the World Rainforest Movement (2004) argues that development should take into account human rights and the environment, rather than creating poverty, social division and disrespect for local culture and customary laws.

The PNG concept of living in forests and in forested landscapes was simply so successful for thousands of years because it involved a humble and pristine live; not many outside influences, low consumption and no (western) capitalism or business ideas while gardening and bartering were fully done though. Yes, trees got cut and burned back then, but in a relatively humble fashion and with taboos embedded in the wider universe. According to PNG believes, human life comes directly out of nature, so one better respects it (Beehler & Laman, 2020; Gillison, 1993). With that PNG was somewhat sustainable in itself, guided by a certain Mother Earth concept and deep spirituality (as found in most other cultures in the world that are sustainable for a very long time; Suzuki, 1993). Can we ever get back to it?

Within reason that is possible and should be promoted and tried more; be indigenous and get de-colonized further. Those processes started already in New Zealand and elsewhere (Zavala, 2013). They are possible to happen if our science institutions allow it to happen and unleash the creativity of people and the best way and knowledge of a sustainable life.

References

Alamgir, M., Campbell, M., Sloan, S., Engert, S. J., Word, J., & Laurance, W. F. (2020). Emerging challenges for sustainable development and forest conservation in Sarawak. Borneo. *PLoS ONE 15*:e0229614. https://doi.org/10.1371/journal.pone.0229614

Alder, D. (1998). The ITTO permanent sample plots in Papua New Guinea: Some results of analysis. *Paper presented at the ITTO Workshop on Permanent sample plots and growth models for natural management in Papua New Guinea, held at the Forest Research Institute*, Lae, 10th–13th November 1998.

Alder, D. (1999). *PINFORM: A growth model for lowland tropical forest in Papua New Guinea.* ITTO Project report PD 182/92—Papua New Guinea.

ATBC. (2011). The Urgent Need to Curtail 'Special Agricultural and Business Leases' in Papua New Guinea. Resolution, Association for Tropical Biology and Conservation.

Australian Network News (ABC). (2014). PNG's land scandal inquiry names an Australian-led company. 11th Feb 2014. https://www.abc.net.au/news/2014-02-12/an-png-land-scandal-inquiry-names-an-australian-led-company/5255528?nw=0&r=Gallery. Accessed May 1, 2022.

Bell, J. A., West, P., & Filer, C. (2015). *Tropical Forests of Oceania. Anthropological Perspectives.* ANU Press.

Baraka, R. (2001). The meanings of independence in post-colonial Papua New Guinea. A Plan B research paper submitted to the faculty of the center for Pacific Islands studies In *The Candidacy for the Degree of Masters of Arts. School of Hawaii, Asia and Pacifci Studies.* The University of Hawai'i at Manoa.

Beehler, B., & Laman T, (2020). *New Guinea: Nature and Culture of Earth's Grandest Island.* Princeton University Press.

Bhagwat, T., Hess, A., Horning, N., Khaing, T., Thein, Z. M., Aung, K. M., Phyo, P., Tun, Y. L., Oo, A. H., Neil, A., & Leimgruber, P. (2017). Losing a jewel—Rapid declines in Myanmar's intact forests from 2002–2014. *PloS one, 12*(5), e0176364

Brack, C. L. (2011). Improving forest inventory in Papua New Guinea: Moving away from the 1% stripline survey. *Native forest management in Papua New Guinea: Advances in assessment, modelling and decision-making*, pp. 69–73.

Brara, R., & Berros, M. V. (2022). A world parliament of rivers. The Rachel Carson center review #1 June. https://springs-rcc.org/a-parliament-of-rivers-legal-personhood-and-beyond/. Accessed July 27, 2022.

Bryan, J., Shearman, P. L., Ash, J., & Kirkpatrick, J. B. (2010). Estimating rainforest biomass stocks and carbon loss from deforestation and degradation in Papua New Guinea 1972–2002: Best estimates, uncertainties and research needs. *Journal of Environmental Management, 19*, 995–1001.

Bun, Y., King, T., & Shearman, P. L. (2004). China's impact on Papua New Guinea's forest industry. Forest Trends. http://www.forest-trends.org. Accessed May 14, 2022.

Cousteau, J.M., & Richards, M. (1999). *Cousteau's Papua New Guinea Journey*. Henry Abraham's Books.

Clark, M. R., & Kozar, J. S. (2011). Comparing sustainable forest management certifications standards: A meta-analysis. *Ecology and Society, 16*(1)

Daly, H., & Farley, J. (2010). *Ecological economics: Principles and applications*. Second Edition. Island press.

Diamond, J, (2011), *Guns, Germs and Steel: The Fate of Human Societies*. Random House.

Demeulenaere, E. S., Yamin-Pasternak, D., Rubinstein, H., Lauren Lovecraft, A., & Ickert-Bond, S. M. (2021). Indigenous spirituality surrounding Serianthes trees in Micronesia: Traditional practice, conservation, and resistance. *Social Compass SAGE Journal*.

Dornelas, M., Antão, L. H., Moyes, F., et al. (2018). BioTIME: A database of biodiversity time series for the Anthropocene. *Global Ecol Biogeogr 00*, 1–26. https://doi.org/10.1111/geb.12729

EIA-Telapak. (2005). The last frontier: Illegal logging in Papua and China's massive timber theft. *Environmental Investigation Agency*. http://www.eia-international.org/files/reports93-1.pdf. Accessed May 1, 2022

ForestPlots.net, Blundo, C., Carilla, J., Grau, R., Malizia, A., Malizia, L., et al. (2021). Taking the pulse of Earth's tropical forests using networks of highly distributed plots. *Biological Conservation, 260*. https://doi.org/10.1016/j.biocon.2020.108849

Filer, C. (2010). The carbon cargo cult in Papua New Guinea. Forests of Oceania Working Session, Association for Social Anthropology in Oceania.

Filer, C. (2011). The political construction of a land grab in Papua New Guinea. Published proceedings paper. In *International Conference on Global Land Grabbing*, 6–8 April, University of Sussex, UK.

Filer, C., & Sekhran, N. (1998). Loggers, donors and resource owners: Papua New Guinea Country Study (Executive Summary). IIED Policy that Works for Forests and People Series No: 2. Meyers J. Ed.

Filer, C., Keenan, R. J., Allen, B., & McAlpine, J. (2009a). Deforestation and forest degradation in Papua New Guinea. *Annals of Forest Science, 66*, 813. https://doi.org/10.1051/forest/2009067

Filer, C., Keenan, R. J., Allen, B. J., & Mcalpine, J. R. (2009b). Deforestation and forest degradation in Papua New Guinea. *Annals of Forest Science, 66*, 813.

Flannery, T. (2002). *The future eaters: An ecological history of the Australasian lands and people*. Grove Press.

Fox, J. C., Yosi, C. K., Nimiago, P., Oavika, F., Pokana, J. N., Lavong, K., & R. I., Keenan,. (2010). Assessment of aboveground carbon in primary and selectively harvested tropical forest in Papua New Guinea. *Biotropica, 42*, 410–419.

Gillison, G. (1993). *Between culture and fantasy: A New Guinea highlands mythology*. University of Chicago Press.

Global Forest Resources Assessment FRA. (2015). Country Report Papua New Guinea Rome, 2014. https://www.fao.org/3/az303e/az303e.pdf. Accessed June 1, 2022

Goldberg, L., Lagomasino, D., Thomas, N., & Fatoyinbo, T. (2020). Global declines in human-driven mangrove loss. *Global Change Biology, 26*(10), 5844–5855.

Gosarevski, S., Hughes, H., & Windybank, S. (2019). Is Papua New Guinea Viable? Political Economic Bulletin 19. https://www.semanticscholar.org/paper/Is-Papua-New-Guinea-viable-Gosarevski-Hughes/6d71c70951c12fa82dea578a36b66923997f3f92. Accessed June 1, 2022

Greenpeace. (2005). Partners in Crime: The UK Timber Trade, Chinese Sweatshops and Malaysian Robber Barons in Papua New Guinea's Rainforests. Greenpeace-UK, London.

Gray, B. (2011). *The boreal herbal: Wild food and medicine plants of the North.* Aroma Borealis Press.

Hannah, L., Roehrdanz, P. R., Marquet, P. A., Enquist, B. J., Midgley, G., Foden, W., Lovett, J. C., Corlett, R. T., Corcoran, D., Butchart, S. H. M., Boyle, B., Feng, X., Maitner, B., Fajardo, J., McGill, B. J., Merow, C., Morueta-Holme, N., Newman, E. A., Park, D. S., Svenning, J.-C. (2020). 30% land conservation and climate action reduces tropical extinction risk by more than 50%. *Ecography, 43*, 943–953. https://doi.org/10.1111/ecog.05166

Hu, T., Su, Y., Xue, B., Liu, J., Zhao, X., Fang, J., & Guo, Q. (2016). Mapping global forest aboveground biomass with spaceborne LiDAR, optical imagery, and forest inventory data. *Remote Sensing, 8*(7), 565–570.

Huettmann, F. (2011). *Protection of the Three Poles.* Springer.

Huettmann, F. (2017). Climate change effects on terrestrial mammals: A review of global impacts of ecological niche decay in selected regions of high mammal importance. *Encyclopedia of the Anthropocene, 2*, 123–130.

Huettmann, F. (2018). Climate change effects on terrestrial mammals: A review of global impacts of ecological niche decay in selected regions of high mammal importance. *Encyclopedia of the Anthropocene* 123–130.

Huettmann, F. (2020). Chapter 24—Investigating Matschie's Tree Kangaroos with 'Modern' methods: digital workflows, big data project infrastructure, and mandated approaches for a holistic conservation governance. In L. P. V. Dabeck, J. Blessington, & K. Schwartzet (Eds.), *Tree Kangaroos: Science and Conservation* (pp. 379–391). Academic Press.

Huettmann, F., & Young, B. D. (2022). The So-called modern 'sustainable forestry'destroys wilderness, old-growth forest landscapes and ecological services worldwide: A short first-hand review and global narrative on the use of 'Growth-and-Yield'as a destructive and even impossible goal. In *Forest Dynamics and Conservation* (pp. 53–82). Springer.

Hyndman, D. (1998). Melanesian resistance to ecocide and ethnocide: Transnational mining projects and the fourth world on the Island of New Guinea. In H. J. Bodley (Ed.), *Tribal peoples and development and issues: A global overview* (pp. 281–298). Mayfield Publishing Company.

Kobayashi, T., Tateishi, R., Alsaaideh, B., Sharma, R. C., Wakaizumi, T., Miyamoto, D., Bai, X., Long, B. D., Gegentana, G., & Maitiniyazi, A. (2017). Production of global land cover data—GLCNMO2013. *Journal of Geography and Geology, 9*(3), 1–15 (2017). https://doi.org/10.5539/jgg.v9n3p1

Karger, D. N., Kessler, M., Lehnert, M., & Jetz, W. (2021). Limited protection and ongoing loss of tropical cloud forest biodiversity and ecosystems worldwide. *Nature Ecology & Evolution, 5*(6), 854–862.

Katewa, S. S. (2009). Indigenous people and forests: Perspectives of an ethnobotanical study from Rajasthan (India). In *Herbal drugs: Ethnomedicine to modern medicine* (pp. 33–56). Springer

Katovai, E., D. D. Katovai, S. G. Laurance, W. Edwards, & W. F. Laurance. (2021). Structural recovery of logged forests in the Solomon Islands: Implications for conservation and management. *Tropical Conservation Science* (https://journals.sagepub.com/ https://doi.org/10.1177/19400829211028125).

Keenan, R. J., Fox, J. C., Brack, C. L., & Saulei, S. (2011). Native forest management in Papua New Guinea: An introduction. *Native Forest Management in Papua New Guinea: Advances in Assessment, Modelling and Decision-Making, 135*, 32–40.

Kingsland, S. E. (2018). Facts or fairy tales? Peter Wohlleben and the hidden life of trees. *Bulletin of the Ecological Society of America, 99*(4), 1–5.

Krishna, P. A., Chand, R., Huettmann, F., & Ghimire, T. R. (2022). Rabies elimination: Is it feasible without considering wildlife? *Journal of Tropical Medicine* (in press)

Ludlam, S. (2021). *Full Circle: Power*, Hope and the Return of Nature. Black Inc.
Laman, T., & Scholes, E. (2012). *Birds of paradise: Revealing the world's most extraordinary birds.* National Geographic Books.
Laurance, W. F. (1998). A crisis in the making: Responses of Amazonian forests to land use and climate change. *Trends in Ecology and Evolution, 13*(10), 411–415.
Laurance, W. F. (2007). Forest destruction in tropical Asia. *Current Science*, 1544–1550.
Laurance, W. F., Kakul, T., Keenan, R. J., Sayer, J., Passingan, S., Clements, G. R., & Sodhi, N. S. (2011). Predatory corporations, failing governance, and the fate of forests in Papua New Guinea. *Conservation Letters, 4*(2), 95–100.
Lawrence, A. (2009). Forestry in transition: Imperial legacy and negotiated expertise in Romania and Poland. *Forest Policy and Economics, 11*(5–6), 429–436.
Leinen, J., & Bummel, A. (2018). *A world parliament: governance and democracy in the 21st century.* Democracy Without Borders.
Lines W. (1999). *Taming the Great South Land: A history of the conquest of nature in Australia.* University of Georgia Press.
Lipschutz, R. D. (2000). Why is there no international forestry law: An examination of international forestry regulation, both public and private. *UCLA Journal of Environmental Law and Policy, 19*, 153–159.
Mace, G. M., Perrings, C., Le Prestre, P., Cramer, W., Díaz, S., Larigauderie, A., Scholes, R. J., & Mooney, H. A. (2013). Science to policy linkages for the post-2010 biodiversity targets. In *Biodiversity monitoring and conservation: Bridging the gap between global commitment and local action.* (Ch. 13, pp. 291–310) Wiley-Blackwell. https://doi.org/10.1002/9781118490747
Mack, A. (2014). *Searching for PekPek: Cassowaries and conservation in the new guinea rainforest.* Cassowary Conservation and Publishing
Mack, A. L., & West, P. (2018). Ten thousand tonnes of small animals: wildlife consumption in Papua New Guinea, a vital resource in need of management.
Mårald, E., & Westholm, E. (2016). Changing approaches to the future in Swedish forestry, 1850–2010. *Nature and Culture, 11*(1), 1–21.
Melick, D. (2010). Credibility of REDD and experiences from Papua New Guinea. *Conservation Biology, 24*, 359–361.
Mongabay. (2022). Papua New Guinea. https://rainforests.mongabay.com/deforestation/forest-information-archive/Papua_New_Guinea.htm. Accessed March 29, 2022.
Montagu, A. S. (2001). Reforming forest planning and management in Papua New Guinea, 1991–94: Losing people in the process. *Journal of Environmental Planning and Management, 44*(5), 649–662.
Narokobi, B. (1983). The Melanesian Way. Institute of Papua New Guinea Studies, Boroko, Papua New Guinea.
Overseas Development Institute (ODI). (2007). What can be learned from the past ? A history of the forestry sector in Papua New Guinea. *Papua New Guinea Forest Sector Studies 1.* www.odi.org.uk
Peng, C. (2000). Growth and yield models for uneven-aged stands: Past, present and future. *Forest Ecology and Management, 132*(2–3), 259–279.
Ploton, P., Mortier, F., Réjou-Méchain, M., Barbier, N., Picard, N., Rossi, V., Dormann, C., Cornu, G., Viennois, G., Bayol, N., & Pélissier, R. (2020). Spatial validation reveals poor predictive performance of large-scale ecological mapping models. *Nature Communications 11*, 1–11.
Prajapati, J., Ghimire, T. R., Regmi, G. R., & Huettmann F. (2020). Chapter 17. Nature and landscape governance in royal times: Experiences from the Shah and Rana regimes in Nepal re-assembled from literature and interview data. In: G. R. Regmi & F. Huettmann (Eds.), *Hindu Kush-Himalaya watersheds downhill: Landscape ecology and conservation perspectives.* (pp. 345–360) Springer.
Pretzsch, H. (2009). Forest dynamics, growth, and yield. *Forest Dynamics, Growth and Yield*, 1–39.
Quaglia, S. (2018). *State of the world's forests 2018—Forest pathways to sustainable development. Food and Agriculture Organisation of the United Nations.* FAO Rome, Italy.

Rajan, S. R. (2006). *Modernizing nature: Forestry and imperial eco-development 1800–1950*. Oxford University Press.
Ramachandra, T. V., & Bharath, S. (2020). Carbon sequestration potential of the forest ecosystems in the Western Ghats, a global biodiversity hotspot. *Natural Resources Research, 29*(4), 2753–2771.
Rangarajan, M. (1994). Imperial agendas and India's forests: The early history of Indian forestry, 1800–1878. *The Indian Economic and Social History Review, 31*(2), 147–167.
Rodriguez, F. C., & Conje J. (2022). The evolution of the dialogue and perspectives on sustainable forest management with special emphasis on the United States of America. *Journal of Sustainable Forestry 5*, 1–45. https://doi.org/10.1080/10549811.2022.2059687
Richards, S. J. (Ed.). (2018). *Rapid biological assessments of Wau Creek, Uro Creek and Lake Kutubu: Documenting biodiversity values to promote forest conservation in the Kikori River basin, Papua New Guinea*. ExxonMobil PNG Limited.
Rumley, D. (2006). Australia's arc of instability: Evolution, causes and policy dilemmas. *The Otemon Journal of Australian Studies, 32*, 37–59.
Saulei, S. M., & Aruga, J. A. (1994). The status and prospects of non-timber forest products development in Papua New Guinea. *The Commonwealth Forestry Review* 97–105.
Samojlik, T., Fedotova, A., Daszkiewicz, P., & Rotherham, I. D. (2020). *Białowieża Primeval Forest: Nature and culture in the nineteenth century* (pp. 223-pp). Springer International Publishing.
Saulei, S. M., & Aruga, J. A. (1994). The status and prospects of non-timber forest products development in Papua New Guinea. *The Commonwealth Forestry Review*, 97–105.
Schoen, J. (2021). *Tongass Odyssey: Seeing the forest ecosystem through the politics of trees*. University of Alaska Press.
Scudder, M. G., Herbohn, J. L. & Baynes, J. (2018). "The failure of eco-forestry as a small-scale native forest management model in Papua New Guinea". *Land Use Policy 77*: 696–704.
Shearman, P., Bryan, J. E., Ash, J., Hunnam, P., Mackey, B., & Lokes, B. (1972). The state of the forests of Papua New Guinea. *Mapping the Extent and Condition of Forest Cover and Measuring the Drivers of Forest Change in the Period, 2002*, 148.
Shearman, P. L., Ash, J., Mackey, B., Bryan, J. E., & Mean, B. (2009). Forest conversion and degradation in Papua New Guinea 1972–2002. *Biotropica, 41*, 379–390.
Shearman, P. L., & Bryan, J. (2011). A bioregional analysis of the distribution of rainforest cover, deforestation and degradation in Papua New Guinea. *Austral Ecology, 36*, 9–24.
Shearman, P., & Bryan, J. E. (2015). *The State of the Forests of Papua New Guinea 2014: Measuring change over the period 2002–2014*. University of Papua New Guinea.
Shearman, P., Bryan, J., & Laurance, W. F. (2012). Are we approaching 'peak timber' in the tropics? *Biological Conservation, 151*(1), 17–21.
Slik, J. W. F., et al. (2015). An estimate of the number of tropical tree species. *PNAS, 112*, 7472–7477.
Simard, S. (2021). Finding the mother tree: uncovering the wisdom and intelligence of the forest. Penguin UK.
Stiglitz, J. (2003). *Globalization and its Discontent*. Norton Paperback.
Suzuki, D. (1993) Wisdom of the elders: sacred native stories of nature. BoD–Books on Demand.
Taber, R. D., & Payne, N. F. (2003). *Wildlife, conservation, and human welfare: A United States and Canadian perspective*. Krieger Publishing Company.
Taylor, A., Zotz, G., Weigelt, P., Cai, L., Karger, D. N., König, C., & Kreft, H. (2022). Vascular epiphytes contribute disproportionately to global centres of plant diversity. *Global Ecology and Biogeography, 31*(1), 62–74.
The Guardian. (2015). The incredible plan to make money grow on trees. 24th November 2015. https://www.theguardian.com/world/2015/nov/24/redd-papua-new-guinea-money-grow-on-trees. Accessed May 28, 2022.
The Guardian. (2017). What David Attenborough should have said at the end of Blue Planet. 12th December 2017. https://www.theguardian.com/commentisfree/2017/dec/12/david-attenborough-blue-planet-bbc. Accessed July 1, 2021.

The Guardian. (2020). Papua New Guinea police accused of gun running and drug smuggling by own minister 10th May 2022. https://www.theguardian.com/world/2020/sep/18/papua-new-guinea-police-accused-of-gun-running-and-drug-smuggling-by-own-minister. Accessed November 5, 2021.

Tree Kangaroo Conservation Program. (2018). *Annual Report 2018*. Woodland Park Zoo, Seattle.

Trends, F. (2006). *Logging, Legality and Livelihoods in Papua New Guinea: Synthesis of Official Assessments of the Large-scale Logging Industry* (Vol. 1). Forest Trends.

Uda, S. K., Hein, L., & Sumarga, E. (2017). Towards sustainable management of Indonesian tropical peatlands. *Wetlands Ecology and Management, 25*(6), 683–701.

Webb, C. O., Boucher, T., Sheppard, S., & Summers, M. (2005). Vegetation of the Adelbert range, Madang Province, PNG. Unpublished Report. https://camwebb.info/doc/pubs.html. Accessed May 3, 2022.

Williams, B. A., Grantham, H. S., Watson, J. E. M., et al. (2022). Reconsidering priorities for forest conservation when considering the threats of mining and armed conflict. *Ambio*. https://doi.org/10.1007/s13280-022-01724-0

Wohlleben, P. (2016). *The hidden life of trees: What they feel, how they communicate—Discoveries from a secret world* (Vol. 1). Greystone Books.

World Rainforest Movement (2004, March 12). Oro Landowners Declaration on Large Scale Commercial Extraction of Natural Resources and the Expansion of Oil Palm Nucleus Estates.

Yen, P., Ziegler, S., Huettmann, F., & Onyeahialam, A. I. (2005). Change detection of forest and habitat resources from 1973 to 2001 in Bach Ma National Park, Vietnam, Using Remote Sensing Imagery. *International Forestry Review, 7*(1), 1–8.

Zavala, M. (2013). *What do we mean by decolonizing research strategies? Lessons from decolonizing, Indigenous research projects in New Zealand and Latin America*. Chapman Publishers.

Ziegler, J., Golay, C., Mahon, C., & Way, S. (2011). *The fight for the right to food: Lessons learned*. Springer.

Chapter 16
Papua New Guinea as a Paradise (of Weaponized Diseases): AIDS/HIV, African Swine Fever, Malaria, Avian Influenza, Weaponized Viruses and Wider Public Health?

Abstract Tropical rural nations like Papua New Guinea (PNG) get easily confronted with diseases and their bad reputations. As PNG is a biodiversity hotspot, it also features naturally many diseases, similar to its tropical neighbor nations. However, tropical diseases can be well managed and treated if a (wider) will and infrastructure is in place. Humans have co-evolved with many diseases, and wilderness tends to help to keep a natural balance, e.g. malaria as an example. While some advanced disease strains are man-made and occur in PNG and its shanty towns and mining and oil & gas hubs, such as cerebral malaria or HIV, many more diseases are benign and can stay within reservoirs for a long time without becoming of large impact. Core questions in PNG's public health center around poverty. The core diseases in PNG are mostly malaria (different strains), tuberculosis, diarrheal diseases, and some acute respiratory diseases and recent pandemics. And all of those can usually be treated. Spectacular disease reports like the kuru disease in PNG are locally isolated cases and essentially extinct but still provide a reputation and legacy. PNG can offer a disease reservoir and be used for more strategic diseases and when those get weaponized (nowadays a reality in Biological Warfare by major nations; a global practice but that PNG never has practiced). Wider public health concepts across the board and ideologies should be considered to bring up health metrics for PNG to a global standard similar and better than Australia, as a major donor for aid and development. Healthcare lessons from other nations like Cuba or China can be helpful rather than ignoring them while PNG had already all basic aspects of modern 'OneHealth' implemented for millennia.

Keywords Papua New Guinea (PNG) · Disease · AIDS/HIV · African swine fever · Avian influenza (AI) · Public health · One Health

16.1 Introduction

Like with many tropical, exotic and remote locations, the western world can often boost its own health record as 'progress' while inflating the tropical disease record as 'inferior' based on 'facts.' Most travel guides are written that way, top down looking upon the tropical nation and its exotic diseases. It's an important argument

F. Huettmann, *Globalization and Papua New Guinea: Ancient Wilderness, Paradise, Introduced Terror and Hell*, https://doi.org/10.1007/978-3-031-20262-9_16

because this scheme is also widely used as a justification for development aid and (sustainable) development, modernity, an assumed progress via urbanization and its plumbing, electricity, taxation and social welfare (e.g. Acemoglu & Robinson, 2013; see also Diamond, 2011 for some examples and discussion). It's a core argument for 'civilization' and what would make other nations lower developed, or 'failed states' (see earlier chapters in this book).

Many western health records and views simply come from an urbanized landscape and infrastructure; it's correlated, whereas rural and wilderness landscapes fundamentally differ, regardless of being in PNG or in 'the west.' Wilderness will always by dusty and likely muddy with a jungle-type vegetation or a desert without much infrastructure; it tends to be a frontier confronting administration with the 'wild and raw' (see Leopold, 1989, etc.). PNG is widely rural, not urban, has much wilderness left and runs a Wantok tribal system with a 'bush' landscape or a similar remote island and coastal society, already referred to as 'savages' if not even cannibals (e.g. Levi-Straus, 1966; Matthiessen, 1987; Mead, 1967). Remote areas in PNG are not well studied but feature those rural metrics, and often a patchy and poorly understood record of diseases and similar, e.g. (Moir et al., 1989). Sorcery remains a strong force in PNG on those topics (as described in Connell 2022, Beehler & Laman, 2020; Gillison, 1993).

And while many diseases are now global and they cross borders, the western and global policies even made it worse (Lerner, 2021; for an example, see Packard, 2007 for Malaria and its advanced strains like *Plasmodium falciparum* as a created man-made disease transported globally by the industrialized world based on colonialism and industrialization; see Carter, 2003 for *Plasmodium vivax*; compare with Trajer 2022

In the western world, the tropics are still widely seen as 'exotic,' the unknown. Diseases from there are automatically declared 'enemies' for modernity and perceived as left over pains from the past and from another world. It's often assumed and promoted—wrongly though—that exotic diseases never occurred in the western world, got wiped out centuries ago, or are primarily a problem of 'the tropics' and of poor nations and poor governance. Again simply looking at malaria shows the flaws of such views, e.g. it was a common and historic problem in Italy, Spain, Greece, Yugoslavia, Central Asia and even subarctic Russia. It's still reoccurring as instances, in France, Germany and beyond. The recent 'pandemics' of 2019 threw another curve ball in this narrative.

Health questions—including vaccinations that brought much global progress and health—now get increasingly fiercely debated and evaluated. A naïve approach then frequently goes: Let the western pharmaceutical industry kill those diseases, we pay for it; the public healthcare service and its commercial entities come to our rescue. Diseases that have experienced such an attitude include traditional ones also found in PNG such as tuberculosis, influenza as well as diarrheal diseases (e.g. neglected tropical diseases like *giardarsis*), and the range of acute respiratory disease (see also Owen 2005 for more disease details). But for a wider medical health view one also need to include here the bad outcomes of more 'traditional' heath concerns, e.g. physical injuries such as from falls, slips, fires, sport-related injuries (Foster, 2006), machete cuts, remote plane crashes, car accidents, as well as abscesses, tumors, depression, etc. Modern diseases like HIV/AIDS further add to the injury (see,

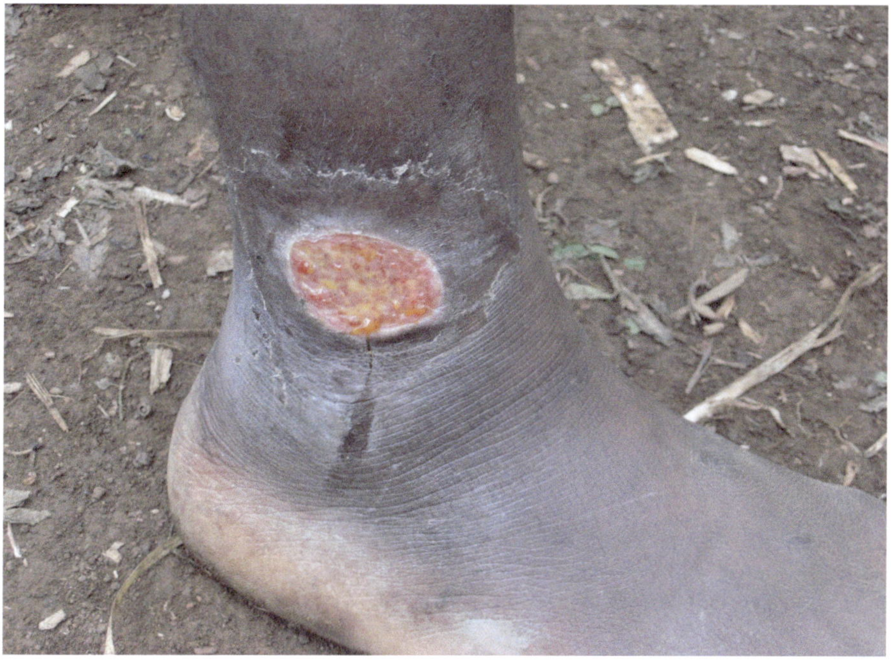

Fig. 16.1 Skin diseases, a common sight in tropical nations

for instance, Luker, 2004). PNG has all of those but lacks the full coverage health treatments (Figs. 16.1 and 16.2).

16.2 PNG as a Tropical Disease Player on the World Stage

Having worked and visited as 'western investigator and traveler' PNG for several seasons across years, I can relate pretty well with the experience of western people, as well as disease treatments in the nation and from abroad (see Table 16.1 for overview). I actually met several health practitioners from abroad in PNG, almost all came from the hospital-pharmaceutical industry angle though. Noteworthy was an experienced outbreak of Cholera in the Sepik region (as treated by a 'Doctors without Borders' campaign '*diseases know no borders*'; see organization details here https://www.doctorswithoutborders.org/). OneHealth efforts—a more holistic effort to disease management—are part of those industrial western schemes. But how effective really is it?

And in all of this, one must not forget the context that tropical diseases occur in the 'tropics' and thus the reality in many nations worldwide around the equator and beyond. The 'tropics' are connected with the northern regions though and diseases get spread; with modern infrastructures and commerce even more. That has been the case

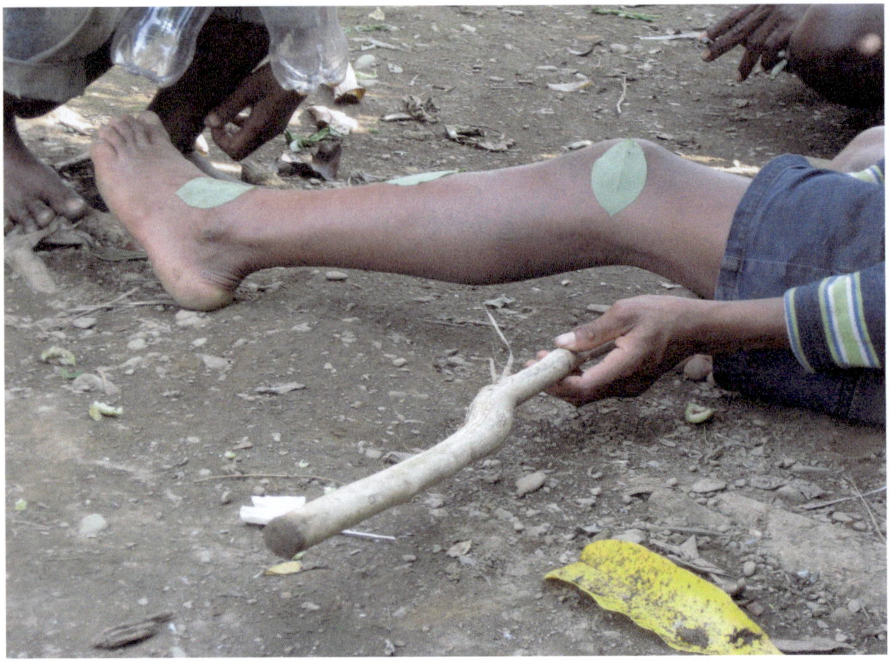

Fig. 16.2 Treatment of skin diseases; use a leaf with natural healing ingredients instead of a plastic band-aid!

Table 16.1 Selection of diseases encountered in PNG and their treatment

Disease	PNG origin?	Citation	Comment
Malaria	No	Packard (2007)	Severe malaria strains are widespread and encroaching in higher altitudes in PNG but essentially a man-made product from elsewhere (see Beehler & Laman, 2020; Packard, 2007)
Tuberculosis	No	Stead (1997)	Widely globally distributed
Rabies	No	NA	Probably a rel. recent disease and primarily linked to dogs and flying foxes/bats. Unknown when it occurred in PNG
COVID	No	NA	Global pandemic, no new for PNG
HIV	No	Grundy et al (2019)	Details not so well known

historically, e.g. pestilence from Central Asia into western Europe, in colonial times, e.g. Malaria brought into the New World (Packard 2007), and it happens nowadays, e.g. Corona (from China worldwide). The economy remains a driver either way (see Gulyaeva et al., 2021 for Avian Influenza in the Pacific region).

It must be stated that the tropics do have their own co-evolved and endemic diseases, but for PNG and its islands, the typical disease load was also brought by colonialists, such as influenza, typhoid, sexually transmitted disease, etc. with devastating effects (Acemoglu et al., 2003; Diamond, 2011 for overview). It's not the case that the western world was 'clean' and the bad diseases pop up and spread from the tropics, or PNG, into the world; far from it (see Acemoglu et al., 2003; Diamond 2011 for a discussion).

In terms of tropical region medicine, Cuba remains an accepted leader in many tropical disease studies and cures (Vos et al., 2006). The president of The World Bank, James Wolfensohn, acknowledged that Cuba had done "a great job" in providing for the social welfare of its people (Lobe, 2001 in Vos et al., 2006). China did some of it too. So why not learning from them for all options available, specifically in a tropical, colonial poverty concept, as applicable in PNG? Australia might not easily approve (Figs. 16.3 and 16.4).

And so, being in the tropics is not automatically a hindrance to good medicine. Being exposed to tropical diseases for such a long time, and running a civilization for over 47,000 years, why then is PNG not a leader? PNG does know well how

Fig. 16.3 Insects and spiders, some can bite indeed (others just look scary), the tiny mites though in forests and grassland can be equally or more impactful though...

Fig. 16.4 Always good to have a mosquito net in the tropics; simple and effective protection

to avoid and how to treat their diseases (details in Gillison, 1993 for instance). Despite such much effort (Lerner, 2021) and wording by the western nations and aid, primarily Australia (see example at https://www.dfat.gov.au/geo/papua-new-guinea/development-assistance), why is the health case in PNG not much better? Arguably, the root cause of diseases is ignored and a pure obsession of treating symptoms remains the action of the day; "*makes money, keeps The West in charge and helps to control the locals*," not? That's totally colonial then!

In modern PNG, the communicable diseases dominate (malaria, tuberculosis diarrheal diseases, and acute respiratory disease; source: https://www.lonelyplanet.com/papua-new-guinea). A HIV epidemic is driven by heterosexual transmission (Grundy et al., 2019).

While the modern times in PNG are still not well under public control when it comes to diseases (see Clements et al., 2006 for immunization), the aspects of climate change and how it moves major diseases like malaria upward by elevation (Beehler & Laman, 2020) are also not well recognized, but those are true facts. It's a real human drama unfolding because the inhabitants of those areas get confronted with new and deadly diseases they did not cause or understand or know how to handle. Presumably more surprises can be expected while pandemics are on the rise. PNG remains a disease hotspot, primarily with PNG citizens to cover the burden and sufferings; foreigners get less affected because they have better options and more means (usually they come well prepared and with own medication, a fact that many people in PNG simply do not have. For instance, in 2017 a national audit showed for PNG a widespread "unavailability of lifesaving and essential drugs").

With many wildlife species widely left unresearched (see Martin, 2005 for tree kangaroos a widely hunted and handled species for human consumption), many aspects of zoonotic diseases are subsequently not so well known or studied neither (see Owen 2005 for reasons, evolution and species), e.g. avian influenza, Japanese encephalitis (Hanson et al., 2004) or COVID. In part, that's because most of the zoonotic aspects are not well known or monitored, which is the vast section of the disease ecology. Without knowing the reservoir it's difficult to treat such diseases (example for rabies in Asia and UN policies discussed in Krishna et al., 2022). PNG has a lot of domesticated ducks around the many villages (see Figs. 16.5, 16.6 and 16.7 for poultry and Muscovy Duck) and which are known national disease reservoirs and vectors but not well accounted for, or managed at all. And then, the pigs play a major role certainly (Diamond 2011). Just like cattle, pigs co-evolved with humans, their dwellings and their diseases and thus those play a big role everywhere in the world (see Diamond, 2011); this is specifically true for PNG where pigs play such a central role in the society and for peace-keeping (Flannery, 1998 for an example). And then, PNG does have many (micro) bats and flying foxes (as widely discussed reservoirs for COVID, etc. Gouilh et al., 2011) and the roosting sites might already present large sources of diseases. Some of those roosts are next to supermarkets, e.g. in Madang downtown. As many local markets are held in the open, under shade trees, it can be generally rather close to roosts, and transmission risks do exist.

Fig. 16.5 A rural village with chicken (think of avian influenza) under your bed…

Fig. 16.6 Muscovy duck dominates the local poultry in Papua New Guinea, and in Asia overall; they often share with humans the same water source also

A third disease risk that should be considered in PNG and Asia are due to pets. Pets from the wild, like cockatoos, cassowaries or opossums and many baby mammals are widespread and commonly found in PNG and its villages (see Mack & West, 2018 for bushmeat examples and large amounts taken). Many of them actually get eaten eventually, e.g. cassowaries once they are app. 1 year old, and a link with the human interface is close and unavoidable. On the one hand, their disease risks are not known nor so well studied. On the other hand, pets are an inherent part of the PNG village lifestyle for millennia and it has co-evolved (see Flannery, 1990, 1998). Pets are living in the adobe houses, with the entire family and interact with the kids too.

Textbox 1: Poultry and pigs: An industrially relevant influenza (poultry) and African swine fever (wild pigs) perspective in PNG

The intense commercial and industrial production of protein, meat, is the basis for 'modern' society. The western world is driven by it (McKenna, 2017). Having meat readily and consistently available in such high quantities and quality in the modernity, and to a low price, e.g. in a concentrated place nearby like a supermarket chain, was not really possible in earlier societies. It's a recent aspect of globalization and its underlying industrialization. This production is usually achieved through a high energy input (Madison, 1997 for E.

Fig. 16.7 A Muscovy Duck, a beauty for disease researchers and the major domestic species in Asia

Odum's famous ecological research on the topic), as well as a use of technology and biochemistry, e.g. painkillers, growth hormones and antibiotics (McKenna, 2017). This is well known for chicken, aquafarming and cattle. All of it comes with a side aspect, which is, contamination. But also other aspects, e.g. waste, and Jevons Paradox, reliance on food, use of soil, and psychological perspectives of obesity and human health in society (Eisenhauer, 2001; see Baraka, 2001 mentioning it as modern lifestyle diseases recently occurring in PNG).

Now, because protein is such an essential issue in society, its functioning and wealth distribution, it becomes a core question where and how it is produced (e.g. Diamond, 2011; Friedmann, 1993). While U.S. has been a global lead in such questions of consumption, within Asia, China and India now play a big role, and so does Australia for the Sahul. Meat production is done, and advised, as an industrial entity. Growth rates and impacts for doing so in nations like PNG can be rather high still, with lots of potential.

Due to the relevance of protein production for an entire city, region, island, or nation, food security becomes easily a strategic item. Any diseases harming food security can quickly turn such a production into a national crisis. As many sources for disease outbreaks remain unclear, much uncertainty can linger around this topic. Some diseases can easily be triggered by humans, and then wipe out the entire food production; the human history has many examples, see Diamond (2011) for potatoes-related diseases in Ireland and affecting North American and UK history. For PNG, this means, production centers of cattle, e.g. Ramu, or chicken, Lae, or industrial farming regions like Mt Hagen can become very vulnerable. That's why diseases like avian influenza, African swine fever or rabies matter a lot for island nations like PNG. These diseases are by now on a global level, and their names can be rather misleading as they are not just confined to Asian or to Africa. As PNG misses the capacity to study and manage those diseases, outside medical capacity sits in the driver seat, once more. Those aspects of commercial control are widely overlooked, also in times of globalization

Textbox 2: Living with snakes in interior PNG

Snakes are a reality in the Sahul region; that's for Australia a serious concern as it is in PNG. Beehler and Latam (2020) describe the relevant species in real life and that they occur primarily in the open grasslands, e.g. several species of adders. Rubber boots are a necessity...to survive.

However, people in PNG lived with such snakes for millennia...barefooted. Arguably, no foot coverage was available, and many people must have died, namely the garden workers.

I heard locals in PNG telling me that snakes observe their targets (humans). They find an activity pattern and then strike accordingly. Snakes also are said to take revenge when treated poorly or disturbed. Some of those snakes are deadly… The key remains to strike a way of living in such environments, and PNG did just that for millennia.

Textbox 3: Corona virus: Not a Papua New Guinea disease whatsoever: PNG on the receiving end, again

Outbreaks of diseases on islands can be devastating. The recent Corona outbreak, which spread from China into Asia and the world, affected PNG wholesale. Corona is a disease linked to mammals, and bats and flying foxes play apparently a bigger role. PNG has many species of flying foxes and of (micro) bats, and the link with COVID remains an easy one; it's a reservoir and disease frontline. However, it's very unlikely that PNG is a global reservoir of tropical diseases like COVID. With that, PNG remains primarily a victim of corona and likely affected by it for a longer time via Asia mainland likely.

16.3 Serious Health Risks in PNG and Their Handling

PNG is world-famous for its serious diseases, namely the classic tropical ones (list provided in the next section). But this is often used to scare people and also to present PNG and its society as a dangerous place, the back woods, as 'savages', to create a hype and exploit it for one own's purpose (examples in Hofffmann, 2015; Mack, 2014; Salak, 2001).

Noteworthy in that discussion is the kuru disease, a deadly prion disease spreading by eating human brains. It's perceived as a sign of predatory cannibalism but which it usually is not (see Lindenbaum, 2001; see also Gingging, 2007 for a discussion of such catchy concepts). Reality shows that these are very local and rare instances though, and that it's part of a spiritual burial practice from the 1950s in a local area mainly with the Fore people. They believe to help the soul (brain) to transfer into the new generation and to avoid the corpse of a relative getting eaten by worms and bugs instead. This is an essential concept in PNG society to care for the well-being of the future soul. Kuru is a disease that can be tracked to Neanderthals etc. and to flesh eating (e.g. Alpers, 2007). Its outbreak is often seen like a monitoring event for ancient, cannibal, practices otherwise hidden in society. But kuru has a delayed incubation and thus cannot always be used to track those practices. It's now virtually extinct though (Liberski, 2013). For those people who hype of kuru, one should be more concerned about other prion diseases in the western food system, e.g. Creutzfeldt-Jakob disease (Mad Cow Disease), etc. which are found in the EU and Australia and can potentially spill into PNG cattle from the outside. Instead, using the concept of 'danger'—a fear factor—obviously works in PR to attract attention. But it stands widely against the concept of mass tourism as it scares people away. It does help development aid though, which then promotes PNG with progress and a good outlook, when there actually is little (Fig. 16.8).

Arguably, most diseases in PNG are not that deadly or 'bad.' That's because they can get self-treated with antibiotics, or in the local hospitals, or patients get flown out of the country, e.g. Australia, within a day. Which tourist really dies from serious diseases in PNG; it's a very low percentage, if at all. In remote areas such as PNG, with very few medication used, antibiotics tend to have a steep success rate (in contrast to western urban areas).

Like with most species in many tropical nations, the diversity of disease variants is also high in PNG. Some of those strains are really dangerous. PNG presents us as a disease pool with incomplete surveys (Lerner, 2021). But what many people likely are not much aware of, PNG offers us many raw materials to be used for pandemics, if weaponized. That is, one can take/collect an endemic and rare disease and make it even stronger in a lab, and then use it as a biological weapon agent on a 'battlefield' (Lerner, 2021; see Rauzon, 2016 for reality applications in the Pacific). Disease hotspots are frequently visited and globally monitored and partly collected by many groups and actors (labs, institutes, NGOs). PNG offers the raw wild materials for it, which are otherwise hard to find! Some PNG diseases can offer a good springboard to weaponize them (see Dudley 2004 and section below). Is that hypothetical? Far from it.

Instead, and as so well presented and introduced to a global audience by Diamond (2011), diseases are part of life (see also Acemoglu et al., 2003). Diseases are another life form. It co-evolved with humans; they are distributed worldwide and most of them are with us, not wiped out by medical campaigns (e.g. the claim that polio is wiped out, just compare for instance with https://www.who.int/westernpacific/emergencies/papua-new-guinea-poliovirus-outbreak). But perhaps some diseases are reduced in spread while the severity remains very high, e.g. cerebral malaria (Packard, 2007).

Arguably, many diseases can be managed, reduced and well treated in humans. And until the pandemics of 2019, that became a safe narrative. Reality somewhat differs though, and always had differed. Weaponized diseases present a dubious albeit unknown deviation; it's uncharted territory for most people and nations (Dudley, 2004; many examples exist, for instance Abbara et al., 2021; Blackburn et al., 2020). This puts the work by Diamond (2011) to another level (but it was not really discussed in Diamond, 2011). Now with the pandemicWeaponized diseases on the rise, we get a reality view and see how the traditional health PR failed us for decades. PNG sits right within that reality, but it is no extreme case, as we learned earlier. Still, PNG is just a pawn in the global medicine game, driven by dominant nations and huge budgets. Arguably, PNG now gets virtually all its medical advice, help and treatment from Australia, which follows a widely western model (the 'techno pill' approach instead of a wider soul and holistic society view that would suit PNG much better).

Fig. 16.8 Another modern disease spreading risk: airports, airplanes, cargo and (international) customers

The other health advisor, the church, usually follows this concept and does surprising little 'anti-vaxer' work in PNG when compared to the U.S., let's say.

For sure, in most nations the more serious diseases got reduced, often can get treated, or even get wiped out, by national efforts and international campaigns. That has worked for public health metrics worldwide and thus is something that gets favored by administrations as they then have metrics to show to funders and stakeholders. It works virtually anywhere on earth. That's the western approach, for those nations and patients ...who can pay for it. The western public mindset still seems to be something like this: We get it done, one gets vaccinated and we wiped out the disease; all else are just small flickers of bad luck. And thus, PNG as a nation can easily be put on that treatment schedule and will get it done in the same way, if there is a will and no commercial block. Remember, PNG features some of the world's largest mines and should have sufficient income to treat diseases and its citizens. As a matter of fact, health care in PNG is somewhat free and most diseases are treatable, avoidable and a public health agenda needs to be done and implemented accordingly (see Clements et al., 2006 for vaccination programs in PNG). That is specifically true for the tropical regions. Nations like Cuba, Singapore, Taiwan or China show us no other.

However, it's clear that many mining and oil & gas hubs in PNG are also the disease hotspots, directly (e.g. Nriagu, 1990) and indirectly (Clark, 1993; Hemer, 2015). That's unlikely to change and when more mining comes online.

16.4 Some Major Disease Characteristics in PNG

From experience, sores, infections, insect bites, parasites, tumors, bruises and cuts might be leading injuries for PNG people and when living in the bush. Scabies and stomach disease should likely be added also. Many of those are not life-threatening but painful and can result into 'sick days.' In the following I will describe some additional major diseases in the PNG context:

Malaria: Malaria is a global tropical disease; it consists of different strains. Malaria strains in PNG are 'serious' due to the tropical strain that develops in the brain. The anopheles is the vector, and its life conditions will determine much of the disease risk. In that regard, suburban development—shanty towns—are major known contributors, e.g. when compared to traditional small dwellings in 'the bush' which could be changed quickly when diseases break out. Compared to other nations, e.g. Cuba, Nicaragua or Florida/U.S. I saw little spraying campaigns in PNG.

Tuberculosis: A widespread disease due to water problems and plumbing issues.

Rabies: PNG has few large mammals and rabies plays a little role. However, some stray dogs and specifically bats and flying foxes are known hosts.

HIV/AIDS: A widespread disease, as typically found now worldwide and in tropical nations.

COVID update: PNG got hid by the COVID pandemic and the shutdowns like most other nations, that is specifically tragic in isolated village and island communities making the public health aspects more prominent. As COVID is a zoonotic disease, the role of (micro) bats and flying foxes, etc. are discussed as a reservoir and risk factor but not well known.

Other diseases: Here is a good place to mention stomach parasites, tumors and cancers, depression and snake bites; national statistics are not really available, but those factors mentioned rank traditionally high, and are likely on the rise.

Of interest are global pandemics of the past, for instance the bubonic plague of 1894–1901, or Influenza 1914 (see Echenberg, 2007), but while those hit hard nearby places and ports, e.g. Polynesians, Hawaii and Australia, the PNG impact appears to be little, or not well recorded. But generally, island populations get hit very hard by those 'invasives'.

Many other diseases are found; some are not hitting humans directly but can be devastating in other ways. A typical example is plant diseases related to food items and cash crops. The pink disease for coffee—a fungi (see online https://perfectdailygrind.com/2019/01/a-guide-to-common-coffee-pests-diseases/) is such a case and for PNG. Others exist, e.g. Ramu sugar smut and the sugar stunt (Waller et al., 1987).

Diseases as invasive species: There is a wide discussion about the origins of diseases, and where, how and when co-evolved. PNG is not necessarily the cradle for many tropical diseases. Instead, those diseases move and adjust and got introduced; they often are associated with human migration waves, of which several can be found for the 47,000 years old human history in PNG. While the science is ongoing, typical examples can be found with ticks on islands of the South Pacific including Hawaii (examples provided by Rauzon, 2016). This raises very relevant, but hardly addressed issues of island biogeography, specifically the core discussion in that discipline about 'island balance' (as part of the biogeography theory and hypothesis presented by Diamond, 1973 for PNG; see Beehler & Laman, 2020 for overview). Adding diseases to that discipline will likely change many of the inferences, claims and facts presented for Biogeography.

16.5 PNG as the Evolving and Endless Global Disease Reservoir?

PNG gets frequently described, and downgraded as an unsafe place, being labeled as something like 'full of diseases waiting to attack for people to die,' or 'I survived PNG and its diseases.' Reality differs, as the life expectancy in PNG is app. 64 years (various sources exist but none are very accurate for a 'bush metric'); that's higher than many Russian men, or what colonial Germany was like in the 1870s with Bismarck at the start of their colonial period! PNG—as a culture—has developed good, and did so with proper actions against the disease fear, they did that for over 47,000 years and they co-evolved. One can live with diseases and have a long life. Melanesians lived a long life, e.g. 70+ years, and that's because—traditionally—they come from a very healthy place to begin with. Pacific Island records and archeology show that no other (Fig. 16.9).

Fig. 16.9 Sticky crowded local airports for domestic flights

16.6 PNG as a Global Reservoir for Weaponized Diseases and a Spreader?

There are good public health measures taken when patients with serious disease get treatment and they are traveling to the best possible nearby hospital (for PNG which usually is located in Australia). Tuberculosis patients in the Torres Strait region make a good case for that topic (e.g. Australian, 2011).

And apart of its international airports and harbors, PNG is located on international bird flyways, marine mammal and fish movement corridors (including sea turtles) and ocean currents, travel routes (land and sea). And thus they can indeed spread disease wider. But has that happened? Probably, but little has been studied and documented. Likely, PNG actually is less a sender of diseases, but more of a recipient. That's because in reality, diseases get spread by intense industrial activities internationally, and when done in a massive commercial level without good oversight and care. Poultry and pig farms are core targets for that (see Gulyaeva et al., 2020). China and similar national operations sit at the forefront on the list to blame on diseases, add capitalism and promotion of economic growth without proper healthcare but much poverty. The impact that western nations had on the death rates of diseases in their colonies—clearly an industrial activity and aim—reaches easily in the 100 millions. Already James Cook and his exploration efforts get credited for many of such subsequent impacts, see Diamond (2011) for assessment and numbers.

16.7 A Public Health Plan Improvement List for PNG: Get Rid of Aided One-Sided Economic Growth for an Assumed Stability, Learn More from Cuba and the Own History, and Roll Out a Sustainable and Effective Public Health Business Plan ('OneHealth')

Like any tropical and other nation, diseases can be handled and dealt with well, if there is a will. Papua New Guinea health needs to be put on a map (Kock & Koch, 2005 for a disease atlas). Public health is a discipline and a craft, and it gets practiced worldwide; usually with a decent, or better, success (some colonial nations were essentially able to increase their citizen's life expectancy by 35% in just the last 150 years). People tend not to do malpractice sanitary medical steps when their own health gets affected.

Cuba is an accepted leader for tropical public health with an incredibly good record on the matter, including COVID. (Christensen 2003). We do not have the luxury to ignore it. Cuba also sends its expertise and doctors worldwide. However, PNG is instead dominated by Australian and western-style health experts and their business model. How 'well' this has worked can be seen in the poverty and disease rankings that PNG finds itself.

Clearly, the Australian health care provision has not yet brought paradise; far from it, Australia simply offered a localized symptom-based western-style unsustainable health care that pushes out some diseases, but made other diseases more marginal and thus more extreme. Already the underlying Australian business model for disease treatment and health care overall is hardly sustainable and not delivering well. That includes to build capacity (compare with CSIRO claims https://research.csiro.au/bismarcksea/) for nurses, doctors and healthcare practitioners (which follow a brain drain and achieved the opposite for PNG, a loss with the PNG health practitioners being drawn to Australia and New Zelaand; Negin, 2008).

A national audit in 2017 documented widespread "unavailability of lifesaving and essential drugs" for PNG; thus sorcery remains (Connell 2022) Who's fault is it?

In PNG, a combination of government, private facilities and church facilities provides a healthcare service. It's meant to be free but usually charged for and primarily located in urban centers. For betterment, the National Health Plan 2011–2020 has placed major emphasis on stronger access in the bush to primary health services; see Trájer 2022 for PNG and malaria).

As diseases play such a big role, those issues should be addressed and confronted head on. Putting more emphasize on the co-evolution of diseases and how those bad impacts can be mitigated with a good culture (instead of leaving it alone or making it all worse). The Guardian (2021) shows some examples how PNG is a reservoir for diseases like COVID and variants.

The popular concept of a holistic approach, 'One Health', has arrived in PNG, and one might roll it out and see its fruits. Having a good landscape available helps to handle many of the disease effects, e.g. to avoid 'slums' and shanty towns and no plumbing (all as frequently related to industry, and mining, oil & gas development). But in reality, we are still far aware from a PNG healthcare system that truly serves the needs of its people (as stated by Chan, 2016).

The treatment 'blockage' comes mostly from the set up of the vaccines and how health and treatments are institutionalized and administered to the people and paid for. In most nations, the cost coverage should come probably from a funded (tax) system and a public health system. PNG has that, conceptually, but apparently this is not working so well at all, and so the cost of available medication and medicine is a real block plunging PNG into deep poverty and associated health crisis. In reality, that means many innocent people die, and thus, the economic system that pharmaceutical companies with a monopoly and their nations have has real-world impacts on the lives of people in PNG. One may honestly ask whether big neighbor Australia - including New Zealand- has done the best job in helping, and how this gets improved?

References

Abbara, A., Zakieh, O., Rayes, D., Collin, S. M., Almhawish, N., Sullivan, R., Aladhan, I., Tarnas, M., Whalen-Browne, M., Omar, M., Tarakji, A., & Karah, N. (2021). Weaponizing water as an instrument of war in Syria: Impact on diarrhoeal disease in Idlib and Aleppo governorates, 2011–2019. *International Journal of Infectious Diseases, 108*, 202–208.

Acemoglu, D., & Robinson, J. A. (2013). *Why nations fail: The origins of power, prosperity, and poverty paperback.*

Acemoglu, D., Johnson, S., & Robinson, J. (2003). Disease and development in historical perspective. *Journal of the European Economic Association, 1*(2–3), 397–405.

Alpers, M. P. (2007). A history of kuru. *Papua New Guinea Medical Journal, 50*(1/2), 10–19.

Australian, T. B. (2011). Doing the right thing for tuberculosis control in the Torres Strait Islands. *The Medical Journal of Australia, 195*(9), 523–524.

Baraka, R. (2001). The meanings of independence in post-colonial Papua New Guinea. A plan B research paper submitted to the Faculty of the Center for Pacific Islands Studies in the Candidacy for the Degree of Masters of Arts. School of Hawaii, Asia and Pacific Studies, The University of Hawai'i at Manoa.

Beehler, B., & Lamann, T. (2020). *New Guinea: Nature and culture of earth's grandest island* (p. 89). Princeton University Press.

Blackburn, C. C., Lenze, P. E., & Casey, R. P. (2020). Conflict and cholera: Yemen's man-made public health crisis and the global implications of weaponizing health. *Health Security, 18*(2), 125–131.

Carter, R. (2003). Speculations on the origins of *Plasmodium vivax* malaria. *Trends in Parasitology, 19*(5), 214–219.

Chan, J. (2016). *Playing the Game: Life and Politics in Papua New Guinea.* University of Queensland Press.

Christensen, A. (2003). Cuba's Jewel of tropical medicine. *Perspectives in Health Magazine, 8*(2). https://www3.paho.org/english/dd/pin/Number17_article4_4.htm. Accessed May 14, 2022.

Clark, J. (1993). Gold, sex, and pollution: Male illness and myth at Mt. Kare, Papua New Guinea. *American Ethnologist, 20*(4), 742–757.

Clements, C., Morgan, C., Posanai, E., & Sakamoto, C. (2006). A qualitative evaluation of the immunization program in Papua New Guinea. *Papua New Guinea Medical Journal, 49*(1/2), 5–13.

Connell, J. (2022). Challenging modernity? COVID-19, sorcery, religion and vaccines in Papua New Guinea.

Diamond, J. M. (1973). Distributional ecology of New Guinea birds. *Science, 179*, 759–769.

Diamond, J. (2011). Guns, Germs and Steel: The Fate of Human Societies. Random House.

Dudley, J. P. (2004). Global zoonotic disease surveillance: An emerging public health and biosecurity imperative. *BioScience, 54*(11), 982–983.
Echenberg, M. (2007). *Plague ports: The global urban impact pf bubonic plague 1894–1901*. New York University Press.
Eisenhauer, E. (2001). In poor health: Supermarket redlining and urban nutrition. *GeoJournal, 53*(2), 125–133.
Foster, R. J. (2006). From Trobriand cricket to rugby nation: The mission of sport in Papua New Guinea. *The International Journal of the History of Sport, 23*, 739–758.
Friedmann, H. (1993). The political economy of food: A global crisis. *New Left Review, 197*, 29–57.
Flannery, T. (1990). *Mammals of New Guinea*. The Australian Museum.
Flannery, T. (1998). *Throwim Way Leg: Tree-Kangaroos, Possums and Penis Gourds*. Atlantic Monthly Press.
Gillison, G. (1993). *Between culture and fantasy: A New Guinea highlands mythology*. University of Chicago Press.
Gingging, F. A. M. (2007). "I Lost my Head in Borneo": Tourism and the refashioning of the headhunting narrative in Sabah, Malaysia. *Cultural Analysis, 6*, 1–29.
Gouilh, M. A., Puechmaille, S. J., Gonzalez, J. P., Teeling, E., Kittayapong, P., & Manuguerra, J. C. (2011). SARS-coronavirus ancestor's foot-prints in South-East Asian bat colonies and the refuge theory. *Infection, Genetics and Evolution, 11*(7), 1690–1702.
Grundy, J., Dakulala, P., Wai, K., Maalsen, A., & Whittaker, M. (2019). *Independent state of Papua New Guinea health system review*.
Gulyaeva, M., Huettmann, F., Shestopalov, A., Okamatsu, M., Matsuno, K., Chu, D.-H., Sakoda, Y., Glushchenko, A., Milton, E., & Bortz, E. (2020). Data mining and model-predicting a global disease reservoir for low-pathogenic avian influenza (AI) in the wider pacific rim using big data sets. *Scientific Reports, 10*, 1681. https://doi.org/10.1038/s41598-020-73664-2
Hanson, J. P., Taylor, C. T., Richards, A. R., Smith, I. L., & Boutlis, C. S. (2004). Japanese encephalitis acquired near Port Moresby: Implications for residents and travellers to Papua New Guinea. *Medical Journal of Australia, 181*(5).
Hemer, S. R. (2015). Breaking silences and upholding confidences: Responding to HIV in the Lihir Islands, Papua New Guinea. *Medical Anthropology, 34*(2), 124–138.
Hoffmann, C. (2015). *Savage harvest: A tale of cannibals, colonialism, and Michael Rockefeller's tragic quest*. William Morrow Paperbacks; Reprint edition.
Koch, T., & Koch, T. (2005). *Cartographies of disease: Maps, mapping, and medicine* (p. 840). Esri Press.
Krishna, P. A., Chand, R., Huettmann, F. & T. R., Ghimire (2022) Rabies Elimination: Is It Feasible Without Considering Wildlife? *Journal of Tropical Medicine* (in press).
Leopold, A. (1989). *A Sand County almanac, and sketches here and there*. Oxford University Press.
Lerner, S. (2021). The Virus Hunters: How the Pursuit of Unknown Viruses Risks Triggering the Next Pandemic, December 28, 2021. https://theintercept.com/2021/12/28/covid-pandemic-virus-hunters-ecohealth-alliance-peter-daszak-wuhan/
Levi-Strauss, C. (1966). *The savage mind*. University of Chicago Press.
Liberski, P. P. (2013). Kuru: A journey back in time from Papua New Guinea to the Neanderthals' extinction. *Pathogens, 2*(3), 472–505.
Lindenbaum, S. (2015). *Kuru sorcery: Disease and danger in the New Guinea highlands*. Routledge.
Lindenbaum, S. (2001). Kuru, prions, and human affairs: Thinking about epidemics. *Annual review of anthropology, 30*(1), 363–385.
Luker, V. (2004) Civil society, social capital and the churches: HIV/AIDS in Papua New Guinea.
Madison, M. G. (1997). 'Potatoes Made of Oil': Eugene and Howard Odum and the origins and limits of American agroecology. *Environment and History, 3*(2), 209–238. (Ecological Visionaries/Ecologised Visions).
Mack, A. L., & West, P. (2018). Ten thousand tonnes of small animals: Wildlife consumption in Papua New Guinea, a vital resource in need of management. https://openresearchrepository.anu.edu.au/bitstream/1885/144062/1/rmap_wp61.pdf

Mack, A. (2014). *Searching for PekPek: Cassowaries and Conservation in the New Guinea Rainforest*. Cassowary Conservation & Publishing.

Martin, R. (Ed.). (2005). *Tree-kangaroos of Australia and New Guinea*. CSIRO Publishing.

Mather, T. N., Wilson, M. L., Moore, S. I., Ribeiro, J. M. C., & Spielman, A. (1989). Comparing the relative potential of rodents as reservoirs of the Lyme disease spirochete (*Borrelia burgdorferi*). *American Journal of Epidemiology*, *130*, 143–150.

Matthiessen, P. (1987). Under the Mountain Wall: A Chronicle of Two Seasons in Stone Age New Guinea. Penguin Classics.

McKenna, M. (2017). *Big chicken: The incredible story of how antibiotics created modern agriculture and changed the way the world eats*. National Geographic Books.

Mead, M. (1967). The rights of primitive peoples: Papua-New Guinea: A crucial instance. *Foreign Affairs, 45*(2), 304–318.

Moir, J. S., Garner, P. A., Heywood, P. F., & Alpers, M. P. (1989). Mortality in a rural area of Madang province, Papua New Guinea. *Annals of Tropical Medicine & Parasitology, 83*, 305–319.

Negin, J. (2008). Australia and New Zealand's contribution to Pacific Island health worker brain drain. *Australian and New Zealand Journal of Public Health, 32*(6), 507–511.

Nriagu, J. O. (1990). Global metal pollution: Poisoning the biosphere? *Environment: Science and Policy for Sustainable Development, 32*(7), 7–33.

Owen, I. (2005). *Parasitic zoonoses in Papua New Guinea*. J. Helminthol, *79*(1), 1–14. https://doi.org/10.1079/JOH2004266

Packard, R. (2007). *The making of a tropical disease: A short history of Malaria*. Randall M. Packard. Johns Hopkins University Press. ISBN: 978-0-8018-8712-3.

Rauzon, M. J. (2016). *Isles of Amnesia*. University of Hawaii Press.

Salak, K. (2001). *Four corners: One Worman Solo's journey into the heart of Papua New Guinea*. Counterpoint Publisher.

Stead, W. W. (1997). The origin and erratic global spread of tuberculosis: How the past explains the present and is the key to the future. *Clinics in Chest Medicine, 18*(1), 65–77.

The Guardian. (2021). Experts warn Papua New Guinea is potential breeding ground for new Covid variants, December 12, 2021. https://www.theguardian.com/world/2021/dec/13/experts-warn-papua-new-guinea-is-potential-breeding-ground-for-new-covid-variants. Accessed May 10, 2022.

Trájer, A. J. (2022). Late quaternary changes in malaria-free areas in Papua New Guinea and the future perspectives. *Quaternary International*.

Vos, P. D., Ceukelaire, W. D., & Stuyft, P. V. D. (2006). Colombia and Cuba, contrasting models in Latin America's health sector reform. *Tropical Medicine & International Health, 11*(10), 1604–1612.

Waller, J. M., Egan, B. T., & Eastwood, D. (1987). Ramu stunt, an important new sugarcane disease in Papua New Guinea. *International Journal of Pest Management, 33*(4), 347–349.

Chapter 17
Papua New Guinea and Climate Change (and REDD/REDD+): The 'Western Carbon Cult' as a New Hostile Relationship that Was Friendly Though Throughout Most of Mother Earth's History

Climate activists are sometimes depicted as dangerous radicals, but the truly dangerous radicals are the countries that are increasing the production of fossil fuels.
—United Nations Secretary General Antonio Guterres (April 5th 2022)
Today, markets value forests more destroyed than standing. Michael Somare, Papua New Guinea's Prime Minister in regards to REDD (as quoted in his public biography https://en.wikipedia.org/wiki/Michael_Somare*)*

Abstract Man-made climate change affects nations like Papua New Guinea (PNG) in very harsh terms. People in PNG often do not really know and understand what caused it and how they can contribute to mitigate the actual cause of such all-encompassing change caused primarily by industrial nations from abroad. Instead, PNG did very well on climate change for over 47,000 years but was already used as a 'pawn' in the carbon sequestration game while global warming and CO_2 release increases essentially unabated. All relevant agreements and COPs have achieved little on that matter; CO_2 and such Greenhouse Gases (GHGs) are not much curbed whatsoever but rising. A 'Carbon Cult,' e.g. carbon sequestration programs like REDD and REDD+, have not been effective in changing these problems, essentially made it worse and thus failed on what they were to deliver in earnest. The ancient old-growth forests pay that price; all relevant metrics show that clearly. In the meantime, PNG and its islands feel the full impacts of sea level rise, ocean acidification, coral reef death and coastal erosion. Many cultures start to get moved, beyond 'just' the Carteret Islands and the Torres Strait region with vast risk planning scenarios on the rise affecting millions of people, villages and cultures.

Keywords Papua New Guinea (PNG) · Carbon sequestration · Carbon stock exchange · REDD+ · Climate change · Climate justice

F. Huettmann, *Globalization and Papua New Guinea: Ancient Wilderness, Paradise, Introduced Terror and Hell*, https://doi.org/10.1007/978-3-031-20262-9_17

17.1 Introduction

As Papua New Guinea ranks at the bottom of the list for per capita carbon emissions (https://data.worldbank.org/indicator/EN.ATM.CO2E.PC), the virgin and ancient forests, the mangroves, wetlands and the peatlands of PNG became globally famous for their carbon sequestration storage trading capabilities (Fox et al., 2010; Melick, 2010). PNG was to safe the world by simply trading the CO_2 emissions away. Using capitalism and continue business as usual, the REDD 'Cap and Trade' scheme by the UN promised to get rid of the access carbon and distribute pollution quotas more equally and 'fairly'! It required that the CO_2 sequestrations are inventorized and known (Schmid et al., 2015 for an example of such applied equations in the tropics).

In December 2010, REDD was rebranded as REDD+ (https://unfccc.int/topics/land-use/workstreams/redd/what-is-redd). That way REDD lost its earlier, narrow focus on reducing emissions and carbon markets and got expanded (Martin, 2020 for REDD+ in PNG). This was to allow for a more holistic approach of the value of forests and the lives of the people who rely on them and live in them. Activities that could now be funded under such a program came to include "*non-carbon benefits*" including "*opportunities for wealth creation and wellbeing.*" While this sounds great initially, more ecological, such win-win rebranding is as neoliberal as it gets and allows to include many aspects unrelated to climate change and greenhouse gases, it remains dubious for progress on climate change (e.g. Spash 2006). Current climate facts of global warming and impacts speak to that effect clearly.

The Sepik river region—the birth area of the first PNG prime minister Michael Somare—became of world relevance for such climate services, and to trade them internationally (The Guardian, 2015). That's because PNG was one of the first tropical nations that received a compensation payment of many million US$s for climate change-related carbon sequestration services from the contaminating western nations (Zhongming & Wei, 2022; see Filer, 2010 for the western 'Carbon Cargo Cult'). It was a record deal and meant to start a process worldwide to distribute carbon more equally among nations and thus to reduce the carbon problem globally once and for all. But both of which has not really happened. Reportedly, Michael Somare was eventually very disappointed with REDD, and the deal fell somewhat apart.

Does anybody remember this, or care, and where did the money go, and the CO_2, and the trees?

The Sepik and PNG's carbon sequestration deals now carry quite a bad label; some money simply went away unaccounted (see The Guardian, 2015 for "*We are not perfect*" by Stephen Hooper, the Australian carbon developer who established the Sepik river REDD+ project,; see also REDD-Monitor, 2015 for details of project 'April Salumei REDD').

Early on, PNG actually became world famous for confronting the U.S. at a climate conference meeting about its inaction within the wider community of concerned climate change nations (The New York Times, 2008; The Guardian, 2015; see for no actions to this very day, e.g. by Australia phrased here with Ludlam, 2021, The Guardian 2022a, 2022b). It was PNG that spoke up, while most other nations remained quite or used 'diplomatic pathways' and ineffective tone, including Australia (a nation otherwise being so concerned about PNG and its well-being)! PNG leaves a mark in the man-made climate change topic that it sees itself increasingly confronted with.

The climate change debate is ubiquitous and affecting many aspects of ancient life. In PNG, it's for instance the question whether humans, or the climate, killed off the species that became extinct in the Sahul region (Flannery, 2002; Wroe et al., 2013). It's a question that Beehler and Laman (2020), Diamond (2011) touched upon, but here comes a more direct inference then: Climate as the main driver (as stated by Flannery, 2002 and elegantly expressed as 'Future Eaters')!'

After colonialism and globalization, man-made climate change marks another new chapter for PNG as a victim from outside forces as it otherwise always adapted well, and lived, with the more or less 'natural' climate for millennia (apart from stochastic events like hurricanes, volcanos and climate-driven landslides due to rain, volcano outbreaks, tsunamis, the universe, etc. let's say).

PNG is among the rainiest and cloudiest place on earth; the relevance of clouds—as part of the global climate discussion—cannot be understated for PNG (Beehler & Laman, 2020). But it's the wider atmosphere that PNG people have no direct control over (Figs. 17.1 and 17.2).

At minimum, with man-made climate change PNG—its people and habitats—will already face known losses due to sea level rise, coastal erosion, coral reef die-off and ocean acidification (e.g. Dixon et al., 2021). A new concept is to be developed and followed for PNG, *sensu* (Robinson et al., 2022; see Pittman et al. 2021for seascape). Like elsewhere in the world, many PNG citizens and Melanesians are to move and lose their local coastal homesteads, etc. (Game et al., 2011; Stone & Obura 2013). It's as if the world, the environment and Mother Earth even turned against PNG once more. In times of man-made climate change, one is to run to the hills; coastal erosion (Figs. 17.3 and 17.4) and spoilage of freshwater in coastal wells leaves no other options. And PNG has those hills, e.g. when compared to Australia! Thus, not all is doomed in PNG with man-made climate change (Figs. 17.5 and 17.6).

17.2 Papua New Guinea Will Be Hit Hard by Industrial Man-Made Climate Change Regardless

While PNG being widely innocent, powerless and a true victim in the global climate change arena, let's look closer what PNG has now to deal with in a changing climate that is man-made by a few powerful and dominating nations of 'the west' and the north (that includes here Australia). PNG is arguably on the receiving end, once again, and it's almost the identical actors than in the colonial and globalization game before. Essentially, Captain Cook hits from the other end, but again.

Progress on the climate change front remains insufficient, as widely noted in the public (see Ludlam 2021 for public protests in Australia and worldwide; The Guardian, 2022b). There is no need to be a cynic in understanding that PNG will not only be hit by a climate change caused primarily by the western/industrial world

Fig. 17.1 Ghosts and spirits are watching whatever happens

(see first law case on that issue of Climate Impact Litigation unfolding with German Energy company RWE vs the people of the Lake Palcacocha in Peru, Latin America; The Guardian, 2022c). But it will also be hit hard by its subsequent mitigation efforts that are set up even more imperialistically to beat climate change (Martin, 2020). Seeing how climate policies stall and are unfolding, the set of latter policies to combat climate change are primarily designed '*by the west for the west*' while the remaining world is left outside, certainly PNG. That's how most modern policy settings have gone now, again. It simply reflects the earlier colonial and global power structure; it unfolds under an increasingly rotten capitalistic framework as the platform (Rich, 1994). It's the track record of power, and it will benefit those who write the policies on behalf of others they are to serve and to deal with in good terms

So to start that discussion, below the first set of facts that knowingly will hit PNG during times of climate change, based on the author's experience first hand in the field:

Fig. 17.2 Clouds from the ocean as an inherent part of Papua New Guinea's climate and for climate change forecasting

Sea level rise: King tides are already having a more serious impact; full island evacuation is becoming reality now, e.g. Torres Strait and Carteret Island (Connell, 2016; see Nunn, 2012 for wider Pacific Island regions).
Ocean acidification: Ocean acidification involves various chemical changes in the ocean ecosystem, namely what is referred to as 'saltwater.' It's an unavoidable consequence of CO_2 increases in the atmosphere and occurs globally, also in waters of PNG and it affects coral reefs and the system overall.
Coral Reef decay: Coral Reef Bleaching (e.g. Foale, 2006), see Dixon et al. (2022) for no 'safe zones' and climate refuge boundaries for coral reefs.
Temperature rise: Due to warming, the single one glacier on the Indonesia side is already melting for many decades (Flannery, 2002), and almost gone now.[1]

[1] The glaciers and snow areas of New Guinea have been studied for many decades and were a piece of research early on, e.g. for British explorations (as described by Beehler and Laman (2020); see citations within). Mt. Hagen was climbed by many Alpine celebrities, e.g. Reinhold Messner. Flannery (1998) has referred to the changes early on. The areas were heavily 'collected'. Still, no relevant climate change actions have come from it, see The Guardian (2020a, 2020b) for public

Fig. 17.3 Coastal erosion, it's for real in Papua New Guinea and elsewhere!

The snow pack on the PNG side disappears in the same fashion. Elevational gradients get pushed up, e.g. for limiting farming species, diseases like malaria, and affecting harvest timings (Beehler & Laman, 2020).

Humidity changes: This is a massive topic in PNG because it affects the ecology, namely disease spread, agriculture, forestry and wider weather patterns. Clouds are very relevant in that discussion but poorly studied.

Carbon sequestration: PNG has tropical peat lands as well as ancient forests. All of which are known to sequester carbon at a record rate.

Invasive Species: One can easily recognize that invasive species are on the rise, usually due to a harmonizing habitat and landscape caused by the overruling

statements. The melting areas play virtually no role for adjacent Australia and its essentially ongoing climate change denial (see Ludlam, 2021 for details).

Fig. 17.4 Coastal erosion also affects king tides making impacts even more severe; that process comes with many surprises including rare but then dramatic impacts as well as spoilage of drinking water in coastal regions and wells

economic regime. As this business plan is fully rolled out more in PNG, it will bring with it invasive species; many examples can be found already, and it is done on the cost of PNG endemic species. For instance, the UN did so for many decades already with devastating effects, e.g. in the Sepik region (see Beehler & Laman, 2020, and citations within).

Diseases: Many diseases are now found in PNG that do not originate in PNG, and/or never were an issue there and for their impact; malaria comes to mind, e.g. moving further in altitude and new strains, including avian influenza, and commercial plant diseases.

Human wealth and poverty gap: In tribal times, village life had its wealth limit, and true poverty was somewhat buffered by the much smaller range between rich and poor and the family structures; there hardly were any rich people or an upper class and caste. Thus, there were fewer poor people in the society (while human power imbalances remained).

Human migration: The refugee crisis on Manus island is a classic indicator. The other aspects are actually the mountains, which are safe from sea level rise. People from the islands will enter the safer lands above sea level, including highlands and mountain villages. Many of those are along the trail system that have been used for bartering between islands and the interior.

Fig. 17.5 Coastal housing in 'real live' including kitchen and outhouses

Human conflict and warfare: It's clear that many parts in PNG were not so peaceful among tribes; aggression plays a big factor, against other tribes and domestically (see Flannery, 2002; Gillison, 1993 for examples). However, international warfare in PNG remains widely unheard of (there is one case of PNG army fighting with Spirito Santo because PNG was asked to engage, based on French and international efforts). PNG has no really big or relevant army to start with, and PNG virtually picks no fights with its neighbors or abroad. However, as climate change changes the set up and fabric of nations, PNG can see such conflicts coming easily, such as climate refugees from islands or adjacent nations. When compared with the Ghurka soldiers from Nepal in WW1 and WW2, the famous PNG warriors never really engaged on the global battlefield abroad.

Lag effects: It can easily be assumed that climate change will not only affect the situations now, but for decades to come, likely centuries. The lag effects can be dramatic, when thinking of the causes being entirely man-made, made by just a few nations (mostly colonial ones) and their leaders.

Synergy effects: Already the list of effects shown here is devastating when taken individually. However, it appears to be much worse when it all comes together, combined, just like real ecology and real live is (Table 17.1). Thus far, an underestimate is reported.

Fig. 17.6 Faces of climate change in Papua New Guinea: the future matters

17.3 There Are no Winners: What Good Has Man-Made Climate Change to Offer for Papua New Guinea?

Not much; simply when judged by the science record. There is no win-win, nor is there an 'opportunity', any good potential or 'winners and losers' (O'Brien & Leichenko, 2003). Instead it's all loosing for PNG—and for most of the world/mankind—when it comes to man-made climate change. As an island ocean nation PNG will pay an incredible bill on the climate change front, for years to come. PNG will not catch up with Australia, or with any other larger Asian nation on techno-solutions. PNG will remain within the Melanesian group, its set of 'failed states' and within its problems an approaches. Melanesia and PNG are certainly planned to be mined regardless (Kirsch, 2014 for 'Mining Melanesia'); see seafloor mining to come. PNG will be pushed into new directions it had not seen before, and the refugee list is likely to get longer for PNG, Australia and New Zealand (see, for instance, Luetz & Havea, 2018; Slee, 2019; the latter already flooded and carry many generic conflicts of integration (e.g. for such ongoing conflicts in the region on islands just see the recent tragic Christchurch bombing; The Guardian, 2021). A major misconception here is that PNG is a mainland with some islands. Instead, PNG is an ocean EEZ and an island nation with a larger block of land in its western side. The climate change impacts for PNG are on a similar state than they are for Torres

Table 17.1 List of minimum effects and impacts of climate change that are known and which can be expected to occur in PNG

Climate change effect	PNG impact	Comment
Global temperature rise	Glaciers and snow packs are melting, and have done so for many years already	Already seen in PNG, e.g. Beehler and Laman (2020)
	Palm trees moving up in elevation and produce more mature fruit there	
	Crop species can be planted in higher elevations	
	Malaria moves upward	
	Sub-alpine fires occurred in El Nino years	
Sea level rise	King tides reach higher	Already seen in PNG
	Coastal erosion PNG-wide	Already seen in PNG
	Carteret Island inhabitants got relocated already	An ongoing situation in Melanesia (Beehler & Laman, 2020, p. 89)
Ocean acidification	Affecting ocean water quality	Widely overlooked and poorly studied but massive impacts
Change of temperature, rain and weather patterns	Seasons are broken up and harvests are more diverse	Already seen in PNG, e.g. Beehler and Laman (2020)
	Rainfall patterns changed	

Islanders or Guam and Kiribati (e.g. Dixon et al., 2022; Stone & Obura, 2013): Even a slight sea level rise will have devastating effects for PNG and its villages, cities, ports and beaches along the coast. It's of national impact and global proportion (sometimes referred to as 'biblical').

Textbox: REDD and REDD+: What it is, what it does, and why it fails (=has virtually no valid open access inventory data, etc.)

REDD is an older and highly inefficient—if not perverted—concept to distribute CO_2 and other Green House Gas (GHG) emissions throughout nations of the world. It's biased because it favors an ideological approach (capitalism but as the root of the initial problem) to trade CO_2, based on the initial idea that one can use money to compensate for CO_2, and that one simply asks others in exchange to pollute less, while the own pollution levels remain 'as is' or can even increase (as we currently experience worldwide, all as approved in a REDD framework).

In such a world, pollution can be traded and bought, it can be transferred into the coin-space (=money). It's like as if your life and death can be bought; one cannot. There is 100% no need entire nations, and the world, fall for such concepts. It's nothing but a common and very harmful paradigm, but as adopted by money-rich nations in their chosen discipline of Environmental Economics (in contrast to Ecological Economics)—promoted at their institutions and outlets. It fails globally. It fails conceptually. And it fails in reality. It has a certain perversity to it, and in the following some simple reasons:

- To get an estimate what the national pollution levels are, one first would need an assessment with a fixed and consistent protocol. That's what REDD achieves. REDD+ is an update from that scheme, a more flexible one but which makes it more complicated for an inventory.

  The way how this relates to PNG is that PNG has vast tracts of carbon sinks, peatlands and old-growth forests that actually, offer such a carbon buy-out for polluting nations, e.g. Norway and Germany (see The Guardian, 2015 for public overview; see also for an Indonesia-Norway payment in Mongabay, 2022). However, neither the world, nor CO_2 quotas, CO2 release or PNG itself have really benefitted from such schemes. Where the money went is less clear also, but the REDD narrative remains.

  As REDD is to be based on data, we are short-founded on those data deliveries; where are they? Kujala et al. (2022) showed the need for such inventory schemes, which are to be linked with databases of global dimensions, as typically done now as best-professional practice (e.g. Huettmann, 2015). But it's already here where the carbon trading scheme falls short: modern book keeping.

- Further, the real money in REDD sits not in the assessment or in the data, but in the actual market trading and CO_2 valuation (which is a large multitude of the actual field work cost, and which is politically assigned but rather large). With that, REDD turns highly political and gets used in that wider framework of nation, global and strategic debt economies. The political economy has a base driving all relevant aspects of life now.
- Lastly, quotas are easily computed and simulated. With just 193 or so nations in the world, and very few main actors and political blocks among them (e.g. U.S., China, EU, G8 and OPEC), any trading negotiation scenarios can easily be captured, quantified, computed and predicted with AI and supercomputers for optimizations beneftting the ones in power. One can easily BUY-OUT entire nations and their quotas, and then control the market. And as if that has not happened and was part of set up to use REDD. That way REDD and its schemes get manipulated, the free CO_2 market falls easily apart and it turns 'perverse' where pollution gets cheaper and CO_2 rises, thanks to the commercial approaches with REDD at the center. Poor nations pay the cost, so does the world.

Simply put, REDD lacks progress, lacks data, lacks a good vision, a shown track record and lacks an achievement and CO_2 and GHG reduction with a real plan and sustainability. Global warming remains on the rise. That is certainly true for PNG where the entire forestry sector is rooted in lack of data and subsequent corruption (see associated chapters in this book; Beehler & Laman, 2020 for repeated details and facts).

17.4 Thinking It Through in More Detail: Climate Change in PNG and What is to Come

Clearly, man-made climate change is driven by industrialization, consumption and globalization with a *laissez-faire* approach to environmental issues (see Stern et al., 2006). Thus, any of those modern items are to be reassessed in that framework. PNG's 47,000 old past has little do with it (Table 17.2).

The list of items in Table 17.2 is quite long and stands in good contrast to what most western people and leading nations should better mitigate and what their governments promote (national well-being, win-win; we do all we can...). The modern world, as we know it, can hardly function any further within the classic paradigm. And so either we set ourselves quotas and accept limits, or stop industrialization as a concept, or engineer our way out of it, find other creative solutions, remain organic, are very lucky, or all of those together. Even the best possible avenues, Steady State Economics and Ecological Economics (Czech & Daly, 2014; Daly & Farley, 2010, Spash 2006) will face massive implementation problems with the realities of climate change to be scaled put (Farley & Kunkel, 2018).

The current level of CO_2 is man-made and caused the global warming, with many implications to come still. It's a global change. Impacts by methane and other GHGs and their feedback loops are hardly mentioned or studied yet; certainly not for PNG.

Table 17.2 List of actions and items that have a large man-made carbon (CO_2) footprint making climate change more severe

Action that increases man-made CO_2	Citations	Relevance for PNG	Comment
Travel of goods from abroad	Klose (2015)	PNG receives a large amount of products from abroad, namely Australia and China	
Travel of goods to markets abroad	Klose (2015)	PNG has a major focus on export products in Asia	
Use of fossil fuel	IPCC.org	PNG does not burn much fossil fuels but produces oil and gas, and mining products	

Whereas PNG has a minor CO_2 footprint and did not cause the global problems. It's the economy, industrialization, as promoted by the colonial nations and a few others that create once more a major headache for PNG. PNG can hardly change in how it was set up by global powers abroad.

17.5 Man-Made Climate Change Adaptation the PNG Way

While a poster child for adaptation, it's easy to see that PNG is a passive player in the world's Climate Change arena. Also, modern PNG has a poor governance structure at hand, and it is locked in into acolonial and British Commonwealth legacy of problems; with Australia, the U.S. and China driving many decisions for PNG, directly and indirectly. Mining, oil & gas and the natural resource extraction model as the prime business scheme for PNG. Harvesting virgin rainforest blocks and stressing coral reefs and the oceans does certainly not help. That leaves not much options then.

So what should be done?

Arguably, the PNG culture stands as a good role model over time to combat climate change. PNG and its tribal culture indeed is rather resilient. But who wants to be, and to live like PNG? Many people of the west do oppose, as already the vast loss of expats shows (see also Lutton, 1981 for university library move from Port Moresby to Perth!).

And such dramatic changes come with costs, costs of human lives either way; difficult to envision any other. There will be suffering, caused by the west again, with nations located literally on the far opposing side of the planet. But we are one, after all.

While I favor cultural life adjustment toward nature, and living close with it and in it, and in the nature of 'now,' there is no good solution in sight to truly deal with man-made climate change benefitting everybody (but see Loewen 2021). Instead we see a one-sided approach with many loosing parties. Be ready for the life boat and run for the hills.

The current policy inaction and lock-in by earlier governance, e.g. mortgages and student loans, debt trading of entire nations by private companies, the International Monetary Fund (IMF) policies, etc. make a relevant change virtually impossible for climate change.

Now where does that all leave us?

It would be great of the wider framework, globalization and corporations and large nations and their cultures if they would save us; unlikely though. Presumably the local efforts and individual work toward survival—for a better life—is where the power sits, bottom up.

References

Beehler, B. M., & Laman, T. (2020). *New Guinea.* Princeton University Press.

Connell, J. (2016). Last days in the Carteret Islands? Climate change, livelihoods and migration on coral atolls. *Asia Pacific Viewpoint, 57*(1), 3–15.

Czech, B., & Daly, H. E. (2014). The steady state economy as the sustainable alternative to economic growth. In *Peak Oil, Economic Growth, and Wildlife Conservation* (pp. 119–129). Springer.

Daly, H., & Farley, J. (2010). *Ecological economics: Principles and applications*, 2nd edn. Island Press.

Dixon, A. M., Forster, P. M., Heron, S. F., Stoner, A. M. K., & Beger, M. (2022). Future loss of local-scale thermal refugia in coral reef ecosystems. *PLOS Climate, 1*(2), e0000004. https://doi.org/10.1371/journal.pclm.0000004

Farley, H., & Kunkel, B. (2018). Ecologies of scale. New left review Jan/Feb 109. https://newleftreview.org/issues/ii109/articles/herman-daly-benjamin-kunkel-ecologies-of-scale

Filer, C. (2010). The carbon cargo cult in Papua New Guinea. Forests of Oceania Working Session, Association for Social Anthropology in Oceania.

Flannery, T. (2002). *The future eaters: An ecological history of the Australasian lands and people.* Grove Press.

Foale, S. J. (2006). The scale and epistemology of coral bleaching in Papua New Guinea. Bridging Scales and Epistemologies.

Fox, J. C., Yosi, C. K., Nimiago, P., Oavika, F., Pokana, J. N., Lavong, K., & Keenan, R. I. (2010). Assessment of aboveground carbon in primary and selectively harvested tropical forest in Papua New Guinea. *Biotropica 42*, 410–419.

Game, E. T., Lipsett-Moore, G., Saxon, E., Peterson, N., & Sheppard, S. (2011). Incorporating climate change adaptation into national conservation assessments. *Global Change Biology, 17*(10), 3150–3160.

Gillison, G. (1993). *Between Culture and Fantasy: A New Guinea Highlands Mythology.* University of Chicago Press.

Huettmann, F. (2015). On the relevance and moral impediment of digital data management, data sharing, and public open access and open source code in (Tropical) research: The Rio convention revisited towards mega science and best professional research practices. In: F. Huettmann (Ed.), *Central American Biodiversity: Conservation, ecology, and a sustainable future.* (pp. 391–418). Springer.

Kirsch, S. (2014). *Mining Capitalism.* University of California Press.

Klose. (2015). https://www.itf-oecd.org/sites/default/files/docs/cop-pdf-06.pdf

Kujala, H., Maron, M., Kennedy, C. M., Bull, J. W., Evans, M. C., Wintle, B. A., Iftekhar, M. S., Selwood, K. E., Beissner, K., Osborn, D., & Gordon, A. (2022). Credible biodiversity offsetting needs public national registers to confirm No Net Loss. *One Earth, 5*, 650–662. https://doi.org/10.1016/j.oneear.2022.05.011

Loewen, R. (2021). *Mennonite farmers . A global history of place and sustainability.* A comparative global history of Mennonites from the ground up. In Young center books in Anabaptist and Pietist studies. https://jhupbooks.press.jhu.edu/title/mennonite-farmers?utm_source=newsletter&utm_medium=email&utm_content=Mennonite%20Farmers&utm_campaign=f21_new-nov-books_hnov_promo

Luetz, J., & Havea, P. H. (2018). "We're not refugees, we'll stay here until we die!"—climate change adaptation and migration experiences gathered from the Tulun and Nissan Atolls of Bougainville, Papua New Guinea. In *Climate Change Impacts and Adaptation Strategies for Coastal Communities* (pp. 3–29). Springer, Cham.

Lutton, N. (1981). From port Moresby to Perth. Archives and Manuscripts, 30–38.

Martin, P. (2020). *The changing climate of development: REDD+ in Papua New Guinea.* Unpublished Doctoral dissertation, UNSW Sydney, Australia.

Melick, D. (2010). Credibility of REDD and experiences from Papua New Guinea. *Conservation Biology, 24*, 359–361.

Mongabay. (2022). Papua New Guinea. https://rainforests.mongabay.com/deforestation/forest-informationarchive/Papua_New_Guinea.htm. Aaccessed March 29, 2022.

New York Times. (2008). Issuing a bold challenge to the U.S. Over Climate. 22 January 2008. https://www.nytimes.com/2008/01/22/science/earth/22conv.html. Accessed July 29, 2022.

Nunn P. D. (2012). Climate change and pacific Island Countries. Asia-Pacific Human Development Report Background Papers Series 2012/07. UNDP Nairobi. https://www.uncclearn.org/wpcontent/uploads/library/undp303.pdf

O'Brien, K. L., & Leichenko, R. M. (2003). Winners and losers in the context of global change. *Annals of the Association of American Geographers, 93*(1), 89–103.

Pittman, S. J., Yates, K. L., Bouchet, P. J., Alvarez-Berastegui, D., Andréfouët, S., Bell, S. S., & Young, M. (2021). Seascape ecology: identifying research priorities for an emerging ocean sustainability science. *Marine Ecology Progress Series, 663*, (pp. 1–29).

REDD-Monitor. (2015). The incredible story of Papua New Guinea's April Salumei REDD project 25 November 2015. https://redd-monitor.org/2015/11/25/the-incredible-story-of-papua-new-guineas-april-salumei-redd-project/. Accessed July 1, 2022

Rich, B. (1994). *Mortgaging the earth: The World Bank*. Earthscan Publishers.

Robinson, S., Bouton, E., Dolan, M., et al. (2022). A new framework for rapidly assessing national adaptation policies: An application to small island developing states in the Atlantic and Indian Oceans. *Regional Environmental Change 22*. https://doi.org/10.1007/s10113-021-01855-2

Schmid, M. S., Baltensperger, A. P., Grigor, J., & Huettmann, F. (2015). Assessments of carbon stock hotspots in Nicaragua and Costa Rica. In: F. Huettmann & F. (Ed.), *Central American biodiversity: Conservation, ecology, and a sustainable future* (pp. 677–701). Springer.

Slee, C. (2019). Inside Australia's PNG refugee prison camp. *Green Left Weekly, 1246*, 7–9.

Stern, N., Peters, S., Bakhshi, V., Bowen, A., Cameron, C., Catovsky, S., Crane, D., Cruickshank, S., Dietz, S., Edmonson, N., Garbett, S.-L., Hamid, L., Hoffman, G., Ingram, D., Jones, B., Patmore, N., Radcliffe, H., Sathiyarajah, R., Stock, M., & Zenghelis, D. (2006). *Stern review: The economics of climate change*. HM Treasury.

Stone G. S., & Obura, D. (2013). *Underwater Eden: Saving the Last Coral Wilderness on Earth*. The University of Chicago Press.

Spash, C. L. (2006). The stern report: The continuing fallacy of global cost-benefit analysis. *European Society for Ecological Economics Newsletter*.

The Guardian. (2015). The incredible plan to make money grow on trees. 24th November 2015. https://www.theguardian.com/world/2015/nov/24/redd-papua-new-guinea-money-grow-on-trees. Accessed May 28, 2022.

The Guardian. (2021). Two years after Christchurch, New Zealand makes plotting a terrorist attack a crime. 29th September 2022. https://www.theguardian.com/world/2021/sep/30/new-zealand-makes-plotting-a-terrorist-attack-a-fixing-legal-loophole. Accessed July 16, 2022.

The Guardian. (2022a). 'Far from adequate': Former Pacific leaders group urges Australia to increase 43% emissions cut. 7th July 2022a. https://www.theguardian.com/world/2022a/jul/08/far-from-adequate-former-pacific-leaders-group-urges-australia-to-increase-43-emissions-cut. Accessed May 15, 2022a.

The Guardian. (2022b). For 50 years, governments have failed to act on climate change. No more excuses. 2nd June 2022b. https://www.theguardian.com/commentisfree/2022b/jun/02/for-50-years-governments-have-failed-to-act-on-climate-change-no-more-excuses. Accessed June 5, 2022b.

The Guardian. (2022c). German judges visit Peru glacial lake in unprecedented climate crisis lawsuit 27th May 2022b. https://www.theguardian.com/environment/2022b/may/27/peru-lake-palcacocha-climate-crisis-lawsuit. Accessed June 4, 2022b.

Wroe, S., Field, J. H., Archer, M., Grayson, D. K., Price, G. J., Louys, J., Faith, J. T., Webb, G. E., Davidson, I., & Mooney, S. D. (2013). Climate change frames debate over the extinction of megafauna in Sahul (Pleistocene Australia-New Guinea). *Proceedings of the National Academy of Sciences, 110*(22), 8777–8781.

Zhongming, Z., & Wei, L. (2022). Huge proposed project to prevent deforestation in Papua New Guinea may sell hot air credits and have misled locals.

Chapter 18
Direct Sustainability Impacts from the Australian Arc of Perceived Terror, Tyranny and 'Development Aid/Help' in Papua New Guinea and the Surrounding Region

> *Australia shows signs of following the United States down the road to kakistocracy,*
> *Ehrlich said. "If it does It is unlikely to survive as a nation."*
> *World-renowned Professor Paul Ehrlich—President of the Centre for Conservation Biology at Stanford University, U.S.; cited by the New South Wales University, Sydney/Australia*
> (Professor Paul R Ehrlich—President of the Centre for Conservation Biology at Stanford University—in an interview for his invited upcoming guest lecture on the sixth mass extinction with the New South Wales University in Sydney, Australia (https://newsroom.unsw.edu.au/news/science-tech/pioneering-ecologist-paul-ehrlich-present-unsw-sydney-sixth-mass-extinction; accessed 17th July 2022).
> *With regional support, Australia led a mission in 2003 to restore law and order. A short-term success, the mission leaves questions about its long-term ability to achieve either well-being for Solomon Islands or security for the region. Its emphasis on shoring up a perennially weak central government, and its inattention to other pillars of Solomons society, threaten to undermine its success and create a crippling sense of dependency. For the mission to succeed, it must empower Solomon Islanders to take charge of their own destiny.*
> *Kabutaulaka* (2004)

Abstract Australia is the strongest nation in the region of New Guinea, Papua New Guinea (PNG), Torres, Vanuatu, Fiji and Solomon Islands. In an 'arc-like fashion,' Australia is the biggest and most influential neighbor—and partner—for those nations. It sets direction and various policies, including policing campaigns to set up 'law and order' or how to deal with biodiversity, science and climate change. However, while Australia has a long history of active engagement in those regions, e.g. as a protectorate power and though economy, aid, development and advising, this arc is plagued with problems. The stronger Australia seems to engage the more those nations seem to show problems, while the concept of development widely failed. In the wider strategy of Australia, the 'arc' turned unstable, unsustainable and frankly dangerous, to be taken-over by other concepts and nations, e.g. China, India and islamist radicals. The impacts for biodiversity and conservation remain

F. Huettmann, *Globalization and Papua New Guinea: Ancient Wilderness, Paradise, Introduced Terror and Hell*, https://doi.org/10.1007/978-3-031-20262-9_18

ignored and worsening while climate change remains widely unabated also. A terror unfolds.

Keywords Papua New Guinea (PNG) · Australia · Australian AID · Embassy · International politics

18.1 Introduction

Early on the Australian administration of Papua New Guinea (PNG) promoted mining and banned cannibalism, and the latter practice was subsequently eliminated in PNG by 1960. It fits the PR image and subsequent perspectives about PNG as 'treacherous savages' presented in Richardson (2006)—a missionary—and such 'ending cannibalism' campaigns get widely perceived by the global community as progress and civilization. Reality—and my own multiyear field experience—widely differs though (just as found by many other scholars also, e.g. Beehler & Laman, 2020; Mack, 2014; West, 2006; Connell 2005).

Based on various foreign policy assessments, Australia was widely described to run an '*Arc of Instability*' and perceived terror (e.g. Henton & Flower, 2006; see May, 1998 on PNG and Solomon Islands, Rumley, 2006). And it's widely now based on neoliberal policies and political power tools (Springer 2015, 2017). This 'arc,' surrounding Australia for its own protection (see Ratnasari & Perwita, 2020 for Australia's Stepping Up Engagement initiative), fully includes PNG (the other nations consist at least of Solomon Islands, Fiji, Vanuatu and Torres Islands/Strait (one should likely add East Timor also, The Guardian, 2019). But PNG is a dominant actor among those nations due to its size, diversity, ocean and well-being, as well as linked with Indonesia via New Guinea and Eastern Timor, etc. (van der Kroef, 1970) Indonesia is a major international actor in SE Asia, in mining and beyond). This arc features some terror fueled by Australia itself (sparked for instance by supporting separatist movements, military aid and build up, policing, promoting ruthless mining in pristine areas, operating land scams abroad on indigenous-owned island land tenure, pushing economic growth on finite resources, providing aid with self-interest and mis-handling the consequential refugee waves as part of many other things and cultural problems like the illegal but occurring runs of gun smuggling and drug transports); subsequently, the state of Australia's Arc is essentially a home-made problem. PNG just reflects that. Already the PNG's currency follows an Australian design template, is printed in Australia, and its exchange rate hinges widely on Australia, e.g. (Garnaut & Baxter, 1984). That way, for PNG, Australia can essentiallyset the price of resources it gets from PNG. Whereas, Australia itself is in a deep and widely acknowledged decade-long crisis of democracy (e.g. Evans et al., 2016).

While China is frequently blamed for massive forest extraction in PNG, e.g. in Beehler and Laman (2020), already the 'largest forestry project in PNG' involves a land scam orchestrated by an Australian company (Australian Network News (ABC), 2014) which results into PNG people losing their land, homes and livelihoods;

PNG locals get essentially evicted from a forest place they lived in sustainably for millennia (see also for an urban example in Port Moresby with The Guardian, 2021a, 2021b). The highly critiqued JANT Forestry project from Japan actually includes Commonwealth Australia also (Beehler & Laman, 2020; Cousteau & Richards, 1999, p. 206/7). The fact that on a finite landmass, a fast-growth plantation forestry is tried all over New Guinea, as fully promoted and supported—again—by the Australian government and allies, for totally exotic species to PNG such as *Pinus caribaea, Araucaria cunninghamii, A. hunsteinii,* Teak as well as fast-growing *Acacia mangium* (Beehler & Laman, 2020) does not help the case. Clearly, refugee problems and biodiversity wilderness loss and loss of home and income are all linked (Afira, 2021). It is easy then to understand the roots of the unrest and terror in the region; the literature for PNG mentions those details clearly, usually related to business practices, with Bougainville as a core example, e.g. Lassett (2012). Bougainville and its mining was an important building block and bootleneck in the PNG nation-building process from the 1970s onward and with support by prime minister Michael Somare (Figs. 18.1, 18.2 and 18.3) It's a core engine in providing funding and foundation of the new modern nation of PNG to be build.

18.2 Papua New Guinea as an Equal and True Australian and Global Partner?

Despite the official claim, Australia and PNG are not on equal footage, and despite the frequently seen wording (e.g. https://www.dfat.gov.au/geo/papua-new-guinea/development-assistance) it's not a mutual or fair partnership whatsoever. That one-way street is easy to see, the bigger neighbor bleeds the weaker one. It's widely expressed by prime ministers Michael Somare and Chan (2016) in the public and for years, as well as by outside observers (e.g. Gosarevski et al., 2019).

The deep structural involvement of Australia in PNG is well and widely documented in all its shapes and forms, and it was done early on, e.g. Nelson (1982), McLennan (1938), Smith (1998), Ward and Ballard (1976), for media; Dinnen et al. (2006), Gosarevski et al. (2019) including its commercial exploration, exploitation, PR image appropriation and settlement done to western terms and policies (Chan, 2016, Nelson, 2016 for a PNG perspective; Dinnen et al., 2006 for policing; and for Economic Growth advising see Gosarevski et al., 2019). It easily goes back to 1850 onward (Flannery, 2002; see Grant, 2013 for WW2, and Gosarevski et al., 2019 for recent times), and PNG is fully caught within that colonial structure and set up that was all decided far away from PNG with Royal courts centuries ago, now playing out through the big neighbor and policies approved through a UN mandate and missions (see Baraka, 2001).

Accordingly, Flannery (1994) stated in his book dedication:

> *…to the Australasians who are trying to forge nations out of the chaos of colonial history.*

Fig. 18.1 Cheap cloths from Australia, re-offered to a PNG city market

This is well spoken indeed. But it remains an understatement, politely leaving out the details of the prime Australian contributions to that very chaos, governance scheme and subsequent decline of biodiversity and conservation, or wilderness for that matter. Without acknowledging the *status quo* and major actors, one can hardly move forward effectively. Connell (2005) shows some of the 'development' problems

Fig. 18.2 Australian mission help and mini-hospital based on a ship that reaches remote villages in coastal Papua New Guinea; this Christian organuzation has a global youth focus overall.

for Australia with PNG (Gounder, 1999 for development concepts): Neither the GDP is a good performance metric nor are the efforts taken effective and help the nation of PNG forward for sustainability; ignoring climate change and its sources cannot achieve. While known for several decades that industrial style development impacts biodiversity and wilderness, Beehler and Laman (2020) are still clear in their assumption that bridges and roads are part of the development mix. While Australia built roads in PNG 1971 onward with a wider funding scheme, it's clear that this will not promote wilderness biodiversity, hardly fixes the poverty problems (as the Australian section in PNG remains utterly poor and is riddled with riots and worse, e.g. kidnappings). It should be really noteworthy that Australia cannot be relied on much as a partner in the first place because it shifts its regional and national policy frequently (details in Wainwright, 2004a).

In the bilateral partnership, an Australian-aided PNG started out with great promises and hopes 100 years ago (at least for the assigned parts of PNG). But the more engagement Australia has in PNG the more PNG seems to decline (May, 1998). It's certainly true for the state of PNG wilderness and where most PNG people live. At best, one can stay here with Bandura (2007)'s moral disengagement harming sustainability, because ignoring the topic and doing nothing here imperils biodiversity conservation further (see The Guardian, 2022 for a generic complaint about Australia from the Arc nations re. climate change, sea level rise).

Fig. 18.3 Relevance of a direct connection to Australia, specifically, remains essential for Australia's influence over Papua New Guinea

But often it is not just a *laissez-faire* by Australia but an active and wanted—explicitly stated—participation of Australia in the destruction, knowingly to occur. That is, the direct involvements of Australia in conservation assessments, in resource destruction, carbon trading deals, or taxonomic species lists and its essentially ineffective debates are among those. For instance, it's found with the project 'Trees of PNG' (see role of Sydney's Royal Botanical Garden https://www.rbgsyd.nsw.gov.au/ in https://www.pngplants.org/PNGtrees/proj_details.html), or Birds of Paradise (BoP, an endeavor started through Australia early on with John Gould as one of the first documentations for these species; https://www.sl.nsw.gov.au/stories/papua-new-guinea-forty-years-independence/birds-paradise). Despite a massive and widely acknowledged forest cover loss and encroachment, BoPs are still labeled and defended as not of being of conservation concern (Beehler & Laman, 2020). Similar can be said for PNG's tree kangaroos. For conservation efforts like done and updated from Alcorn et al. (1993; Faith et al. 2000) onward using prioritization algorithms promoted by Australian universities, Beehler and Laman (2020) stated that those are often just an endless game of biologists drawing circles on maps. What is the real-world conservation protection outcome for people in PNG?

PNG has been widely credited with tribal conflict. It follows a certain pattern (as described by Banks, 2008 for instance). However, an entirely different dimension of terror came to PNG from the outside, namely through colonial powers, Christian missions, WW1 and WW2 and then the Australian mining, oil and gas, fisheries, forestry, and the 'Arc of Terror' all done in the form of ongoing foreign aid and help. To see the magnitude, already the bombing raids in Rabaul involving former colonial powers are the largest in the entire Pacific (war) Theatre and WW2 (= 'center of fire,' Cousteau & Richards, 1999). And much of those historic colonial powers unleashing such terror remain in charge to the very day 'due to the global realities' (e.g. in Melanesia; see The Guardian, 2021a, 2021b). The bad role of Japan in PNG remains somewhat unaccounted for.

As early as 1922, parts of New Guinea was made a Mandated Territory of Australia by the League of Nations (a failed predecessor of the United Nations; Baraka, 2001). In the same year, a mining ordinance was put in place to legalize prospecting in PNG. Like done in many nations, that took PNG itself widely out of the equation and its resources were essentially put in the hands whoever can afford it to grab them from abroad. This were Australians, the next neighbor, and to no surprise, many of those Australians walked out of PNG as rich men, if not even millionaires. The scheme is a repeat scheme from South American gold, from many colonies, and other bad examples (see Hochschild, 1999 for King Leopold in Congo, Palmer 2016 for Cecil Rhodes; see Eichstaedt, 2011 for ongoing problems in Central Africa). It's neocolonialism in almost its pure form found today, e.g. done via international corporations and with Australia as a platform and willing actor under a UN oversight.

Only due to the WW1 rulings made in Europe for a territorial transition of PNG areas, Australian Gold mine owner Cecil John Levien (1874–1932) took out 1923 the first miner's right issued in the 'Territory of New Guinea.' In that year, Levien took out the now infamous leases at Bulolo flats. Arguably, that land and subsequent resource became available to Australia due to the WW1 events, initiated in Europe with PNG being the pawn.

It then followed a typical pattern: Making use of his insights and job tasks at hand, C. J. Levien, was a District Officer in Buka and Morobe, then a miner. And so he was very important in the process of making New Guinea significant in gold mining between the First World War and the Second World War. He was also instrumental in New Guinea's commercial aviation history, which tends to be linked to modern mining (details in Sinclair, 1978, e.g. Lae region operated mines from air service with Germany's Junker playing a central role; and in recent times Ok Tedi mine had one of 'the busiest airport in the world'). It's clear that a connection and interaction between the socio-economic sector, PNG, Australia, and outside lead to massive wealth accumulation by just a few knowing 'the game' (as Julia Chan, 2016 called it) but leaving the majority of people empty-handed, behind and with the clean up costs. The Australian approach came hand-in-hand with 'pacifying' savage people, with bush patrols and using missionaries which consequently resulted into the vast loss of a culture and sustainable lifestyle and expertise (Beehler & Laman, 2020). This is a typical set up in colonial nations, leaving the populations behind, in the dirt, poor and as an unstructured society; it creates a mob. It also speaks to the subsequent resource curse using mining, which benefits the west the most, if even that (e.g. Kirsch, 2014; Langton & Mazel, 2008).

Michael Somare—the long-serving prime minister of PNG supported by Australia—never saw a real problem with PNG's funding scheme of a 'trickle down economy' based on natural resource extraction (Gosarevski et al., 2019; just later he regretted the REDD aspect for carbon sequestration). And the autobiography of another leading PNG politician—Julius Chan (2016)—presents explicit details on that topic for PNG. It fully lays out the consistent and dominating Australian involvement, their plan and strategy, as well as Chan's experiences; it calls for a change, all elaborated on in terms of tropical governance. It's a set up for PNG and where the conspiracy theory was shown to be correct; bad powers to steer and control nations to exist and get applied (see Sandline mercenary affair in PNG helped by England/South Africa and described first-hand in Chan, 2016 for his decisions, more context provided in McCormack, 1998).

The author has witnessed, observed and experienced first-hand such type of situations and discussions—in Australia as well as in PNG and outside—on the issue of Australian conduct and attitudes toward PNG and abroad. Already the flying to PNG is a forced Australian act, with administrative payment hurdles. And then 'passenger conduct' policies are along the same lines (see also experiences shared in Henton & Flower, 2006 screening out PNG travelers at will).

The earlier prime minister of PNG, Mike Somare, has repeatedly reported similar policies and complained of Australian dominance for a long time, with a famous diplomatic incident and statement at the airport on a flight to Australia for not having his shoes inspected by the air traffic security guards (ABC, 2005).

Textbox 1: Embassies and Diplomacy in an Internet Age and hacking: Passports, visa, independence and embassy income *ad absurdum*

Embassies are important national representations abroad. It's the established way of doing business and to address and resolve conflicts between nations. Embassies have usually close connections with the host government and they would lobby for their citizens, and on behalf of their nation abroad. Embassies are to operate in national interest, and they tend to be highly political locations as they are the center of disputes between governments, etc. trying to intervene and to resolve conflicts. Arguably, embassies do much more than providing visas and passport replacements; they are political.

However, there is reality also, technology and globalization. There remains an essential question of how are embassies funded, and linked, with the host nation and its government? In a democracy, the embassies, and their leadership—the ambassadors—are not necessarily in-line with their government or with the leading party of its time. Embassies are not to be paid or corrupted in their daily affairs. They thus need some form of independence. How that is achieved remains somewhat unclear though, and different models of liability and funding exist. It's clear though, embassies are money hungry, e.g. visa processing costs and for their stamps. It's a widely heard and experienced complaint that embassies lack any liability, and are not even contactable behind walls. One reason is that an ambassador may work in secret and carry a level of immunity. A typical example is parking tickets, speeding tickets and even traffic accidents, or inspection of diplomatic suitcases and materials. But it's commonly seen that ambassadors and such staff get removed from the official mission; they can operate instead in another sphere. It's done to reduce a political escalation and for being pro-actively.

While ambassadors and embassy employees are often government-employed (with a pension), not all really are. In reality, embassies operate on their own terms and have their own policies and procedures. The day-to-day work of embassies is then left to lower level employees or contractors, which often causes many problems for the citizen and beyond.

In times of globalization and an international society, I think it's time to look at those problems closer and ask whether such a concept, secrete negotiations and independent embassy actions, still have their place in a modern society with the internet serving citizens, nations or sustainability? For instance, embassies are the natural point of contact in most international fisheries conflicts, e.g. to get confiscated fisheries vessels and crew back home, somewhat unpunished and with public money buy-back. The state of such a global fisheries speaks to the success or failure of such steps. Ambassadors also play a key role in preparing environmental summits and policies, as seen in the climate agreements; all virtually failed. Embassies are usually great for industrial and economic growth schemes, including national culture and the arts, but less so for sustainability. By now, many ambassadors come out of established and quite wealthy diplomat families.

Clearly, there is a large international clique of ambassadors around in the world, traveling and working the world. That's what they do, but it's not really trackable. Arguably, they are often well paid and have a high pension; wider family employment (crony'ism?) has been observed, but details are pretty secret and not well exposed. Sometimes the diplomatic immunity gets waived, e.g. by accident, and then some facts show. Drug smuggling and the transfer of important documents are part of the game and have been observed by the public. WIKILEAKS exposed some of those details, online and otherwise, but more is to be done to end, change and improve bad diplomacy, harmful to the environment, the nation itslef and people. The world can easily suffer from bad diplomats. Based on experience, interactions and discussions with the international community of embassies, the diplomatic circles are often not even aware, skilled on the environment, certainly not willing, to change and give up their status, lucrative jobs and meddling in policy. Much of it is based on visa and legal monopolies given, and networks. Arguably, ambassadors might be effluent speakers and celebrate culture—aka their national (colonial) civilization—and know the local power brokers, but most of them are not trained on biodiversity, on conservation or PNG for that matter. Most of them still cater economic needs and—at best, argue on a win–win where business and the environment both can win (which is widely untrue). That's exactly what the Sustainable Development Goals (SDGs) by the U.N. with wide diplomatic buy-in and negotiations promote though; to no surprise. While widely overlooked, and not well known, the clique of ambassadors has global influence and it has been shown to act on self-interest. The ancient beaurocratic power concept of signatures, or restraining citizenship, use of hardcopy passports, is widely outdated, presents a designed bottleneck with a monopoly and does not allow modern progress for all citizens of the world. Such embassy webportals expose themselves for hacking, as widely shown. That must be exposed and improved in times and realities of the internet and when serving a global audience for the wider public good, as the aim and mandate. Whereas, serving earlier family cliques with ancient bureocratic power means remains less useful.

18.3 How Does Australia Achieve Terror, Outside of Australia[1]?

The leading groups and the government of Australia have been strongly promoting an image of the clean, just, fair and modern Australia where everybody is happy (see online https://www.dfat.gov.au/development/australias-development-program) operating in mutual agreement with everybody, a well-liked partner for assisting in development (https://www.dfat.gov.au/geo/papua-new-guinea/development-assistance). Australia is to be part of the Pacific family. It's presented as a good Australia that celebrates the rule of law and ethics—even human rights—to everybody, and just as it is to come straight out of Oxford and Cambridge for centuries: The royal order is confirmed by the eloquent academy, its self-cleaning culture and agency, diplomacy and embassies, and then using a global bill of human rights everybody in their society is happy with. It appears something like this: Australia the shining beacon of civilization in the south Pacific, you can come to Australia to start a new life, while PNG happily agrees and supports it, and the late Cecile Rhodes sits on the side lines and one can make a huge profit which is to triple down to everybody, using great sciences, e.g. based in Oxford and Cambridge and its affiliates and friends. The biodiversity and the environment are to benefit (but do not; hardly PNG is).

Metrics show the reality clearly instead

Arguably, that shiny world and Australian world view is not happening much for most people and it's far from reality, e.g. most people in the world earn less than 4$ a day,

---

[1] Australia has sparked massive conflicts onto its own land as well, as seen in Tasmania and the old-growth forest protests, Great Barrier Reef disputes, and the land claims with Aboriginees. For details, see Ludlam (2021) and citations within.

Table 18.1 App. GDP metrics of nations in Australia's Arc of Instability; Henton and Flowers 2007, see also Browne (2006)

Nation	GDP	GDP per capita	Comment
Solomon Islands	$1.51 billion	$2,258	
Papua New Guinea	$23.59 billion	$2,636	PNG ranks in the middle
Vanuatu	$8.54 million	$2,782	A very small national GDP
Fiji	$4.376 billion	$4,881	One of the highest metric per capita
Timor-Leste	$1.821 billion	$1,381	One of the lowest metric per capita

and that is certainly so for many tropical nations, for Australia's Arc and certainly for PNG. After being in 'that game' with Australia as a partner for over 100 years, PNG features some of the worst poverty on the planet while Australian gold mining and development is ongoing in parallel, the party still runs (see NME, 2021 for Ok Tedi mine finance scandal involving a leading part of the Australian entertainment industry; see Sydney Morning Herald 2015 for known paths and patterns of corruption between PNG and Australia) (Table 18.1).

The alleged engine of this mindset and machinery is the economy, economic growth and its currency and promoters, backed up by military might; the case of Solomon Island and the Australian military intervention and help shows no other [see for instance Wainwright (2006) of policing there and in PNG (Dinnen, 2006)]. Timor-Leste received military and similar police force guidance, so did PNG. It's a template narrative and which tries eventually always to put neocolonial mindsets back in the driver seat using foreign policy and their diplomacy; one may call it imperialist and racist as it does not allow much for diversity other than favoring existing elites (which tend to be white, english speaking, close to the royal institutions, and established for generations; see Jayasuriya et al., 2003, Smith-Khan, 2015 for Australia, and Tuffin, 2008 for New Zealand). As it gets criticized, the white colonizers still tend to win, as they act as the assumed superiors using science, law, governmental seats, money and force (=power tolls of modernity). Universities and education provide quite a central role in this context (de Carvalho & Flórez Flórez, 2014). It's a typical argument and profile shown and critiqued by the decolonialists for decades and who favor an indigenization. Part of that old profile to move on still is the use of 'modern' medicine, then technology and science, including flying to space. One must be superior when flying with rockets to the moon, not? As initiated by James Cook, this scheme comes as a template onto the people in the Pacific (Diamond, 2013; Flannery, 1994), for the latter virtually all of them have dark skin. In the region, New Zealand shows similar patterns on the white superiority aspect, and the Maori have reported on that for a long time with aggressive but successful moves for a change (see Wehi et al., 2021 for Maori management for some Antarctica waters; see Haertel, 2015 for indigenous concepts and differences to current dominating western approaches). New Zealand and Australia do present a lion-share for PNGs mining efforts, including subsequent 'development.'

Now how is it really done, to roll out an imperialistic model worldwide against the citizen of the world in real live of 2023? Ten simple ways doing so; a policy that is promoted easily for over a century and can be easily seen in the 'arc' and PNG as well when using:

- UN security council—permanent seats—for all global decisions and have members with a permanent seat, all enforced by UN Blue Helmets
- use a UN mandate
- apply development aid to former colonies and protectorates
- industrial activities
- mining, banking
- western education
- science
- carbon sequestration deals
- media
- health care and pension plans
- taxation, as well as
- military aid/power.
- airlines (tourism and cargo)
- shipping

Arguably, Australia is a democracy (according to P. Ehrlich moving into an kakistocracy; see footnote 1 in title quote page 407) operating with a UN mandate (Baraka, 2001) to help very poor nations (e.g. Hezel, 2012), and all what happens there is to be done by an elected parliament but locally as part of the dominion (Belanger, 2019; Twomey, 2006 for the global mining empire using that framework). So who is at fault? Not the politicians, so they say (Ludlam, 2021), but in a democracy it's the people who vote: the citizens.[2] While this is true conceptually, a large resource eating mass or mob (Flannery, 1994), in reality, there is an elite clique that promotes things like the Australian policy terror for money; often done by CEOs and companies, headquartered elsewhere though. All done with a U.N. approval. This assumes the Australian citizen is a benign being caring for its neighboring island mates.[3] This claim is hopefully true, but one may easily agree that Australian mining and the policy apparatus fueled up the instability and subsequent terror in those instances, whereas the UN mandate does not help it much last 100 years (Lasslet, 2012; Henton and Flower, 2006 for real world examples in PNG). There is also real terror spilling from Australia into PNG, but presumably that's part of a wider crime

[2] This line of thought and argument got elaborated and shown wrong already much more advanced in North America, as well as for Africa. See, for instance, Churchill and Ryan (2017), Hughes (2006).

[3] While many Australians perceive themselves as friendly and social, -based on experience they are- in reality, the Australian policy -based on a claimed democracy with votes—is far from it. Domestically and internationally that can be seen nowhere better then with Climate Change denial and in Australia's Arc. And for most parts, there is not much outcry about that in Australia (see Ludlam 2021 for details).

global scene (see for drug smuggling and illegal guns and weapon transfer often also involving Asia; RNZ, 2005, The Guardian, 2020b).

A typical example how Australia officially operates, and which affects natural resources and the set up of nations and in the Australian arc is shown in the handling of the Sepik region, beyond tourism. There is for instance an Australian involvement in a land carbon sequestration deal that went bust (The Guardian, 2015). The net effect of that involvement has been land scams, habitat destruction, loss of income and livelihood, and certainly loss of faith in Australia (apart the fact that climate change was not helped). The Australian-helped scheme of SABLS has been discussed in other chapters already but are fully adding to that argument presented here.

Another typical case is found with port facilities and associated naval bases:

> "'The Australian' (2018) reported that Australia and Papua New Guinea were discussing providing port facilities to the Royal Australian Navy and US Navy on Manus Island. [17] Australia and the United States would help expand Lombrum Naval Base, so there would be facilities for Australian naval vessels there.[18] The newspaper reported that Australia was countering interest China had placed in expanding Papua New Guinea's port facilities at Wewak, Kikori, Vanimo and Manus Island. Manus Island is the most important of these four ports, as it is a deep-water port near important shipping lanes. The RAN operated a naval base on Manus Island from the 1950s until transferred to the Papua New Guinea Defence Force in 1974. [19]" (see also ADF News, 2022)

By now it almost comes as a side-fact that the year-long APEC (Asia–Pacific Economic Cooperation) 2018 meeting was hosted by Papua New Guinea in Port Moresby (https://en.wikipedia.org/wiki/APEC_Papua_New_Guinea_2018). However, major operations were designed from the outside and daily flights from Brisbane/Australia etc. occurred during the meeting. PNG citizens had little direct engagement or gain (a few new roads were build in Port Moresby from that budget to look good in public world media). This was the first time PNG was the host of an APEC meeting but Australia provided app. ¼ of the cost to host the meetings, and it also helped with logistics and security. An independent nation would arguably not operate that way. PNG became a tool, a toy to look good in public—window dressing-, driven for exchange deeds to act in support of Australia and the U.S. against ...China?

18.4 What Terror Has Australia Already Inflicted onto Others?

Australia perceives itself as a global peace keeper. But PNG was already almost put into a secession of the entire nation and certainly into a prolonged island civil war, with deaths, economic dependence and financial crisis (see Chan, 2016 for details). But beyond itself (The Guardian 2021c, ABC News 2019) the Australian policy also threw PNG into an underlying business model with ongoing inequality, poverty and destroying ongoing nation-building in PNG; virtually all nature metrics in PNG show a decay. May it be called a genocide for ancient forest and island tribes? And beyond

PNG, Australia played already a massive bad role elsewhere (see for instance Ayson 2007). As it almost exclusively deals with islands and finite space affecting the natural riches of Melanesia, in the following I will present a short list for evidence:

Fiji conflict : Fiji has a long and deep history, and much effort is dealing with its colonial past; it is going through hard times culminating 2006 in a coup, and Australia was not short of help but came with conditions (Hayward-Jones, 2009, 2011). The Fiji government remains not happy with Australia's efforts and rather moves to Chinese options. In the meantime, the ecosystems of Fiji and its people pay the prize.

Torres Island: Australia has a difficult time to justify its actions in colonial times and after to this very day (https://australian.museum/learn/first-nations/genocide-in-australia/). It makes for a poor common ground to start a trust partnership. Many examples can be provided. This further gets enforced with man-made climate change, where Australia is not acting with enough progress to save Torres Strait Islanders (e.g. https://www.theguardian.com/australia-news/2021/oct/26/fearful-of-losing-their-homelands-islands-are-taking-australia-to-court-over-climate).

Timor-Leste: As a near neighbor, Australia was initially very receptive to Timor as an independent state from Indonesia and as a new nation. Timor is one of the youngest but also poorest nations in Asia, and the struggle of Timor was bloody, deadly and intense with many refugees, large suffering and international outcries. Nature there paid the costs and got spoiled in that course. Australia provided initially leading military build-up support and training, and was in support. But that then changed when Australia spied in governmental negotiations and leaked them to oil companies who had claims in the region; consequently relationships between both nations flipped (The Guardian, 2019 for details). Timor considers now linking with China as a partner. Any of those relations affect land and sea ecosystems, because Timor has diverse and rel. large ocean resources and matters as a shipping and travel partner for the region overall.

The Australian policy has not helped the conflict on either side. It certainly affects how Australia interacts with Indonesia, a Muslim nation. And now oil and gas resources are to be grabbed by Australia also suddenly changing policy and trust onmore.

Solomon Islands: The Solomon Islands are now referred to as a failed state, and Australia plays no innocent role as a bystander in that path (e.g. Kabutaulaka, 2004 Connell 2006, McDougall 2007). Australia was already heavily criticized on the Bougainville topic, but Solomon Islands is equally and even more complex as a colonial legacy to be resolved well. Environmental issues for this region can easily be taken from Steadman (2006) as well as Mayr and Diamond (2001) and show many extinctions.

After PNG, the Solomon Islands was somewhat given up for the bigger problem and player in the region: Indonesia (see also Chauvel, 2004; Wainwright, 2004b). But as Indonesia shares a border with Papua New Guinea, East Timor might become the new Tibet in that game. Access to oil and mining resources remain core items of negotiations. Of course Australia does all it can to avoid those problems and to mitigate them, but arguably, Bougainville, Solomon Islands and Indonesia are linked. Now heavily interfered with by China. (e.g. Firth, 2018). China is close to be directly involved in the state politics of Solomon Islands; moving from there to the environmental land—and seasscape failureis a minor step.

Vanuatu: The island nation of Vanuatu never was much happy with its colonizers, or with Australia (e.g. Lini, 1982). However, like most nations in the Pacific, the Chinese influence adds to complexities, e.g. Atkinson, 2007 for Taiwan). The current discussion centers on Australia's internationally recognized failure to act on man-made climate change (e.g. https://www.theguardian.com/world/2022/jun/20/vanuatu-calls-on-australia-to-back-its-un-bid-to-recognise-climate-change-harm). Australia remains the climate bully for those nations.

Global: A first set of the wider Australian foreign policy issues are outlined by Bergin and Wall (2021). Securing Australia and starting to build navy ports with diplomatic missions in the arc are a key scheme. Refugee problems are now the topic of the day. But it does not end there (see ABC 2020 for wider terror links).

Textbox 2: Manus Island dilemma: It's not extraterrestrial but Australian-made

Based on some international exchanges, political lock-ins and agreements, Papua New Guinea (PNG) allowed the government of Australia to operate and run a controversial offshore immigration detention system, the Manus Regional Processing Centre. It was a refugee camp situated on adjacent Los Negros Island, operating for over 16 years, from 2001 to 2017. It was meant to improve the refugee problem but created major outcries, in PNG, in Australia, the Pacific region and beyond, but created a drama (e.g. The Guardian 2021c, Human Rights Watch 2017)

The Supreme Court of Papua New Guinea ruled in 2016 that the detention of asylum seekers on Manus Island was illegal, and Prime Minister of PNG Peter O'Neill announced that the center would be removed. After a camp stand-off that involved the PNG military and police, all remaining men were removed in 2017 to new accommodations at the East Lorengau Refugee Transit Centre, Hillside Haus and West Lorengau Haus (more details in Dastyari & Sullivan, 2016).

In late 2019, the still remaining asylum seekers were moved to Port Moresby, and upon request by the PNG government, the Australian Government terminated the contracts of the service providers for the detention center and other facilities in 2019.

While Australia is not really taking any liability, responsibility or support for its Journalist and citizen Julian Assange or associated Wikileaks (Australia had no major rejections to deliver him to the U.K., and thus, toward a potential extradition to the U.S.), it has participated in virtually any war efforts done by the western nations: from WW1 to WW2 and then Vietnam as much as Iraq and Afghanistan, and many other missions out there. Australia is a stench supporter of the U.S. approach, including the war on terror and immigration policies involving PNG, but it lacks a good and meaningful track record of success and world peace—just as most other nations who engage in such 'missions.' This does affect biodiversity and the environment, certainly wilderness because those are all linked. Already looking at refugees, and squatters shows that clearly.

18.5 How to Maintain, or End the Terror Toward, Peace in the wider Australian Sector

There is a widely shared concept of world peace (e.g. https://www.un.org/en/un-chronicle/world-peace-one-hour). It comes from many groups and entities, but Australia stays far from it.

Arguably, maintaining world peace remains a key question for humankind; with climate change this might get more relevant than ever. But how to get there is not a simple feat. For some people, it comes wrapped in religion, others are more spiritual keeping it with Mother Earth. While others might even push for a benevolent dictator, the track record for that option shows us no good though. Same must be said for democracy with a social welfare spin, but which usually results more into a higher-end resource eating mass and excluding others from the 'pie' not achieving sustainability (China as a classic example; see Elvin, 2008; see Connolly 2020 for a PNG application. Those masses might well be the REAL Future Eaters (*sensu* Flannery 2002 but that was not mentioned and inferred in his assessment. The key inference was widely left out: number of people and their entitled consumption covered by an approving ideology and religion). Many examples exist for it also.

So where to go from here? While we cannot just go on MARS or the MOON, we need a solution down to earth, and I see no valid one that is truly discussed or available in the western world; certainly not by Australia though (see Ludlam, 2021 offering some approaches for Australia usually involving the Greens; Dobell 2012 for a strategy). All awhile, PNG has already lived on the planet for over 47,000 years without a global crisis, so did the aborigines. Thus, some relevant answers do sit there in the wider Sahul indeed.

References

ABC. (2005). https://www.abc.net.au/am/content/2005/s1340930.htm

ABC. (2020). Australia's Future Fund 'in bed with Adani' after freedom of information request reveals $3.2 million investment. 15th Dec 2020. Australia's Future Fund 'in bed with Adani' after freedom of information request reveals $3.2 million investment. Accessed November 30, 2021

ABC News. (2019). Queensland mine deaths heighten terror of the unknown for families, industry vows to 'reset'. 10th July 2019 https://www.abc.net.au/news/2019-07-10/queensland-mine-deaths-emergency-meeting/11294770. Accessed April 30, 2022.

ADF News. (2022). Australia to invest in PNG port infrastructure. January 27th 2022. https://adbr.com.au/australia-to-invest-in-png-port-infrastructure/. Accessed July 17, 2022.

Afira, M. R. I. P. (2021). The role and impact of Australian refugee policy in Asia-Pacific Region: From Australia-Cambodia Agreement to Rohingya Crisis. *Mandala: Jurnal Ilmu Hubungan Internasional, 4*(1).

Atkinson, J. (2007). Vanuatu in Australia–China–Taiwan relations. *Australian Journal of International Affairs, 61*(3), 351–366. https://doi.org/10.1080/10357710701531529

Alcorn, J. B., Beehler, B. M., & Swartzendruber, J. F. (1993). *Papua New Guinea conservation needs assessment*. (Vol. 1). Biodiversity Support Program.

Australian Network News (ABC). (2014). PNG's land scandal inquiry names an Australian-led company. 11th Feb 2014. https://www.abc.net.au/news/2014-02-12/an-png-land-scandal-inquiry-names-an-australian-led-company/5255528?nw=0&r=Gallery. Accessed May 1, 2022.

Ayson, R. (2007). The 'arc of instability'and Australia's strategic policy. *Australian Journal of International Affairs, 61*(2), 215–231.

Beehler, B., & Lamann, T. (2020). *New Guinea: Nature and Culture of Earth's Grandest Island*. Princeton University Press.

Bandura, A. (2007). Impeding ecological sustainability through selective moral disengagement. *International Journal of Innovation and Sustainable Development, 2*(1), 8–35.

Banks, G. (2008). Understanding 'resource' conflicts in Papua New Guinea. *Asia Pacific Viewpoint, 49*, 23–34.

Baraka, R. (2001). The meanings of independence in post-colonial Papua New Guinea. A Plan B research paper submitted to the faculty of the center for Pacific Islands studies. In *The candidacy for the Degree of Masters of Arts. School of Hawaii, Asia and Pacifci Studies*. The University of Hawai'i at Manoa.

Brown, M. A. (2006). *Development and security in the Pacific Island region*. International Peace Institute.

Bélanger, P. (2019). *Extraction empire: Undermining the systems, states, and scales of Canada's global resource empire, 2017–1217*. MIT Press.

Bergin, A., & Wall, J. (2021). The fix: Solving Australia's foreign affairs challenges. *Australian Foreign Affairs, 12*, 89–94.

Chan, J. (2016). *Playing the game: Life and politics in Papua New Guinea*. University of Queensland Press.

Chauvel, R. (2004). Australia's strategic environment: The problem of Papua. *Agenda: A Journal of Policy Analysis and Reform* 39–52.

Churchill, W., & Ryan, M. (2017). *Pacifism as pathology: Reflections on the role of armed struggle in North America*. PM Press.

Cousteau, J.M., & Richards, M. (1999). *Cousteau's Papua New Guinea Journey*. Henry Abraham's Books.

Connell, J. (2005). *Papua New Guinea: The struggle for development*. Routledge.

Connell, J. (2006). 'Saving the Solomons': A new geopolitics in the 'Arc of instability'? *Geographical Research, 44*, 111–122.

Connolly, P. (2020). The belt and road comes to Papua New Guinea: Chinese geoeconomics with Melanesian characteristics? *Security Challenges, 16*(4), 41–64.

Dastyari, A., & Sullivan, O. (2016). Not for export: The failure of Australia's extraterritorial processing regime in Papua New Guinea and the decision of the PNG Supreme Court in 'Namah' (2016). *Monash University Law Review, 42*(2), 308–338.

de Carvalho, J. J., & Flórez Flórez, J. (2014). Knowledge encounter: a project to de-colonize the eurocentric university knowledge. *Nómadas 41*, 131–147.

Diamond, J. (2013). *Guns, germs and steel: A short history of everybody for the last 13,000 years*. Random House.

Dinnen, S., McLeod, A., & Peake, G. (2006). Police-building in weak states: Australian approaches in Papua New Guinea and Solomon Islands. *Civil Wars, 8*(2), 87–108.

Dobell, G. (2012). From “arc of instability” to “arc of responsibility.” *Security Challenges, 8*(4), 33–46.

Eichstaedt, P. (2011). *Consuming the Congo: War and conflict minerals in the world’s deadliest place*. Chicago Review Press.

Elvin, M. (2008). *The retreat of the elephants*. Yale University Press.

Evans, M., Stoker, G., Halupka, M. (2016). A decade of democratic decline: How australians understand and imagine their democracy. In *From Abbott to Turnbull: A New Direction? Australian Commonwealth Administration 2013–2016* (pp. 23–41). Echo Books.

Faith, D. P., Nix, H. A., Margules, C. R., Hutchinson, M. F., Walker, P. A., West, J., Stein, J.L., Kesteven J. L., Allison, A., & Natera, G. (2000). The BioRap biodiversity assessment and planning study for Papua New Guinea. *Pacific Conservation Biology, 6*(4), 279–288.

Firth, S. (2018). *Instability in the Pacific Islands: A status report*. Lowy Institute For International Policy. http://hdl.handle.net/11540/8550

Flannery, T. (1994). *The future eaters: An ecological history of the Australasian lands and people*. Grove Press.

Flannery, T. (2002). *The future eaters: An ecological history of the Australasian lands and people*. Grove Press.

Garnaut, R., & Baxter, P. (1984). Exchange rate and macro-economic policy in independent Papua New Guinea. Report to the Government of Papua New Guinea, Reprinted as Pacific Research Monograph No. 10, National Centre for Development Studies, The Australian National University, Canberra, Australia.

Gosarevski, S., Hughes, H., & Windybank, S. (2019). Is Papua New Guinea Viable? Political Economic Bulletin 19 https://www.semanticscholar.org/paper/Is-Papua-New-Guinea-viable-Gosarevski-Hughes/6d71c70951c12fa82dea578a36b66923997f3f92. Accessed June 1, 2022.

Gounder, R. (1999). Modelling of aid motivation using time series data: The case of Papua New Guinea. *Oxford Development Studies, 27*, 233–250.

Grant, L. (2013). Operations in the Markham and Ramu Valleys. *Australia 1943: The Liberation of New Guinea*, 233.

Haertel, C. E. (2015). Indigenous management styles. In *International Encyclopedia of the Social & Behavioral Sciences* (pp. 784–787). Elsevier.

Hayward-Jones, J. (2009). *Fiji: The failing state*. Lowy Institute for International Policy.

Hayward-Jones, J. (2011). *Policy overboard: Australia’s increasingly costly Fiji drift*. Lowy Institute for International Policy.

Henton D., & Flower, A. (2006). *Mount Kare Gold Rush. Chapter 10 and 22 Solomon Islands*. Mount Kare Publisher.

Hochschild, A. (1999, first edition). King Leopold’s ghost: A story of greed, terror, and heroism in colonial Africa. *The Wilson Quarterly, 23*(1), 103–105.

Hezel, F. X. (2012). *Pacific Island Nations: How viable are their economies?* East-West Center.

Hughes, L. (2006). *Moving the Maasai: A colonial misadventure*. Springer.

Human Rights Watch. (2017). Australia/PNG: Refugees Face Unchecked Violence. 25th October 2017. https://www.hrw.org/news/2017/10/25/australia/png-refugees-face-unchecked-violence. Accessed February 1, 2021.

Jayasuriya, L., Walker, D., & Gothard, J. (2003). *Legacies of white Australia: Race, culture, and nation*. UWA Press.

Kabutaulaka T. T. (2004). “Failed State” and the war on terror: Intervention in Solomon Islands. Asia Pacific Issues 72: https://scholarspace.manoa.hawaii.edu/handle/10125/3798. Accessed February 5, 2022.

Kirsch, S. (2014). *Mining capitalism*. In Mining Capitalism. University of California Press.

Langton, M., & Mazel, O. (2008). Poverty in the midst of plenty: Aboriginal people, the ‘resource curse’ and Australia’s mining boom. *Journal of Energy & Natural Resources Law, 26*(1), 31–65.

Lasslett, K. (2012). State crime by proxy: Australia and the Bougainville conflict. *British Journal of Criminology, 52*, 705–723.

Ludlam, S. (2021). *Full Circle: Power*, Hope and the Return of Nature. Black Inc.

Lini, W. (1982). Australia and the South Pacific: A vanuatu perspective. https://doi.org/10.1080/10357718208444779

Mack, A. (2014). *Searching for PekPek: Cassowaries and Conservation in the New Guinea Rainforest*. Cassowary Conservation & Publishing.

May, R. J. (1998). From promise to crisis: A political economy of Papua New Guinea. *Governance and reform in the South Pacific* 54–73.

Mayr, E., & Diamond, J. M. (2001). *The birds of northern Melanesia: Speciation, ecology and biogeography*. Oxford University Press on Demand.

McCormack, T. (1998). The 'Sandline Affair': Papua New Guinea Resorts to Mercenarism to End the Bougainville Conflict. *Yearbook of International Humanitarian Law, 1*, 292–300. https://doi.org/10.1017/S1389135900000167

McDougall, D. (2007). "Intervening" in the Neighbourhood: Comparing Australia's Role in East Timor and the Southwest Pacific. *International Journal*, *62*(4), 867–885. http://www.jstor.org/stable/40204341

McLennan, A. (1938). The population problem in Australian New Guinea. *The Australian Quarterly, 10*, 44–52.

Nelson, H. (1982). *Taim Bilong Masta, The Australian involvement in Papua New Guinea* Australian Broadcasting Commission, Sydney. ISBN 9780642975669, 0642975663. Google Books. https://www.google.com/books/edition/Taim_Bilong_Masta/y2MhAAAAMAAJ?hl=en

Nelson, H. (2016). Black, White and Gold: Goldmining in Papua New Guinea 1878–1930. ANU Press, Google Books https://www.google.com/books/edition/Black_White_and_Gold/8M0PDQAAQBAJ?hl=en&gbpv=1&printsec=frontcover

NME. (2021). Sheppard band members' father and former manager arrested in Papua New Guinea for alleged fraud. 22nd January 2021. https://www.nme.com/en_au/news/music/sheppard-band-members-father-and-former-manager-arrested-in-papua-new-guinea-for-alleged-fraud-2862298. Accessed July 18, 2022

Ratnasari, N. N. A. T., & Perwita, A. A. B. (2020). The Implementation of Australia's "Stepping-Up Engagement" with the Melanesian States as the Defense Strategy of Australia to Respond to the Rise of China in the South Pacific Region (2013–2018) (Doctoral dissertation, President University).

RNZ. (2005). Australian jailed in PNG for smuggling guns in exchange for drugs across the Torres Strait. 27th January 2005. https://www.rnz.co.nz/international/pacific-news/152953/australian-jailed-in-png-for-smuggling-guns-in-exchange-for-drugs-across-the-torres-strait. Accessed July 3, 2022.

Rumley, D. (2006). Australia's arc of instability: Evolution, causes and policy dilemmas. *The Otemon Journal of Australian Studies, 32*, 37–59.

Sinclair, J. (1978). *Wings of gold: How the aeroplane developed New Guinea*. Pacific Publications. ISBN 0-85807-033-2.

Smith, R. (1998). The Role of the Media in Relations between Australia and Papua New Guinea. *op. cit*, 305–326.

Smith-Khan, L. (2015). Discrimination by any other name: Language tests and racist migration policy in Australia. researchonline.mq.edu.au

Springer, S. (2015). *Violent Neoliberalism: Development, Discourse and Dispossession in Cambodia*. Palgrave MacMillan.

Springer, S. (2017). The Age of Perplexity: Rethinking the World we Knew. Neoliberalism and Antiestablishment Movements. Madrid, BBVA, OpenMind.

Steadman, D. W. (2006). *Extinction and biogeography of tropical Pacific birds*. University of Chicago Press.

The Guardian. (2015). The incredible plan to make money grow on trees. 24th November 2015. https://www.theguardian.com/world/2015/nov/24/redd-papua-new-guinea-money-grow-on-trees. Accessed May 28, 2022.

The Guardian. (2019). Australia's history with East Timor isn't pretty but it must be told truthfully. 12th November 2019. https://www.theguardian.com/australia-news/postcolonial-blog/2019/nov/13/australias-history-with-east-timor-isnt-pretty-but-it-must-be-told-truthfully. Accessed June 28, 2022.

The Guardian. (2020). PNG's biggest drug bust: the plane crash, the dead man and the half tonne of cocaine. 1st Aug 2020b. https://www.theguardian.com/world/2020b/aug/02/pngs-biggest-drug-bust-the-plane-crash-the-dead-man-and-the-half-tonne-of-cocaine. Accessed November 3, 2021.

The Guardian. (2021a). PNG residents evicted a decade ago for casino development face violence and poverty: report. https://www.theguardian.com/world/2021a/nov/12/png-residents-evicted-a-decade-ago-for-casino-development-face-violence-and-poverty-report. Accessed March 17, 2022.

The Guardian. (2021b). 'Is it to satisfy you or to satisfy us?' Why New Zealand's Pacific colony doesn't want independence. 29th October. https://www.theguardian.com/world/2021b/oct/30/is-it-to-satisfy-you-or-to-satisfy-us-why-new-zealands-pacific-colony-doesnt-want-independence. Accessed January 15, 2022.

The Guardian. (2021c). Australia sends troops and police to Solomon Islands as unrest grows. 25th November. https://www.theguardian.com/world/2021c/nov/25/honiaras-chinatown-targeted-as-violent-protests-break-out-for-second-day-in-solomon-islands. Accessed January 15, 2022.

The Guardian. (2022a). 'Far from adequate': former Pacific leaders group urges Australia to increase 43% emissions cut. 7th July 2022. https://www.theguardian.com/world/2022/jul/08/far-from-adequate-former-pacific-leaders-group-urges-australia-to-increase-43-emissions-cut. Accessed May 15, 2022.

The Sydney Morning Herald. (2015). Steering corrupt cash into Australia from PNG: a how-to guide. June 24th 2015. https://www.smh.com.au/national/steering-corrupt-cash-into-australia-from-png-a-howto-guide-20150623-ghv1sx.html. Accessed June 4, 2022.

Tuffin, K. (2008). Racist discourse in New Zealand and Australia: Reviewing the last 20 years. *Social and Personality Psychology Compass, 2*(2), 591–607.

Twomey, A. (2006). *The chameleon crown.* Federation Press.

van der Kroef, J. M. (1970). Australia and the West Irian problem. *Asian Survey, 10*, 483–500.

Wainwright, E. (2004a). *Our failing neighbour: Australia and the future of Solomons Islands.* Australian Strategic Policy Institute, Canberra, Australia

Wainwright, E. (2004b). Solomon Islands, Papua New Guinea and Australia's Policy Shift. *Sydney Papers, 16*, 122–132.

West, P. (2006). *Conservation is our government now. In Conservation Is Our Government Now.* Duke University Press.

Ward, R. G., & Ballard, J. A. (1976). In their own image: Australia's impact on Papua New Guinea and lessons for future aid. *Australian Journal of International Affairs, 30*(3), 439–458.

Wehi, P. M., van Uitregt, V., Scott, N. J., Gillies, T., Beckwith, J., Rodgers, R. P., & Watene, K. (2021). Transforming Antarctic management and policy with an Indigenous Māori lens. *Nature Ecology & Evolution, 5*(8), 1055–1059.

Chapter 19
Papua New Guinea's Education, Science and Space Exploration: A Primer, and the Bad Role Models of Christianity and Oxford/Cambridge, Again, Calling for an Improved and Globally-All Inclusive STEM-Type Cybereducation for 'the Bush' and for Global Sustainability

We don't need no education,
no dark sarcasm in the classroom,
hey teacher, leave us kids alone
We are just another brick in the wall"
Lyrics of 'Another brick in the wall' by Pink Floyd "The Wall" album
... did not possess the big ego that today apparently is the prerequisite for a career in Science...
P. Hiepko life review Paul Hiepko (1932–2019) (bioone.org)

Abstract Throughout generations, Papua New Guinea (PNG) has taught its young people in a very sophisticated way how to stay sustainable and remain within the bounds of Mother Earth. This 'school' was not institutionalized nor supported by the international missionaries or by Australia who both happen to dominate PNG education now. While 'modern' education today is done in English, or at least in Melanesian pidgin, tokples and its many unique variations get lost and thus the cultural lifestyle, habitats and species with them. That has been known for decades but still English-style education is promoted. Here, it gets outlined how a 'modern' science-based education on a national level is pretty inefficient, tends to make PNG poor and not competing much with global society. Instead, it will further result into a brain-drain, and decay of the 'bush village' and sustainability skills and good lifestyle that evolved over millennia. It's a cultural destruction unabated eradicating cosmologies that co-evolved over millennnia but happening in front of our eyes and knowingly now for decades. Alleviation of poverty and wealth re-distribution remain questions of the day, and how achieved using modern tools like computing, education and the www? An outlook is provided how PNG can move forward and what can be learned from ancient PNG concepts for a relevant global sustainability education and subsequent governance instead that serves everybody well.

Keywords Papua New Guinea (PNG) · Students · Education · New generation · Generatonal justice

F. Huettmann, *Globalization and Papua New Guinea: Ancient Wilderness, Paradise, Introduced Terror and Hell*, https://doi.org/10.1007/978-3-031-20262-9_19

19.1 Introduction

If '*education is everything*' and a future investment is an indicator, then PNG might have a tough road ahead.[1] That is because PNG is already '*on the brink*' (Windybank & Manning, 2003). Internationally, the education levels in PNG rank rather low, and also when considering that the public school system is widely in the hands of conservative religious groups and sects, e.g. from America (http://www.pngembassy.org/religion.html; Beehler & Laman, 2020), as well as Australian-supported and approved (Megarrity, 2005; Papoutsaki & Rooney, 2006 for colonial education in PNG through Australia; O'Donoghue, 2009 for British Empire perspective). While the religious groups in PNG stick to their bible dogma (and that is primarily the old testament including such topics like family planning, role of females in society, lack of abortion, the ancient notion of 'eye for eye' and national patriotism), the Australian reform of PNG schooling is ongoing in parallel. That's done to put PNG on a modern industrial path and give the youth a better outlook and job opportunities: "*why not becoming an astronaut ? The world is yours...Dream big!*". For PNG, that might mean to land a job in Australia and New Zealand and live in a suburb but hardly more. It's a programmed braindrain for PNG towards Australia, New Zealand etc, typically found in the high-paid jobs. If Australia supports education in PNG, it supports itself.

Well, the narrative of education sound all common and great to western ears; except that the job market in PNG is not industrial, nor large or high tech, certainly not unionized or paying high wages. And the adjacent Australian job market is not that big neither, nor that welcoming, or easy to navigate, all inclusive and accessible for PNG citizens in the first place (Colic-Peisker & Tilbury, 2007). Australia's job and living niches are globally flooded and have been for a long time already. PNG citizens might have a hard time to compete there nowadays; this is even more so for females in and from PNG (Fox, 1999). And obtaining highly educated workers for jobs in Australia from PNG easily contributes to a PNG brain drain of 'the best and brightest' (Gibson & McKenzie, 2012; Negin, 2008) for nurses and medical practitioners, etc.), which is not what national education should support, much. A common phrase heard with such type of education from Oxford and Cambridge says "*Wanna some fries with that* ?". So why all re-applied in PNG then?

If one were highly educated in PNG, where to go other than to work abroad and make real money? Keep in mind, PNG citizens are usually dark skinned, which makes it still hard on them to blend in, in Singapore, in U.S. or in Australia. And if you ever have a great or highly paid job in PNG, e.g. in engineering, medicine, or

[1] For long time, the wealthiest nation was the U.S. Whereas soon, the global leader in education, and as judged by the graduation rate, will likely be India; a nation and culture with an entrenched cast system of poverty. India will not only soon be the nation with most people in the world, but also with most students and with PhDs and M.Sc.s. The implication of this situation is not well understood and addressed yet by most western nations. For instance, many U.S. universities have essentially a quota, a limit, on how many foreigners they can accept so that the local students are not outcompeted, creating a skew to access and output, who gets educated.

the court system, the pressures of tribalism will be felt strongly on those employed as the tribal members will come asking for help, tribal village support and for favors with the highly educated 'upper ones.' In the tribal system, one is to serve the tribe, as the prime unit of concern. That's what a Big Man does and where your home is. Many examples elsewhere in PNG show us no other.

But if the Australian model is 'PNG's educational aim' and the rule now (Klaus, 2003; Malone & Paraide, 2011 for adjustments to local languages, and Kulik, 2019; Aikhenvald, 2004 for subsequent language extinction), let's also agree that most Australians themselves will not get, or take, a job with the space agencies, or become Astronauts or work as a surgeon in a hospital, in an American software company or be a manager with a Japanese car manufacturer (Note: The Australian Space Program and such jobs are actually next to non-existent. Australia as a middle-power is not relevant enough and not part of the 'space game' to start out wit). And that is then even more true for PNG citizens and immigrants. If Australians do not get such jobs (see Redmond, 2015 for Australian inclusion issues and Saunders et al., 2016 for national poverty rates), then PNG citizens will be ranked much lower and have virtually no relevant opportunity whatsoever. That's reality. Keep in mind again, in just a few years from now most graduates will be produced by mainland Asia, namely China and India, as the world education powerhouse of jointly app.2.2 billion people. Adjacent Fiji for example sees those influences and effects already for decades (Gillion, 1977), and virtually, all of the Pacific experiences the upcoming China dominance struggle, certainly Solomon Islands (e.g. Aqurau, 2021).

In the meantime, the costs of education in PNG are still going through the roof, and it breaks down the initial village and society system, as described by Cousteau and Richards (1999) in the community management of coastal ocean resources where a much stronger emphasis on 'money making' overharvests the ecosystem. In that example, the 'money grab' commercialized the ocean resources of the community—the 'common good' of giant clam gardens. The change was driven by fear of lower education for their own children, and thus, the giant clam gardens were overharvested for monetarization, paying school costs of the Australian educational business model imposed onto PNG, its people and resources. It's done now to a degree never seen before and destroying an ancient concept around giant clams—the ocean commons—that co-evolved successfully for millennia but not anymore; why?[2]

Just looking at the gender equality statistics, family planning and sexually transmitted disease statistics (STDs) for PNG show the long-running ineffectiveness of the religious approach to education which dominates PNG and its society for over two centuries (see in contrast Richardson, 2005). There has been little progress (see Watkins, 2018 for feminism in global context instead). PNG citizens are not really making educated western-style decisions, and that's because there is little relevant effective education on the topic available to them nor does it fir their live context. PNG is a young nation after all, and the age pyramid is dominated

---

[2] This reminds of the insane student loans in Canada and the U.S. literally enslaving entire generations of university graduates. For instance, the president of the U.S., Barack Obama, paid back his student loans eventually during his first office term in the White House.

by young people also. But what is the progress with the church and Australia, both are given the upper hand for the public good in PNG such as education and public opinion (Foster, 2002)? Are we seeing any astronauts from PNG in space yet? Or do we see many PNG citizens as software engineers, CEOs or as leading international fashion models (see https://www.businessadvantagepng.com/papua-new-guineas-fashion-queen-design-and-hard-work/). And so, just compare that with socialist Mongolia for instance which had a cosmonaut in space with Russia already as early as in the 1960s (public information here: https://en.wikipedia.org/wiki/J%C3%BCgderdemidiin_G%C3%BCrragchaa). So why not PNG by now?[3]

The Australian educational model does not send you to space. All what The West and its society and workplace really does is to run around in an ever-incomplete unsustainable spiral, with a dangling dream and hyped hope, but from one event and problem to another, and trying to pseudo-resolve problems that are consistently unresolved and popping up in a re-occurring cyclic fashion, exacerbated and hyped up by the media (owned by vested big money; Cockburn, 2013). From the eco-crisis over world hunger to climate change and overconsumption, pollution, warfare and human crisis and pandemics. Major programs and projects get done and approved—also for PNG—but which resolve next to nothing longterm while the wider unsustainable bubble explodes further, globally (Czech, 2000). So after 100 years of very intense education (Megarrity, 2005), some might call it industrial brainwashing, where are we now? Are people and humanity, world peace, any better by now, is the world and its resources better off? Are we all 'free' or enslaved?

19.2 A Better World Through Better (Globally) Institutionalized and Certified Education?

Public institutional education remains a big and unresolved scheme though, virtually anywhere in the world. It receives a global assault, e.g. it's way too expensive and with such narrow cost metrics, and it's work undone. In earnest, western education is an anti-experience for many (drop-out rates etc will show no other, teacher resignations speak the same language). That's certainly true for PNG. As I can attest from my own work, and as shown in Beehler and Laman (2020) for New Guinea, schools they rarely have books, or even pencils for their students. The WWW is widely absent there. Chalkboards rule, and memorization and recitation dominate—bible school style; modern pedagogy is widely lacking but follows older schemes, often even with a whip (see for instance Papoutsaki & Rooney, 2006). The notion of a cyber infrastructure, remote or digital online education—as per STEM by the National Science Foundation U.S. (https://ncses.nsf.gov/interest-areas/stem)—sounds rather funny-tragic for PNG realities, and the bush (=majority of PNG)

[3] The Indian astronauts in the U.S., both as the major global nations, were a big deal and widely celebrated.

As per Beehler and Laman (2020) teachers in PNG are not so well educated neither (while I cannot assess that argument, I found great skilled and and curious PNG teacher friends all over during my work in PNG). They lack a modern pedagogy, continued training, and are paid poorly, with a delay or not at all; often they leave their assignment and do not come back. The salary of teachers remains a major discussion item loaded with ideology. Arguably, from my observation—as backed up by metrics and expert opinions (Beehler & Laman 2020)—teachers in PNG are not only poorly paid but their employment terms are adventurous; that's even more so in rural areas (PNG is to 86% rural), for females, and where they are not even speaking the local language; they are not part of the tribe and culture there. So how to educate the children then effectively? Not many teachers even want to be in villages or rotated around often for job contracts all over PNG and its bush and islands. Many teachers then are '*gone for Xmas*' (a term presented by Beehler and Laman 2020 for those teachers working in the remote bush, in a foreign language and cultural tribe of PNG without much support, and then left for good).

With over 700 languages, PNG has the massive language maintenance and cultural issue. Whereas I would call it a very advanced and living concept of languages in PNG—many locals speak easily 3–6 dialects. Those are precisely adjusted to local living conditions—but now this is to go away and become unified, as per latest education decrees imposed by Australia to be 'modern.' As per Beehler and Laman (2020), some schools were teaching in local language, others in Neomelanesian Pidgin, or English. Whereas the Australian concept of "Bridging to English'—as supported by AusAID (http://www.education.gov.pg/TISER/documents/in-service/in-service-primary-unit8-bridging-to-english.pdf) and widely imposed onto PNG (Litteral, 1999)—is essentially a 'big vacuum cleaner' to turn everybody into an Australian and western mindset and push them from the bush into hubs and cities; suburbs rule. The paying jobs are in Port Moresby (POM), cities along the road system, and in Australia; unless one stays with the gardening concept, so-called 'uneducated'. Such an institutional education wipes out and impoverishes rural areas, just as it did in the western world decades ago (e.g. Leon, 2005 for the EU, FH pers. com.). It's like a train: Parents essentially have to cover the education costs of such a ride. It starts with the lower primary classes (grades 3–5), there is a starting bilingual education (including English), and at the higher primary (Grades 6–8), it's fully English then (http://www.education.gov.pg/TISER/documents/in-service/in-service-primary-unit8-bridging-to-english.pdf). Loose your roots in full with a degree towards a sub-urban city live

In official terms, that means 'tokples' is replaced with pidgin English (e.g. Klaus, 2003; see Kulik, 2019) for subsequent extinction of languages and then ancient lifestyle and sustainable culture. And for western educated people, even the latter solution is not much appreciated because bush English is actually not much regarded in effluent urban areas of 'The West'. According to that, PNG remains in the dark ages (Richardson, 2005). In other words, the vast and globally unique and globally recognized diversity of languages and cultures in PNG is toyed with, given up, and everybody is unified for a single 'civil' tongue to satisfy 'progress' metrics from Australia, the churches and 'the western industrial society,' that is, the tongue of the industrial mind toward efficiency. Money talks. It's social engineering at its finest toward a global neoliberalism in the tropics failing most aspects, including biodiversity and wilderness (see Huettmann, 2015 for Central America)!

And that tokples tongue is still not the tongue of the world leaders, its essentially perceived as a lower form of English which is frowned upon once you use it outside of PNG. It's second class. Melanesian Pidgin is far away from Oxford or Cambridge English; the latter is a style favored by Australia. It sets PNG people up for being 'outsiders' and not to fit it, simply by speaking. In real world, that means loss of income and to be excluded, as a people, as a nation.

It must be stated here in bold (and just like Jerry Diamond did in his books): Despite a wide lack of western education, PNG people are obviously highly intelligent and easily outsmarting 'average' industrial and urbanized—so-called civilized—western citizens—on issues of biodiversity and wilderness maintenance, and living in PNG. They did so for over 47,000 years, why not now?

The western model of a 'modern,' subsequent industrial and neoliberal education fails in the bush, while in PNG other values and skills matter more. Already the use of several languages and dialects achieves that, and so does the cultural intelligence and expertise being a 'gardener,' often making use of over perhaps 100 species of plants and animals for a subsistence use, applied on a steeped sloped soil and remote bush in a village framework with deep social context (see Lawrence, 1984 for deep and wider knowledge). Where is that taught, and for free, and why changing it with such a high cost (see Beehler & Laman, 2020 for impacts of a New Guinea culture loss)?

From a perspective of a university teacher, I was always amazed about the PNG skill sets and education level in real live, and its realities in remote villages. Of course, PNG people are very smart indeed, just like we all are! They know how to live and to make a living where no westerners would though. One can easily see that fact by looking where the highly educated western people live, including the missionaries and teaching advisors: Not in the bush; most have left PNG, and if they stay in PNG they are in the cities of course.

Textbox 1: Impact Factors in Science and nothing in Papua New Guinea? Politics beats easily the Sciences

Papua New Guinea (PNG) has a surprising small market for science employment. Whereas the pool of people who want to do research in PNG is much larger, specifically those who want to collect species, find new ones, and export specimen, all done from areas of 'uncontacted tribes.' It coincides with a stereotype featured by The Western World and their institutions and media.

People who are experts of PNG topics hope for an academic position—any position on that matter—so that they can continue their business. It just follows an uninspiring mindset that comes from the colonial time and represents a wider powergrab, e.g. 'me and the last wild' but leaving out people of PNG.

Table 19.1 Features where PNG dominates in absence of the so-called modern society

Item	Rank	Impact	Modern context
Pisa study (Math, Reading, Writing) Source: https://factsmaps.com/pisa-2018-worldwide-ranking-average-score-of-mathematics-science-reading/	PNG is not included in the assessment	PNG ranks extremely low for reading, writing and math skills in the population	PNG gets widely behind and has little options
Mandatory years of education Source: https://education.stateuniversity.com/pages/1166/Papua-New-Guinea-EDUCATIONAL-SYSTEM-OVERVIEW.html	Not well stated; primary education only (7–12 years, attendance is not compulsory; 70% finish a formal education)	PNG has many citizens not well trained and a high rate of illiteracy	Like above
Number of universities Source: https://www.commonwealthofnationsorg/sectors-papua_new_guinea/education/universities_and_colleges/	App 8 (and colleges and technical schools)	PNG left with little graduate-level education	Like above, little science power possible
University ranking	NA	Low attraction rates to be there and get high quality researchers	Like above Weak graduate education
Software companies	None	Little employment and options to shape the future	Without software one can hardly participate
Car manufacturers	None	Modern transportation means hardly available	Cars are essential for modern/western society

Despite the huge topic of PNG with relevance to humankind, PNG scientists and experts are rel. few. Like in the western world, science has a low standing and pay in PNG also, so why to engage there? Consider the actual need for a PhD and elaborated employment paths. Consequently science is not an attractive job option for most people, not in PNG neither. Becoming a scientist, and working hard at it, is not a good vision for most PNG people; it's not affordable nor so meaningful.

That matters when other jobs are easier and more lucrative, e.g. engineering, nursing. But the lack of a proper science platform in PNG harms PNG, and it's done on purpose. For such essential things PNG must rely on Australia, and U.S. to get it done (beyond earlier missionary efforts, the role of the EU in PNG remains surprisingly tiny in the educational sector; how come?) To get it accomplished, that's true for health research, museums with specimen and industry and impact assessment research; and it happens to dominate income and wealth for a few people. PNG is not really to be funded, developed and included in such things for a good reason. The development by the international community that we see in PNG is not allowing for a well-done and sound research and academic model. Apparently, it was overlooked?

Rest assured that such a PNG science—done by outsiders instead -, and as we widely see in PNG, then has an 'impact factor' (usually the Hirsch h-index, or Google Scholar downloads) which benefits the western world and outside sources the most but leaves PNG further behind. It's here where the change in effort and emphasis is still needed, and where it can inform the western world on their mis-guided science model for betterment (Table 19.1).

19.3 Science Education in Papua New Guinea: The 'Modern' College and University Debt Conundrum Revisited with Public Sustainable Gardening

Looking at the metrics, the top 300 impact researchers are not in PNG or from PNG (https://www.webometrics.info/en/hlargerthan100); politics drives science which is 'money only', chasing funding to make a few rich people richer! The wider public good is ignored and essentially discriminated against.

Would it not be nice to be western educated and then to work as an academic with easy tenure and teaching in PNG at a college or university to less fluent locals for a good cause and money?

Well, many people had that concept in their mind; it worked only rarely though. Not many public research and science institutions in PNG are in a good shape or have long-lasting budgets, or can compete with the west for faculty members, tenure, social welfare and pension plans. There are only a few universities in PNG and while attractive and at adventurous study locations none are ranked highly internationally or even much certified; see Lutton (1981) for the PNG library moving to Perth.

And how should a college or science curriculum really look like for PNG? Traditional PNG societies tend to have very little individual and material wealth, just common wealth. Modern 'western' school education is trying to flip that: turning public ownership individuals into modern effluent members of the western capitalistic society, not? Best to forget gardening as a skill. From public entities to materialism and the '*homo economicus*' as the citizen template for everybody.

It's also worthwhile to ask what the costs are and what a human model the modern sciences and universities really promote, and what they have promoted (see Vetta, 2021 for modern science and their statistics tools actually used to support racism, etc.). The recent mobbing discussion in academia and the loss of positions and talent speaks to that model and its failure. Modern metrics of a so-called progress show us no other. It's the direct real-world outcome of Cecile Rhodes and his message and scholastic legacy in the British Commonwealth and Dominion, and through Cambridge and Oxford (Knudsen & Andersen, 2019) delivered globally; what did it really bring to the table and for people of PNG, or to tackle poverty?

There are perhaps three concepts with university science education to go by:

- –elite university (Ivy League),
- –good quality modern university (middle class) and
- –people's university (bottom up, city college or rural campus concept).

Harvard has been involved in PNG work, just like Oxford and Cambridge have, or many universities who wanted to be at the frontier, and at the cutting edge of civilization and colonial research. But arguably, PNG itself is not really part of the elite university system and cannot make it there any time soon, nor is anybody pushing for a powerful PNG university and PNG research really. Staying in the traditional imperial research and teaching paradigm is not healthy or sustainable for costs and outcome. There is very little what ivy league universities really have contributed to PNG and its people, period. They are happy to use PNG for their own university and career agendas (usually just money-related), but less so to help PNG or to ask for that.

And so, other options come to play for moving forward in good terms: middle class college. That's essentially the public Australian or Canadian university model. As middle powers such nations cannot provide a top ivy-league institution or even maintain it, instead they try to offer a certain middle-class hybrid striving for 'the best' but using and selectively picking such world-leading metrics as they can. Their universities can still be 'decent,' humble and sustainable but usually are short of money, short of long-term commitments, poorly advised for sustainability, and heavily in a neoliberal path they tend to pursue, e.g. for their private funders. Patent production is a good metric for them, but hardly for the wider public good!

In reality, the Australian and Canadian etc universities rely now on outside money and that often tends to be stock market, nuclear industry, mining and gas and oil funding the most; their forestry and fisheries funds tend to be subsidized and/or highly industrial and not supporting much sustainability (the national state of forestry and fishery speak for themselves). There are major problems in such a university system with health care costs (see Archibald & Feldman, 2014 for cost schemes and a very poor reality). Thus, under such a business scheme those universities can have an engineering and military-type section as a money maker, but widely lack many other details, including well-funded arts, humanities or philosophy (and should PNG not have a major philosophy department, and a history one, and an anthropology and a social sciences one?). A well-balanced Humanity Science is widely diminished or absent there in such a "modern" university business plan also, but while the Oxford /Cambridge link is promoted (and where most good students then are to go and try to apply for; try for a Rhodes Scholarship...). Money talks in western education while student debt is on the rise. The Anthropology Department in such a system invokes some relevant questions on the topic, as science and Anthropology have some rather colonial underlying concepts, to study the 'other', exotic people, and measure, describe and experiment with them (a concept that has been widely dismissed as biased and faulty in Psychology and elsewhere for decades already, e.g. Garrison, 2009). PNG has still been a feast and platform for such antiquated thinking (Meade & Leaves, 2010; West, 2006, etc.). Be aware, PNG hardly has a middle class to start with. PNG is not really a class society. The otherwise typically found entrenched middle class of bureaucrats is mostly missing in Wantok PNG. Most people in PNG are rural, gardeners, and cannot afford a solid western college education, and the few high-income PNG citizens go abroad, or get jobs to qualify them for leaving.

The third option, people's university (education for the general public), is widely unused and somewhat absent from PNG's Australian advisor lead. Educate to put power towards PNG is threatening Australia and the church. Perhaps some church efforts can be perceived as a public schooling? But overall, it's a model that I find attractive and powerful though, similar to a community college. As the churches apply it, it's just for their own gain and remains in that sphere. This system stands powerful but runs a long way against the Elite University system, and also against the industrial-funded university because industry is not in charge and costs and gains are 'public,' not private (Figs. 19.1, 19.2, 19.3, 19.4 and 19.5). At minimum, it can help out PNG.

PNG remains a 'young' nation, also judged by its population pyramid. And that's specifically where the lack of education hits the hardest: And thus, PNG will remain behind for the next generation also. PNG's modern education outlook remains pretty grim, so then does its generic development outlook. But that's just one side of the coin. In terms of sustainability, gardening, PNG has a great outlook and it can sustain PNG well into the future! But how taught and what degree and certification?

Education is a key ingredient though when it comes to putting a nation ahead, and it's not really happening in PNG. The key discipline for survival in PNG simply remains 'Gardening.' But exactly that is certainly not promoted or taught with the Australian model or the church model, or the western university scheme. Hundred years of trying shows us no other, while 47,000 years of inter-generational informal education lead to sustainability instead. The current education scheme in PNG destroys gardening and sustainability, village life.

Fig. 19.1 Papua New Guinea heroes: who wants an education to fly to the moon?

So how good really fares the education model that comes out of Oxford and Cambridge and the Ivy League?

In the global discourse, violence and chaos drives the wider public conception of PNG. But upon a closer look, the European and EU history and legacy are far from non-violent itself, see for instance Die Welt (2021) for Giles de Rais alisa Blaubart as a serial killer in some core aspects of Central European history. And see subsequent disputes including the church (https://www.pbs.org/newshour/show/pope-francis-apologizes-for-abuse-at-indigenous-schools-but-the-pain-remains-for-many) and their PNG involvement (see for PNG artefacts by F. Kirschbaum from the 1920s https://news.artnet.com/art-world/pope-francis-canada-2151805). In reality, and like with much of western society, brutality and all its aspects are an inherent part of their own underlying model. Kindness was hardly promoted by its cultural leaders (see for instance biographies of recent western prime ministers like German H. Kohl (by his son Kohl, 2011), F. Mitterrand for France or T. Blair for UK and Iraq war entry; compare with Buckley and Lama (2021) for Dalai Lama, Asia, etc.). Claiming the moon, the space and the universe show us again no other than an imperialism and its aggression (compare with Gillison, 1993, Suzuki, 1993). So why educating then astronauts?

Now, providing a good civil strive (*sensu* Gosarevski et al., 2019), that's where education is claimed to be a good measure and to actually create a civil society. But just look at the details for the western nations and you will see, modern educationa; ethics and values are widely absent (see for instance Bandura, 2007). How can Civic Strife develop when the teachers themselves come from a highly elite and

Fig. 19.2 Huge skill and globally acclaimed talent and expertise sits in the Papua New Guinea culture and society, as exemplified in the ancient and highly-valued art and carving

competitive, ruthless and fierce education system, but often with jobs simply handed over throughout generations (see typical examples in German President Weizsaecker and family and generational links including the Nuremberg Trail, Diplomatic career, industry contracts, Vietnam war and Catholic Church https://en.wikipedia.org/wiki/Richard_von_Weizs%C3%A4cker, contrast with the lack of job and hiring opportunities in academia for working class children in colonial Germany as a generic social reality critique, Frisch, 2020)?

The western education has an obsession with narratives of modernity and success. It is to look good, to serve the industrial employment, making money, starting with the obsession of teaching children very early (2–5 years) trivial 24 character alphabet (which is not rocket-science, e.g. when compared to learning biodiversity species in 'the bush' and subsistence lifestyles). The global society is getting obsessed with

Fig. 19.3 A superior sustainability and food security skill: egg and protein production, as shown here with Megapode eggs harvested from a long-term use of a forest mound

'reading and writing' performance metrics in the earliest age classes possible; as if that matters so much or speaks to the high quality of the educational system, money well spent and teachers…or as it would provide you with a job as a professor or astronaut, or engineer or economist for that matter. The root of human society, of cultural success, remains in sustainability, wilderness resources and benign gardening. And that's really all about that; hardly more. Relax. PNG shows us no other.

The current 'modernity' and its obsession and drill with 'education' and its hype is a western mindset imposed onto the world. But it helps little and does not do good justice to the world's diversity and sustainability needs, e.g. geographically adjusted local life and income skills that are sustainable. And they better be sustainable as one cannot fare any other, globally.

Fig. 19.4 "*Where is a plumber when you need one*"? Just like many other nations with a western scheme and on the coast, urban Papua New Guinea has a serious sewage problem, which relates to an education system problem, e.g. contamination handling, trained plumber and sewage craftsman, policy and good awareness and management; photo taken in a coastal hostel by the author

19.4 Some Options for a PNG Education Model from Looking Backwards to Move Forward at least another 47,000 Years: Indigenization of the PNG Curricular for the inclusive Online Delivery of Sustainable Gardening

As stated by Conner (2009) and others (e.g. Vetta (2021; see references within), one may agree that the classic sciences failed widely on the wider common public good, certainly in PNG. Pushing forward for educating hardcore scientists in PNG might be far off and not even modern anymore (see recent debate on STEM education in the U.S. and in Australia; Carter, 2017).

In addition to public healthcare, or some specific engineering, civic, administrative and self-sustenance skills, what should the education for PNG really be like, and for whom? Already computing puts problems into the PNG world where electricity is not available throughout the nation. Still, PNG is a world civilization located in a world class natural resource setting of strategic global relevance. And so it cannot be they are found running and lagging behind. They should lead instead and teach

Fig. 19.5 A typical sight in Papua New Guinea and its remote landscape and trails: what is unsustainable and not educated?

us, not vice versa. As suggested by Montgomery and Bishop (2006), conservation should be put on the curriculum, as well as local PNG sustainability expertise (which is vast and essentially world class).

In terms of teaching materials, PNG has so much to offer, and it will keep many universities busy with material and knowledge for decades to come. So why not using it? The big value of deeper ecological knowledge, how we relate with the universe and in a sustainable fashion, is widely acknowledged in the western world, e.g. Næss and Jickling (2000), but still not used well yet.

The indigenization of the university curricular is a worldwide scheme now, done in New Zealand as much as in Alaska and Canada (Table 19.2). There are many of those discussions in Australia also. So this offers a new hope and can be used stronger. As shown in Suzuki (1993), the earth's knowledge is deep and wise.

Table 19.2 Selected examples to indigenize the curriculum and other activities

Location	Topic	Example
New Zealand	Maori knowledge	Kukutai and Walter (2015)
Alaska	Fisheries	Whyte (2017)
Canada	Process overall	George (2019)

But for modernity, online learning will remain a deal breaker. While it becomes a global standard and can be done virtually from any angle and location, PNG in the bush does not have a strong eLearning (distance learning) component. That's because they have little internet and less computers and no vision really. In the absence of secure classrooms, the computers in schools easily get stolen out of adobe housing with palm leaf roofs and no doors.

With such a reality view, there is still no real public need flying to the Moon, or explore Mars, hardly to dive to the seafloor, for anybody, nor for PNG citizens (where most people cannot really drive a car or have no driver's license let alone a proper car insurance). But a basic fluency on issues of the modern world, and embedded into sustainability will help it, in a decent and humble fashion. That's a good model to start from.

So where has over 200 years of colonial, industrial activity and Australian oversight really left us? How much progress was provided, e.g. Imboun (2007), Dorney (2016). The infamous Rhodes scholars—the ones that are essentially paid by exploitive mining money supposed to lead the Commonwealth and trickle down a good economy to everybody—arguably seem to be at the end, and their arguments appear rather hollow by now (e.g. Walker, 2016). What does such a system really lead and contribute to? And what education, for what and for whom,STEM-style (https://www.nsf.gov/news/news_summ.jsp?cntn_id = 243,502) education has not really reached PNG yet, nor does it perform well in the first place and when computers are added. Instead cybereducation—in the cloud—can only work when it has a sustainable business model underneath. After all, good Earth Stewardship and World Peace matters and the connection with Deep Earth as stated by the elders of this world (Suzuki, 1993, Chapin et al., 2011; https://www.wisdomweavers.world/).

One would argue the world needs simply more PNG teachers, parents and their traditional knowledge spreading bush knowledge online to overcome the western 'Nature Deficit' (Louv, 2011) invoking a wider teaching reform that goes to the core of the problem: poverty, society wealth and a strong nation being sustainable and ready for a global roll-out.

References

Aikhenvald, A. Y. (2004). Language endangerment in the Sepik area of Papua New Guinea. *Lectures on Endangered Languages, 5*, 97–142.

Aqorau, T. (2021). Solomon Islands' foreign policy Dilemma and the switch from Taiwan to China. In *The China alternative* (Vol. 319).

Archibald, R. B., & Feldman, D. H. (2014). *Why does college cost so much?* Oxford University Press.

Beehler, B., & Laman, T. (2020). *New Guinea: Nature and Culture of Earth's Grandest Island*. Princeton University Press.

Bandura, A. (2007). Impeding ecological sustainability through selective moral disengagement. *International Journal Innovation and Sustainable Development, 2*, 8–35.

Bialostok, S. (2006). Literacy campaigns and the indigenization of modernity: Rearticulations of capitalism. *Anthropology and Education Quarterly, 37*, 381–392.

Brydon, K., & Lawihin, D. (2016). Melanesian visions? Some preliminary thoughts about social work education and practice in Papua New Guinea. *International Social Work, 59*, 192–204.

Buckley, M., & Lama, D. (2021) In *This fragile planet: His Holiness the Dalai Lama on environment*. Sumeru Press

Carter, L. (2017). Neoliberalism and STEM education: Some Australian policy discourse. *Canadian Journal of Science, Mathematics and Technology Education, 17*(4), 247–257.

Chapin, F. S., III., Power, M. E., Pickett, S. T., Freitag, A., Reynolds, J. A., Jackson, R. B., & Bartuska, A. (2011). Earth Stewardship: Science for action to sustain the human-earth system. *Ecosphere, 2*(8), 1–20.

Cockburn, A. (2013). *A Colossal Wreck: A road trip through political scandal*. Verso Publishers.

Cousteau, J. M., & Richards, M.(1999). *Cousteau's Papua New Guinea Journey*. Henry Abraham's Books.

Colic-Peisker, V., & Tilbury, F. (2007). Integration into the Australian labour market: The experience of three "visibly different" groups of recently arrived refugees 1. *International Migration, 45*(1), 59–85.

Conner, C. D. (2009). In *A people's history of science: Miners, midwives, and low mechanicks*. Hachette UK.

Czech, B. (2000). In *Shoveling fuel for a runnaway train*. California University Press.

Diamond, J. (2011a). In *Guns, germs and steel: The fate of human societies*. Random House.

Diamond, J. (2011b). In *Collapse: How societies choose to fail or succeed*. Revised Edition. Penguin Books.

Dorney, S. (2016). The embarrassed colonialist. https://policycommons.net/artifacts/1345734/the-embarrassed-colonialist/1957881/. Accessed May 1, 2022.

Foerstel, L. (1994). In Confronting margaret mead: Scholarship, empire, and the South Pacific. Temple University Press.

Foster, R. J. (2002). *Materializing the nation: Commodities, consumption, and media in Papua New Guinea*. Indiana University Press.

Fox, C. (1999). Training in Papua New Guinea. In *Gender, education, and development: Beyond access to empowerment* (Vol. 33).

Frisch, K. (2020). Tear down this invisible wall: We need to talk about West German Hegemonies in German Academia. Elephant in the Lab. https://doi.org/10.5281/zenodo.4312885

Garrison, M. J. (2009). In *A measure of failure: The political origins of standardized testing*. SUNY Press.

George, C. T. (2019). Decolonize, then indigenize: Critical insights on decolonizing education and Indigenous resurgence in Canada. *Antistasis, 9*(1), 73–95.

Gibson, J., & McKenzie, D. (2012). The economic consequences of 'brain drain' of the best and brightest: Microeconomic evidence from five countries. *The Economic Journal, 122*(560), 339–375.

Gillion, K. L. (1977). *The Fiji Indians: Challenge to European dominance 1920–1946*. Australian National University Press.

Gillison, G. (1993). *Between culture and fantasy: A new Guinea highlands mythology*. University of Chicago Press.

Gosarevski, S., Hughes, H., & Windybank, S. (2019). Is Papua New Guinea Viable? *Political Economic Bulletin 19* https://www.semanticscholar.org/paper/Is-Papua-New-Guinea-viable-Gosarevski-Hughes/6d71c70951c12fa82dea578a36b66923997f3f92. Accessed June 1, 2022.

Huettmann, F. (2015). On the relevance and moral impediment of digital data management, data sharing, and public open access and open source code in (Tropical) research: The Rio convention revisited towards mega science and best professional research practices. In F. Huettmann (Ed.), *Central American Biodiversity: Conservation, ecology, and a sustainable future* (pp. 391–418). Springer.

Imbun, B. Y. (2007). Cannot manage without the, significant other: Mining, corporate social responsibility and local communities in Papua New Guinea. *Journal of Business Ethics, 73*, 177–192.

Klaus, D. (2003). The use of indigenous languages in early basic education in Papua New Guinea: A model for elsewhere? *Language and Education, 17*, 105–111.

Knudsen, B. T., & Andersen, C. (2019). Affective politics and colonial heritage, Rhodes must fall at UCT and Oxford. *International Journal of Heritage Studies, 25*(3), 239–258.

Kohl, W. (2011). *Leben oder gelebt werden: Schritte auf dem Weg zur Versöhnung*. Integral.

Kulik, D. (2019). A death in the rainforest: How a language and a way of life came to an end in Papua New Guinea, Algonquin Books. (9781616209049).

Kukutai, T., & Walter, M. (2015). Recognition and indigenizing official statistics: Reflections from Aotearoa New Zealand and Australia. *Statistical Journal of the IAOS, 31*(2), 317–326.

Lawrence, P. (1984). *The Garia: An ethnography of a traditional cosmic system in Papua New Guinea*. Manchester University Press.

Léon, Y. (2005). Rural development in Europe: A research frontier for agricultural economists. *European Review of Agricultural Economics, 32*(3), 301–317.

Litteral, R. (1999). Four decades of language policy in Papua New Guinea: The move towards the vernacular. SIL Electronic Working Papers, February.

Louv, R. (2011). In *The nature principle: Human restoration and the end of nature-deficit disorde*r. Algonquin Books.

Lutton, N. (1981). From port Moresby to Perth. Archives & Manuscripts (pp. 30–38).

Malone, S., & Paraide, P. (2011). Mother tongue-based bilingual education in Papua New Guinea. *International Review of Education, 57*(5), 705–720.

Meade, M., & Leaves, F. R. (2010). Mead as anthropologist: "To Study Cannibals". (Margaret Mead: The Making of an American Icon, pp. 113).

Megarrity, L. (2005) Indigenous education in colonial Papua New Guinea: Australian government policy (1945–1975). History of Education Review.

Montgomer, S. Y., & Bishop, N. (2006). *Quest for the tree kangaroo: An expedition to the cloud forest of New Guinea*. Houghton Mifflin Company.

Mufwene, S. S. (2015). Colonization, indigenization, and the differential evolution of english: Some ecological perspectives. *World Englishes, 34*(1), 6–21.

Murray, T., & White, J. P. (1981). Cambridge in the bush? Archaeology in Australia and New Guinea. *World Archaeology, 13*, 255–263.

Næss, A., & Jickling, B. (2000). Deep ecology and education: A conversation with Arne Naess. *Canadian Journal of Environmental Education (CJEE), 5*, 48–62.

Negin, J. (2008). Australia and New Zealand's contribution to Pacific Island health worker brain drain. *Australian and New Zealand Journal of Public Health, 32*(6), 507–511.

O'Donoghue, T. (2009). Colonialism, education and social change in the British empire: The cases of Australia, Papua New Guinea and Ireland. *Paedagogica Historica, 45*, 787–800.

Papoutsaki, E., & Rooney, D. (2006). Colonial legacies and neo-colonial practices in Papua New Guinean higher education. *Higher Education Research and Development, 25*(4), 421–433.

Redmond, G. (2015). Social inclusion in Australia-what has it achieved?. *Australian Journal of Social Issues, The, 50*(2), 115–117.

Richardson, D. (2005). Peace child: An unforgettable story of primitive jungle treachery in the 20th century. Regal Publishers.

Saunders, P., Wong, M., & Bradbury, B. (2016). Poverty in Australia since the financial crisis: The role of housing costs, income growth and unemployment. *Journal of Poverty and Social Justice, 24*(2), 97–112.

Suzuki, D. (1993). *Wisdom of the elders: Sacred native stories of nature.* BoD–Books on Demand.

Vetta, A. (2021). Comparison of the scientific developments of first 2 decades of 20th century and 21st century. Academia Letters. https://doi.org/10.20935/AL1127

Walker, G. (2016). 'So much to do': Oxford and the Wills of Cecil Rhodes. *The Journal of Imperial and Commonwealth History, 44*(4), 697–716.

Watkins, S. (2018). Which feminisms. New Left Review 109 Jan/Feb. https://newleftreview.org/issues/ii109/articles/susan-watkins-which-feminisms

Welt, D. (2021) Gilles De Rais Alias Blaubart: Hinter den Burgmauern begann er aus purer Lust zu morden, 21st November 2021. https://www.welt.de/geschichte/article202504788/Gilles-de-Rais-Blaubart-Bevor-sie-starben-trieb-er-mit-ihnen-Sodomie.html?wtrid=kooperation.reco.taboola.free.welt.desktop. Accessed August 1, 2022.

West, P. (2006). *Conservation is our government now. In Conservation Is Our Government Now.* Duke University Press.

Whyte, K. (2017). Indigenous climate change studies: Indigenizing futures, decolonizing the anthropocene. *English Language Notes, 55*, 153–162.

Windybank, S., & Manning, M. (2003). *Papua New Guinea on the brink.* Centre for Independent Studies.

Chapter 20
Case Study: The YUS (Yupno, Uruwa and Sam Rivers) Landscape, Huon Peninsula, in Papua New Guinea as an Ongoing and World-Leading Conservation Area Success Story Despite International Coffee and National Park Model Landgrabs

Abstract Like for most of Melanesia, and as commonly found with many indigenous peoples throughout the world, Papua New Guinea (PNG)'s land is mostly owned in a complex public land tenure system. A simplistic western-style biodiversity wilderness top-down land protection, e.g. a typical National Park with a single ownership and governance, does not really exist nor will it likely be effective, e.g. due to tribal conflicts, deep legacy, heritage, PNG sustainability management and spiritual and enforcement problems across generations to come. An alternative is the community-owned and people-driven approach ("bottom-up") with a conservation aim, such as Wildlife Management Areas (WMAs) and similar concepts. The rugged and ancient landscape of 'YUS' (encompassing the watersheds of the Yupno, Uruwa and Sam rivers) is one of those unique examples of living landscapes in a community ownership where regional sustainability can be found. It's a certain role model for landscape protection alternatives having used endemic tree kangaroos, cloud forests, 'ridge-to-reef' concepts and rangers, One Health and a social emphasize (see Freeman et al. in Bulletin of the British Ornithologists' Club 133:4–18, 2013 for birds). While it is unique in PNG for its long-term success, it's also unique globally because YUS has a missionary legacy and involves a somewhat balanced outside support that actually works over time locally. The YUS region has many conservation species, including the iconic Matschie's Tree Kangaroo found in the virgin mountain forests of over c. 1500 m elevation. YUS has much to study left, and research data are few. YUS is an ancient landscape with coastal trails connecting the coast into the highlands making a case for ridge-to-reef watershed approaches while a commercial export coffee production is carried out in YUS, partly perhaps to make the operation more viable with a western input.

Keywords Papua New Guinea (PNG) · YUS (Yupno, Uruwa, Sam Rivers) · Sustainable landscapes · Organic coffee · Wildlife-friendly

F. Huettmann, *Globalization and Papua New Guinea: Ancient Wilderness, Paradise, Introduced Terror and Hell*, https://doi.org/10.1007/978-3-031-20262-9_20

20.1 Introduction

Papua New Guinea (PNG) offers many fascinating topics and living landscapes (see Loney Planet Guide https://www.lonelyplanet.com/papua-new-guinea, Flannery, 1995, p. 18, or Beehler, 2020 for a selection and overview). While most of PNG is 'in use' and a livelihood area for people, only a small fraction of them have been described yet with some relevant and organized science detail (Table 20.1 for a selection). Many other details remain unknown due to lack of access, or open data shared, lack of peer-reviewed publications or awareness even (see Huettmann, 2020 for tree kangaroo examples). Arguably, PNG is a wilderness place, but roads, bridges and other developments such as mining and tourism come in the way (Hayan, 1990; West, 2006). On a landscape scale, those then create a mélange of cultures; the ancient meets the west and globalization. These cultures and their areas do not have to directly overlap, but they reach far and beyond.

According to Beehler and Laman (2020), PNG carries no National Park really (New Guinea just has one, and details on the land management are not so clear as authors describe), but the public record for PNG states 8

Table 20.1 List of selected landscapes in Papua New Guinea with some available science subject descriptions

Landscape	Science description details	Citation
PNG border northern region (~Telefomin)	Mammals, tree kangaroos	Flannery (1995, 2000)
Adalbert mountain range	Plants/trees	Webb et al. (2005)
Finisterre range	Linguistics	https://en.wikipedia.org/wiki/Finisterre_languages
YUS	Usually a Tree Kangaroo focus	See chapter and citations within
Fly river	Specific aspects covered as part of the Ok Tedi impact assessments	Kirsch (2014)
Sepik river	Specific ecology aspects	Cousteau and Richards (1999)
Mountain forests	A small selection of mountain forest aspects in PNG	See also Crater mountain citations
Cloud forests	A specific selection of mountain forest aspects in PNG	Montgomery and Bishop (2006); Tree Kangaroo Conservation Program (2018)
Crater mountain region and landscape gradient	Mostly a biodiversity, bird focus as well as wider Anthropology	Mack (2014); see also Gillison (1993); Gillison (2002)
Wau region	Wildlife (e.g. birds and mammal aspects), usually in response to impact assessments for oil and gas	Richards (2018)

National Park designations (https://en.wikipedia.org/wiki/List_of_protected_areas_of_Papua_New_Guinea); many of them are disputed by the land tenure holders though; see Hyndman, 1998 for deeper land conflicts in modern Melanesia). Like in many places in Melanesia, the western-style protected area concept in PNG is arguably inefficient and lacks an assessed outcome and with indigenous people involved; it can be outright destructive and harmful. Landscape components are often maintained by the PNG society. One reason is that those national park areas are hardly untouched by humans, considering PNG's deep history of over 47,000 years with a human occupation on much of the land, including fire and big animal pursuit. It also appears that PNG is somewhat overlooked and ignored for such questions. It's hardly listed in World Heritage Site lists (just one listed and seven are tentative: https://whc.unesco.org/en/statesparties/pg); virtually over 95% of PNG is instead in public land trust tenure (Baraka, 2001). With that, the third category of managed/protected areas in PNG, the Wildlife Management Areas (WMAs), become a major item. That's because WMAs are managed through an elected committee that consists of customary landowners. The WMAs are similar to other community conservation and comanagement areas in the world. Likely, they are a success story overall because they are less static and allow humans to exist in a somewhat benign and indigenous fashion. However, they are also relatively "liberal" and allows various industrial usage options, including industrial ones. The latter is a major lobby for land designations in the region and how the landscapes really look like, affecting people!

In the meantime, for many people the real beauty of PNG remains in the areas that are still widely undescribed, hardly well discovered yet.[1] Arguably, PNG offers many opportunities and contributions to mankind and it asks for new concepts to be tested and applied, all the time. Confronting the western world with PNG remains full of insights; it experiments the west.

There are many 'remote' places in PNG but where people make a living already for millennia (see Matthiessen, 1987; Flannery, 2000). A certain human footprint can often be found, directly and indirectly. A few of those places, usually based on access from the coast or rivers, carry a western history from explorers, warfare and prospectors with little documentation (e.g. Nelson, 1982; Flannery; 1995; Wilson, 2019, see Goroka in the highlands as a typical example, e.g. described in Cousteau & Richards, 1999; West, 2006; Mack, 2014).

Many expeditions into the so-called unknown of PNG have been made and written about, e.g. 'AN EPIC JOURNEY:The 1930 Expedition of Michael Leahy and Michael Dwyer across New Guinea via the Purari River' (Willis, 1969).

While the Huon Peninsula was explored and crossed by 'western' explorers (see Table 20.3 for a selected list of YUS explorers), many areas still remain to be investigated and to be understood more.

The remote landscape centered in the uplands of the Huon Peninsula around the Yupno, Uruwa and Sam rivers (YUS) is one of the unique examples where regional sustainability can be found (Montgomery & Bishop, 2006). The area has app. 50

[1] This statement stands in deep opposition to what Jerry Diamond claimed and that all birds of Papua New Guinea have been described and nothing new can really be found there, far from it.

villages ('hamlets') and 12,000 people, is virtually free of roads, and just a few airstrips exist. It attracts explorers to this very day, e.g. for pursuing Bird of Paradise found at it slopes (Laman & Scholes, 2012).

YUS is located north from the famous Bulolo river connecting with a road to the wider Markham tributaries where an initial set of PNG's early gold fields were located. However, YUS is not really impacted from roads and has its own dynamics and development going on. Gold and mineral exploration is likely not on the agenda there any time soon.

Despite this remoteness, in the discipline of Landscape Ecology it is understood that all landscapes are truly embedded, linked and telecoupled with other processes, namely human ones (Liu et al., 2018). Over the last 100 years or so, the area received its fair share of missionaries trying to make the area 'peaceful'—god's eden—but then filling it with their own ideology leading to cultural loss and such an onslaught (Beehler & Laman, 2020 for general details).

Living in the Anthropocene, such examples where those connections among the community at large actually work out are quite rare these days (see for instance the CAMPFIRE project in Africa but that ran dry though, Taylor, 2012; or see the Annapurna Conservation Area in the Hindu Kush-Himalaya; Prajapati et al., 2020); PNG seems to have here one of those rel. stable areas while wider infiltrations occurred such as missionaries, explorers, trading and warfare (local, WW1 and WW2; Table 20.2).

Table 20.2 List of selected landscape aspects in Papua New Guinea that await their detailed description contributing to world's unique places and conservation

Landscape	Description details and justification	Citation
Coastal orchid-scape	Orchids and their conservation in the landscape	Orchid society of Papua New Guinea (2006)
Steep slopes	Likely widely unsurveyed due to access constraints	See Laman and Scholes (2012) for birds of Paradise focus
Seagrass beds	Major site for spawning, fish and sea turtles	
Cloud forests	Major site for endemics	Tree Kangaroo conservation program 2018 for an example
Grasslands	Major sites of human civilization	See for instance Matthiessen (1987)
Sand beaches	Erosion monitoring due to sea level rise	Huettmann pers. com
Old-growth rainforest	Last areas left for sustainability widely untouched by physical disturbance	Beehler and Laman (2020)
River headwaters	Essential ridge-to-reef aspects	Huettmann per. com

Table 20.3 List of selected explorers of Huon Peninsula where YUS is located

Exploration	Topic	Citation	Comment
Leahy brother	Internal southern Papua New Guinea, widely unexplored at its time	Willis (1969)	Touched on southern Huon Peninsula watersheds but leaving most of YUS unexplored
Colonial Germany	Explorations	NA	Little effort, just coastal and almost none in the interior; likely rivers nearby
C. Schmitz	Anthropology	Schmitz (1960)	German research
J. Diamond	Birds	Diamond (2011 and earlier citations within)	Several initial collection and study sites

20.2 What is YUS: A Baseline

The Huon Peninsula is a high mountainous area (Finisterre Range) with steep slope habitats connecting to the coast (Laman & Scholes, 2012). It had a human presence for thousands of years (Groube et al., 1986). YUS has received western contact for over 400 years, namely by the Australians and later by some Japanese and allied soldiers during WW2 (mostly coastal and along major rivers; Duffy, 2016). Interest in gold remains a driver to this very day, specifically in the wider area close to YUS like the connected Markham River watershed. While YUS was little driven by colonial Germans, it still was a bit perforated from the German-dominated coast through the trails to the jungle and highlands. During WW1, it was handed over from the Germans to Australia but invaded by Japan. And YUS became an area of major interest during WW2 during the wider 'Finisterre Range campaign' where the Japanese-Australian battle played a big role for the control of the 'Pacific Theater'. The Markham River villages near Lae got described by Robinson (2017) and others through (escaped) soldiers and with 'spies' on either side (Japan and allied troops; mostly Australia and U.S.). Some traveled through YUS.

After WW2, it received the construction of the largest bridge at the time (1955). But till then, not so much has happened in YUS, which must be seen as amazing and considering the booms-and-busts ongoing in other parts of PNG and nearby, e.g. Wau area (see Sinclair, 1978 for relevance of air traffic for those gold mining operations). A few airstrips were done for village exchange, later for coffee trade.

YUS is also characterized by many earthquakes, sometimes more than 10 earthquakes a day and which one can easily feel when being there. A certain danger comes with it, as landslides can get triggered by them and people have been killed by them at slopes. I have personally witnessed those events and impacts.

YUS and its mountainous landscapes and the cloud forests were already studied by J, Diamond and got some fame for the Matschie's Tree Kangaroo (e.g. Montgomery & Bishop, 2006). Nowadays, YUS is essentially a missionarized "conservation area with a small coffee production pursued and likely growing" (details in Montgomery & Bishop, 2006; Beehler & Laman, 2020).

What made YUS famous is that the conservation area consists of landowners who donated and pledged their land for tree kangaroo protection and habitat (details in Dabek & Wells, 2021; Montgomery & Bishop, 2006). This creates a unique, otherwise globally rarely found wildlife protection cooperative type. It's fueled by people of PNG and some NGOs and the coffee production within. As the tree kangaroos live primarily in the higher ranges and peak areas, it's an area where most people are not really going. Apart of climate change, it thus remains a naturally protected area, consisting of a unique and large endemic habitat and species.

The area then got mapped out ("gazetteer") and later was re-done with a revision and then that concept got put to work (Dabek & Wells, 2021). The efforts benefitted from missionaries who presented already the hunting of tree kangaroos as a sin (something like "*Jesus never hunted tree kangaroos and so should not you*"). This gospel is now also taught in local school curricular (=replace math with conservation); it has a public impact on the landscape (details in Montgomery & Bishop, 2006).

The area is actually monitored using 'rangers' which regularly walk the trails and record any hunting and potential poaching problems in the area and villages (Delie, 2015; Montgomery & Bishop, 2006). The rangers are vested. This makes for a unique, somewhat ancient patrolling concept applied to wildlife conservation; apparently, it's rather successful.

20.3 Where YUS Currently 'is'

The YUS landscape is somewhat hidden and thus a bit overlooked, and that's likely one reason for its protection. It's a certain 'gem' hidden from major developments and out of sight. It's currently on the edge of developing, and many avenues of making money are tried and pursued, most do not work yet. Still, YUS and its geography have a unique position for 'ridge to reef.' That's because YUS is not only connected well with the coast, it's also part of a mountain range and its plateaus, valleys, steep slopes, grasslands and extensive cloud forests. Much of the vegetation remains vast and in a natural state; indeed, it's one of its last pristine places on earth (Laman & Scholes, 2012) (Fig. 20.1).

20.4 A Unique but Quite New Connection with YUS, Religion, Science and Coffee

Like many places in PNG, YUS lived for over 97% of its time untouched by the western world. But upon contact, it got missionized early on. By now, missionary groups like Seven Days Adventists dominate many villages and the flying schedules; the area is pretty peaceful but affected by wider gang conflicts in the city regions.

Fig. 20.1 Rugged terrain with virgin forests, gardening patches, a network of villages, trails and some grasslands. The YUS landscape is occupied for millennia and a rather sustainable human high civilization

Western people enter frequently, namely medical doctors, nurses, linguistic scholars, zoo scientists and Australian researchers.

But the advent of coffee has changed the lives of PNG and many farmers lost their skill and expertise for a cash crop which made everybody poorer at the end (Cousteau & Richards, 1999, p. 90; Baraka, 2001). This pattern is also unfolding in YUS. The YUS area is prime recipient area for organic product companies looking for pristine supply chains abroad but based in the U.S. such as Café Vita (https://www.caffevita.com/). While YUS is near the coast, coffee is still flown inland to Goroka for the highland sellers and traders; YUS itself is hardly growing in this deal but its resources taken out and the older (mining) hubs get re-inforced; a colonial game.

20.5 How About the Role of Remoteness, Spirituality, the Cost of Fuel as Well as the Global Market and Its Repercussions? Can the Pink Disease and Coffee Bean Borer Destroy It All?

Perhaps a reason why YUS was so peaceful is simply because it is so rugged and inaccessible, and thus of little relevance to The West? It's easy to be spiritual in

Fig. 20.2 A human-dominated but quite sustainable landscape with gardens, hunting zones, edge fragments, fire patches and some virtually untouched wilderness taboo areas

Fig. 20.3 Fire plays a role in virtually any human society

Fig. 20.4 A village in YUS

YUS. Protection by remoteness is for real. Those areas are often a National Park (but which is unlikely a suitable concept for PNG though with living lands in public land tenure). But remoteness is not everywhere equal, and travel routes exist in the area already for millennia; bush airport hubs are now the YUS centers. Those tend to bring in goods, trading items and people; not all come without dispute. One can watch it on facebook.

Another factor for the conservation success could be the dominating religion, which is a peaceful but dominating one wiping out ancient cosmologies. It argues for instance that the bible, and Jesus, never hunted tree kangaroos; thus, nor should any Christian in YUS! In the meantime, remote pockets remain with the usual PNG nature believes, sorcery included. YUS offers many facets of live. Hunting dogs remain to be used; some run free and leave a 'footprint'.

And third, what do global markets do? Are they helping YUS? Is the coffee bean and fuel price in support of good sales, income, local wealth and access? As global experiences for PNG commodities showed us, those forces are not to be trusted much (see various chapters of this book). Arguably YUS follows a wider destructive trend but is just behind in time and thus impacts are not so obvious, yet but are to come. A good fact on this trend are the mentioned coffee diseases which are threatening YUS coffee now.

Fig. 20.5 An ancient trail connecting coastal areas of Northern Papua New Guinea with interior Huon Peninsula and the highlands

20.6 YUS Outlook

Coffee has its issues; one is diseases (e.g. coffee borer) another is seasonal labor, and another one is the land demand. It also connects remote PNG directly with downtown Seattle/U.S. or Hamburg/Germany, where PNG coffee is often roasted and consumed for much higher value and money. It can be an organic independent roaster product. More demand for coffee will push pressure on hiking and road development which will enter this 'pristine' paradise and affect it.

And likely it will not end with the coffee once a good infrastructure is in place. Although not well studied, man-made climate change will knowingly bring changes and push up vegetation and diseases. It will likely result into the decay of the cloud forest habitat sooner or later. That's because many fitting examples

Fig. 20.6 Water, forests and people; what can be better than taking a 'splash' as done for thousands of years; it's ancient

of village development after the gold rush can be found—good or devastating (see chapters of this book) but all with massive changes into the modernity—see for instance https://www.businessadvantagepng.com/how-one-town-in-papua-new-guinea-found-life-after-the-gold-rush (Figs. 20.2, 20.3, 20.4, 20.5, 20.6, 20.7, 20.8, 20.9 and 20.10).

Fig. 20.7 Moss-loaded cloud forests in the upper section of YUS, app. 3000 m above sea level; a world heritage

Fig. 20.8 Chasing tracks of a Tree Kangaroo

Fig. 20.9 Influence of the church on landscapes has been, and remains, rather big in YUS; a well-maintained religious rectangle in the ancient wild landscape

Fig. 20.10 A sunday afternoon past time: village rugby with competing village teams

References

Baraka, R. (2001). The Meanings of Independence in Post-Colonial Papua New Guinea. A Plan B Research Paper Submitted to the Faculty of the Center For Pacific Islands Studies in the Candidacy for the Degree of Masters of Arts. School Of Hawaii, Asia and Pacific Studies. The University Of Hawai'i At Manoa.

Beehler, B., & Laman, T. (2020). *New Guinea: Nature and Culture of Earth's Grandest Island*. Princeton University Press.

Cousteau, J. M., & Richards, M. (1999). *Cousteau's Papua New Guinea journey*. Henry Abraham's Books.

Dabek, L., & Wells, Z. (2021). Creating the first conservation area in Papua New Guinea to protect tree kangaroos. In *Tree Kangaroos* (pp. 135–151). Academic.

Delie, J. (2015). Indigenous rangers as stewards for the environment. In A case study of the YUS conservation area rangers, Papua New Guinea (Doctoral dissertation, Oregon State University).

Duffy, J. P. (2016). War at the end of the world: Douglas MacArthur and the forgotten fight for New Guinea, 1942–1945. Dutton Caliber.

Freeman, B. G., Class, A., Mandeville, J., Tomassi, S., & Beehler, B. M. (2013). Ornithological survey of the mountains of the Huon Peninsula, Papua New Guinea. *Bulletin of the British Ornithologists' Club, 133*(1), 4–18.

Flannery, T. (2000). *Throwim Way Leg: Tree-kangaroos, Possums, and Penis Gourds—on the Track of Unknown Mammals in Wildest New Guinea*. Grove Press.

Flannery, T. F. (1995). *Mammals of the south-west Pacific & Moluccan Islands*. Cornell University Press.

Gillison, G. (1993). *Between culture and fantasy: A new Guinea highlands mythology*. University of Chicago Press.
Gillison, D. (2002). *New Guinea Ceremonies*. Harry N Abrams Inc.
Groube, L., Chappell, J., Muke, J., & Price, D. (1986). A 40,000 year-old human occupation site at Huon Peninsula. *Papua New Guinea. Nature, 324*(6096), 453–455.
Diamond, J. (2011). *Guns, Germs and Steel: The Fate of Human Societies*. Random House.
Hayan, D. M. (1990). *Road through the rain forrest: Living anthropology in highland Papua New Guina*. Waveland Press.
Hyndman, D. (1998). Melanesian resistance to ecocide and ethnocide: Transnational mining projects and the fourth world on the Island of New Guinea. In H. J. Bodley (Ed.), *Tribal peoples and development and issues: A global overview* (pp. 281–298). Mayfield Publishing Company.
Huettmann F. (2020). Chapter 24—Investigating Matschie's Tree Kangaroos With 'Modern' Methods: digital workflows, big data project infrastructure, and mandated approaches for a holistic conservation governance; In Dabeck L. P. Valentine, J. Blessington & K., Schwartz (Eds.), *Tree Kangaroos: Science and Conservation* (pp. 379–391). Academic Press.
Kale, E., Armstrong, K. N., Amick, P., & Woxvold, I. (2018). Mammals of the Wau Creek proposed wildlife management area, Gulf Province, Papua New Guinea. In S. J. Richards (Ed.), *Rapid biological assessments of Wau Creek, Uro Creek and Lake Kutubu: Documenting biodiversity values to promote forest conservation in the Kikori River basin, Papua New Guinea*. ExxonMobil PNG Limited. Port Moresby.
Kirsch, S. (2014). *Mining capitalism*. University of California Press.
Laman, T., & Scholes, E. (2012). *Birds of paradise: Revealing the world's most extraordinary birds*. National Geographic Books.
Liu, J., Dou, Y., Batistella, M., Challies, E., Conno, T., Friis, C., Millington, J. D. A., Parish, E., Romulo, C. L., Bicudo Silva, R. F., Triezenberg, H., Yang, H., Zhao, Z., Zimmerer, K. S., Huettmann, F., Treglia, M. L., Basher, Z., Chung, M. G., Herzberger, A., … & Sun, J. (2018). Spillover systems in a telecoupled Anthropocene: Typology, methods, and governance for global sustainability. *Environmental Sustainability, 33*, 58–69. https://doi.org/10.1016/j.cosust.2018.04.009
Montgomery, S. Y., & Bishop, N. (2006). *Quest for the tree kangaroo: An expedition to the cloud forest of New Guinea*. Houghton Mifflin Company.
Mack, A. (2014). *Searching for PekPek: Cassowaries and Conservation in the New Guinea Rainforest*. Cassowary Conservation & Publishing.
Matthiessen, P. (1987). Under the Mountain Wall: A Chronicle of Two Seasons in Stone Age New Guinea. Penguin Classics.
Nane, D., Pontio, M., & Holbrook, T. (2021). Land-use planning for a sustainable future in Papua New Guinea. In *Tree Kangaroos* (pp. 153–163). Academic.
Nelson, H. (1982). *Taim Bilong Masta, The Australian involvement in Papua New Guinea Australian Broadcasting Commission*. ISBN 9780642975669, 0642975663. Google Books https://www.google.com/books/edition/Taim_Bilong_Masta/ y2MhAAAAMAAJ?hl=en
Orchid Society of Papua New Guinea Inc. (2006). Spatulata Orchids Papua New Guinea. Moore Printing Ltd. ISBN 9980-86-070-7
Prajapati, J., Huettmann, F., Ghale, T. R., & Regmi, G. R. (2020). Chapter 24. The Annapurna conservation area project (ACAP): Towards a success story in landscape feature and watershed conservation management. In G. R. Regmi & F. Huettmann (Eds.), *Hindu Kush-Himalaya watersheds downhill: Landscape ecology and conservation perspectives*. (pp. 473–496). Springer Gland.
Richards, S. J. (Ed.). (2018). Rapid biological assessments of Wau Creek, Uro Creek and Lake Kutubu: documenting biodiversity values to promote forest conservation in the Kikori River basin, Papua New Guinea. ExxonMobil PNG Limited. Port Moresby.
Robson, S. K., Inkster, T. E., & Krockenberger, A. K. (2012). Bats of the YUS Conservation Area Papua New Guinea.

Robbins, R. G., Haantjens, H. A., Mabbutt, J. A., Pullen, R., Reiner, E., Saunders, J. C., & Short, Webb, Boucher, Sheppard, Summers Adelberts Vegetation, K. (1976). Lands of the Ramu-Madang Area, Papua New Guinea. Land Research Series 37. CSIRO, Australia, Ding.
Schmitz, C. A. (1960). *Historische Probleme in Nordost-Neuguinea; Huon Halbinsel. Studien zur Kulturkunde*, (Vol. 16).
Sinclair, J. (1978). *Wings of gold: how the aeroplane developed New Guinea.* Pacific Publications. ISBN 0-85807-033-2.
Souter, G. (1963). In *New Guinea: The last unknown.* Angus & Robertson Publishers. ISBN 0-207-94627-2.
Tree Kangaroo Conservation Program. (2018). Annual Report 2018. Woodland Park Zoo, Seattle.
Taylor, R. (2012). The performance of CAMPFIRE in Zimbabwe: 1989–2006. In *Evolution and innovation in wildlife conservation* (pp. 219–240). Routledge.
Webb, C. O., Boucher, T., Sheppard, S., & Summers, M. (2005). Vegetation of the Adelbert range, Madang Province, PNG. http://camwebb.info/files/pubs/webb2005_adelbert_veg.pdf. Accessed July 30, 2022.
West, P. (2006). *Conservation is our government now. In Conservation Is Our Government Now.* Duke University Press.
Willis, J. (1969), An epic journey: The 1930 Expedition of Michael Leahy and Michael Dwyer across New Guinea via the Purari River. A $ubthesis prepared by I. J' Willis in part completion of requirements for the M.A. Preliminary Course, University of Papua and New Guinea, Port Moresby. https://library.ucsd.edu/dc/object/bb40968273/_1.pdf
Wilson, M. (2019). *A True Child of Papua New Guinea: Memoir of a Life in Two Worlds.* McFarland.
Woodland Park Zoo. (2010). Tree Kangaroo Conservation Report. Annual Report. YUS report https://www.zoo.org/document.doc?id=1724. Accessed December 10, 2021.

Chapter 21
Case Study on Outsider Impacts by Exploration, Mining, Science and western NGOs: An Honest Look at 'Crater Mountain' in Papua New Guinea

The Professional Society of [xxx] is a nonprofit membership-supported organization dedicated to the protection and improvement of the quality of life for present and future generations.

Type of a typical mission statement scheme found *with 1000s of NGOs* and Professional Societies, many of them operating international, tax-exempt and some also active in Papua New Guinea, based on temporary visas (whereas the long-term PNG metrics for socio-economics and biodiversity wilderness conservation speak for themselves, for their progress and achievements, and now, lack there of). It simply makes for the wider international community of the AID and Development scene and what they can, do, achieve, and what they totally miss and destroy.

Abstract Crater Mountain is a world-famous research site near Goroka in the highlands of Papua New Guinea (PNG). It served as international conservation inspiration and as a study site for many studies, thesis and PhD dissertations—local and from abroad. Anthropologists, photographers and naturalists alike visited in order to study in remote tropical virtually untouched wilderness areas surrounded by 'ancient tribes.' Birds of Paradise got intensely collected and pursued, Cassowaries and many other aspects got studied with year-long efforts and supported by NGO efforts. This image of perceived harmony changed dramatically though when mining exploration got announced and started to unfold on the landscape and its social fabric. Subsequent tribal and other conflicts became intense and violent, people got injured, resulting into a departure of NGOs and researchers making this 'conservation' area a bad showcase how external forces shape and affect outcome in an 'original' setting initially little touched by the western world, apart from Australian mining efforts, colonialism and their missions 100 years ago. While many people in the bush are worse off, just like the cassowaries and Bird of Paradise, Crater Mountain presents a small conservation area but a wider and horrific example for the realities of western conservation and research efforts operating in the PNG fabric all created by, and with, a western oversight. This was more complicated than western people had initially thought, and Crater Mountain is now a major proposed mining place waiting for a wide and deep landscape modification to come.

F. Huettmann, *Globalization and Papua New Guinea: Ancient Wilderness, Paradise, Introduced Terror and Hell*, https://doi.org/10.1007/978-3-031-20262-9_21

Keywords Papua New Guinea (PNG) · Conservation · Mining · Tribal warfare

21.1 Introduction

Crater Mountain is another unique and relatively well-described landscape in Papua New Guinea (PNG). It's located in the Eastern Highlands of PNG near Goroka; this region has received some major interruptions and disturbances, e.g. after finding gold near Goroka and when missionaries as well as 'conservation issues' entered that landscape. It now is considered a mining site for scheduled production ("*Crater Mountain is shaping up as PNG's next large scale, gold and copper discovery.*"; http://www.cratergold.com.au/irm/content/crater-mountain.aspx?RID=211). Hick ups occurr...

This plays a bigger role for PNG and beyond in regard to NGO-lead conservation work being affected by subsequent exploration and tribal infighting, as a certain role model to learn from (Hayan, 1990; Mack, 2014; West, 2006a, 2006b). In constant search for great study sites, e.g. for specimen collection and to describe new species for eager institutes and their directors and employees, many investigators and NGOs have since shifted away to other, easier and less disturbed places by industry but still pursuing same minds (see Diamond, 2011; Beehler & Laman, 2020).

Parallel with a photographic exploration in the 1970s onwards (Gillison, 2002) and with a long-term PhD thesis in Anthropology (Gilison, 1993), Crater Mountain reached a certain global fame for its pristine set up, research infrastructure, species collections, socio-economics and research outcome (Mack, 2014 for major description; see West, 2006a, 2006b for conservation and Beehler & Laman, 2020 for context and some photos; see GBIF.org for sparse data for the region despite such a major exploration, science and international effort). Birds of Paradise (BoP) and their conservation, as understood and promoted by the Bird Curator of the New York Zoological Society (Pearl, 1994) and others (Laman & Scholes, 2012; Mack, 2014), played a central role in the development of this area, but also in its subsequent problems (see report by Mary Pearl, 1994 in West, 2006a, 2006b). The planned building of a wilderness lodge there for high-paying clients—including scientists—wanting to see and collect BoPs (Beehler & Laman, 2020), all connected with helicopter transport, was not much rewarded by clients and eventually failed (Laman & Scholes, 2012). Certainly, local communities were affected as they provided much of the workforce required. It makes for a classic example of that 'development' template, e.g. using the eco-argument and its failures (West, 2006a, 2006b). The western world is not sustainable, certainly not high-level eco-tourism and when hacked out in downtown New York.

Like often found for research stations in wilderness regions, most people speak in high admiration about this area and its people. And it remained until several years later when prospective exploration and mining inflicted just another war on this 'paradise' (detailed descriptions by West, 2006a, 2006b; Mack, 2014 based on InterOil). It affected the landscape and its people who lived there 'a live', as it was initially appreciated and studied by Anthropologists and biologists alike as it was a treasure trough of inspiration, perceived 'harmony' and nature wilderness for decades

(Mack, 2014; West, 2006a, 2006b). To some people, it ranked like a 'Shangri La', while the research studied essentially the 'deck chairs' of the Titanic, specifically Birds of Paradise and cassowaries and the many specimen collections. Still, neither the ecology nor the conservation for those species collected is achieved or understood in Crater Mountain. Data are few. Instead, the industrial destruction—linked with the well-known socio-economic decays—now plays a big role affecting the landscape and being the dominant force.

One could also say, based on several visits and available research information—formal and informal—Crater Mountain is not short of tragedy. That is very true throughout its history (Gillison, 1993 for a compilation), but certainly for the role that outsiders, including NGOs, have had on the Crater Mountain area. As it is well described by West (2006a, 2006b) and Mack (2014). 'Crater' plays a certain—kind of textbook—exemplary role model for outsider impacts, for NGOs operating, in wilderness settings and how a local fabric can get destroyed easily with quite fatal consequences for everybody involved. International NGO funding remains a key in this scheme and pattern (see also Lawrence, 2008).

Goroka as the hub provides a hotspot for international NGOs in PNG. This raises some serious questions about conservation, and what it does to local communities and wider sustainability in nations, and in set ups like PNG, for biodiversity wilderness, and in remote sites (but compare for instance with Prajapati et al., 2020 for Annapurna Conservation Area in the Hindu Kush-Himalaya for an alternative).

It's quite a typical example and follows a certain script as mentioned by Macintyre and Foale (2004). While conservation has often tried to fight off mining, it actually failed, and it also failed in conserving people and landscapes alike nor did the mining industry really go away (compare with Ludlam 2021 for some success stories in Australia for same mining companies). So what was really gained in PNG? Conservation certainly presented PNG with a massive game changer of the lifestyle (West, 2006a, 2006b), that otherwise stayed stable for many thousands of years. Seeing its reality, the NGO-led conservation efforts have hardly helped Crater Mountain conservation, or its people!

21.2 Where and What is 'Crater Mountain'?

Crater Mountain is part of an Eastern Highland mountain range that is in the landscape located not that far from Goroka. It's connected with the coastal zones through ancient trails. One can get there by foot or bush plane from Goroka and crossing and walking in many rivers. The highlands landscape that Crater Mountain is part of overall must be seen as an incredible location and inspiration for global humanity and as a resource—spiritual and otherwise (Gillison, 1993). It's an area of humanity indeed, certainly an area that does not deserve to be messed up by anybody. If given a choice, everybody I know of would protect it, of course. But live took its course... (Figs. 21.1 and 21.2).

Fig. 21.1 A basic Papua New Guinea cabin in the dark might well be and remain "romantic"

21.3 Something Special About Crater Mountain?!

As part of the PNG highlands, Crater Mountain perhaps summarizes what this wider area all has to offer: biodiversity, forests, geology, people and their unique dialects (e.g. Gimi; see Gillison, 1993 for deeper legacy). On the first look, Crater Mountain was seen by a few insiders as a perfect experiment in nature where all comes together creating a paradise for all societies, ancient and modern alike. It's an internal human experiment, too. And the wider role of 'New York' cannot be denied for this part of very remote wilderness so exotic to the western eye, certainly from effluent urbanites and America (Details in Gillison, 1993; Pearl, 1994; Gillison, 2002; West, 2006a, 2006b; Mack, 2014). Perhaps it was seen as a study lab for those who want to learn more about humanity, wilderness and its synergy in times of the eco-conservation wave? Sounds all great, but in a way, it's done on cost of the locals and of biodiversity wilderness, and those paid truly the bill also (West, 2006a, 2006b).

21.4 Nature, Harmony and Win–Win at Crater Mountain?!

One might argue whether Crater Mountain is a paradise, or not (see Gillison, 1993 for lifestyle, inherent aggression and domestic abuse and industry; see also Wilson,

Fig. 21.2 Goroka, and its administration: a major hub for the Eastern Highlands and an important place in Papua New Guinea's modern history. Goroka has app. 19,000 inhabitants and is a 'lovechild' of western NGOs, including missionaries and mining operations. This road is located just opposite to a large airstrip connecting directly and c. 3 times a day with the capital of Pt Moresby.

2019). One might also argue whether Crater Mountain was relatively free of serious conflict among humans/tribes, or not. But clearly, tribal killings were described a lot, and the wider area around Goroka was tried to be pacified for over 100 years by western people, namely Australians (Nelson, 1982). It was part of their mission, while running a wicked mining Australian operation with an international workforce in parallel that benefitted them the most with a few to become millionaires (Nelson, 1982).

However, different human set ups resulted into some better lifetimes than others, a matter of context. Already bush plane helicopter traffic, supported by missionaries and mining, affects those areas quite a bit and is not benign. But at least Crater Mountain was free of the otherwise typical set of problems that are found in the western and industrialized world, e.g. no world war, no nuclear power and no serious financial pressures (as people lived a subsistence lifestyle). And very few roads. Let's celebrate that. The pressure of investment banks came later, parallel with the troubles (details in Mack 2014).

One may see Crater Mountain perhaps as a control site for humanity and its conflicts. It contrasts western life for a longer while, but got fully affected by it from the outside eventually: industry and NGOs alike leaving behind a '*mélange*' in its wake (West, 2006a, 2006b). Goroka also remains a coffee hub and an investment hotspot for missionaries, and so does the wider landscape overall, e.g. road system, airport and hydrodams. It's essentially an industrial production area in the PNG highlands now serving global interests. It's far from a standalone region living in a vacuum whatsoever, and now it gets affected by another influx wave of 'modern' people, beyond the gold miners, missionaries and coffee traders that are already there for long time. Crater Mountain never was really untouched wilderness or pristine last 300 years when colonialists entered the north coast and infiltrated the highlands of PNG. The global framework matters!

Of specific note is the initial concept for Crater Mountain, and it was meant to be a showcase in the first place where conservation can turn into a decent business, a win–win. That western idea came from the initial conservation NGO and straight out of New York and got written into stone (see P. West in Steward & Strathem, 2005 for those Crater Mountain details). But perhaps that was precisely the problem: A table-drawn win–win with foreign business plans from New York underneath 'using' native people in the bush remains dubious, or outright impossible and unethical. One will have to pay the bill.

21.5 Baseline Conflicts, and Intense Conflict, and Why It Broke Loose in Crater Mountain?

Life at Crater Mountain was certainly never really short of struggles and conflict (Gillison, 1993). But one may also easily argue that western influence—the outsiders from the west—brought a major imbalance to Crater Mountain and likely made it worse; it then further spilled into regional hubs and centers, certainly to Goroka and also Port Moresby and Manang (see Mack, 2014). They did it with their tools and approaches, namely western culture, the phone, camera, science and internet, funding, banks, loans and resource extraction plans: the mine and its anticipation. It's clearly described in Mack (2014) and by West (2006a, 2006b). Putting so much money in front of local communities, many not well-trained with such modern issues and problems will almost by definition change their sophisticated co-evolved fabric and thus harm and destroy them. It's a generic rule found in PNG (examples shown in Henson & Flower, 2007; Kirsch, 2014) and beyond (see Ross, 2015 for resource curse).

To nobody's fault, geological activity resulted for Papua New Guinea into resources of gold, copper and silver due to the stark uplift over time from the deep earth to the surface. Resources are then for the taker. Some oil and gas resources are sometimes associated with such a geology it also. The negative impacts of gold mining in PNG are well described in Cousteau and Richards (1999) and beyond. Human right violations are widely documented, certainly in PNG's mine sites, and those occupy the media and courts of this world, e.g. see The Guardian (2021). See Wilson (2019) for human changes described from a first-hand account for the highlands, e.g. due to gold.

And so, what had happened is that the mining exploration created 'wants' and 'needs.', broken dreams Western mining with a hunger for gold is something that is not really entrenched in the PNG society. For 97% of its time, the PNG society lived just well without a massive hunger and exploration for gold. In the long and vast PNG history, gold was never much accumulated and nothing really to fight or to kill for in PNG. But modern gold exploration—individualized wealth—created a steep rift among people and the community in a public land tenure. It created major mismatches and wrong expectations. It enabled a few tribal leaders to get wealthy and use it in their own way, unconstrained, unevenly and non-traditionally. That then created dynamics and demands, and it involved banks, and the police as well as the wider human fabric in the landscape, in Goroka, PNG and beyond (see Mack, 2014 for how it was handled and managed). It is a perfect mix for resulting violence and decay.

With that in mind, it's almost a side-fact now of history that "*Macmin Silver Limited, a subsidiary of New Guinea Gold, had signed a joint venture agreement with Celtic Minerals Limited, which gave Celtic a 75% interest in the Crater Mountain project*" (West, 2006a, 2006b) and so mining at Crater Mountain started the wider process and subsequently documented *havoc* there (Mack, 2014; West, 2006a, 2006b) and ongoing.

21.6 What is Crater Mountain Today

Considering its deep past, earlier (Gillison, 1993) and recent (Gillison, 2002), the good parts of Crater Mountain fell again off the map. After decade-long study, the cassowary is hardly better off now, nor are the Birds of Paradise (BoPs, Beehler & Laman, 2020) or most of the indigenous people of the area (Gillison, 2002). Summary work describes its new status (e.g. West, 2006a, 2006b; West & Kale, 2015) and the industrialized plans (Johnson, 2000 for a Wildlife Management Area WMA). But lacking the wider excitement and subsequent buy-in, not much new science is really done there, and the area lives in its ancient ways but now widely disturbed by recent events with an apparent trauma. Healing takes time while more industrial plans unfold. Crater Mountains do not hit much the mainstream media anymore or any news really. Ever heard about it? It's just part of the global collateral now. And PNG has many of those by now, many triggered by The West and through remote places like New York. It's now a designated Wildlife Management Area and listed with European Commission funds (https://econservation.jrc.ec.europa.eu/site/6085). Looks very good, on paper, does it not?

Perhaps Crater Mountain and its legacy can get over the trauma of the impact, starting 1970s from onwards? And that takes energy, time and skill…

21.7 Crater Mountain Tomorrow?

As Crater Mountain had a very long and a very sustainable history, literally for millennia, with massive tremors just the last two decades, the future remains unclear though. It clearly was a mess up for the last 15 years or so. As found worldwide, the sheer brutality of the western world, the so-called western civilization, NGOs and development entering other geographies, hit Crater Mountain very hard. Genocide was documented for all major leading and so-called civil nations, e.g. Russian purge in the Old World, Japanese atrocities in Korea and China, German Nazis, US and Canada atrocities on indigenous people, or with the Danish in Greenland and the Spanish and Portuguese in their colonies, e.g. Mozambique, Brazil. Crater Mountain experienced a soft version of it, but so many people got hurt, assaulted and many died.

Can that heal well, and how long does it take?

The new world, COVID, wider aspects of global change and new climate regimes will likely add more stress to the fabric; the wounds of Crater Mountain—the initial paradise, so-called, —are not left alone much. And once open, it can get worse; the track record of modernity and human civilization shows us no other (Diamond, 2011; Acemoglu and Robinson, 2013). One may easily think of Facebook and its infuration across parties and villages. Many of such examples exist worldwide. Whether PNG can take it, and can take more, as another Crater Mountain-many Crater Mountains can actually be found in PNG and in the world—is a good question, but likely it will, in a PNG way.

References

Acemoglu, D., & Robinson, J. A. (2013). Why nations fail: The origins of power, prosperity, and poverty paperback.

Beehler, B., & Laman, T. (2020). *New Guinea: Nature and Culture of Earth's Grandest Island*. Princeton University Press.

Cousteau, J. M., & Richards, M. (1989). *Cousteau's Papua new Guinea journey*. Abrahams Publishers New York.

Cousteau, J. M. & Richards, M. (1999) Cousteau's Papua New Guinea Journey. Henry Abraham's Books.

Diamond, J. (2011). In *Collapse: How societies choose to fail or succeed*. Revised Edition. Penguin Books.

Hayan, D. M. (1990). *Road through the rain forrest: Living anthropology in highland Papua New Guina*. Waveland Press.

Henton, D. & Flower, A. (2007). Mount Kare Gold Rush: Papua New Guinea, 1988–1994. Published by Mt Kare Gold Rush, Brisbane Australia. ISBN 10: 0646482815; ISBN 13: 9780646482811.

Gillison, G. (1993). *Between culture and fantasy: A new Guinea highlands mythology*. University of Chicago Press.

Gillison, D. (2002). *New Guinea ceremonies*. Harry N Abrams Inc.

Johnson, A. H. (2000). Monitoring and evaluation of an enterprise-based strategy for wildlife conservation in the Crater Mountain wildlife management area (Papua New Guinea) (Order No. 9996895). Available from ProQuest Dissertations & Theses Global; ProQuest One Business. (304637002). Retrieved from http://uaf.idm.oclc.org/login

Kirsch, S. (2014). *Mining Capitalism*. University of California Press.

Laman, T., & Scholes, E. (2012). *Birds of paradise: Revealing the world's most extraordinary birds*. National Geographic Books.

Lawrence, R. (2008). NGO campaigns and banks: Constituting risk and uncertainty. In *Hidden hands in the market: Ethnographies of fair trade, ethical consumption, and corporate social responsibility*. Emerald Group Publishing Limited.

Ludlam, S. (2021). *Full Circle: Power, Hope and the Return of Nature.* Black Inc.

Mack, A. (2014). *Searching for Pekpek: Cassowaries and conservation in the New Guinea rainforest.* Cassowary Conservation and Publishing.

Mack, A. L., & Wright, D. D. (1996). Notes on occurrence and feeding of birds at Crater Mountain biological research station. *Papua New Guinea Emu, 96*(2), 89–101.

Macintyre, M., & Foale, S. (2004). Politicized ecology: Local responses to mining in Papua New Guinea. *Oceania, 74*(3), 231–251.

Nelson, H. (1982). Taim Bilong Masta, The Australian involvement in Papua New Guinea Australian Broadcasting Commission. ISBN 9780642975669, 0642975663. Google Books https://www.google.com/books/edition/Taim_Bilong_Masta/y2MhAAAAMAAJ?hl=en

Prajapati, J., Huettmann, F., Ghale, T. R., & Regmi, G. R. (2020). Chapter 24. The Annapurna conservation area project (ACAP): Towards a success story in landscape feature and watershed conservation management. In G. R. Regmi & F. Huettmann (Eds.), *Hindu Kush-Himalaya watersheds downhill: Landscape ecology and conservation perspectives* (pp. 473–496). Springer Gland.

Pearl, M. C. (1994). Chapter 8 local initiatives and the rewards for biodiversity conservation: Crater mountain wildlife management area, Papua New Guinea. In D. Western, M. Wright & S. Strum (Eds.), *Natural connections: Perspectives in community-based conservation* (pp. 193–214).

Ross, M. L. (2015). What have we learned about the resource curse? *Annual Review of Political Science, 18*, 239–259.

Stewart, P. J., & Strathern, A. (Eds.). (2005). Anthropology and consultancy: Issues and debates (Vol. 1). Berghahn Books. https://www.google.com/books/edition/Anthropology_and_Consultancy/1AGt1K5BNP0C?hl=en&gbpv=1&dq=crater+mountain+photographer+papua+new+guinea&pg=PA65&printsec=frontcover

The Guardian. (2021). Villagers file human rights complaint over plan for giant PNG goldmine. 9th December 2021. https://www.theguardian.com/world/2021/dec/10/villagers-file-human-rights-complaint-over-plan-for-giant-png-goldmine. Accessed May 15, 2021.

West, P. (2006a). *Conservation is our government now: The politics of ecology in Papua New Guinea.* Duke University Press.

West, P. (2006b). Environmental conservation and mining: Between experience and expectation in the Eastern Highlands of Papua New Guinea. The Contemporary Pacific 18. https://google.com/ps/i.do?p=AONE&u=googlescholar&id=GALE|A148482167&v=2.1&it=r&sid=googleScholar&asid=901e69b7

West, P., & Kale, E. (2015). The fate of crater mountain: Forest conservation in the eastern highlands of Papua New Guinea. In J. Bell, P. West & C. Filer (Eds.), *Tropical forests of Oceania: Anthropological perspective's. Asia-Pacific Environment Monograph* (Vol. 10, pp. 155–178).

Wilson, M. (2019). *A True Child of Papua New Guinea: Memoir of a Life in Two Worlds.* McFarland.

Woodland Park Zoo. (2010). Tree Kangaroo Conservation Report. Annual Report. YUS report https://www.zoo.org/document.doc?id=1724. Accessed December 10, 2021.

Chapter 22
"*Pacifism as a Western Pathology*"? A Case Review of Bougainville Terror, Civil War and Independence, Civil Unrest, Missing Mining Impact Studies and Public Data, and Why the Nation Construct of Modern Papua New Guinea is Unsustainable and Could Fall Apart Soon Due Its Colonial Creators

The civil rights movement happened because there was civil disobedience,
U.S. President B. Obama's ABC News Interview Transcript on Praising Missouri Protesters, with G. Stephanopoulos) (Source: FIRE November 15, 2015. www.thefire.org/obamas-abc-news-interview-transcript-on-missouri-protesters/(accessed 3rd July 2022))
It seems clear that the mine has a profound effect on the freshwater and marine environment below it…Its damages are borne entirely by local people and the environment that has long sustained them.
Cousteau and Richards (1999, p. 122).
'The European Shareholders of Bougainville Copper' Bougainville Copper Ltd shall become a role model for forward-looking, advanced and environmentally harmless mining in the developing world, a project to the benefit of a whole nation. […].
Well meaning, right-minded, and unexpected. But they have a curious, faintly colonial motive, it seems, when they say, "It is not only limited to the fact that Bougainville was a German colony back in the nineteenth century why in particular the German members of the group of The European Shareholders of Bougainville Copper want to stress their active support for the people of Bougainville." (http://www.bougainville-copper.eu/pageID_5753635.html).
Sullivan (2015)

Abstract In society, there is a general understanding that 'nobody gets hurt' and all works smoothly and peaceful; a harmony. This also applies to the workplace, corporations and business abroad. However, reality is quite far from it. Most nations have extracted natural resources with a conflict; civil war is not unheard of in the so-called civilized world or outside. Bougainville in Papua New Guinea (PNG) is part of the Solomon Island group and shares its ecological and colonial history. It was - in part - a German colony and got affected by all global events during the last 300 years,

F. Huettmann, *Globalization and Papua New Guinea: Ancient Wilderness, Paradise, Introduced Terror and Hell*, https://doi.org/10.1007/978-3-031-20262-9_22

including a very strong Australian involvement last 100 years, now including China. Like PNG, the Solomon Islands are widely referred to as 'failed states,' but mining operations made it worse. The Bougain Island conflict reached world headlines when hired UK mercenary contractors—Sandline International—added to the war-like dispute in PNG and threatening the entire nation fabric, including the Port Moresby capitol. It's clear to any scholar of Australian, North American and world (conservation) history that being peaceful does not always obtain wanted objectives and that a new, effective and safe approach is to be found and used to achieve a balanced wider peace and justice for humankind and Mother Earth.

Keywords Papua New Guinea (PNG) · Bougainville · Independence · Federalism · Civil war

22.1 Introduction

Bougainville is an island and an inherent part of Papua New Guinea (PNG), the modern PNG nation, and of Melanesia. There is a lot of colonial and outside history in that region (see for instance impacts brought by forced labor, Fitzpatrick, 1980), and Melanesians have tried to resist that pressure in various ways (see Hyndman, 1998 for ecocide and genocide). Many of the Bougainville Island inhabitants in PNG closely associate themselves with the Solomon Islands (app 7 km away) instead of mainland PNG (app. 500 km away). In my discussions in PNG, many people of Bougainville actually feel little connected with the nation construct of PNG, with PNG mainland or with Port Moresby. And that's even if the annual 'PNG Day' celebration in fall might indicate otherwise. On the matter of nation, and like stated for Solomon Islands, PNG as a national governance might easily be seen as dysfunct (Wainwright, 2004; see also Filer, 1990; Riley, 2004; see for instance Kabutaulaka, 2004 for Solomon Islands). PNG's leading prime minister at the time, Julius Chan and an islander himself, has described those details in his biography in great length (Chan, 2016).

As shown in Wainwright (2004), 'Bougainville' is actually a group of islands, arguably part of the wider Solomon Island group. It certainly features the same culture, heritage and legacy than the nation of Solomon Islands (for details and history, see http://speedysnail.com/pacific/bougainville.html). It was initially claimed by colonial powers, became British, turned over via Spain as a German colony, then changed to Australia, invaded by Japan and 'freed' by the allies (part of operation Cartwheel) and then assigned to PNG (The Guardian, 2019a for public summary and status). (Bougainville was also part of The Republic of Northern Solomons but it had no status and fell apart after 6 months; however, it sets a precedence).

The deep culture of Bougainville remains unfathomed and an ongoing topic of research (e.g. Burton, 2012; Sheppard & Walter, 2006[compare with Naess 1988 for a recent western concept but rarely applied. Bougainville was settled for many thousand of years and had contact with several waves of human cultures from Asia. But after the PNG assimilation, in recent times Bougainville made primarily the news for its reported terror and the violence, a nine-year-long civil war-like state (The Guardian, 2019a, 2022a).

Such style, a certain island war of independence—a secession fight— and a 'Pacific Crofters's war' (MacIntosh 1990) is though a complete '*no no*' in the mind of western world nations, certainly in Australia, UK and associated funders who seem to perceive it as an affront, culminating in mindsets like '*we did nothing wrong in Bougainville*' or '*we are the good outpost, the beacon of western civilization*' (more details in Henton & Flower, 2007; Lasslett, 2012; Woodbury, 2014; Abbondanza, 2021; Underhill, 2021; see Heinecke et al., 2008 for Australia's Samatarian dilemma & Kolova, 2015 for an open critique of associated Australian aid).

Whereas the Bougainville conflict is a people's war (MacIntosh, 1990), the western world and its govrnance widely condemns such 'uncivil' actions involving to question authority or to fight and to displace them, e.g. Mietzner and Farrelly (2013) or Churchill and Ryan (2017). This is surprising when knowing the French Revolution (= the foundation of modern France and democracy) or the American fights for independence resulting into the 'free world' (= essentially an act of terror by a militia with physical violence and destruction against the empire; e.g. Boston tea party. Examples in Table 22.1).

But as most indigenous societies can share, as part of subsequent struggle and warfare when contacted by the western world (Hyndman, 1998), it's the 'western model' (imperialism) that came with force upon them, usually unwanted (see for James Cook and Christoph Columbus in Diamond, 2011a,b; see Berzunza-Sanchez et al., 2013 for ecological coral reef changes since then in PNG which represents an

Table 22.1 A selection of famous revolutions and Civil War conflicts

Conflict	Act of terror	Outcome	Comment
India	Riots, hunger strike	Independence	A hunger strike can be a rather brutal act and has obviously vast impacts
South Africa	Bombing	End of Apartheid	Nelson Mandela was listed by the U.S. as a terrorist byt then
North Ireland	Bombing	Peace Accord	Likely a certain win for North Ireland and Ireland overall in regards to British Royal dependence
Basque	Bombing, kid-napping	Remains part of Spain, joining the EU	Unclear outcome and win, but stated recognition of the problem and more independence within Spain
Timor-Leste	Civil war	Better living conditions	Ongoing and a struggle

essential change of life caused by the environmental change due to an abusive global framework). Many contracts got actually broken by the western world (see examples by Venne, 2007 for North America). And as it is well documented and known (Flannery, 2002; Diamond, 2011a,b), it was eventually the indigenous societies who got raped and killed, easily in the thousands and millions, and who died by the diseases (famous arguments presented by Diamond, 2011a,b). It was not the western ones who got affected much and who just came to take the land (but not to give land or resources in exchange).[1]

So the wide-spread narrative about peace, good obedience, following law and order, the nation state, world harmony, and subsequent well-earned prosperity—full employment—provided by colonial powers and Australia and its supporters—including the mining industry, missionaries or earlier Australian Peace Controls and village courts in PNG (Dinnen, 2000)—might have to be adjusted to better include the intimidation, terror, massacre, bias and ongoing impacts (e.g. Henton & Flower, 2007; see The Guardian, 2022b for not stating reality; see Flannery, 1998, 2002 for details on the ground).

The western world is far from being peaceful or benign itself, e.g. Kavanagh (2022) for Ireland and the British Empire. In the western world, killings are on the order of the day and an integral part of their history (see Bergen & Rothenberg, 2014 for Drones). Already the Australian and western recent history itself shows it clearly to this very day: Fighting and war are the mother of the nationhood and widely celebrated, remembered by governance support and rewarded, e.g. Mackenzie (1987); Grant (2013), see for strategic policy which is not short of tragedy or death (Ayson, 2007; Connell, 2006; Wainwright, 2003). PNG impacts of war brought by the outside world are widely shown in Cousteau and Richards (1999).

Why does it matter?

It's because when the mine project of Bougainville started—fully promoted by Australia from the start and till today (The Guardian, 2022a)—it initiated the open conflict there and a looming and intense PNG secession dispute which resulted into a civil war-like terror and many deaths. It certainly had strong environmental impacts, e.g. Cousteau and Richards (1990) and had wide repercussions in PNG, in the capitol of POM, in Australia and beyond (Chan, 2016 for details): Mercenaries from abroad—Sandline International—were brought in to control the situation, and it had direct conflicts in Port Moresby even resulting into an international event and subsequent global outcry (more context and details explained by Chang, 2016 who made many decisions and was involved first hand). There was no other terror of this kind and magnitude before in PNG or after; it's truly inflicted by western nations, Australia and its mining industry included. It's part of Australia's Arc of Instability (May, 2006).

---

[1] Mayr and Diamond (2001) show some interesting statistics on western beachcombers and escapees living on Solomon Island beaches 'on their own'; many of them died or got killed in that process.

The international Rio Tinto corporation was involved in the Bougainville mine early on (but walked away eventually https://news.mongabay.com/2017/04/rio-tinto-walks-away-from-environmental-responsibility-for-bougainvilles-panguna-mine/), and so was the PNG Central Bank (Baraka, 2001). Thus, Bougainville and its resources and money are part of the initial fabric of the PNG nation construct, as designed and pushed for by Australia, all approved with a UN mandate and started with Prime Minister Michael Somares' lead. In that process 1970s onwards, the people of PNG were indifferent and/or not really pushing for independence itself, much. Instead, it was primarily Australia that moved PNG towards it. Thus, in PNG there was little struggle or fighting for their nation, until the Bougainville conflict (compare that with Timor-Leste for instance, The Guardian, 2019b). The actual PNG nationhood came essentially without bloodshed.

Well, there was wider 'terror' in PNG, and that is, the initial PNG nation creation imposed on people against their will, hardly with their knowledge or with agreed terms, consent (Baraka, 2001 for many highlanders being neutral to opposed to an independent PNG). Also, much terror sat with the colonial powers as far back as to James Cooke and then with numerous others (see Fitzpatrick, 1980; Mayr & Diamond, 2001 for recruiting workers, slavery), including the Australian gold miners and explorers that started the western run into PNG and its islands and mining wealth, including Australia's Bush Control patrols Similar can today be said for seafloor mining projects. As an effect, all of it can be seen in Bougainville first hand. The widely upheld argument from Australia that mining also 'pacifies' the wild tribes and puts down tribal conflict—adding to law and order, and a well-to-do nationhood—can likely be put to rest here; it had the opposite effect, and many people died and suffered in that proces ongoing from colnialisms; a civil war unfolded instead. So the question then is: 'Mining Melansia' (Kirsch, 2014) for what and for whom? Hyland (1998) provides the answer referring to ecocide and genocide (as shown in Diamond, 2011a also).

As the history of mining Bougainville clearly shows: Australia's governance stated and insisted that 'Bougainville' would get nothing in that deal. And it also used aspects of foul play—accusations of internal industrial sabotage remain upheld—and it ended in lawsuits and subsequent conflicts.

So how could it come that economic growth, essentially subsidized mining, fueled by the outside/Australia and in the wider British Commonwealth with their lust for money and control abroad—the U.S. oversight in the Pacific Theater—could have such global repercussions? Well, promotion of economic growth and underlying mining has that effect virtually everywhere else too (see Eichstaedt, 2011 for mining the Congo; see Buckley, 2020 for Tibetan watersheds). And imposing a western, industrial model achieves usually the same. The western world remains imperialistic, and it wants to make and use money from elsewhere as an explicit aim and policy structure. Their own business model is insufficient to cover a social welfare model for all but it got promised. And that's what makes people upset. In the meantime, The West likes to celebrate the narrative and concept of it being a peaceful, pacifist and egalitarian society, and thus any conflict then gets silenced, coined as 'pathetic.' But that narrative itself can easily be taken to rest as a myth with many examples showing completely otherwise (e.g. Churchill & Ryan, 2017). China, as a nation active in Asia and PNG, that promotes 'harmony' is widely described showing the same flaws, e.g. documented lack of freedome of speech, and suppression of Tibet and many minorities.

22.2 A Quick Recall of a Recently Independent Bougainville Island

By 1976, Bougainville was added as a province to PNG within the construct of that nation in 1975. It remained a remote and relatively peaceful island till the 1980s (Golup, 2014) when the Paguna mining site started and conflicts became obvious mid-1980s onwards (see MacIntosh, 1990 for an ecocide). Afterward, conflicts started, and from 1989 onwards, it left the island in a nine-year-long war-like situation with environmental spoilage and human loss (Cousteau & Richardson, 1996). Bougainville is part of the PNG construct, and the question of a split, separation, independence and/or joining Solomon Islands was not on the PNG agenda but got promoted onwards once mining money hit the set up (Chan, 2016). PNG stays strong as a unit and it operates that way. Bougainville voted for independence, and it became an independent territory 2002. From a PNG perspective, a Bougainville split is not for discussion and never truly was, but with concessions made. Arguably, obtaining some funds toward a better supported autonomy is part of that discussion and of the current solution, e.g. how the Bougainville mine and its income is distributed, compared to PNG overall (see The Guardian, 2022a on reopening the mine for that purpose).

The question from there is whether Bougainville is joining Solomon Islands, whether it is a fully independent nation, or whether it remains an Autonomous Region and what that all means in the real world? Conceptually, Bougainville can walk through the entire PNG nation-building experience itself, just as a small version of it. In reality though, an independent Bougainville will have a tough road ahead to be a global actor in an international marketplace and how to remain viable, with a set of problems just like PNG faced them (Gosarevski et al., 2019). How small can a nation be to be viable and meaningful (see Gosarevsky et al., 2019)? But in that discussion, it should not be forgotten that Bougainville is not just a 'land' discussion but that the ocean, the global fabric and the "Pacific Theater' really matter. With so much global unrest ongoing, Bougainville plays a relevant role (Fig. 22.1).

22.3 A Closer Look at the PNG Fabric At-Large: A Global Showcase in Local Complexity for Governance of Land

PNG was a British and German colony, and its subsequent brutal path is widely exposed in public (see The Guardian, 2019). MacIntosh (1990) referred to the Bougainville dispute and its terror as a Pacific Crofter's War. The fierceness of the

Fig. 22.1 Bougainville map scheme (see arrow) and its fit within the Solomon Islands (large box) and PNG (smaller box; blue line indicates the Exclusive Economic Zone (EEZ) of Papua New Guinea)

Bougainville conflict might sit in the land rights of the PNG region and affected by mining (Mining Watch Canada 2022). Baraka (2001, pp. 8/9) states:

> *As landholders, they feel secure when they are still in possession and in control of their land. But when they realize that control has been relinquished, or the returns are not what they expected, or the secondary effects of these development projects disturbs their livelihood (e.g., the pollution of vital streams and rivers caused by mining) they find their basic security threatened. Subsequent resistance to such state sanctioned initiatives is often fierce.*

PNG is obviously a complex governance system with a diverse land fabric and mosaic of people, history, cultures and attitudes. With over 700 languages, perhaps it is one of the most diverse places in the world? PNG is a country of many nations (Baraka, 2001). If it can be done in PNG it should be possible anywhere in the world. PNG lived by itself fine for 47,000 years. As a 'unique' nation—non-feudal-it was essentially overpowered from the outside, created way too late by colonial powers for independence, and it was essentially let go to Australia and now cannot break out of such mindsets and geography, much. The PNG nation was designed actually after the fact, it was an afterthought and a bad compromise made, and relevant details were given up, by Australia and colonial powers, but with leashes attached, e.g. via development aid, policing help and education and training of its leaders and leading 'talents.' (Northern) PNG still shows some Polynesian influence, a culture that came to the southeast Pacific by app. the twelfth century (Diamond, 2011a). The Asian link is also obvious, specifically in the islands, such as in the Solomon culture. And in earnest, PNG is somewhat a divided nation to start out with, by birth. PNG is really diverse, and thus, it defies the principle of top-down and homogeneity. PNG means to discuss and debate. This sets the stage for the perceived western 'failure of the nation construct' (Kabutaulaka, 2004; Reilly, 2004; see also Otter, 2002; Mietzner & Farrelly, 2013; Gusarevsky et al., 2015).

PNG itself is just not really top-down, and it lacks a feudal and royal court. PNG is bottom-up and the Anti-West!

22.4 A Closer Look at the Bougainville Conflict

The online version of Encylopedia Britannica states essentially that Australia had established the mine in Bougainville in order to lower the new state's dependence on foreign aid (https://www.britannica.com/place/Papua-New-Guinea/The-colonial-period). Done on a finite space and with an ancient local fabric of society and biodiversity (Cousteau and Richards, 1990), the Bougainville conflict in the 1980s turned bloody, deadly and ugly (MacIntosh, 1990; Golup, 2014). It looked awful from the outside and was very tragic and bad at the inside (Chan, 2016), having long-lasting shattering impacts to this very day. Table 22.2 shows some summary metrics. No western nation would accept such a reality in their own borders. Details are widely covered and described by Chan (2016) as a first-hand account and also covered by Filer (1990), Henton and Flower (2007), The Guardian (2019); public knowledge shown here (https://en.wikipedia.org/wiki/Bougainville_conflict). Based on Australian intervention (Lasslet, 2012), it dragged down the entire PNG fabric till now (The Guardian, 2019, 2022b). But Bougainvillers did not cause it really. It came from the outside and was fueled there. Some milestones of that conflict can be described the following way:

- till 1740 ~ precontact, indigenous world
- 1770 Capt James Cook contact and 'discovery'
- 1899 German Colony
- 1914 WW1 Australia claim to the island
- 1939 WW2; Japanese invasion 1942 and allied troops 'freed' the island 1945[2]
- 1964 Australian geologists discover copper and gold; one of the largest reservoirs in the world
- 1975 PNG independence
- 1975 Australian mine starts; 2nd biggest income for PNG nation
- 1989 serious conflict and terror starts
- 1990–1991 PNG imposed an embargo on media and aid on the island
- 1997 Sandline affair
- 1998 Peace treaty signed in New Zealand.

Various sources emphasized that during the Bougainville conflict females and their suffering played a major role, e.g. https://ramumine.wordpress.com/2011/10/18/women-call-the-shots-on-panguna-mega-copper-mine/.

[2] For indigenous perspectives of the war and terror, please see Kwai (2017)

Table 22.2 Some metrics in regards to the Bougainville conflict

Metric	Detail	Comment
Value of Panguna mine	c. $60 billion	A theoretical number
People that died in conflict	15,000 or more	Some estimates are as high as 50,0000 people; this includes civilians and soldiers of PNG, Australia, Fiji, etc
Length of civil war	c. 9 years	

Overall, Bougainville follows the classic model of colonial impacts and decay: It was a colonial hand-me down - German, than Solomon Islands got British - but Bougainville actually remained German till WW1. Before WW2, it became Japanese for three years till it returned during WW2 to Australia and with a UN oversight till independence. Bougainville makes the tragic poster child for mining gone awful, as a template for local community disruption (Macintyre & Foale, (2004)).

But it's the landscape and biodiversity angle that I am discussing here the most, and it's linked with the struggle for land, the island and the nation overall. That's the point I am trying to make here, and all facts are in support.

22.5 Role of the Bougainville Environment

Arguably, the environment of Bougainville is the playground and ultimate reason for the Bougainville conflict inflicted by mining. The environment and the associated land are the life-support system, the income (Baraka, 2001). Bougainville Island consist of hills, rainforests, palm plantations, beaches, reefs and the ocean system with islands nearby. The reason for the conflict lays in the geology, the island set up, the ocean and readily available resources on the surface, with a proximity to other islands, Australia and Asia. The geology brings deep-earth resources readily to the surface, and mining easily affects the tribal land ownerships, the river systems and the coastal zones, including the island-mainland dynamics. Mining cannot be benign in such a setting. The specific handling how mining interferes was widely ignored or at least done poorly.

As Bougainville is an island region, many of the dynamics then come predetermined. Bougainville is beautiful but has many deadly aspects to it.

Like elsewhere in the world, the forests are refuges and those harbor many of the rioters and it becomes a reservoir and frontline. Remote and inaccessible forests play a role in this conflict.

22.6 The Mine from Australia to Bring Bougainville and PNG Overall to Explosion?

In just its brief history as a nation locked-in within the international community, the GDP of PNG consists mostly of mining, oil and gas, fisheries and timber export.

That means, PNG stands for hardcore mining and resource extraction—done by white people from abroad—just as it is promoted and well greenwashed by Australia, U.K. funders, Asia and banking investment. Latest approaches to seafloor mining in the region add fuel to the fire (The Guardian, 2022).

But as discussed by Lasset (2012), Australia remains widely liable for the conflict and terror (The Guardian, 2019, 2022a).

And so, Bougainville was proposed to receive a mine site (mineral deposits of copper and gold).

The mine at Paguna—one the largest in the world once more—was opened in 1972, and it was majority-owned by Rio Tinto. The mine was vitally important to the economy of Papua New Guinea as a national jump-start but it is one of the largest man-made holes on Earth, and it totally spoiled the social fabric of the island, tribes and beyond (Golup, 2014). How could it not be? The PNG national government received an app. 20% share of profit from the mine, of which the Bougainvilleans received app. 5–1.25% share of the total profit. And here sits part of the inherent conflict.

22.7 Where Are Valid Impact Statements, Studies, Data and Policies?

As mining gets pushed, a reflective and consented approach to the problem, with impact studies that are meaningful and open access data does not really exist (https://miningwatch.ca/news/2018/9/12/bougainville-mine-or-not-mine). Any mine needs an industrial impact assessment; it's part of a legal requirement and certainly best professional practice. The mine on Bougainville is a major example for it: if not done here well then where else?

Cousteau and Richards (1999, pp. 119–123) show clearly water pollution due to the mine, e.g. Jaba River. There are some environmental impact assessments but the pollution has not stopped.

As this conflict turned into a civil war affair, with outside troops sent in, the impacts of the mine got obviously ignored, assessed wrongly and were not realistically phrased. Where is the information and the data (see for instance GBIF.org as a mandatory data repository for many nations, including Australia and South Africa [as core nations of origin of relevant gold mining companies])?

22.8 Pacificism as a Pathology?

Ideally, one perhaps wants to settle conflicts in peace, by peaceful means and with diplomacy and expertise; no harm done but just a good change. It's a rel. quick and easy concept, and everybody can go on from there. The western world hangs on to that narrative. But if the conflict sits very deep and large, involving many people on a platform exposed nationally, or even globally, then this concept is tricky to achieve, if at all. As a matter of fact, pacifism sounds like a great means, e.g. attributed to Mahatma Gandhi, Nelson Mandela and others. But the reality shows us a somewhat different picture (e.g. Churchill & Ryan, 2017). Terror played a role in many civil unrests and changes. Mandela (Olesen 2015) was indeed classified as a terrorist and achieved a change, and Gandhi participated in warfare, e.g. South Africa (both cases involve the British Colonial Powers and their concepts btw; e.g. Gupte 2012). One may argue that a hunger strike -as used by Gandhi - is an act of terror, as most governments would probably label it, and they really do not like any disagreement, or public expression of disconsent. Nevertheless, the change often comes through such actions, and it is even celebrated that way, e.g. in the violent 'Boston Tea Party' in the U.S. starting independence from the British.

There can be no doubt that pacifism remains the tool of choice for most objections. It's powerful because one can easily become a victim in good standing, simply by saying 'No' to an event (see Ludlam, 2021). But is it effective or the only way ?

Many revolutions were not really achieved by pacifist means only. Violence almost always occurred (Table 22.1), and on either side. No sides really truly played it benign or were non-violent. The heroic part was usually added by the winner, who wrote the history!

Beyond Bougainville, PNG has uprisings and some are violent, for instance in the southern highlands of PNG or Milne Bay (a former Australian claim and affected by Australia to this very day), related to earthquakes, mining, oil, elections and other problems (https://www.cnn.com/2018/06/18/asia/papua-new-guinea-riots-intl). This area—as a former Australian protectorate—remains one of the poorest areas in the world.

22.9 Bougainville: Plusses, Minuses, the Environment and the Future (Climate Change)

Many people died in Bougainville, and many more had a bad life and lots of fear and stress. Whose fault is it, and can it be avoided?

There are pluses: Bougainville got recognition and more independence.

There are minuses: Bougainville conflict was painful for all sides involved; no real winners can be named.

- *Today, Bougainville—an autonomous region in Papua New Guinea—is devastated by damages wrought by both war and the mine.*
- *The Bougainville government is considering re-opening the mine in order to fund a cleanup*

The future? Bougainville moves closer to independence, and PNG looses some standing and islands. But then, Bougainville is not sustainable by itself, and the people might loose even more. Australia still has plans for Bougainville and remains a core actor (Underhill 2022).

As stated in Beehler and Laman (2020, p. 353) "*Large-scale resource development in Eastern New Guinea never happens without fight.*" And this will likely get acerbated by climate change. Now what a future is that and was it all worth it going a modern way?

References

Abbondanza, G. (2021). Australia the 'good international citizen'? The limits of a traditional middle power. *Australian Journal of International Affairs, 75*, 178–196. https://doi.org/10.1080/10357718.2020.1831436

Ayson, R. (2007). The 'arc of instability' and Australia's strategic policy. *Australian Journal of International Affairs, 61*(2), 215–231.

Baraka, R. (2001). The meanings of independence in Post-Colonial Papua New Guinea. A Plan B research paper submitted to the faculty of the Center for Pacific Islands Studies in the candidacy for the Degree of Masters of Arts. School of Hawaii, Asia and Pacific Studies. The University of Hawaii at Manoa.

Bergen, P. L., & Rothenberg, D. (Eds.). (2014). *Drone wars: Transforming conflict, law, and policy*. Cambridge University Press.

Berzunza-Sanchez, M. M., Cabrera, C. G., & Pandolfi, J. M. (2013). Historical patterns of resource exploitation and the status of Papua New Guinea coral reefs. *Pacific Science, 67*(3), 425–440. https://doi.org/10.2984/67.3.9

Buckley, M. (2020). A first-hand narrative account on Tibet's Paper Parks: How China's Greenwashing in Tibet Flies under the Radar. In G. Regmi, & F. Huettmann (Eds.), *Hindu Kush-Himalaya Watersheds Downhill: Landscape ecology and conservation perspectives* (pp. 197–212).

Burton, L. J. (2012). Building on living traditions: Early childhood education and culture in Solomon Islands. *Current Issues in Comparative Education, 15*(1), 157–175.

Chan, J. (2016). *Playing the game: Life and politics in Papua New Guinea*. University of Queensland Press.

Churchill, W., & Ryan, M. (2017). *Pacifism as pathology: Reflections on the role of armed struggle in North America*. PM Press.

Connell, J. (2006). 'Saving the solomons': A New Geopolitics in the 'Arc of Instability'? *Geographical Research, 44*, 111–122.

Dinnen, S. (2000). *Law and order in a weak state: Crime and politics in Papua New Guinea*. University of Hawaii Press.

Eichstaedt, P. (2011). *Consuming the Congo: War and conflict minerals in the world's deadliest place*. Chicago Review Press.

Filer, C. (1990). The Bougainville rebellion, the mining industry and the process of social disintegration in Papua New Guinea. *Canberra Anthropology, 13*, 1–39.

Fitzpatrick, P. (1980). Really rather like Slavery: Law and labour in the colonial economy in Papua New Guinea. *Crime, Law and Social Change, 4*(1), 77–82.

Golub, A. (2014). *Leviathans at the gold mine: Creating indigenous and corporate actors in Papua New Guinea*. Duke University Press.

Grant, L. (2013). Operations in the Markham and Ramu Valleys. In *Australia 1943: The Liberation of New Guinea*, p. 233.

Gupte, P. (2012). *Mother India: A political biography of Indira Gandhi*. Penguin Books India.

Heinecke, D., Dollery, B., & Fleming, E. (2008). The Samaritan's dilemma: The effectiveness of Australian foreign aid to Papua New Guinea. *Australian Journal of International Affairs, 62*, 53–71. https://doi.org/10.1080/10357710701842686

Henton, D., & Flower, A. (2007). Mount Kare Gold Rush: Papua New Guinea, 1988–1994. Published by Mt Kare Gold Rush, Brisbane Australia. ISBN 10: 0646482815, ISBN 13: 9780646482811

Hyndman, D. (1998). Melanesian resistance to ecocide and ethnocide: Transnational mining projects and the fourth world on the island of New Guinea. In H. J. Bodley (Ed.), *Tribal peoples and development and issues: A global overview* (pp. 281–298). Mayfield Publishing Company.

Kabutaulaka, T.T. (2004). "Failed State" and the War on Terror: Intervention in Solomon Islands. Asia Pacific Issues 72: https://scholarspace.manoa.hawaii.edu/handle/10125/3798. Accessed February 5, 2022.

Kavanagh, J. (2022). *The Irish Assassins: Conspiracy, revenge and the murders that stunned an empire.* Grove Press.

Kirsch, S. (2014). *Mining capitalism.* University of California Press.

Kolova, S. (2015). Critique of Australian aid to Bougainville. *Contemporary PNG Studies, 23*, 83–92.

Kwai, A. A. (2017). *Solomon Islanders in World War II: An indigenous perspective.* Australian National University Press.

Lasslett, K. (2012). State crime by proxy: Australia and the Bougainville conflict. *British Journal of Criminology, 52*, 705–723.

Mackenzie, S. S. (1987). The Australians at Rabaul. The capture and administration of the German possessions in the Southern Pacific, Official History of Australia in the War of 1914–1918 (Vol. 10, p. 412). University of Queensland Press. Angus and Robertson Ltd., Sydney, Australia (1941).

Macintyre, M., & Foale, S. (2004). Politicized ecology: Local responses to mining in Papua New Guinea. *Oceania, 74*, 231–251.

May, R. J. (2006). Papua New Guinea: Disorderly democracy or dysfunctional state? In *Australia's Arc of Instability* (pp. 151–170). Springer.

McIntosh, A. (1990). Origins of the Bougainville Crisis PNG: The Bougainville Crisis, A South Pacific Crofters' War. https://www.alastairmcintosh.com/articles/1990_bougainville.htm. Accessed August 5, 2022.

Mietzner, M., & Farrelly, N. (2013). Mutinies, coups and military interventionism: Papua New Guinea and South-East Asia in comparison. *Australian Journal of International Affairs., 37*, 342–356. https://doi.org/10.1080/10357718.2013.788128

Mining Watch Canada. (2022). Bougainville: To mine or not to mine? (Source: Jubilee Australia) (September 12, 2018) https://miningwatch.ca/news/2018/9/12/bougainville-mine-or-not-mine. Accessed May 15, 2022.

Naess, A. (1988). Deep ecology and ultimate premises. *Ecologist, 18*, 128–131.

Otter, M. (2002). Development planning for a divided society in a failed state: The case of Solomon Islands. *Dev Bull, 60*, 46–49.

Olesen, T. (2015). Global political iconography: The making of Nelson Mandela. *American Journal of Cultural Sociology, 3*(1), 34–64.

Reilly, B. (2004). State functioning and state failure in the South Pacific. *Australian Journal of International Affairs, 58*(4), 479–493. https://doi.org/10.1080/1035771042000304742

Sheppard, P. J., & Walter, R. (2006). A revised model of Solomon Islands culture history. *The Journal of the Polynesian Society, 115*(1), 47–76.

Sullivan, N. (2015). Introduction to the prospects for sustainable development in Papua New Guinea. https://www.academia.edu/4492757/Introduction_to_The_Prospects_for_Sustainable_Development_in_Papua_New_Guinea. Accessed July 3, 2022.

The Guardian. (2019a). The brutal history of Bougainville: A look back at the archipelago's history, from German colony to civil war, and from reconciliation to independence referendum. November 23, 2019a. https://www.theguardian.com/world/gallery/2019/nov/24/the-brutal-history-of-bougainville-in-pictures. Accessed May 15, 2022.

The Guardian. (2019b). Australia's history with East Timor isn't pretty but it must be told truthfully. November 12, 2019. https://www.theguardian.com/australia-news/postcolonial-blog/2019/nov/13/australias-history-with-east-timor-isnt-pretty-but-it-must-be-told-truthfully. August 8, 2022.

The Guardian. (2022a). Panguna mine at centre of bloody Bougainville conflict set to reopen after 30 years. February 10, 2022a. https://www.theguardian.com/world/2022/feb/11/panguna-mine-at-centre-of-bloody-bougainville-conflict-set-to-reopen-after-30-years. Accessed May 10, 2022.

The Guardian. (2022b). Not calling a massacre a massacre is ridiculous': A model of truth-telling at Myall Creek. May 2, 2022. https://www.theguardian.com/australia-news/2022/may/03/not-calling-a-massacre-a-massacre-is-ridiculous-a-model-of-truth-telling-at-myall-creek. Accessed May 15, 2022.

Underhill, B. (2022). The Post-War Australian objective in Papua-New Guinea: Liberal ambition or colonial control? *Australian Journal of Politics & History*.

Venne, S. H. (2007). Treaties made in good faith. *Canadian Review of Comparative Literature/Revue Canadienne de Littérature Comparée, 34*(1).

Wainwright, E. (2003). Our failing neighbour: Australia and the future of Solomons Islands. Australian Strategic Policy Institute, Canberra, Australia.

Wainwright, E. (2004). Solomon Islands, Papua New Guinea and Australia's Policy Shift. *Sydney Papers, 16*, 122–132.

Chapter 23
Case Study of Urban Ecology: Who Runs the Shots in Port Moresby (POM)? An Attempted Exposure and Review of Cliques and Their Profiles in a Global Crime Hotspot

We are not a banana republic where anyone can pick up a plane and just come into PNG unannounced," he said. "We will have no place for those who think they could peddle drugs in this country.
in The Guardian (2020a)

Abstract Port Moresby (POM) is a major colonial port and the capitol of Papua New Guinea (PNG). Having an international airport makes it a hub for PNG, including its crime. PNG crime in Port Moresby is widely described and has a unique character as it is driven by many factors such as Asia, Australia and 'rascals' (usually young men from the bush organized in gangs). The rural society and modern PNG simply cannot provide them with a suitable social net and an income so they get pushed into the capital, where they are stranded and are to survive on any way they can. That's one narrative. But reality sits deeper and links with the social structure, poverty, colonial system and education, social welfare and lack thereof in modern PNG designed by Australia. The crime situation in Port Moresby shows how unsustainable the 'modern' PNG construct is. In many cases, the crime in Port Moresby links with Australia, e.g. via Brisbane, Darwin, Sydney, parts of Asia and beyond, as done via speed yachts, cargo ports, banks and airports. Crime in Port Moresby is not all local but a wider modern and global construct as a cultural dominant phenomenon also found elsewhere in the world and globalization with its underlying economy model and administration.

Keywords Papua New Guinea (PNG) · Port Moresby · Global crime · Urban planning

23.1 Introduction

While Melbourne, Australia, ranks as one of the best cities in the world to live, crime in neighbouring Papua New Guinea (PNG), and specifically in the capitol of Port Moresby (POM), is widely out of control (Pitts, 2002). It's even worse for females and minorities (e.g. BBC, 2018; but see Sparks, 2021). POM ranks at the bottom of world's cities to live (The Guardian, 2004; Asian Development Bank 2012). It's

F. Huettmann, *Globalization and Papua New Guinea: Ancient Wilderness, Paradise, Introduced Terror and Hell*, https://doi.org/10.1007/978-3-031-20262-9_23

more than just a social phenomenon (Lasslett, 2018). Apart from studying crime and associated human behavior for its own sake, PNG is a sustainable nation after all, so how does POM fit into that picture?

The emerging discipline of urban ecology (Chen & Wu, 2009; Forman, 2008, 2016) also includes sociological and governance questions of cities, in Asia and beyond (e.g. Koczberski et al., 2001). Many Asian cities are now looked at and well covered with those type of studies. A specific emphasize can be seen with national capitols, e.g. for Nepal Kathmandu's biodiversity (Hansen & Huettmann, 2020), Caistor (2022) for Mexico City history ,and Ludlam (2021) for Kenya's Nairobi and its policing problems killing members of the public (many even underaged).

23.2 PNG Cities

In PNG, several 'hubs', and two big cities can be found, Lae (~105,200 inhabitants) located at the North coast and with a port and airport and then the capitol Port Moresby (POM) facing south toward Australia with 'c. 360,000 to 380,000 inhabitants (Mongabay, 2022). POM is still growing and has the key ports and airport (Korstanje 2015 for airport dynamics and patterns). But internationally, and from an Asian perspective, those PNG cities are relatively small cities still, and it speaks to the fact that PNG consists to 86% of a rural population (for comparison, POM's population is still more than the entire population of the nation of Iceland, as another island nation but then with a much wealthier per capita GDP!).

POM makes for a unique national capitol during globalization, but which has little study recognition (see Piits, 2002, Goddard, 2017 and Lasslett, 2018). However, as a location and relevance for humanity it easily ranks on the same level than Mexico City, let's say (see Caistor, 2022 for describing it as a cradle of nations and with 20,000 years of legacy; app. 8.8 million inhabitants). POM is over 47,000 years populated and one city that finds itself consistently in the bad press due to one of the world's highest crime rate and other bad metrics (Table 23.1). Accordingly, it gets described by travel guides and those who seek adventures (Salak, 2014). Crime there is often explained superficial but not with relevant causes and justifications. One may easily blame then 'the rascals' (raskols; Dupont, 2012), young males leaving the rural areas and villages, unemployed (Levantis, 1997) and trying to survive in a lost new world without much of an appropriate structure (Salak, 2014). But reality is a bit different and it has changed, too, and for everybody to see and to know, e.g. Levantis (2000), The Guardian (2020a, b). Black young men can quickly be blamed by white colonialists, industrialists and females, a classic western misangony. But instead, the story sits wider, in the world society and who made it (see Macintyre 2008 for policing). And that's arguably Australia, the colonial and global powers with the U.N. (as shown in Baraka, 2001; Gosarevsky et al., 2019). Globalization is now in the driver seat (Stiglitz, 2003; Yiu, 2010; Chan 2016 for a PNG and island perspective (Fig. 23.1).

Table 23.1 Some metrics of Port Moresby (POM) capitol of Papua New Guinea

Metric	Detail	Comment
Carjackings per year	"High" (numbers not available)	Details found in the public record, likely more than 2 per week.
Crime-related death per year	Above nation average (numbers not available)	Despite the bad reputation of POM on that question, the author is not able to find those official and reliable numbers
Domestic violence per year	Above nation average (numbers not available)	Details found in the public record, however, many cases are likely underreported
Average life expectancy	Unknown	No specific records known for Port Moresby

Fig. 23.1 Port Moresby has many things to offer

Textbox 1: Rascals: A classic case study in sociology, racism and globalization

Arguably, rascals are a key component for the high crime rates in POM and PNG cities. The cause of the PNG rascal phenomenon sits deeper though (Levantis, 1997, 2000). It has to do with poverty, with a broken down social structure and the rural exodus, the nation state construct and framework, as well as with drug crime and existing long-term lack of employment or hope (e.g. Laslett, 2018). If one wants to understand it, it clearly runs deeper than described in Salak (2014) and in the tourist guidebooks. A good start is to look at the colonial impoverishment starting 300 years ago, the resource curse, and the neocolonial approach by Australia and other nations claiming 'to help' and sending policing and aid and development for decades (see Gosarevski et al., 2019), whereas, those efforts widely failed for betterment. In the current mindset it's not getting better, neither for the people nor for rural areas or for the metrics…or for POM.

As most policemen can probably share, crime is directly associated with roads and travel paths, in and out of PNG, and with POM as the hub. The nearest neighbor matters. Arguably, much of the crime goes through Brisbane, Sydney, Darwin etc. then, a fact that one can easily detect in POM. At minimum, problems in PNG are connected to Australia and its northern coast (see for Torres Strait; McFarlane, 1998; Alpers, 2005), and they have been for a long time (see Sharman, 2017 as well as The Guardian, 2021a for casino, housing and poverty problems). Asia provides the other link and many crimes link global.

23.3 POM

Based on numerous trips and stay-overs in POM (Port Moresby), I realized when looking closer at the cultural phenomenon 'Port Moresby' that there are relevant patterns, main actors and dynamics and major drivers from the outside though. POM is a certain international mining hub, a port and airport entry into PNG, which one can easily see when staying in the hotels with a nightly rate of easily $180 and above; other housing choices are few (see for instance https://www.lonelyplanet.com/papua-new-guinea). Already due to safety concerns, one tends to stay at those expensive venues; it's the standard there. My taxi experiences were not that impressive, but got me where I needed to be. A way to learn about PNG and how it functions.

POM had many good things going for it initially. It's at a good and well-proven shipping location. It's in a relatively dry spot of PNG, and it has the location of the University of Papua New Guinea opened in 1965, with an initial library (Lutton, 1981). The development of the national PNG government started in POM, and some nice indicative Melanesian architecture can be found there also. POM had an elected city council from 1971 to 1980, but then the national cabinet suspended it due to financial mismanagement. Still, POM has a museum, a zoo also and is famous for some of its diving sites. POM was the event of the recent APEC meetings (see other chapters in this book).

POM has many other aspects and which I will present here next: the port, the gangs, the suburbs and the native setting of POM. And there is a new angle, and that is the social media of POM.

To start out with Baraka (2001) it was expressed that unlike in the urban areas where people are isolated from their land and thus have to depend on the cash economy to meet their needs, the rural population only needs land to sustain their livelihoods. POM cannot break out of this reality, and thus rural life drives it. In PNG, POM is unique and not a typical experience for the nation (Fig. 23.2).

After all, POM is a city created, designed and made big by western colonialists again (Flannery 1994 and Beehler and Laman 2020 for New Guinea's history of western exploration). POM—as a PNG entry point from abroad—is the place where most products come and leave PNG. It was established for that purpose app 1873 and named after captain John Moresby, a British admiral and explorer. POM is a 'port,' a major airport hub as well as a mining industry's lovechild, and that comes with all its problems.

POM has several aspects to it all in one: local ownership of the land, political ownership, PNG capitol, port, airport, shipping hotspot, tourist destination, colonial structure, etc. (Table 23.2).

Many cities in the world are driven by certain groups. Traditionally, POM is part of two indigenous tribes (Motu and Koitabu) but those are now in the vast minority; POM represents the microcosmos of all of PNG as well, and thus, most tribes and their families are represented there; some can be very powerful; and POM has many tribal groups present these days. But economy-wise, POM is mostly driven by a Western and Asian society group; usually, these are the ones who operate the port and its

Fig. 23.2 A typical airport transport in Papua New Guinea

Table 23.2 Features of Port Moresby

Feature	Detail	Comment
Main language	Motu, Pisin, English	
Tribes	Motu, Koitabu and many others	Because Port Moresby is a national hub, virtually all PNG tribes are represented there, as well as the international community
People	360,000–380,000	Exact number not known due to being a hub, and high amount of squatters and seasonal workers
Railway stations	0	
Subways	0	A subway is not on the agenda for years to come
International airports	1	It's a relative small airport space
Ports	1	A major international cargo hub

business with international links, dominated by nearby Australian cities (Brisbane, Darwin and Sydney with their banks and lawyers). I was told some European families are rather influential in POM but details are not well presented.

Nowadays, the airport also matters and overrules. But the cargo container port plays a large role as it connects PNG with Lae, with PNG islands but also with nearby Australia and with Asia and the world: from Singapore to Taiwan, Hongkong, Shanghai and the Atlantic including Hamburg and London. The list of major companies that operate in Port Moresby is diverse and consists of banks, phone companies, supermarket chains, container operators, etc. Of course, POM has many embassies also.

With that comes an Asian and major international port culture and ethics into PNG and certainly into POM; while the EU link fades out Asian groups increased their influence there, with the top 10 container ports being of Asian origin now, mostly China.

Asian crime can operate in POM also and it becomes a heaven for accessing Australia, as the bigger market. And thus, POM crime is not all local and the rascal narrative becomes an easy scapegoat (Laki, 2006; see NME 2021 for a wider Australian corruption scheme example unfolding in POM) (Tables 23.3 and 23.4).

As PNG has major natural resources, so are the ports reflecting it, namely mining, oil and gas, timber, fish, farming and other natural resources. Many ships deal with such commodities. It's the typical colonial structure and patterns, and those resources are to be shipped abroad, out of PNG, away from its people quickly; time is money. It's a typical scheme found all over the world impoverishing local areas, done with in containers with a computerized efficiency across borders, and worse (many containers actually get lost; see for details https://www.joc.com/maritime-news/international-freight-shipping/ap-moller-maersk-group/lost-and-stolen-containers-trackable-database_20171127.html).

Table 23.3 Details of Port Moresby (POM) ports and hubs (Fig. 23.1 for international airport connections to POM, e.g. Seoul. South Korea)

Metric	Relevance	Comment
Tons cargo per year	High	Exact numbers are not known but distributed among just a few companies
Containers per year	Very high	
Tourists per year	210.000 (2020)	https://www.papuanewguinea.travel/annual-visitor-arrivals
Number of ships per year	High	Details can be tracked here: https://www.marinetraffic.com/en/ais/details/ports
Number of flights per year	Low	Airport details here: https://www.port-moresby-airport.com/

Table 23.4 Selection of social media entries of interest for Port Moresby

Facebook entry	Topic	Comment
https://www.facebook.com/www.museumpng.gov.pg/	Art	All capitols are to have 'art'
https://www.facebook.com/groups/671084006396203/	City news	
https://www.facebook.com/portmoresbydaily/	News	

The 'container box' is a major defining feature for POM, if not even for most of coastal PNG and larger cities, globally and for neocolonialism (Klose, 2015). The 'in and out' concept remains the key issue for PNG, a reversed form of cash-and-carry; just like it ever was in colonial times. Now with globalization, it just got scaled up globally and made look better, and so are the problems hidden in window dressing, often not even hidden at all. It's noteworthy that PNG has no railroad or a road system connecting PNG internally. Ships and planes remain essential for PNG and for POM. All what is done in PNG tends to meet in POM, and one uses the domestic planes to do so; it becomes part of the fabric but which is driven by Australian maintenance (airlines).

But beyond the port and airport, POM has an additional level and dynamic. And that is, the trend of Motu becoming the main national language, next to Tok Pisin and English. PNG mainlanders take over the local operation (while islanders and remote groups have little influence in PNG). Talk Pisin has real-world implications.

And beyond Australia itself, POM is part of the wider Asian and global trend of urbanization. Seen on a PNG-scale, human population growth is biggest in POM and so is the surrounding deforestation, e.g. for fuel wood in the growing suburb and shanty towns (Beehler & Laman, 2020).

The road connections around Port Moresby are essential, and the roads just got paved as part of the APEC summit deal (https://en.wikipedia.org/wiki/APEC_Papua_New_Guinea_2018). And to no surprise, carjacking is a big deal in PNG (as it is in other nations and cities of the world, e.g. Johannesburg and Cape Town in South Africa). That is specifically the case for Luxury SUVs (see online movie by All Jazeera (2021). A car hijack has a value of app US$10,000.

23.4 Crime is Quite Simple…

Daily crime in PNG and in POM is usually not a rocket science, hardly sophisticated. It's a simple affair, blunt theft done with violence; humans might get in the way. While the POM crime tends to happen at night, e.g. house break-ins, much occurs also throughout the day, on the street and in your house. Crime seems to have some seasonal patterns also. It's an open crime scene in most of POM, with many (sad) stories to tell. Many of those crimes are simply done with a machete or with rifles made of cut-off pipes and barrels (see Al Jazeera, 2021); physical violence is dominating.

Fig. 23.3 Provided connections to POM for some offer opportunities for others; a typical night flight

But crime also has 'simple' motivations underneath, e.g. to eat well or to have fun with the quick wealth obtained. And for have-not's, what else is there? Poverty remains the key item unresolved in this social detective game, it's a resilient feature now.

23.5 POM Moves Forward Either Way

There are clear lock-ins in POM, the social fabric and crime, such as the geography, the history, the tribes, the PNG set up with POM as the center and hub, the road network, the mines and the shanty towns; urban planning could help, a bit, but one will have already a hard time to get a good road map for POM; the architecture and banking financing sets other limits. Still, the wider framework of POM and PNG, the modern PNG construct remains the true key for improvement.

How to move POM into betterment (see Jones, and Lea 2007 for Pacific examples)? To end crime is a poverty and a cultural issue, for humanity, and thus a post-colonial issue and a vast social issue involving rural areas, village structures, Big Men and the Wantok system in modern times. As a key item in this discussion, see The Guardian (2021b) for including females as full PNG citizens for improving POM and PNG. The recent PNG election had two female politicians included, for the first time! Still, things are not changing much in POM any time soon, and it can hardly be changed from within or from outside. Australia tried it. Local community control sounds nice and helps—as per Pitts (2002)—but it's not being the major delivery of improvement. PNG needs to grow from within, revise the Australian model and have a functional governance, but that has not much happened because PNG is a colonial construct, now runs neocolonial and got deeply pushed into globalization.

References

Asian Development Bank. (2012). *The State of Pacific towns and cities: Urbanization In ADB's cacific developing member countries.* Asian Study Series. Metro Manila.

Al Jazeera. (2021). *The Gangs of Port Moresby*, 30th April 2021. https://r.search.yahoo.com/_ylt=AwrOqq7bkfxiRXEFd49XNyoA;_ylu=Y29sbwNncTEEcG9zAzEEdnRpZAMEc2VjA3Ny/RV=2/RE=1660748380/RO=10/RU=https%3a%2f%2fwww.aljazeera.com%2fprogram%2f101-east%2f2021%2f4%2f30%2fgangs-of-papua-new-guinea/RK=2/RS=zkq1cyL_i_BH2QEyJqyNZaeeEhs. Accessed August 15, 2022.

Alpers, P. (2005). *Gun-running in Papua New Guinea.* Small Arms Survey, Graduate Institute of International Studies, Geneva, 1–6.

Baraka, R. (2001). *The meanings of independence in Post-Colonial Papua New.*

BBC. (2018). *Port Moresby: The world's most dangerous city to be a woman?* 27 September 2018 https://www.bbc.com/news/av/stories-45654549. Accessed August 2, 2022.

Caistor, N. (2022). *Mexico City: Cradle of Empires.* Reaktion Publishers, 240 pp.

Chan, J. (2016). *Playing the game: Life and politics in Papua New Guinea.* University of Queensland Press.

Chen, X., & Wu, J. (2009). Sustainable landscape architecture: Implications of the Chinese philosophy of "unity of man with nature" and beyond. *Landscape Ecology, 24*(8), 1015–1026.
Dupont, S. (2012). *Raskols: The gangs of Papua New Guinea.*
Forman, R. T. (2008). The urban region: Natural systems in our place, our nourishment, our home range, our future. *Landscape Ecology, 23*(3), 251–253.
Forman, R. T. (2016). Urban ecology principles: Are urban ecology and natural area ecology really different? *Landscape Ecology, 31*(8), 1653–1662.
Goddard, M. M. B. (2017). *The unseen city: Anthropological perspectives on Port Moresby, Papua New Guinea.* Pandanus Books.
Gosarevski, S., Hughes, H., & Windybank, S. (2019). Is Papua New Guinea Viable? *Political Economic Bulletin, 19.* https://www.semanticscholar.org/paper/Is-Papua-New-Guinea-viable-Gosarevski-Hughes/6d71c70951c12fa82dea578a36b66923997f3f92. Accessed June 1, 2022.
Guinea. *A Plan B Research Paper Submitted to the Faculty of the Center for Pacific Islands Studies in the Candidacy for the Degree of Masters of Arts.* School of Hawaii, Asia And Pacifci Studies. The University of Hawai'i at Manoa.
Hansen, L., & Huettmann, F. (2020). Chapter 18. Swallows and sparrows in the human street-market interface of Urban Nepal: Towards a first open access GIS data and model inference on the role of religion and culture in bird distribution. In G. R. Regmi, & F. Huettmann (Eds.), *Hindu Kush-Himalaya watersheds downhill: Landscape ecology and conservation perspectives* (pp. 361–399). Springer Gland.
Jones, P., & Lea, J. P. (2007). What has happened to urban reform in the island Pacific? Some lessons from Kiribati and Samoa. *Pacific Affairs*, (pp.473–491).
Korstanje, M. E. (2015). The anthropology of airports, criticism to non place theory. *Advances in Hospitality and Tourism Research (AHTR), 3*(1), (pp. 40–58).
Klose, A. (2015). *The container principle: How a box changes the way we think.* MIT Press.
Koczberski, G., Curry, G. N., & Connell, J. (2001). Full circle or spiralling out of control? State violence and the control of urbanisation in Papua New Guinea. *Urban Studies, 38*(11), 2017–2036.
Laki, J. (2006). *Non-traditional security issues: Securitisation of transnational crime in Asia.* Nanyang Technological University.
Lasslett, K. (2018). *Uncovering the crimes of urbanisation: Researching corruption.* Routledge.
Levantis, T. (1997). Urban unemployment in PNG—It's criminal. *Pacific Economic Bulletin, 12*(2), 73–84.
Levantis, T. (2000). *Crime catastrophe-reviewing Papua New Guinea's most serious social and economic problem.*
Ludlam, S. (2021). *Full circle.* Forward Press.
Lutton, N. (1981). From Port Moresby to Perth. *Archives & Manuscripts*, 30–38.
Macintyre, M. (2008). Police and thieves, gunmen and drunks: Problems with men and problems with society in Papua New Guinea. *The Australian Journal of Anthropology, 19*, 179–193.
McFarlane, J. (1998). *Torres Strait: Policing the open border* (pp. 1–6). Australian Institute of Criminology.
Mongabay. (2022). https://rainforests.mongabay.com/deforestation/forest-information
NME. (2021). *Sheppard band members' father and former manager arrested in Papua New Guinea for alleged fraud.* 22nd January 2021. https://www.nme.com/en_au/news/music/sheppard-band-members-father-and-former-manager-arrested-in-papua-new-guinea-for-alleged-fraud-2862298. Accessed July 18, 2022.
Pitts, M. (2002). *Crime, corruption and capacity in Papua New Guinea*, Asia Pacific Press at The Australian National, University, Canberra.
Salak, K. (2014). *Four corners: A journey into the heart of Papua New Guinea.* Restless Books.
Sharman, J. C. (2017). Illicit global wealth chains after the financial crisis: Micro-states and an unusual suspect. *Review of International Political Economy, 24*(1), 30–55.
Spark C. (2021). *The New Port Moresby.* University of Hawaii Press.
Stiglitz, J. (2003). *Globalization and its discontent.* Norton Paperback.

The Guardian. (2004). *Raskol gangs rule world's worst city*. 21 September 2004. https://www.theguardian.com/world/2004/sep/22/population.davidfickling. Accessed June 30, 2021.

The Guardian. (2020a). *PNG's biggest drug bust: The plane crash, the dead man and the half tonne of cocaine*. 1st Aug 2020a. https://www.theguardian.com/world/2020a/aug/02/pngs-biggest-drug-bust-the-plane-crash-the-dead-man-and-the-half-tonne-of-cocaine. Accessed Nov 3, 2021.

The Guardian. (2020b). *Papua New Guinea police accused of gun running and drug smuggling by own minister*. 10th May 2022. https://www.theguardian.com/world/2020/sep/18/papua-new-guinea-police-accused-of-gun-running-and-drug-smuggling-by-own-minister. Accessed Nov 5, 2021.

The Guardian. (2021a). *PNG residents evicted a decade ago for casino development face violence and poverty: Report*. https://www.theguardian.com/world/2021/nov/12/png-residents-evicted-a-decade-ago-for-casino-development-face-violence-and-poverty-report. Accessed March 17, 2022.

The Guardian. (2021b). *If we want Port Moresby to rise in the liveability rankings, start by protecting its women*. 16th July 2021b. https://www.theguardian.com/commentisfree/2021b/jul/16/if-we-want-port-moresby-to-rise-in-the-liveability-rankings-start-by-protecting-its-women. Accessed August 1, 2022.

The Guardian. (2022). *PNG prime minister denies connection to suitcase full of cash found as voting starts*. 6th July 2022. https://www.theguardian.com/world/2022/jul/06/png-prime-minister-denies-connection-to-suitcase-full-of-cash-found-as-voting-starts. Accessed July 6, 2022.

Yui, M. (2010). Transnational organised crime and the threat to strategic stability. *Asia-Pacific Defence Reporter, 36*(7), 17–20.

Chapter 24
'RAMU Sugar' in Papua New Guinea from Fiji and Australia: 'Take-Home' Lessons from Another Perverted Project Based on the Neoliberalism Principle and Industrialization

Abstract The Ramu valley watershed presents a wider fertile landscape around the Ramu river located in northern Papua New Guinea (PNG) near the Bismarck Sea and the Finisterre Range. It's close to the cities of Lae and Madang. Ramu valley has been occupied by humans for a long time and was affected by several cultural waves of immigration over the last thousands of years. Low elevational forests got turned into grasslands but the area overall remained fertile and sustainable. It was an easy and logical place for colonialism to start, and the region was affected accordingly. In the last 100 years, commercial entities started to develop there and the region has coconut palm plantations and sugarcane and cattle operations, for instance. A company called Ramu Sugar became dominant in the PNG market and built infrastructure such as a hydrodam needed for their operations. However, none of those efforts catering international commodity markets offered stable income, and a certain boom-and-bust cycle occurred affecting the region and its sustainability overall. Following neoliberal policies, various subsequent patterns and actions occurred to maintain viable income but instead the region experiences a generic environmental and social decay requiring policing from Australia and other outside measures, non really showed success long-term or relevant signs of sustainability.

Keywords Papua New Guinea (PNG) · Ramu Valley · Ramu Sugar · Sugar Plantations · Natural resources · Pacific boom-and-bust

24.1 Introduction

The Ramu valley is a global legacy of mankind; it's very productive and fertile. It has been sustainable in its own ways for easily over 47,000 years and received several waves of human migration and cultures (details in Beehler & Laman, 2020). It connects with Asia, and together with the Sepik region, the Ramu region stands out as one of the highest language diversity in PNG.

Beyond the island linkages within Melanesia, many people might not know that PNG links genetically with ancient Siberia and that a Chinese connection can be found also along the north coast of PNG. PNG consists of a large mix of tribes. The current state of the Ramu valley might well reflect a destructive human enterprise

F. Huettmann, *Globalization and Papua New Guinea: Ancient Wilderness, Paradise, Introduced Terror and Hell*, https://doi.org/10.1007/978-3-031-20262-9_24

Fig. 24.1 Ramu Sugar reality image of a cattle grazing field and barren rotation in an otherwise would-be coastal rainforest and modified grassland of the ancient Ramu valley

of the colonial times, and specifically of the last 50 years of Australian, industrial and global influence and pressures. It stands as a record in time for such impacts and political powers. In colonial times, the Ramu river was explored by the Germans (Sauer, 1915; Gibb et al., 2016). A quick look at the paved/official road system in PNG (Fig. 24.1) shows us a direct link with the industrial (western) investment and where the exploitation sits, where products come from and go to. It represents a highly uneven development (examples in the wider region by Beer, 2017; Beer & Chruch, 2019). People and villages—as the core of PNG's economy for millennia—play no real role.

The north coast of PNG, where Ramu valley is located, is relatively accessible to humans from the ocean and through rivers; ancient trails lead into the highlands. Many international explorations and contacts consequently occurred in that region (Gibbs et al., 2016), including a massive WW2 operation and a western search for resources, namely gold. It's not much surprising then that the wider Ramu valley has one of the longest efforts of mining over time in PNG and subsequent industrial activities; most of them failed though. It's a boom-and-bust operation and as unsustainable as it gets, ignoring PNG's deep civilization and associated humanity. As per metric and experience, there is no sustainable modern industrial society, or education or prosperity there, hardly peace (as the frequent Australian police missions in the area show us, e.g. for Madang and Lae, etc., e.g. https://asopa.typepad.com/asopa_people/2017/03/chaos-for-australian-police-as-png-mentoring-mission-imploded.html). What one can easily track instead are the boom-and-bust cycles, and associated unrest, with PNG citizens dragged in. Similar changes have been documented, for instance in the Markham Valley (Beer, 2017; Beer & Church, 2019; Church, 2019).

The legacy of 'RAMU Sugar' (https://www.destinationpng.com/section-3/ramu-sugar-limited/), now Ramu Agri Industries (https://www.dnb.com/business-directory/company-profiles.ramu_agri_industries_limited.0cd986efa15e4c2e95c10b22487c896e.html; Fig. 24.2), has virtually all the aspects of a colonial misadventure (compare for instance with the orchestrated Massai move in East Africa, Hughes, 2006). It's a logical step that continues with the legacy of the failed colonial industrial period and the subsequently widely failed globalization undertaking—highly unsustainable—all occurring in a Dominion of the British Commonwealth and global commodity markets. Like with the coconut enterprise in the area and as reported earlier (Beehler & Laman, 2020; see Fig. 2.11 in Huettmann, 2015). RAMU sugar (I am using this term loosely as several companies and constructs are involved over time) went essentially through all major typical phases of development in tropical colonial top-down styles for industrial neoliberal and investment aided resource development: exploitation of natural space and resources, harvest, intensification, global crashes and debt, diversification, strategic selling of branches, subsidized spiral into decay and aspects of bankruptcy leaving the clean-up with the global public and PNG. The only game changer in that generic template is that this area was frequented by several cultures and colonial powers. And they changed due to warfare—conflict started in, and was fueled among, a power clique in remote Central Europe. It then went its path of globalization (as per Stiglitz, 2003).

And no doubt, the initial intention of RAMU sugar and PNG's sugar supply to be independent of outside sources—a truly PNG-driven enterprise, starting in the 1970s—remains as noble as it can get (see details provided by https://postcourier.com.pg/how-a-game-of-golf-sweetened-ramu-sugar/). Earlier, PNG imported its sugar from Australia, Colonial Sugar Refinery (CSR). Who would not agree to have it in PNG instead, where the ancient sugar cane plant essentially came from?

But what came from this?. The wealth did hardly enter PNG or arrived to its people; a classic example that the trickle-down economy promoted by many, and in PNG by its leader, the prime minister Michael Somare, is not working. Neoliberalism is not a win–win, and never has been (MacLaran & Kelly, 2014). Rather, PNG is left with a vast burden from this enterprise that is ongoing to fail. The recent portfolio of Ramu Agri that include oil palms is as devastating as it gets but a consequence. The leading corporation of the region remains a decaying image of its own promise, and that's exactly how the Ramu valley looks like now: industrial destruction with social problems. It's a certain global template of such business models and frameworks.

Based on several site visits and public data, this chapter presents the case that essentially from the date of its initiation Ramu sugar is widely driven by outside powers and investors, always with latest 'modern' methods, using science, but simply to reap profits from PNG, now with globalization on the rise and capitalist models becoming pointless, adding more pain to the injury in a global arena.

Table 24.1 List of corporations with a bad colonial-national or neocolonial legacy

Name	Location	Nation or Colony	Issue of critique	Citation
VW	Various	Nigeria, Mexico, China	Nigeria, China (Uyghurs), VW has a deep link with the German NAZI party	University of Michigan. Center for European Studies (2000)
Japanese	Various	Asia, e.g. Korea, China and Taiwan	Colonial industrial labor	Zhen and Chen (1984)
NIKE	Several	SE Asia	Cheap labor in SE Asia	Discussed in Business textbooks etc.

Keep in mind, PNG is a cradle of sugar cane genetics and one of its first domesticated farming sites in the world. It did not have to come to this! It found itself in a position early on where it exactly did not want to be from the start (=dependent on outside and international markets) (Table 24.1).

24.2 The Start of the Industry

PNG has some of the most ancient sugar species in the world. It's a natural place for sugar production then. But until 1970s, PNG had no real sugar production of its own, a key issue for any nation to be independent. Thirty years ago, the sugar development and industrialization in PNG were started in earnest. It roughly coincides with the establishment and independence of PNG. RAMU Sugar was to turn into a showcase. However, this growth and so-called independence are a pure consequence of the colonial history and approach, leaving PNG's environment and people behind.

24.3 The Boom-and-Bust Starts

It clearly is encouraging to see when the business starts to grow, it moves into a boom phase. During that phase one tends not to change anything because 'it runs' and it cannot be changed now… But the trend cannot go on, and global markets and other issue start to become complex and a problem. Boom turns into bust; banks ask to pay back the loans, the earlier 'product' is no *en vogue* anymore, and the profit declines. The spiral down starts in the boom phase, which is a poorly managed growth.

24.4 Diseases Enter the Scene

Ramu Sugar went early into a down cycle, app in the early 1980s. Workers had to be mobilized. Soil became degraded, and the plants (monocultures) developed diseases. The disease of sugar stunt (1986) and sugarcane smut (entering the scene later in

2016, Tom et al. (2017); see also Waller et al. 1987, Croft et al., 2008). The sources of the sugarcane diseases are not exactly known, outbreaks are local and depend on the sugarcane variety, and not all cases are severe. For some of those there seems to be connections with adjacent Australia and Fiji. Still, it can threat harvest and national product and sugar supply, which can become strategic. It now becomes a complex problem of decay.

24.5 Sugar Diversifies with Cattle

The traditional product approach—sugar—showed its limits. The PNG sugar market demand is not that big. Many nations seek sugar—a human love product for centuries ('sugar cubes'), often produced with slaves (see legacy in Haiti, the wider Caribbean and southern U.S.; Greenfield, 2018). Many ways exist to produce sugar, from honey bees to sugar beet and sugarcane in various shapes and forms of refining However, the market for those sugar sources is equally international and complex already for a long time. Sugar gets refined and was easily a global core product in the slavery times, affecting civil wars in U.S. and slavery wars in Haiti, in Brazil and elsewhere. It shaped entire nations, and was a globally relevant trade item, starting in the Atlantic and Caribbean region and then extending more. In the meantime, the market by now got globally saturated in all aspects: from cheap human labor over refinery industry, and market access and subsidies. Those modern issues come with the industrial sugar production, Cuba onwards. Nowadays, sugar products cover white sugar, brown sugar, organic sugar, export-grade sugar etc.

As a solution in Ramu PNG, sugarcane got de-emphasized and the production—the portfolio—got diversified. It's a widely used strategy by companies and once they have support and infrastructure available. From 'RAMU Sugar,' one went into beef cattle, etc. The province is app. the second leading producer of copra and cattle, and third in cocoa. Ramu sugar and the woodchip mill are among the biggest employers now, including the tuna cannery. From a PNG perspective, the Ramu valley is quite industrialized. But this turned into its own dilemma and did not really solve the problem: 'RAMU Sugar,' what for and for whom, and how it is done?

24.6 Add Some Water and Energy to the Mix for the Overall Non-Sustainable Footprint in an Otherwise Ancient Valley of Sustainability and Humanity

Sugar is a plant-based energy source. And when concentrated, it's even more so of an energy pack; widely unhealthy to humans also. But sugar needs manpower to be concentrated, and it needs to be refined. This refinery energy is essentially an energy transition. Sugar is concentrated energy, after all. But where does all the energy really come from? 'RAMU sugar' essentially got its own electrical dam (Yonki; https://en.wikipedia.org/wiki/Yonki_Dam), creating energy for the industry with a smaller spin-off for the local region (although it was barely argued that way, with people supposed to benefit the most. Figs. 24.3, 24.4 and 24.5). Impacts are rather big, e.g. fish and watertable for food production are affected, and this is far from green energy (Huettmann et al., 2020)! Another dam project in the area stalled but can come forward soon.

The Ramu river got dammed to provide electricity; and that's a major game changer for that ancient riverflow and its people. And it did not follow principles outlined by Rasul (2019). This leaves a larger water and energy footprint on a production plant and the watershed that should have been sustainable and providing natural food instead, a sustainable wilderness for millennia. Clearly, many lines got crossed, all awhile the actual profit is also debated and a problem. It's a typical case of wider

Wilderness

Humans

Nomads and Farming

Colonialization

Industrialization

Problems occur, e.g. labor, introduced labor and diseases, wars

Market Expansion and Economic Growth

Product Improvement, e.g. through science

Global Commodity problems

Fierce Competition

Marketing intensity increases

Portfolio Expansion

Internet

Pandemics

Fig. 24.2 Typical sequence of modern enterprises and events on ancient to developed landscapes

Fig. 24.3 Ramu dam

losses that comes with such an operation but where people instead have already lived and produced well and more or less sustainable in the same region for millennia across various cultural waves. So what went wrong now?

24.7 Global Commodity Markets Turn Violent

By now, and due to the global market and world order, virtually any commodity turned into a global mess of liability, book keeping styles, tax evations, money loss and speculation risk, profitable only to a few and who can afford insurance and who know 'the game'. Same applies then for RAMU products, and specifically for any small and inexperienced new nation: modern PNG is no exception. Going global was not a good idea for Ramu Sugar and the environment, or the people. While export markets like the U.S. get catered, and an own airport (GUSAP) is maintained at the facility to do so, the environment and the social fabric all pay the costs.

24.8 Oil Palms Added to Sugarcane and Cattle: How Bad Can It Get?

As the story unfolds, 'Ramu Sugar' added palm oil plantations to their portfolio mix. This might well excite investors, and outside business people and some

Fig. 24.4 Ramu river

consumers, e.g. ones that are removed from PNG realities, but it's totally unsustainable for the locals. The oil palm footprint in 'Ramu Sugar' is on the rise, and it creates many problems indeed. Habitats are transitioned further, land management got intensified. It's part of a global wave but hardly helps biodiversity or PNG. In the wider perspective of time and sustainability, it will stand out as another major problem in the list.

24.9 Where RAMU Sugar is Going?

RAMU Sugar tries to paint a good picture to the public, with a school built, high education levels of the students, local malaria levels dramatically reduced etc. But simply based on investment history, lock-in and globalization, 'RAMU Sugar' will follow the future like the wider western enterprise. New things come and go, and an endless tide and ebb of assumed—but highly unsustainable—efforts and make-overs are popping up from everywhere. Flying to the moon, using a new earth, is not an option.

In that regard, Fiji should be considered because it was an initial sugar supplier for the region and for PNG, and it still is a major provider for the global market and thus competing Parts of northern Australia do the same (remember Australia driving

Fig. 24.5 Ramu dam reservoir

aid to PNG, as well as the actual currency and the regional trade regime). So where does that leave us now, what's new for a problem?

24.9.1 Some Easy Take-Home Lessons

Ramu valley presents us with what the human enterprise is capable of. Like for most of PNG, the people of PNG did a decent—great-sustainable job for Ramu valley for 97% of the time. As a matter of fact, those initial types of inland valleys were probably representative where a large part of people in PNG lived (Beehler & Lamn, 2020, p. 25).

But when the colonial misadventure started, it went downhill with the Ramu valley, typical industrial problems occurred, and people suffered in vast quantity, and stressors show a large burden for mankind and its habitats in the region and wider. It easily spills into cities of Port Moresby and certainly Madang, Lae and villages that are connected by the road, and for the coastal regions. There is no real end of suffering in sight.

The neoliberal concepts, and with aid and experts, simply failed. Looking back, Ramu Sugar and its CEOs have a lot of explaining to do when assessed ecologically by 2030 and beyond.

References

Beer, B., & Church, W. (2019). Roads to inequality: Infrastructure and historically grown regional differences in the Markham Valley, Papua New Guinea. *Oceania, 89*, 2–19.

Beer, B. (2017). The intensification of rural-urban networks in the Markham Valley, Papua New Guinea: From gold-prospecting to large-scale capitalist projects. *Paideuma: Mitteilungen zur Kulturkunde, 63*, 137–158.

Beehler, B. & Laman, T. (2020) New Guinea: Nature and Culture of Earth's Grandest Island. Princeton University Press.

Church, W. (2019). Changing authority and historical contingency an analysis of socio-political change in the colonial history of the Markham Valley (Papua New Guinea). Paideuma: Mitteilungen zur Kulturkunde, 65.

Croft, B. J., Magarey, R. C., Quinn, B., Royal, A., & Kerkwyk, R. (2008). Sugarcane Smut in Queensland. *Proceedings of the Australian Society of Sugarcane Technologists, 32*, 52–62.

Gibbs, P., Aiwe, M., Tia, A. D., & Wangihama, E. (2016). Quality of life and development challenges in a Middle Ramu community. *Contemporary PNG Studies, 25*, 1–12.

Greenfield, S. M. (2018). Plantations, sugar cane and slavery. In *Plantation Societies in the Era of European Expansion* (pp. 1–36). Routledge.

Huettmann, F., Shrestha, M., Puri, G., & Regmi, G. R. (2020). Chapter 32 Hydrodams Again: A closer look at expensive Gharials, lost dolphins, dissected fish migration, and destructive energy for non-sustainable mining and societies abroad fueling a non-democratic governance for the designed '*homo electricus*' In G. R. Regmi, & F. Huettmann (Eds.), *Hindu Kush-Himalaya Watersheds Downhill: Landscape ecology and conservation perspectives* (pp. 649–661). Springer Gland.

Huettmann, F. (2015). If leaders like Alexander Von Humboldt, Ronald Reagan, and Xi Jinping would have been environmental justice radicals using public internet and drones: A short political ecology of Central America with Regard to Global Agenda Setting in 2014. In: F. Huettmann (Ed.) *Central American Biodiversity: Conservation, ecology, and a sustainable future* (pp. 25–76). Springer.

Hughes, L. (2006). *Moving the Maasai: A colonial misadventure.* Springer.

MacLaran, A., & Kelly, S. (2014). Neoliberalism: The rise of a bad idea. In *Neoliberal Urban Policy and the transformation of the city* (pp. 3–19). Palgrave Macmillan.

Rasul, G. (2019). Development beyond hydropower: Towards an integrated solution for water, energy and food security in South Asia. *International Journal of Water Resources.* https://doi.org/10.1080/07900627.2019.1579705

Sauer, C. O. (1915). Exploration of the Kaiserin Augusta River in New Guinea, 1912–13. *Bulletin of the American Geographical Society, 47*(5), 342–345.

Stiglitz, J. (2003). *Globalization and its Discontent. Norton Paperback.*

Tom, L., Braithwaite, K., & Kuniata, L. S. (2017). Incursion of sugarcane smut in commercial cane crops at Ramu, Papua New Guinea. In *Proceedings of the Australian Society of Sugar Cane Technology* (Vol. 39, pp. 377–384). https://www.researchgate.net/publication/316967031_INCURSION_OF_SUGARCANE_SMUT_IN_COMMERCIAL_CANE_CROPS_AT_RAMU_PAPUA_NEW_GUINEA

University of Michigan. Center for European Studies. (2000). The "Goldhagen effect": History, memory, Nazism--facing the German past. University of Michigan Press.

Waller, J. M., Egan, B. T., & Eastwood, D. (1987). Ramu stunt, an important new sugarcane disease in Papua New Guinea. *International Journal of Pest Management, 33*(4), 347–349.

Zhen, J., & Chen, C. C. (1984). *The Japanese colonial empire, 1895–1945.* Princeton University Press.

Chapter 25
Role of Outside Nations in Papua New Guinea's Sustainability Model: China, U.S., UK, the EU, Russia, Australia, NZ and Japan Need Bob Marley and the Dalai Lama Coming to the Rescue

It is entirely possible—perhaps even likely—that much of what intelligence agencies have done and are currently doing is morally wrong, at the bar of the account I defend here
Fabre (2022, *on undercover security services working abroad)*

Too bad Aussie politicians are unaware of recent research showing the huge and crucial role that forests play in the global water cycle ...
Pioneering ecologist Paul Ehrlich to present 11th Nov 2019 at UNSW Sydney on the sixth mass extinction
(https:Newsroom.unsw.edu.au News Science-tech Pioneering-ecologist-paul-ehrlich-present-unsw-sydney-sixth-mass-Extinction; accessed 3rd May 2022)

Here comes the conman
Coming with his con plan.
We won't take no bribe;
We've got (to) stay alive.
We gonna chase those crazy -
Chase those crazy baldheads -
Chase those crazy baldheads out of the yown.
Bob Marley song "Crazy Baldhead"

Abstract Papua New Guinea (PNG) as a 'modern' (industrial) nation was fabricated, primarily a compromise by the colonial powers and a deal carried out by Australia serving their own purposes, e.g. to save money and maintain control and income, all approved by the U.N. In such a world fabric, driven by just a few but permanent security council seats in charge of world power, there is little that nations like PNG can do and truly operate in. They are to 'participate and playing the game,' as Prime Minister Julius Chan expressed. However, people of PNG cannot fairly participate in the global markets, nor can PNG as a nation and democracy embedded in the British Commonwealth lead by the late queen from London. Many colonial and powerful nations determine the shape of PNG, its set up, industry and business, none are really doing so well or are sustainable themselves. Manmade climate change remains unresolved. That is certainly true for mining, forestry, farming and fisheries sectors. A similar situation can be stated for PNG's neighbors making the region overall desperate and affecting its wilderness of world proportions. While rural PNG is widely sustainable the outlook for PNG and its role in the

F. Huettmann, *Globalization and Papua New Guinea: Ancient Wilderness, Paradise, Introduced Terror and Hell*, https://doi.org/10.1007/978-3-031-20262-9_25

'modern' global agenda remains miniscule. Consequently, the people of PNG will retreat into their globally passive but well-proven action of the Wantok tribal system within ecological bounds available to them.

Keywords Papua New Guinea (PNG) · Globalization · International politics · Global justice

25.1 Introduction

It is often forgotten or actively ignored and denied that most poor nations, certainly PNG and its great environment and culture, do not stand in a vacuum (Stiglitz, 2003). They come from different times and contexts and are not independent whatsoever in the Anthropocene and its universe (Anderies et al., 2004). PNG is a nation construct designed by others, namely Australia to safe costs from a U.N. mandate (Baraka, 2001). Thus, nations like PNG cannot really self-determine their own fate (examples for mining and timber projects in PNG provided by Blazey & Perkiss, 2016 [see Baraka, 2001 for PNG governance pushed by Australia; Hyndman, 1998 for wider Melanesian issues of resisting the subsequent ecocide and genocide brought from the outside]). Political advisors and facilitators play a big role in this, certainly when money is to be made and administered (examples in Rich, 1994; Stiglitz, 2003; Gosarevski, 2019 for PNG; see The Guardian, 2015 for Carbon Sequestration projects and their (Australian-U.S.) advisors within the UN Framework) (Figs. 25.1, 25.2, 25.3, 25.4, 25.5, 25.6, 25.7, 25.8, 25.9, 25.10 and 25.11).

As an administrative construct, all nations and their agencies operate within a context and legacy, under certain frameworks, with a goal, with allowed and permitted actions, all embedded in certain wider dynamics, synergies and intense lag effects, including the culture they initiated from and receive patronage (usually that means here an underlying European colonial concept, e.g. Westphalian Peace, Patton, 2019). And those details can hinder good action and affect outcome; it commonly does (Ostrom, 2009). It usually impacts the environment and society (examples for Melanesia provided in Hyndman, 1998). That is nowhere so obvious than in international markets and with global climate change impacts reaching easily across borders. The people in the remote bush, rural landscapes—most of them highly sustainable—pay though the bill. We see that certainly in PNG very clearly (Gosarevski et al., 2019).

Historically, after western contact, the people of PNG have always been driven by a small clique from the outside (Gray, 1996); details vary by topic. As shown in Beehler and Laman (2020), it's for instance a small clique of paleaobiologists -none of them from PNG—trying to tell the PNG history story without PNG citizens and elders involved (and that narrative gets revised and fine-tuned all the time). Similar is the story of the languages: Foreign anthropologists are literally hanging on the tongue of PNG citizens—who ARE the speakers—to 'document' like from a zoo displayed to the outside world (Gillison, 1993, 2002) what languages are spoken,

Fig. 25.1 The national currency of Papua New Guinea —the Kina— as an international construct with vested interests elsewhere is in trouble once more

what words they are using and when, and how the languages are 'lived' (e.g. Kulik, 2019, Z'Graggen, 1980 and citations within). Like done in a Human Zoo. Arguably, nobody truly appreciates outsiders/foreigners to describe their personal and private lives, tongues and tones, and PNG does not really need (religious) linguists to tell them who they are and what they do and speak. Thus far, those linguists do not really make their lives better or bring back the language (examples in Z'Graggen, 1980, etc.), data are hardly shared and provided for the global public; reality speaks for itself.

Same can be said with the mammals of PNG (Flannery, 1994): usually a powerful group of species that have a high livelihood and cultural value in any nation and thus are to be controlled by the nation itself. For PNG the wholesale description, specimen set and classification of that complete species group is entirely in the hands of Australian institutes, e.g. Flannery (1990), Martin (2005). And it does not end there. Birds of Paradise (BoP) were an early obsession of western people and they collected and shipped them out of New Guinea *en masse* (see details in Laman & Scotland, 2012, Beehler & Laman, 2020 for annual numbers easily in the 100,000 s). Australia and the U.S. dominate much of that bird and BoP narrative, e.g. Australian James Gould—referred to as an 'obsessive' bird collector—creating an Atlas of Papua New Guinea Birds (https://www2.sl.nsw.gov.au/archive/discover_collections/society_art/png/index.html, Diamond, 1973, https://australian.museum/learn/collections/museum-archives-library/john-gould/ Laman & Scholes, 2012).

One may easily continue with the plants; virtually every 'good' and big herbarium will feature PNG-collected specimen (already the German colonists collected for many years specimen as part of their territorial explorations, e.g. Hiepko, 1987). It's

Fig. 25.2 Entering Papua New Guinea requires a visa; it's a strategic bottleneck and played that way from all sides

an important fact that so many collection expeditions were done, most are virtually not known to the global public and hardly published well, and if so, then usually with a long delay (GBIF.org for details, see Hiepko, 1987 for an example for Berlin-Dahlem/Germany). Japanese collections are widely inaccessible to most people, certainly not for PNG citizens. Arguably, most specimen of national collection expeditions are not worked up well, carry no metadata, are not analyzed well, or even known and published internationally [in English] or available in a goof format online for a global and PNG audience; they are hardly well referenced in space or time and thus difficult to use, if at all. Examples of an ongoing discussion for Germany in collected artifacts from abroad found here: https://www.facebook.com/berlinpostkolonial/). While plants and botany might be perceived more as a marginal matter, it's not. Just think of bioprospecting, ethnomedicine and copyrights of plant DNA and crops. It's a big deal and PNG sits on major resources, none in its own hands though.

Lastly, same can be said for the diseases and their medical records of PNG. People of PNG are not in the river seat but on the receiving end for AID. PNG cannot steer its own path…

Fig. 25.3 Food from abroad, e.g. sugar-charged and colorful with plastics in a supermarket, is a big ticket item for many Papua New Guineas, while local garden food, domestic poultry and some (canned) fish remains the choice of the day for the vast majority

25.2 Framework Changes and Shifting Baselines as the Rule

In addition, the initial framework that those nations are to operate under has changed a lot too (example for some of the Australian policy shifts already here: Wainwright, 2004). It sets even further limits for steps possible and for options, and for policies available to move forward in good and sustainable terms for its citizens and everybody else. Already seen from that angle, PNG—as a colonial construct (Papoutsaki & Rooney, 2006)—has little options really and was already heavily battered in the last 60 years during its 'modern' existence when it was tried to propel PNG into the global community of a market economy (Srijinda, 2019; for metrics and policies see Gosarevsky et al., 2019). It's a place where PNG has never been before and likely does not want to be in the first place really.

In the modern global perspective and value system, any land in public ownership—such as the Melanesian lands (Lea, 1993)—gets seen as undeveloped, unclaimed and available for anybody who can buy it. And thus it gets easily grabbed by outside developers and planners, e.g. for investment reasons and privatization (a typical example is the Special Agricultural and Business Leases SABLS; ATBC, 2011).

Fig. 25.4 Architecture of Melanesia: elegant, robust, very suitable and adjusted for the environment and its people (protecting against tsunamis, earthquakes and tornados; note the solar panels)

That's obviously entirely opposite from the PNG perspective and public land tenure in tribal hands and leads to fierce conflict (Baraka, 2001).

Same applies to ocean resources, e.g. giant clams, sharks or thuna (Havice & Reed, 2012). But PNG is a signatory of the (abusive?) South Pacific Tuna Treaty of 1988 and bound to those terms. In the meantime, many of the canneries it is to cater are not safe nor good for employment (Sullivan et al., 2003 for an example). The PNG labor force needs effective protection, unions!

But to cater and align with national economic growth—to be 'modern'—the state of PNG moved into the rural areas all just done in search of natural resources it needed to raise tax revenues, ideally from industrial sources. The government put special emphasis to the mining and forestry industries—farming is ongoing now-(Baraka, 2001) but with fatal consequences (see for instance chapters of this book and citations within). Global corporations were to drive the local agenda with no consideration of the deeply co-evolved fabric in PNG and its people, its land, or of any mutual form of democracy or sustainable governance.

Initially, and back to deep times, PNG was mostly on its own and stayed there for millennia, being a more or less sustainable entity in the wider global picture (Flannery, 2004). PNG was mostly left alone and then part of the wider universal earth and ocean fabric and cosmology (the human population worldwide was probably then much less than a billion). PNG evolved till then, but at least ecologically it virtually never stepped out of its bounds. The global footprint that PNG life had the

Fig. 25.5 Roads as an essential piece for any industrial nation (Note: Good tires are essential, but shoes are not a requirement necessarily)

last 47,000 years was widely unrecognized, and it was without much effect (Flannery, 2002 for a discussion). And that 'independence'—subsistence within ecological limits; a steady-state economy (Daly & Farley, 2011)—is also part of the wider global context where PNG and other nations simply did not exchange much when compared to globalization now. And thus they all were on their own but together on a wider planet under the wider universe. They essentially operated within what was available and provided by the earth, the ocean and the sun. It's a life. And for over 97% of the time of that PNG existence it has been good for PNG and its sustainability and for the world. Most of its time PNG lived well without much direct outside influence, without colonial nations, without world wars, without industrialization and globalization and even without the potatoes, pigs, mining, beer or Coca Cola and supermarkets, for a matter of fact.

When the first globally organized—so-called—new-world order was defined in Europe, at the Treaty of Tordesillas, the Westphalian Treaties, the Berlin (Congo) Conference, and then WW2 onwards, including the subsequent Bretton Woods conference, G7 & G20 sessions as well as APEC meetings, the western nations, the U.N.—and specifically Australia and the U.S.—were very thankful to PNG and made full use of PNG and its post-colonial construct. App. 37,000 conscripts from PNG joined the U.S. troops to fight during WW2 (see Cousteau & Richardson, 1999). In a way, PNG allowed the colonial nations to keep up their structure and interests worldwide to this very day. Global trade companies developed and reached out, all

Fig. 25.6 Straight roads: most western funders will be impressed. Straight roads in an ancient landscape are a total industrial artefact though of the Anthropocene; it's usually provided by an outside funder and comes with major strings attached, e.g. export quotas for mining products, rice exports and automobile imports

the way to PNG. And in WW2 this was defeating the Japanese invasion; the PNG cargo cult developed around it (Diamond, 2011). A globally relevant battle in the Pacific Theatre was fought in PNG for the 'free world': that was the outcome and value of the infamous Kokoda trail battleground and of other areas like Lae and Markham valley, Port Moresby, and some islands of PNG (Chan, 2016 for firsthand examples). The western world owes PNG a lot.

Subsequently, the remote but powerful nations play a large role in PNG, historically and now. Nowadays, this is China (Smith, 2013) and the U.S. But earlier, it was Germany, Holland, England, even Russia, France, Spain and certainly Australia with New Zealand and Japan. Many of those nations are the prime investors for PNG even today.

And within that, the Australian influence on PNG is all dominant (well documented by Australia itself, Nelson, 1982; see Gosarevsky et al., 2019 beyond 'just' mining). It's a long-established tradition going back to the early 1800s (Flannery, 1994, Nelson, 1980, 2016). It's very obvious in the culture, infrastructure, administrative and policy legacy. For instance, places like the 'Papua Club' in Port Moresby are a striking example and this has been a central policy powerhouse for colonial PNG for a long time. As described by Cousteau and Richards (1999, p. 205), it actually makes PNG politics and policies, and yes, it is strongly driven by Australians and the British, with females having traditionally only a very limited access (allowed to enter

Fig. 25.7 Road support from international donors comes usually with self-interest (It's self-evident that roads and bridges promote car sales)

four times a year)! Or in other places like in Madang and Lae, one can detect such things in the Yacht Club and Gulf Clubs, where mostly white men tend to meet and discuss matters of sailing, fishing, gulf, Ramu Sugar, the locals, Australia and PNG overall. And who can own a yacht in the first place in a nation where many people make less than 4$ a day? Gulf Clubs might carry those powers also (as described with a real-world outcome for 'Ramu Sugar' in https://postcourier.com.pg/how-a-game-of-golf-sweetened-ramu-sugar/). If that is how decisions are made one must be scared; it's not democratic and does not serve PNG, its people or the environment.

PNG made it to the world stage through its portrayal as savages and cannibals in the remote jungle not to be trusted (Richardson, 2015), Birds of Paradise (e.g. foreign expert photographs and text by Laman & Scholes, 2012 for National Geographic etc), cannibalism (Hoffman, 2015) and ubiquitous Penis Gourds (Flannery, 1994, 2002; see bare naked photographs in Flannery, 1990, p. 119, 134, 216 and Martin, 2005, p. 13). Through the colonial wars and subsequent world wars, PNG was made famous when Australia fought its first battles in the Great War against the Germans and their coconut fiefdom and commercial enterprise, e.g. taking over the 'wireless' stations in Yap and Nairu (New South Wales State Library, 2015). Finally, PNG became a friendly contributor in the 'Pacific Theatre' when the allied troops fought off the Japanese there for a 'free world'

All those nations remain with direct and deep ties to PNG and have vested interest on several political levels, e.g. fisheries, resources or culturally (see Germany

Fig. 25.8 The logic step from roads over selling vehicles to gasoline is already an easy link from the 1950s

as a part-owner in the devastating Ok Tedi mine (Kirsch, 2014) as well as in the Bougainville mine: http://www.bougainville-copper.eu/comment-kommentar02-3.html).

One can usually track back those colonial power links through the sciences, specifically Anthropology (see chapter of this book), Ornithology (see for instance German bird collections described by Stresemann, 2917), plants (see also German collections in Berlin-Dahlem as described by Hiepcko, 1987) and collection of artifacts and specimen, as found in large museums abroad (Knoefel, 2021 for a German discussion), including the Vatican (Pietrangeli, 1982; Metzler, 1990). In reality, and more linked to money, one can find these frameworks of operations and their tools in the Asian Development Bank, The World Bank (MacWilliam, 1986; Rich, 1994), and also linked with Kyoto-type agreements and UN Climate Change policies. This is not too surprising for the traditional western nations to see. Now China is a rel. new international player also, and so are other nations (see for The World Bank investor

Fig. 25.9 Scare the bad ghosts away, PNG-style; it can be done

Fig. 25.10 Village harmony (mind the coconuts falling down)

Fig. 25.11 Beauty remains (a torch ginger *Etlingera elatior*, widespread over Pacific Islands and an ornamental plant species in many tropical regions including Asia and Central America)

information here inviting virtually anybody who can afford it: https://tcdata360.worldbank.org/countries/PNG).

The actual role of China for PNG is not so new, it goes back easily till 1920s (see book by Chan, 2016 for those direct and early links and exchanges), but also archeologically it is found that PNG has a close link with mainland and island Asia for many thousand years for its coastal and islandic people, (Cousteau & Richardson, 1999; Flannery, 2002; see Beehler & Laman, 2020 and citations within for Lapita culture). PNG actually still is a mix of people, adding to the country of 1000 nations…

There were major events with many of those outside nations and with relevance for PNG; see Table 25.1. This allows for a comprehensive understanding of PNG's international role and actors in PNG (Table 25.1 & 25.2).

25.3 Modern PNG is Virtually not Independent nor a Democracy: Examples

A typical example of 'modern' use of power is shown in Boyden et al. (1975). It uses for PNG a science argument through a world-recognized institution for one of the biggest money makers there: a mine. Can a Cecile Rhodes mind get more re-invoked than that?

Table 25.1 Selected list of outside nations driving major events and instances in PNG, many more can be named such as due to the World Trade Organization (WTO) impacts

Event	Detail	Comment
Treaty of Tordesillas, Westphalian Treaties, Berlin (Congo) Conference	Set up of the world's colonies by European nations and their churches all done from remote Central Europe with just a few family-linked actors and competitors and their naive understanding of the world	This has not only deep impacts today, but was essentially a massive global real estate grab, including resources disowned for very sophisticated cultures who lived there for millennia in a somewhat very sustainable fashion
WW1	Breakup and re-arrangement of the colonial ownerships	Meant to be The Big War; short and resolving all wars
WW2	Determining the set up of the world order as we know it	A true world war based on war declarations of Germany and Japan against 'the world'; driven by a dictator and an emperor
Australia signs protectorate	Australia essentially extends its zone of influence, its territory, and makes more money, getting elevated to a global stage	While officially ended, this situation has essentially not changed till today and continues as political policies, treaties and financial 'tools'
Russian anthropologist Nicholas Miklouho-Maclay	Operated in northern PNG	Argued for humanitarian considerations
Australian—PNG scandal https://www.afr.com/companies/energy/asx-oil-firm-mired-in-15m-png-bribery-scandal-20200207-p53ypj	Bribery and corruption of Australian oil firm	Several others exist
China Nickel mine and pipeline (Ramu Nickel)	Brought a big push of direct Chinese impact to the region and PNG overall	Major environmental crisis and shortcomings
Manu island Refugee policy	Pacific refugee crisis to be resolved through PNG	Australian efforts backfired
APEC summit 2019 in Port Moresby	Link Papua New Guinea with Pacific and global business	Has not brought much gain, and PNG leadership got dragged further into corruption. but Port Moresby got some new roads

As shown in Table 25.1, even in the last 50 years, PNG is dominated by the outside, essentially just like it was before and in the colonial times (see Baraka, 2001). PNG—as a nation construct—actually comes from abroad, it literally started in Europe, helped by Australia, the U.N. and the U.S. The case of Bougainville (part of northern PNG and Solomon Islands, a German colony given away to Britain (Ohff, 2008) but Bougainville remained, then got added to PNG and with an Australian mine

Table 25.2 Selected list of items from abroad in Papua New Guinea

Item from abroad	Origin	Relevance	Comment
Computer	Usually IBM, Apple MacIntosh, etc	Modern nation cannot function without it	A classic issue that now also gets more relevant in industrial nations, e.g. the EU and Germany have quite a limited software development and market when compared to the U.S. etc
Phone Network	Australia	Modern nations need an effective and cheap phone service	Privatization or government-owned; both concepts have problem
Cars/trucks	Usually Japanese and Asian producers	Transportation, public transport	Import, left-sided steering wheel. PNG has no train system
Mobile phones	Australia, Asia	A 'must' for any modern nation	Problematic on many accounts

but once more with some German engagement via their EU[1] bringing soon a civil war and a fundamental PNG crisis including UK mercenaries to Port Moresby). And the western nations, many in their core civilizations based on the western mindset, are fully rooted and operating outside of PNG, including Canada and South Africa. PNG was to play to that tune, but did not. Instead, PNG is clearly driven by tribal structures and local dynamics regardless (details for the Porgera mine dispute in Golub (2014).

In the meantime, PNG delivers to, and relies on, outside markets: For instance mother-of-pearls (Simard et al., 2022), or crocodile farms at Angoram, Sepik River as a good example where crocodile skins are sold to EU and Japan. Similar applies to butterfly farms (e.g. Beehler & Laman, 2020, p. 135), and with PNG orchids (Orchid Society, 2006) and the Birds of Paradise (BoP) trade. All of this keeps an impact footprint in PNG but money is made elsewhere underlying massive global market variations.

Another good example of how the outside shapes PNG from the inside is shown with PNG's road system, as a core infrastructure needed for any 'modern' nation to function (just think of the car industry, trucks and 'goods'). The road from Mendi to Mount Hagen was actually motivated and widely built by Chevron, but not by the PNG government or planned for its people (Cousteau & Richards, 1999, p. 102). The inequality of road access and associated development has been shown by Beer (2017) and Beer and Church (2019). Japan has traditionally provided roads and bridges worldwide (see Huettmann, 2015 for examples in Central America to obtain fish rights in exchange), as also done in PNG. Such an investment helps getting at resources, e.g. when leading to ports, and it further results in a demand for cars and

[1] :http://www.bougainville-copper.eu/comment-kommentar02-3.html .

oil, as well as rice; the classic model unfolds once more driving many basic human needs. How more can on control a nation?

Exotic species can be added to that list. Already, the invasive species of *Salvina molesta* was not there in rivers 1960s and 1970s but came then from outside of PNG. It quickly took oxygen off the river and fish starved, people lost their income from the river and had to move (Cousteau & Richards, 1999, p. 145). More of those vast outside effects are brought by the U.N.-supported fish species and aqua-farming introduced in the wider Sepik Region (Beehler & Laman, 2022), As a matter of fact, PNG deals with over 21 exotic species that create local problems.

Of course, even larger outside effects came with the missionaries destroying PNG cultures on a landscape scale. This is well-documented – in PNG and outside. Beehler and Laman (2020) p. 325 re. religion, these authors refer to it as a crime against traditional humankind). Main actors are the London Missionary, the German Lutheran Church, the German Catholic Mission, and many North American groups (some details shown here http://www.pngembassy.org/religion.html). It's a bit surprising because PNG is a British Dominion lead by the late queen and thus in the hands of the Anglican church, not? But many believes act in PNG instead. Despite the devastating cultural losses, all brought from the outside, sorcery remains (Beehler & Laman, 2020, p. 326–327; Gillison, 1993 and see real-world example in villages for witch hunts with The Guardian, 2021a).

Another factor fully out of the hands and control of PNG are the word's economy during globalization, as well as associated recent 'world wars' (those are large-scale ongoing international wars involving many nations and stakeholders such as in Afghanistan, Iraq, Jemen, Sudan, the global war on terror, the war on drugs, Ukraine, etc.). For instance, already the economy chaos in 1997 in Asia resulted in intense forest exploitation in PNG and is still ongoing (Beehler & Laman, 2020). PNG pays the price of such a situation and from the spill-offs. Already the WW2 brought one of the world's largest use of dynamite through its massive bombing raids. It left behind PNG as a collateral, which then used the leftovers for dynamite fishing destroying unique reefs, fish stocks etc. It did not only killing and injuring fishermen but it also destroyed long-term precious ecological services (details in Cousteau & Richardson, 1999).

The recent enterprise in seafloor mining, where—once more—a firm from remote Canada runs bankrupt in PNG (Mining Watch Canada 2019) shows us no other: outside influence now runs its course even in PNG's untouched seafloor where no human has even been before; that is the curse of mining and natural resources. Many more of those examples exist, e.g. see The Guardian (2021c) where already the headline makes the PNG problem clear, now as a truly international one:

> *"Frieda River mine proposed by Chinese-owned PanAust sparks appeal to government* in Australia *where company is registered".*

It's almost a side detail that this happens to be—again—one of the largest mine projects in PNG.

Textbox 1: Contrasting deep spirituality, cosmology and the Melanesian cargo cult with the colonial Western Coconut and Carbon Cults: How PNG handles and implements outside influences as a ritual
Islands are known for their diversity of cultures and biodiversity. They 'drift', e.g. genetically, and are microcosmos of the universe and mankind. Cultures are evolving in response to the outside environment. The fast and alive culture of PNG was easily seen in action during WW2, when 'strange objects came from sky and unloaded products/cargo.' It was perceived in PNG as outsiders, gods sent from heaven, bringing goods and leaving them behind. The Cargo Cult was born; different islands and tribes pursued it in various fashions, often in secret. To anthropologists, it's a great indication of culture and human evolution in real life, caused by global and industrial events. In this case, military planes got worshiped.

Those things and evolutions have been described earlier and in (fiction) books like 'The Lord of the Flies' by Nobelprize winner W. Goulding, from U.K. Within the German colonies, a German coconut cult was started by August Engelhardt (as a public topic and covered by poetry see details here https://en.wikipedia.org/wiki/August_Engelhardt). It had it many Central European followers, but most died or came back. Perhaps those trends remain—as a cult—in the western obsession with pure and organic food?

In modern times, a so-called Carbon Cargo Cult evolved (Filer, 2010), somewhat unnoticed by its participants (the global society, trying to fight industrial society impacts but living fully in it for their own benefit). It's based on the increasing use of fossil fuel encroaching into virtually all aspects of life but CO_2 release getting treated as 'holy' and untouchable; simply 'trade it away.' Clearly, no human society is free from outside influences and its adjustment and evolution. The global culture drifts indeed, always based on the best-available knowledge of its time; and new machines can create new facts, a (techno-) cult ongoing. While the computer, online and mobile phone cults receive less attention by people, in reality, it's part of a similar phenomenon; PNG has such events ongoing for centuries; it's part of human governance also found in all other human societies (and just await it to happen in MARS missions, let's say)

The so-called modern nation and its evolved lifestyle—perceived by some as the crown of human society—stand in good contrast to the PNG life and sustainability of the last millennias. As most urban areas show, there is unsustainable living in such population centers, as part of a failed western governance scheme and framework imposed onto PNG. Like many urban areas, PNG cities are in certain freefrawl chaos, e.g. unplanned and not safe (see Lae's large-scale prisoner escape; The Guardian, 2017). As a result, areas around those centers are overused and described as drowning in crime and lack a stable social network for betterment (Lasslett, 2018). It follows the generic trend of slums (see Davis, 2004), as typically found in nations that run a mining curse, receiving 'aid' from international donors to boost an economic growth. It's nothing unusual that oil, gas and mining are linked with this and initially promised' massive wealth when actual poverty and decay of wilderness biodiversity is the outcome. A world with most people earning less than $4 a day shows us just that.

25.4 Conservation Designed by the West Failing in the West, Failing also in PNG

The way how nature and the environment are now understood, globally, is widely driven by a western mindset, as promoted by their media and the sciences to please the funders (a typical example found for New Guinea in Beehler and Laman 2021 exposing how protected areas are designed around mines so that business as usual can go on while ecological integrity is truly sacrificed). It usually comes with a business model and promotes a western-style sustainability. But in reality, it always means: money first and nature second (Taber & Payne, 2003). Biodiversity conservation has no chance and is not an equal partner in that. It's never a fair win–win there (typical example shown in Gulob 2014 for the indigenous group of Ipili). Slogans heralded then by the world agencies run something like this: "*National parks and protected areas are to save the environment; just set aside 15% and we can keep running industrial society as 'business as usual'.*" In such a mindset, protected areas and conservation are something liked a micromanaged tiny fenced-in park. The failure of this model is logical, fact-based and consequential. It can easily be seen on many indicators; for instance, the rise of invasive species and the decline of the North American avifauna for nations that are rather rich but still cannot get the environment under control and for a sustainable perspective (Rosenberg et al., 2019a, 2019b). Under such a regime, with tiny and fragmented protection areas, wilderness maintenance and the unleashing of large-scale ecosystem processes - including migratory species conservation and healthy and pristine watersheds - are hardly possible. Ecosystem processes do not stop at a fence or a national boundary. Climate change has shown the problems more than no other (Lawler et al., 2011; but see The Guardian, 2022 for upcoming legislation pursuing industry).

Western Nations are great in using their self-centered methods and reasoning; a typical example is found in legal, contractor and university-based Conservation Assessments, Impact Studies and Optimizations; see for PNG with Alcorn, 1993 and Hamilton et al. 2019 for oceans. Beehler and Laman (2020) refer to the optimization efforts for protected areas as just drawing circles on a map. Arguably, such a concept with context fails in PNG and Melanesia where th underlying property concepts are not private but a common good and widely shared (one reason why in Hawaii beaches are public and so appreciated; for the U.S. that widely differs with New Jersey or the Carolinas, let's say).

If one leaves it to the dominating western science perception, one will find oneself outside the serious reality and natural world – as in 'The West' all else but science would not be so serious. And thus, taxonomy is usually the assumed root discipline required before any of their science-based conservation is to happen, to be based on 'mechanistic' research with the need to 'understand' everything first and significantly one-by-one (which is not happening and impossible in the first place for humans to achieve). Endless discussions with science entities make clear that conservation is seen as a dogma but taxonomy and DNA work are seen as objective, pure, unbiased and worthwhile (while an achieving conservation and sustainable culture is ignored in that process; see Steiner & Huettmann, 2021 for an example). This is nowhere so clear than in the specimen collections that are ongoing in PNG (aka "*PNG remains significantly undercollected*" as promoted by Beehler & Laman, 2020 p 132). The conservation progress for PNG's wilderness biodiversity is quite minor, the taxonomic disagreements are vast and remain unsettled for often more than a century (see in GBIF.org for taxonomic uncertainty in specimen holdings, see also Lowe 2004). But the taxonomy concept (Steiner & Huettmann, 2021), and the mandated underlying science or conservation concept, is so widely not reflecting what PNG all has to offer, and what it does, or wants and needs, or how it operates. While much is at decay, PNG is more and beyond what the western world understands and sees (Gillison, 1993, 2002).

In PNG, the concept of 'owning,' of having own materials wealth and property, e.g. a specimen, is quite foreign. Best to share it and to give it away as a present and to put it to good use with the wider world. Do not get in the way of others. PNG has a complex publicly owned land tenure system (Baraka, 2001; Beehler & Laman, 2020). It's community-based instead; as a typical feature in Melanesia (e.g. Lea, 1993). Many PNG people simply cannot relate to the 'modern' understanding of nature, and associated privatization or protection. That's not how they know, experience and see it. Their views on nature differ. Nature is a living world, spiritual, with ancient ghosts and taboos, and fully embedded within the universe. Nature in PNG can be terrifying, but it is also a workplace to make a living. Nature is the garden that provides; if treated well. Nature is a common good and shared with the villages and their communities. Those are the sustainable concepts taught 'in the bush' (Baraka, 2001; see also Gillison, 1993 for an example).

Instead, now we see major outside influences on the species level even; typical examples are found with the infamous potatoes introduced by Portuguese from Latin America and which changed the ecology and balance in PNG, and in many other parts of the world including central Europe or mid-west U.S. forever (Smith, 2012[2]). Many other examples can be named, e.g. highly breed coconuts (see details Cousteau & Richardson, 1999 or Beehler & Laman, 2020), sugar plans, and then, of course, coffee. As a cash crop, it was introduced in the 1950s by rich (Australian) miners from abroad changing much of the economy and fabric of PNG (Cousteau & Richardson, 1999; Baraka, 2001). Coffee as major cash crops for PNG, grown for export, which

[2] There is some discussion that the sweet potato was introduced to PNG earlier; perhaps several waves of introduction occurred and different subspecies, etc.

then put PNG, its products, landscapes and people under direct outside influence from abroad, once more. How imperial can it get? And of course, PNG citizens are not used much to drinking coffee—or tea—anyways; it's not indigenous. Coffee was widely promoted by the native African, and later the Arab world; it got used in Spanish colonies

Invasive diseases like the coffee borer present other examples how PNG is interwoven and linked with the context; the outside world. Diseases like the sugarcane smut (Tom et al., 2017) show the same pattern. Invasive species are additional examples, e.g. cane toads (van Winkel & Lane, 2012).

25.5 Subsequent Environmental Policy Laws Designed by the West, Failing in the West, Failing in PNG also

Arguably, many 'modern' policies and the law in modern PNG are not really made by, nor come from, deep PNG and its people. The law, as we tend to know it, comes usually from western sources and western philosophy, ethics and values; Westminster style and Federalism; for PNG and Dominions that includes the queen even., based in remote London as the head of the state. Clearly that's rather ambiguous to roll this out globally then and for PNG (where a legal practiced community law existed for over 47,000 years; Baraka, 2001 and citations within. PNG features now a legal system commonly done in the Dominions, with the Queen of England as the legal as well as the spiritual lead. A reference to the ancient Greek, 'the Romans'—or subsequent Christians—must fail already on the grounds that these initial nations failed over 2000 years ago themselves; just like the Egyptians none play relevant roles in modern global society or in PNG back then. And we just see their old artifacts decaying further (Acemoglu & Robinson, 2013). So why to follow them, why to herald them, why are these the baselines now, and why to build an entire new nation-state from it? And why still done now that we have computers to deal with during globalization and all can be on the www in a better way, open and transparent? Already an easy link made to ancient cultures and indigenous governances like the ones found in Mongolia, China, Vietnam or Amazonia would be equally or more relevant for PNG (see Chappell, 2005 for Africanization of the Pacific). A western, British system of law (Twomey, 2006), with consistent Australian enforcement and policing (Gosarevsky et al., 2019) has no good cultural footing in PNG, it's hardly a success or sustainable whatsoever

In modern times, PNG has been seen as a resource reservoir free for the taking. That is true for its natural resources, likely for some PNG labor too (see for brain-drain Negin, 2008). But also many museums and 'private' collections abroad are full of PNG art and specimens, and vast collection efforts were carried out in PNG to fill museums, collections and books (a German review on the topic of artifacts shown here but which does not have PNG mentioned yet: Der Spiegel, 2021). To show the impact of wealthy people on PNG's research, Ted Turner supported Cousteau

Table 25.3 Selection of laws that fail in the west as well as in PNG

Law	Topic	Failure	Comment
Climate mitigation	Global climate change	Widely unconstrained CO_2 and GHG emissions for over a century	Essentially totally unresolved, and unlikely resolved well any time soon
Clean water	Water use and selling	Contamination	A public good; PNG is one of the wettest places in the world
Wildlife conservation	Sustainable hunting quota setting	Poaching, ecological impacts, sustainable management, lack of resources, missing MSY most of the time	A marginalized problem but widely unresolved
Human rights	Immigration, education access, domestic abuse, fairness, right of work	Refugees, world peace	A topic defining world wars, sustainability and global unrest

and Richards (1999) work in PNG, and so did Statoil's Archebold expeditions for a long and excessive time (https://www.archbold-station.org/documents/publicationspdf/lohrer_2019_11ArchboldExp.pdf), including M. Rockefeller (who received a tragic death in a 1970s boating accident during a PNG expedition but which then got flipped around again and adding to PNG as a mythical place of cannibals with foreigners cashing in further (see Hoffman, 2015) (Table 25.3).

25.6 Climate Change is not an Apparent Game-Changer for Power, but makes it Worse

PNG got world-famous for speaking up in Bali 2007 at the climate conference to the U.S. and asking this nation for good action (https://www.nytimes.com/2008/01/22/science/earth/22conv.html). "Bravo" to the PNG representatives centered around Kevin Conrad, an American born in Sepik region (now living in New York). Still, PNG gets bullied down and has not much to say in an economy where power is the only metric that matters. Instead, climate change seems now to change global dynamics for no change: the powers in charge remain, thus far. This applies unless they get destroyed themselves by the climate they create, such as by sea level rise, e.g. Shell in Holland or BP in UK (Dunlap & McCright, 2011; Grasso, 2019; Jansen et al., 2020).

25.7 The Global Powers Reflected in PNG and That Never Allowed PNG to Grow

25.7.1 England

At some stage in the recent times England dominated the world or was close to it. The Anglo-Saxons, initiating from northern Germany and southern Denmark region, developed into a world-leading nation and culture - land and sea (Morris, 2021). However, that has changed (see Srinivasan, 2005 for declines). But in the meantime, the British Dominion remains and has its wealth centered in London, and the late queen/royals remain the power in charge. PNG was heavily influenced by the UK, and its royal court features regular visits to PNG (e.g. George & Millet, 1984). The essential link of PNG with UK can easily be seen in the Sandline scandal—a mercenary company from Britain that was called in to handle the Bougainville conflict. More can be said, e.g. through the 'city' of London as the non-sustainable finance hub for the Dominions (Norfield, 2016).

25.7.2 U.S

The U.S. is a spin-off and reservoir of the failed European feudal system. It's relatively recent and established itself just in the last 100 years or so in the Pacific Theatre. The recent history of Hawaii reflects that clearly, and even more so, after WW2 the establishment of U.S. bases in Guam and other places of the Pacific (e.g. Okiniwa/Japan, South Korea). In WW2, PNG played their part for the Pacific Theatre, among Singapore and Hong Kong. PNG was a U.S. battle zone in WW2 and General MacArthur set up his war office in the infamous Port Moresby hotel (Papua Hotel). The won-war resulted into a wide U.S. oversight for the Pacific. One reason for that is because England during the WW2 had to make many concessions of its earlier colonial empire and transferred it to the U.S. This U.S. influence can be seen beyond military power, as Chevron exploration in PNG was pursued in the Leonard Murray Mountains to Kikori for a pipeline and port. The mountain landscape development and impacts from the oil can be seen here (https://www.chevron.com/-/media/chevron/investors/documents/png_factsheet.pdf).

Exon Mobile then just works along the same lines, with a dubious concept (Main, 2021) and business model; and their LNG plant was already coined as '*economic parasitism.*' (The Guardian, 2018).

And this path of influence is not an isolated case. Powerful nations like the U.S. can easily dominate other and smaller nations; for instance described by Azie (2021) for Cameroon. It's just a global scheme (see Blum, 2003 for global list, patterns and overview). Either way, it's difficult to deny that U.S. is the powerhouse for PNG politics, as implemented locally through Australia and the U.N.

25.7.3 Australia

The link of PNG with Australia and its influence has received its own assessment (see literature and book chapters with citations); it's all institutionalized. It's also documented elsewhere and stated in widely used books and media, e.g. Nelson, 1982, Wilson, 2019. Australia itself struggles, e.g. Evans et al. (2016). For that, the Australian concepts of 'help', aid and engagement come as a template of failure and tried and achieved dominance, long-term (e.g. Barak 2001). A typical example is Paga Hill in Port Moresby. The Paga Hill Development Company (PHDC) is a company registered in PNG but it has strong and direct ties to Australia. In 2012, it announced its plan to create the Paga Hill Estate (resort, casino and an aquarium). And so, 2012 onwards the police, backed by bulldozers, began moving into the settlement of Paga Hill in Port Moresby, to clear the area. This was widely done against the citizens, all just possible due to outside/Australian funding efforts (details described in The Guardian, 2021a).

Generic drug runs and smuggling are almost 'standard' by now in the world, e.g. The Guardian (2020) for Australia and PNG. Many more of such links can be named starting from Australia and spilling into PNG and back (see for instance The Sydney Morning Herald, 2015 and NME, 2021).

Still, the Australian case at Bougainville will stand out as a textbook example of outside interference shaking up an entire nation done for money and self-interest (Filer, 1990; Lasslett, 2012). It was devoted a chapter of its own in this book for more details and history; see citations and studies listed within.

While those types of impact appear as 'isolated cases' and petty reporting, they can be of major impact and shock PNG finances. As a matter of fact, the largest reported fraud done in PNG thus far was done by an Australian, and it was done repeatedly (The Guardian, 2021b). This citation reads "*Police allege 268 m kina was misappropriated from the Western Province People's Dividends Community Mine Continuation Agreement (WPPD CMCA) trust fund and improperly paid into the accounts of the Ok Tedi Fly River Development Foundation and Sheppard's law firm.*" It goes further and then reports "*Sheppard was also the manager of Brisbane indie-pop group, Sheppard, of which three of his children are members. The group had a No 1 Australian hit single, Geronimo, and performed at the AFL grand final last October.*"

It's pretty clear that Australian business—including some of the entertainment industry—is linked with PNG, benefits from PNG, and that it affects both nations and even the culture, e.g. in an unsustainable fashion. If that is not enough, see the 'How to Guide' for corruption between PNG and Australia '(The Sydney Morning Herald, 2015).

As a matter of fact, and as outlined by Baraka (2001) the independence of PNG came less pushed from PNG citizens themselves—some highlanders were widely opposed—but was more of a cost-saving, power grab and income measure by Australia as the paying Protectorate power in charge under the U.N. mandate. PNG suffered from that concept ever since (Guvaretsky et al., 2019). Claiming a partnership between Australia and PNG makes it worth, as this relationship is not equal or even, and thus, turns quickly abusive. Many examples are provided in this book and elsewhere.

25.7.4 EU (Germany, France, etc.)

The EU is driven by colonial powerful nations such as Belgium, France, Germany, Denmark, Sweden, Portuga, Italyl and Sweden. The EU includes 27 nations but itself is widely dominated by Germany and France which provides a vast amount of money and a large power set up and administration for other EU member nations - an entire continent really with global spill-over; governance formulas within the EU make sure how those funding distributions and voting rights are institutionalized. Also, Germany is a former colonial power in PNG and either way it plays a role from the outside; example shown here https://www.eeas.europa.eu/papua-new-guinea/european-union-and-papua-new-guinea_en?s=150 and here for Church Aid with a large legacy: Hempenstall (1975).

Already by geography one may happily link Bougainville's problems -in part - with Germany's colonial efforts (The Guardian, 2019). Despite Germany starting and loosing WW1 and WW2, each, this was not ended and is ongoing (see German role in the mine: http://www.bougainville-copper.eu/comment-kommentar02-3.html).

But beyond money, Germany left its long-lasting footprint in PNG. Already many geography features and names carry Germanic terms, e.g. Bismarck Sea, Hindenburg Wall, Mt Hagen, etc. And it continues with biology, like the taxonomic naming of some Birds of Paradise and Matschie's Tree Kangaroo, as well as Ernst Mayr (a German biologist later moving to U.S. but still working in Melanesia, e.g. Mayr & Diamond, (2001), simply remaining with two global powers throught history).

25.7.5 Japan

In the global public, eye the Japanese perspective to PNG is widely missing, see Flannery (2002), Diamond (2011) and Beehler and Laman (2020) for a wide lack of the Japanese science relevance and contributions. It's not surprising though to see Japanese airlines flying to PNG so frequently, and with tourists, many of them are curious to learn about their Japanese legacy and army losses there (see also for a Japanese apology to PNG regarding war atrocities: https://www.asianews.it/news-en/The-Japanese-Church-apologizes-to-Papua:-Peter-To-Rot%E2%80%99 s-death-is-our-fault-25276.html).

While in the Western World perspective, it might appear that Japan just plays a little role in PNG, in reality, Japan is a strong actor for PNG and the region. It's already big donor of aid and has close and frequent ties with PNG, namely fisheries, rice, some research and humanitarian help, e.g. volcanoes and earthquakes. In the meantime, finding data openly and useful for PNG expeditions and explorations, or aid, are hard to track down.

25.7.6 China

China is a major expanding nation and culture worldwide (Smith, 2011). Bun et al. (2004) report on a simple but deep link of China with the PNG forestry sector. While this appears to be sectorial, it actually links deeper and includes many of the political sectors, as well as mining and the wider human society and well-being (Laurance, 2011). But Chinese impact is now much bigger, e.g. Conolly, (2020) for the Road and Belt initiative, and it reaches into the entire region (see example for Solomon Island as well as seafloor mining).

25.7.7 Other Major Factors From Abroad: Western Religion but Taken Out by Bob Marley and Dalai Lama

Beyond national policy impacts, a major outside impact is brought by religion. PNG has one of the highest Christian mission efforts in the world. A brief religious overview for PNG is presented with the U.S. embassy (http://www.pngembassy.org/religion.html).

While Jesus was never in PNG, not even remotely, the influence of Christians onto PNG must remain mind-boggling. The transforming impacts of western religions, many of them by many denominations, have been rather devastating for PNG landscapes, cosmologies and nature. (the naive example of tree kangaroo hunting not mentioned by Jesus and the bible and thus forbidden to PNG villagers is shown in Montgomery & Bishop, 2006).

It should not be forgotten that in PNG, Islam also exist, e.g. in POM as well as near the border to Indonesia, but has a lower abundance in PNG society.

Instead, the extremely popular Bob Marley in PNG—the Jha (Toynbee, 2007)—found all over PNG, as well as Buddhism (Buckley & Lama, 2021) or Hinduism have good and alternative messages to bring to the table: World Peace. So why not promoting it, and for PNG? At least for Bob Marley, this is already in existence.

25.8 The Old-New Colony Hardwired: Will PNG Ever Be a Self-sustained Democracy Safeguarding Its Trusted Wilderness Resources?

To answer this question, I think the facts show us easily a NO whether PNG will be an equal player in the global fabric any time soon (Govarestky et al., 2019). At least PNG will not make this transition within the next 20 years, likely not the next 50 years. PNG is so locked in with the UK system, Australia and the U.S. and China that PNG cannot compete and cannot really be viable. The bilateral and multilateral treaties allow for none of it. Some examples are shown in Babbage (1987). Small cliques seem to run the most relevant decisions in PNG (selected examples shown in Green et al., 2017).

In addition, PNG is obliged to follow international laws, directly or indirectly (see Table 25.4 for a short overview). Those make it very difficult for PNG to operate according to PNG terms, or in a traditional Melanesian way (Narokobi, 1983). Those laws are designed and made for the western world and benefit those the most, as cemented in the U.N. power structure with their administrative culture, membership fees, and permanent seats in the Security Council.

As a Dominion of the Commonwealth PNG is exposed to an overwhelming Australian-helped international powerhouse, with U.N. mandates (Baraka, 2001) and U.S. dominance. PNG plays a major role in the 'Pacific Theatre' for U.S. security and business.

> Perhaps the attitude can simply be summarized with a quote from the recent U.S. Presiden D. Trump "Fuck the law. I don't give a fuck about the law. I want my fucking money" (https://twitter.com/emilybobrow/status/1094280529434820608).

If all else fails then, perhaps it would be good for all actors to consider alternative concepts (James et al., 2012). Out-of-the-box thinking can help. On a purely international level, and while Bob Marley is widespread in PNG but apparently already forgotten by PNG leaders, for instance the Dalai Lama brought already forward a lot of new and innovative conducts and thinking for international policy, diplomacy and achievements (Buckley & Lama, 2021)?

It's time the sustainable PNG approach hits the global agenda; what can the damage be, considering the western dominance failed everybody so much?

Table 25.4 Selection of international laws and frameworks of relevance for Papua New Guinea or where PNG is a signatory

Policy	Citation/URL	Relevance	Comment
Convention of biological diversity (CBD)	http://www.biodiv.org/	Biodiversity management and protection	Biodiversity is poorly known and studied and suffers from poor management and failing agency
Convention on international trade in endangered species of Wild Fauna and Flora (CITES)	http://www.cites.org/	International trade	Poaching and collection in PNG remain very high and poorly documented
Convention to combat desertification (UNCCD)	http://www.unccd.int/	Re-forestation	PNG features a massive forest loss
Framework convention on climate change (UNFCCC)	http://unfccc.int/	Carbon sequestration	PNG as the largest forest wilderness tract in the Pacific as well as ocean carbon sinks. Buta relevant carbon sequestration deal was not much successful
The World heritage convention	http://whc.unesco.org/	World heritage sites	PNG as a world civilization and wilderness place has just c. 2 World Heritage Sites
RAMSAR convention	https://www.ramsar.org	Wetlands of global relevance	PNG as a world endemic bird site and with many international migrants has just 2 RAMSAR sites

References

Acemoglu, D., & Robinson, J. A. (2013). Why nations fail: The origins of power, prosperity, and poverty paperback.

Alcorn, J. B. (1993). Papua New Guinea needs assessment. A USAID-funded consortium of world wildlife fund. The Nature Conservancy, and World Resources Institute Washington, D.C. with the Government of Papua New Guinea Department of Environment and Conservation, Boroko, Papua New Guinea. https://www.sprep.org/attachments/VirLib/PNG/png-conservation-needs-assessment-alcorn-1993.pdf. Accessed June 2, 2022

Anderies, J. M., Janssen, M. A., & Ostrom, E. (2004). A framework to analyze the robustness of social-ecological systems from an institutional perspective. *Ecology and Society, 9*(1).

ATBC. (2011). The urgent need to curtail 'Special Agricultural and Business Leases' in Papua New Guinea. Resolution, association for tropical biology and conservation.

Azie, N. (2021), Impact and prospects of the trump/biden administrations on Cameroon's political economy. Academia Letters https://doi.org/10.20935/AL3445. https://www.academia.edu/51164856/Impact_and_Prospects_of_the_Trump_Biden_Administrations_on_Cameroon_s_Political_Economy. Accessed March 17, 2022

Babbage, R. (1987). Australia and the defence of Papua New Guinea (PNG). *Australian Journal of International Affairs, 41*, 87–94.

Baraka, R. (2001). The meanings of independence in post-colonial Papua New Guinea. A Plan B Research Paper Submitted to the Faculty of the Center for Pacific Islands Studies in the Candidacy for the Degree of Masters of Arts. School of Hawaii, Asia and Pacifci Studies. The University of Hawai'i At Manoa.

Beer, B. (2017). The intensification of rural-urban networks in the Markham Valley, Papua New Guinea: from gold-prospecting to large-scale capitalist projects. *Paideuma: Mitteilungen zur Kulturkunde, 63*, 137–158.

Beer, B., & Church, W. (2019). Roads to inequality: Infrastructure and historically grown regional differences in the Markham Valley, Papua New Guinea. *Oceania, 89*, 2–19.

Blazey, T., & Perkiss, S. (2016). The empowerment of Papua New Guinea's tribespeople: Overcoming the challenges of foreign investment projects. *Australasian Journal of Regional Studies, 22*, 206–2024.

Blum, W. (2003). Killing hope: US military and CIA interventions since World War II. Zed Books.

Buckley, M., & Lama, D. (2021). In *This fragile planet: His Holiness the Dalai Lama on environment*. Sumeru Press Inc. ISBN 10-1896559735

Boyden, C. R., Brown, B. E., Drucker, R. P., & Tuft, S. I. (1975). Ok Tedl environmental study. Report of the 1974 Cambridge expedition to the Western District, Papua New Guinea.

Bun, Y., King, T., & Shearman, P. L. (2004). China's impact on Papua New Guinea's forest industry. Forest Trends. http://www.forest-trends.org. Accessed April 28, 2022.

Chappell, D. (2005). "Africanization" in the Pacific: Blaming others for disorder in the periphery? *Comparative Studies in Society and History, 47*(2), 286–317.

Cocounut cult https://en.wikipedia.org/wiki/August_Engelhardt.

Connolly, P. (2020). The belt and road comes to Papua New Guinea: Chinese geoeconomics with Melanesian characteristics? *Security Challenges, 16*(4), 41–64.

Daly, H. E., & Farley, J. (2011). Ecological economics: Principles and applications. Island Press.

Davis, M. (2004). Planet of slums. *New Left Review, 26*, 5.

Der Spiegel. (2021). Schluss mit der Verharmlosung. by U. Knoefel. 28th May 2021. https://www.spiegel.de/kultur/koloniale-raubkunst-schluss-mit-der-verharmlosung-kommentar-a-338f7afd-8f91-4da2-98e6-272cec7cf2e1. Accessed May 15, 2022.

Diamond, J. M. (1973). Distributional ecology of New Guinea birds: Recent ecological and biogeographical theories can be tested on the bird communities of New Guinea. *Science, 179*(4075), 759–769.

Dunlap, R. E., & McCright, A. M. (2011). Organized climate change denial. *The Oxford Handbook of Climate Change and Society, 1*, 144–160.

Evans, M., Stoker, G., & Halupka, M. (2016). A decade of democratic decline: How australians understand and imagine their democracy. In *From abbott to turnbull: A new direction? Australian Commonwealth Administration 2013–2016* (pp. 23–41). Echo Books.

Fabre, C. (2022). *Spying through a glass darkly?* Oxford University Press.

Filer, C. (1990). The Bougainville rebellion, the mining industry and the process of social disintegration in Papua New Guinea. *Canberra Anthropology, 13*, 1–39.

Filer, C. (2010). The carbon cargo cult in Papua New Guinea. Forests of Oceania Working Session, Association for Social Anthropology in Oceania.

Flannery, T. (1990). *Mammals of new Guinea*. The Australian Museum.

Flannery, T. (1998). *Throwim way leg: Tree-Kangaroos*. Atlantic Monthly Press.

Foster, R. J. (2002). *Materializing the nation: Commodities, consumption, and media in Papua New Guinea*. Indiana University Press.

George, B., & Millett, P. (1984). Papua New Guinea: A South Pacific democracy. *The World Today, 40*(8/9), 377–385.

Gillison, G. (1993). *Between culture and fantasy: A new Guinea highlands mythology*. University of Chicago Press.

Gillison, D. (2002). *New Guinea ceremonies*. Harry N Abrams Inc.

Golub, A. (2014). In *Leviathans at the gold mine: Creating indigenous and corporate actors in Papua New Guinea*. Duke University Press.

Gosarevski, S., Hughes, H., & Windybank, S. (2019). *Is Papua New Guinea Viable? Political Economic Bulletin* (Vol. 19). https://www.semanticscholar.org/paper/Is-Papua-New-Guinea-viable-Gosarevski-Hughes/6d71c70951c12fa82dea578a36b66923997f3f92. Accessed June 1, 2022.

Grasso, M. (2019). Oily politics: A critical assessment of the oil and gas industry's contribution to climate change. *Energy Research and Social Science, 50*, 106–115.

Gray, G. G. (1996). 'The next focus of power to fall under the spell of this little gang': Anthropology and Australia's postwar policy in Papua new Guinea. *War and Society, 14*, 101–117.

Green, P., Lasslett, K., & Sherwood, A. (2017). Enclosing the commons: Predatory capital and forced evictions in Papua New Guinea and Burma. In *The Routledge handbook on crime and international migration* (pp. 329–350). Routledge.

Guglya, L. (2021). The Wto trade and environmental sustainability structured discussions (Tessd): An introduction and pre-assessment. 2021, Academia Letters https://www.academia.edu/50307465/THE_WTO_TRADE_AND_ENVIRONMENTAL_SUSTAINABILITY_STRUCTURED_DISCUSSIONS_TESSD_AN_INTRODUCTION_AND_PRE_ASSESSMENT?email_work_card=view-paper. Accessed May 12, 2022.

Hamilton, R., Green, A., & Almany, J. (Eds). (2009). Rapid ecological assessment: Northern Bismarck Sea, Papua New Guinea. Technical report of survey conducted August 13 to September 7, 2006. TNC Pacific Island Countries Report No. 1/09.

Havice, E., & Reed, K. (2012). Fishing for development? Tuna resource access and industrial change in Papua New Guinea. *Journal of Agrarian Change, 12*, 413–435.

Hembernstall, P. J. (1975). The reception of European missions in the German Pacific empire: The New Guinea experience. *The Journal of Pacific History, 10*(1), 46–64.

Hiepko, P. (1987). The collections of the botanical Museum Berlin-Dahlem (B) and their history. *Englera, 7*, 219–252.

Huettmann, F. (Ed.). (2015). *Central American biodiversity: Conservation, ecology, and a sustainable future*. Springer.

Hyndman, D. (1998). Melanesian resistance to ecocide and ethnocide: Transnational mining projects and the fourth World on the Island of New Guinea. In H. J. Bodley (Ed.), *Tribal peoples and development and issues: A global overview* (pp. 281–298). Mayfield Publishing Company.

James, P., Nadarajah, Y., Haive, K., & Stead, V. C. (2012). In *Sustainable communities, sustainable development: Other paths for Papua New Guinea*. University of Hawaii Press.

Jansen, R., vn der Vlugt, M., & Zukowska, O. (2020). Shell and climate change reputation. University of Amsterdam Corporate Communication, Strategic Communication of Organizations Piet Verhoeven, Holand https://melissavdvlugt.nl/wp-content/uploads/2021/05/%C2%A9-Melissa-van-der-Vlugt-Shell-Climate-Change.pdf. Accessed July 15, 2022.

Knoefel, U. (2021), Debatte über koloniale Raubkunst: Schluss mit der Verharmlosung. Der Spiegel, Germany 28th May 2021. https://www.spiegel.de/kultur/koloniale-raubkunst-schluss-mit-der-verharmlosung-kommentar-a-338f7afd-8f91-4da2-98e6-272cec7cf2e1. Accessed May 11, 2022.

Kulik, D. (2019). *A death in the rainforest: How a language and a way of life came to an end in Papua New Guinea*. Algonquin Books.

Laman, T., & Scholes, E. (2012). *Birds of paradise: Revealing the world's most extraordinary birds*. National Geographic Books.

Lasslett, K. (2012). State crime by proxy: Australia and the Bougainville conflict. *British Journal of Criminology, 52*, 705–723.

Lasslett, K. (2018). *Uncovering the crimes of urbanisation: Researching corruption*. Routledge.

Laurance, W. F. (2011). China's appetite for wood takes a heavy toll on forests. Yale Environment 360 Magazine. http://e360.yale.edu/feature/chinas_appetite_for_wood_takes_a_heavy_toll_on_forests/2465/. Accessed February 20, 2022.

Lawler, J. J., Wiersma, Y. F., & Huettmann, F. (2011) Using species distribution models for conservation planning and ecological forecasting. In *Predictive species and habitat modeling in landscape ecology* (pp. 271–290). Springer.

Leah, D. R. (1993). Melanesian axiology, communal land tenure, and the prospect of sustainable development within Papua New Guinea. *Journal of Agricultural and Environmental Ethics, 6*, 89–101.

Lowe, C. (2004). Making the monkey: How the Togean macaque went from "new form" to "endemic species" in Indonesian's conservation biology. *Cultural Anthropology, 19*(4), 491–516.

MacWilliam, S. (1986). International capital, indigenous accumulation and the state in Papua New Guinea: The case of the Development Bank. *Capital and Class, 10*(2), 150–181.

Main, M. (2021). Absence as immoral act: The PNG LNG project and the impact of an absent State. The Absent Presence (Vol. 141).

Martin, R. (Ed.). (2005). *Tree-kangaroos of Australia and New Guinea.* CSIRO Publishing.

Mayr, E., Diamond, J. M., & Diamond, J. (2001). In *The birds of northern Melanesia: Speciation, ecology and biogeography.* Oxford University Press on Demand.

Mining Watch Canada. (2019). Nautilus seeks financial protection: Key players looked after at the expense of small shareholders and Papua New Guinea 3th March 2019. https://miningwatch.ca/news/2019/3/13/nautilus-seeks-financial-protection-key-players-looked-after-expense-small. Accessed August 10, 2022.

Montgomery, S. Y., & Bishop, N. (2006). *Quest for the tree kangaroo: An expedition to the cloud forest of new Guinea.* Houghton Mifflin Company.

Morris, M. (2021). In *The Anglo Saxons: A history of the beginnings of England* (pp. 529. Hutchinson).

Negin, J. (2008). Australia and New Zealand's contribution to Pacific Island health worker brain drain. *Australian and New Zealand Journal of Public Health, 32*(6), 507–511.

Nelson, H. (1982). *Taim Bilong Masta, The Australian involvement in Papua New Guinea.* Australian Broadcasting Commission, Sydney . ISBN 9780642975669, 0642975663. Google Books https://www.google.com/books/edition/Taim_Bilong_Masta/y2MhAAAAMAAJ?hl=en

Nelson, H. (2016). Black, white and gold: Goldmining in Papua New Guinea 1878–1930. ANU Press, Google Books https://www.google.com/books/edition/Black_White_and_Gold/8M0PDQAAQBAJ?hl=en&gbpv=1&printsec=frontcover

New South Wales State Library. (2015). Papua New Guinea and World Wars. 15th September 2015. https://www2.sl.nsw.gov.au/archive/discover_collections/society_art/png/wars/index.html. Accessed July 10, 2022.

NME. (2021). Sheppard band members' father and former manager arrested in Papua New Guinea for alleged fraud. 22nd January 2021. https://www.nme.com/en_au/news/music/sheppard-band-members-father-and-former-manager-arrested-in-papua-new-guinea-for-alleged-fraud-2862298. Accessed July 18, 2022.

Norfield, T. (2016). In *The city: London and the global power of finance.* Verso Books.

Ohff, H. J. (2008). *Empires of enterprise: German and English commercial interests in East New Guinea 1884 to 1914 (Doctoral dissertation).* University of Adelaide, School of History and Politics.

Orchid Society of Papua New Guinea Inc. (2006). Spatulata Orchids Papua New Guinea. Moore Printing Ltd. ISBN 9980-86-070-7

Ostrom, E. (2009). A general framework for analyzing sustainability of social-ecological systems. *Science, 325*(5939), 419–422.

Papoutsaki, E., & Rooney, D. (2006). Colonial legacies and neo-colonial practices in Papua New Guinean higher education. *Higher Education Research and Development, 25*(4), 421–433.

Patton, S. (2019). The Peace of Westphalia and it affects on international relations, diplomacy and foreign policy. *The Histories, 10*(1), 5–15.

Pietrangeli, C. (1982). *The Vatican collections: The Papacy and Art. Metropolitan museum of art.* Rome, Italy.
Rich, B. (1994). *Mortgaging the Earth: The World Bank*. Earthscan Publishers.
Richardson, D. (2005). *Peace child: An unforgettable story of primitive jungle treachery in the 20th century.* Regal Publishers.
Rosenberg, K. V., Dokter, A. M., Blancher, P. J., Sauer, J. R., Smith, A. C., Smith, P. A., & Marra, P. P. (2019a). Decline of the North American avifauna. *Science, 366*(6461), 120–124.
Simard, N. S., Militz, T. A., Kinch, J., & Southgate, P. C. (2022). Shocks within a Pacific island fishery: A historic study of events impacting the mother-of-pearl fishery in Papua New Guinea. *Marine Policy, 143*, 105173.
Smith, A. F. (2012). *Potato: A global history*. Reaktion Books.
Smith, G. (2013). Beijing's orphans? New chinese investors in Papua New Guinea. *Pacific Affairs, 86*, 327–349.
Srijinda, P. (2019) Papua New Guinea: Problems of economic integration at the beginning of the twenty-first century. In *Actual economy: Local solutions for global challenges* (pp. 223–227).
Steiner, M., & Huettmann, F. (2021). Justification for a taxonomic conservation update of the rodent genus Tamiasciurus: Addressing marginalization and mis-prioritization of research efforts and conservation laissez-faire for a sustainability outlook. *The European Zoological Journal, 88*(1), 86–116. https://doi.org/10.1080/24750263.2020.1857852
Srinivasan, K. (2005). *The rise, decline and future of the British commonwealth.* Springer.
Stiglitz, J. (2003). *Globalization and its discontent*. Norton Paperback.
Sullivan, N., Warr, T., Rainbubu, J., Kunoko, J., Akauna, F., Angasa, M., & Wenda, Y. (2003). Tinpis maror: A social impact study of proposed RD Tuna Cannery at Vidar Wharf, Madang. Madang, Papua New Guinea: Nancy Sullivan and Associates. Academia. Accessed June 18, 2022.
Taber, R. D., & Payne, N. F. (2003). *Wildlife, conservation, and human welfare: A United States and Canadian perspective.* Krieger Publishing Company.
The Guardian. (2015). The incredible plan to make money grow on trees. 24th November 2015. https://www.theguardian.com/world/2015/nov/24/redd-papua-new-guinea-money-grow-on-trees. Accessed May 28, 2022.
The Guardian. (2017). Papua New Guinea mass jailbreak: suspects awaiting trial among 17 shot dead. 15th May 2017. https://www.theguardian.com/world/2017/may/15/papua-new-guinea-mass-jailbreak-suspects-awaiting-trial-among-17-shot-dead. Accessed April 19, 2022.
The Guardian. (2018) ExxonMobil's project in PNG is economic parasitism 1st May 2018. https://www.theguardian.com/commentisfree/2018/may/02/exxonmobils-project-in-png-is-economic-parasitism. Accessed April 19, 2022.
The Guardian. (2019). The brutal history of Bougainville: A look back at the archipelago's history, from German colony to civil war, and from reconciliation to independence referendum. 23rd Nov 2019. https://www.theguardian.com/world/gallery/2019/nov/24/the-brutal-history-of-bougainville-in-pictures. Accessed May 15, 2022.
The Guardian. (2020). PNG's biggest drug bust: the plane crash, the dead man and the half tonne of cocaine, 2020. 1st August. https://www.theguardian.com/world/2020/aug/02/pngs-biggest-drug-bust-the-plane-crash-the-dead-man-and-the-half-tonne-of-cocaine. Accessed April 20, 2022.
The Guardian. (2021a). Two women tortured in latest sorcery-related attack in Papua New Guinea. 26th April 2021a. https://www.theguardian.com/world/2021a/apr/27/two-women-tortured-in-latest-sorcery-related-attack-in-papua-new-guinea. Accessed August 2, 2022.
The Guardian. (2021b). PNG residents evicted a decade ago for casino development face violence and poverty: Report. https://www.theguardian.com/world/2021a/nov/12/png-residents-evicted-a-decade-ago-for-casino-development-face-violence-and-poverty-report. Accessed March 17, 2022.
The Guardian. (2021c). Australian lawyer arrested for third time in Papua New Guinea over alleged $100m fraud. https://www.theguardian.com/world/2021b/jul/29/australian-lawyer-arrested-for-third-time-in-papua-new-guinea-over-alleged-100m. Accessed March 17, 2022.

The Guardian. (2021d). Villagers file human rights complaint over plan for giant PNG goldmine. 9th December 2021d. https://www.theguardian.com/world/2021d/dec/10/villagers-file-human-rights-complaint-over-plan-for-giant-png-goldmine. Accessed July 3, 2022.

The Guardian. (2022). German judges visit Peru glacial lake in unprecedented climate crisis lawsuit 27th May 2022. https://www.theguardian.com/environment/2022/may/27/peru-lake-palcacocha-climate-crisis-lawsuit. Accessed June 4, 2022.

The Sydney Morning Herald. (2015). Steering corrupt cash into Australia from PNG: a how-to guide. June 24th 2015. https://www.smh.com.au/national/steering-corrupt-cash-into-australia-from-png-a-howto-guide-20150623-ghv1sx.html. Accessed June 4, 2022.

Tom, L. E. K. A., Braithwaite, K. A. T. H. Y., & Kuniata, L. S. (2017). Incursion of sugarcane smut in commercial cane crops at Ramu, Papua New Guinea. In Proc. *Aust. Soc. Sugar Cane Technol, 39*, 377–384.

Toynbee, J. (2007). *Bob Marley: Herald of a postcolonial world?* Polity Press.

Twomey, A. (2006). In *The Chameleon crown.* Federation Press.

Rosenberg, K. V., Dokter, A. M., Blancher, P. J., Sauer, J. R., Smith, A. C., Smith, P. A., Stanton, J. C., Panjabi, A., Helft, L., Parr, M., & Marra, P. P. (2019b). Decline of the North American avifauna. *Science, 366*, 120–124.

van Winkel, D., & Lane, J. (2012). The invasive cane toad (Bufo marinus) in West New Britain, Papua New Guinea: Observations and potential impacts on native wildlife. *Biological Invasions, 14*(10), 1985–1990.

Wainwright, E. (2004). Solomon Islands, Papua New Guinea and Australia's policy shift. *Sydney Papers, 16*, 122–132.

Wilson, M. (2019). In *A true child of Papua New Guinea: Memoir of a life in two worlds.* MacFarland Publishers.

Z'graggen, J. A. (1975). The languages of the Madang District, Papua New Guinea. *Pacific Linguistics, Research School of Pacific and Asian studies*, The Australian National University.

Zhang, D. (2019). Perceiving China's influence in the Pacific-the case of Solomon Islands. The Diplomat 18 Oct. http://hdl.handle.net/1885/205944 Accessed August 19, 2022.

Chapter 26
Papua New Guinea from a Modern Anthropology Perspective: PNG 1, Western World 0. Bob Marley and the Tribes Remain the Judges to Follow

We were brought up fighting, fighting is like food to us
J. Waiko (in Cousteau & Richards, 1999, p. 75)
Life changes along the Sepik, and life does not change at all
Cousteau and Richards (1999)
Though television has reached some parts of the country, most youngsters still find their entertainment in natural materials and events, as well as the dramatic stories told by elders
Cousteau and Richards (1999, p. 203).

Abstract Papua New Guinea (PNG) has been a hotspot for Anthropologists* studying 'ancient' and traditional societies. Usually that is seen as progress for global society and allows learning and excitement. However, considering the global crisis and problems, in PNG—or the wider world—with poverty and claimed decay the outcome of this research work for PNG and society at large has been less than achieving/impressive—and is not reaching out well while the overall message remains not even agreed on. Already the tragic top-down discussion about 'savages' makes that clear. Here a short overview of Anthropology expeditions and other human-related research in PNG is provided and how it fits in the wider context of PNG culture and society, colonialization, globalization and the Anthropocene for 2023 and beyond. Despite one of the highest international Christian missionary efforts in the world done now for over two centuries, in PNG sorcery, the tribal Wantok system and its policies remain widespread and dominating; the late Bob Marley, the rasta movement and associated styles are very popular and chosen by many PNG citizens. It's found that unsustainable practices in modern PNG governance have not been improved through Anthropology, or any western engagement and its money scheme; rather the opposite! Considering 'development,' man-made climate change and massive industrial 'aid' ongoing such as mining, oil and gas, as

* The term 'Anthropology' is used loosely here in its international sense. It's acknowledged that interdisciplinary approaches are commonly applied now in Anthropology, and that the discipline has widely moved forward, out of its private donor constraints into a more public and diverse funding landscape. Urban Anthropology and social, equity and sustainability aspects might dominate now, but are still low on the global achievement side. Still, many shortcomings can be found regarding its sustainability, and certainly for the traditional Anthropology concept, e.g. one that was used in Papua New Guinea and that dominates the global discourse for PNG.

F. Huettmann, *Globalization and Papua New Guinea: Ancient Wilderness, Paradise, Introduced Terror and Hell*, https://doi.org/10.1007/978-3-031-20262-9_26

well as one-sided commercial forestry, farming and fisheries, and the discipline of Anthropology hardly mitigating it, a grim reality outlook is provided for PNG, its people, biodiversity wilderness conservation and the global community at large. This is to change for betterment, within Anthropology, its institutions, funders and PNG the most

Keywords Papua New Guinea (PNG) · Anthropology · Neocolonialism · Social Studies · Environmental Justice

26.1 Introduction

Science—as we know it—is a western dominated construct happening nowadays mostly in funded institutional frameworks of nations and thus it carries that footprint and easily is biased (e.g. for Karl Popper and scientifically objective knowledge creation see Wilkinson, 2013; for a Papua New Guinea example see Golub, 2014 and a critique). It's meant to serve society and nations at large (see Narokobi, 1974, 1975 and 1983 for a PNG self-identity). But as national science budgets easily show (some of the biggest research budgets are located in the U.S., as well as Switzerland, Canada, UK, Australia and in some EU nations; now challenged and topped by international corporations) not all nations contribute and gain equally, nor can all nations and their people access it well, and have the same science institutions and research options to work from.

Anthropology is not a god-given discipline or automatic self-creation. It's a paid and funded activity designed and carried out by humans with humans for humans under a wider platform of global engagement. Anthropology—as a discipline and its focus—has actually changed a lot over time and can be defined as the study of human societies and cultures and their development (e.g. Sanjek, 2019). But it's much more. Anthropology can also help to describe myths (Gillison, 1993), rituals (Gillison, 2002), or historically, human race theory (for PNG see for instance Levit & Hossfeld, 2020) and how it all changed (Mead, 1967). Anthropology can describe its actors and their views, as well as describing 'humans' coming from various perspectives and with underlying motives. The question of '*science for whom*' and '*how done unbiased and objective*' remains widely debated though, certainly in the disciplines of economy, statistics, conservation, psychology and Anthropology (e.g. Rooney & Papoutsaki, 2004 for Papua New Guinea PNG). Universities themselves get widely and frequently debated for their best structure, funding and good and fair impacts too (e.g. Miller et al., 2016; see Lutton, 1981 for PNG and the underlying university library located to Perth/Australia etc); it's an ancient discussion in the institutionalized world and virtually all institutions now lack funding for their tenured staff, employees, infrastructure and libraries; it's not a sustainable model at all.

Much of the 'modern' western sciences have their roots in just a few European employers, namely royal courts and their perspectives, biases and needs (contrast with Conner, 2009 for people's history of science and their consistent exclusion). As it will be discussed here for PNG, Anthropology offers us a good example on those topics. But also other disciplines of the western core scheme—and globally relevant ones—are easily affected. For instance 'Economics' (= biased toward making already empowered entities richer quickly instead of equal rent paying for everybody and eliminating poverty for good; Gaffney & Harrison, 1994). Or 'Statistics' (= stuck in frequency distribution and parametric assumptions by C. Gauss from the 18th century and R.A. Fisher from the 1940s instead of machine learning and AI progress; Humphries et al., 2018 and citations with) is another very vivid and so obvious case. Salsburg (2001) showed how that discipline is still being occupied with gambling questions (= to increase the odds of winning and making money for a leisure gambler, e.g. at the royal court, using 'science', instead of dealing with modern advanced questions for the wider public good of the world, climate change included). Other cases can easily be found in medicine, psychology, taxonomy, etc., or man-made global warming.

Western science has also widely engaged in promoting certain views about the feudal system or race theory (e.g. applications of Darwin's Evolutionary principles to humans; see also Vetta, 2021 for science emphasize; or see reviews in Ostrom et al., 2002), or it has widely left them out in its research and inference (e.g. Gulop, 2014; Beehler & Laman, 2020).

26.2 PNG Breaks Anthropology?

All of this matters for PNG, Anthropology and beyond because a science just driven by the royal courts, the wealthy—international corporations—and their political powerhouses of the west for self-interest tends to be linked with colonialism and thus 'enlightenment,' industrialization, a ruthless resource-grabbing, subsequent globalization and with such a human reasoning and justification (see examples for PNG with British and German efforts featuring research views as well; Ohff, 2008; for land grabs in the Commonwealth see for instance https://www.cbc.ca/news/canada/new-brunswick/madawaska-land-claim-patricia-bernard-1.5984019). Whereas many deeper but relevant aspects are ignored and side-lined. For PNG, the entire colonial enterprise hardly was profitable (see Firth, 1972 for German colony bankruptcies), and it remains so for much of the modern day (http://www.bougainville-copper.eu/comment-kommentar02-3.html). The latest use of science, technology, education, mathematics (STEM) education is a consequential outflow of such an approach; a not well-achieving one (see Sanders, 2008 for a review and critique).

Arguably not all things in the world can be studied, explained, understood or said and expressed by humans, or institutions (see Gillison, 1993 for a PNG example). Humans have sensory limits and are easily biased and thus widely incomplete in perception and for outcome (= human-centered), whereas the world and universe are much wider (Naess, 2009; see Nakorobi, 1983; Baraka, 2001; Gulp, 2014 for the Melanesian way being cosmological (see Forsyth and Eves 2015 for witch craft and sorcery). Such points can already easily be seen in the western theory of relativity, in quantum physics, in questions relating to the universe and its origin (many 'big bangs' are now discussed, beyond neutrinos and black mass), or when trying to understand some base aspects of ecology, evolution or advanced human medicine, life-and-death (Le Bellac, 2011). There is more to live than what meets the eye; already simple computer arrays can easily outcompete humans (Humphries et al., 2018), and so can most animals (e.g. De Waal, 2016) and plants (Wohlleben, 2016). Thus, why being so human-centric then when the latest science shows us otherwise and better? There is no need to hail 'the expert' when those are known to be widely biased, often just inappropriate and plain wrong. The same should apply to PNG then, and specifically to PNG, which is one of the deepest and most complex subject areas to study.

Seen from that perspective, "Anthro"—the discipline of Anthropology in scholastic slang—carries a heavy western burden. It is one that is changing like human society is, but its problems are not going away. Anthropology, by definition, comes with a lense to look at things, and that lense is murky 'at best' (see Narokobi, 2000 for PNG); alternatives exist.

For Melanesia and PNG, Mead (1935, 2002) showed how the study of 'primitive people' (also referred to as savages, e.g. Hurley, 1924; Spitz, 1935; Levi-Strauss, 1966; see Mead, 1967 for an updated change) affected our own (biased) understanding of humans, childhood and sexuality changes over time (in part reviewed with a discussion by Lipset, 2003, etc.). Another early Anthropologist of impact for PNG is Nicholas Miklouho-Maclay from Russia but mostly based in Australia (Tutorsky et al., 2019); his work was 'modern' and promotes many humanity aspects from the nineteenth century but it is still less known though (his work is widely appreciated in PNG though; see discussions in Levit & Hossfeld, 2020). A rather relevant Anthropology scholar for PNG—based in PNG—was the late Bernhard Bullo Narokobi as it provided base principles of the new PNG nationhood (see review in Bablis, 2010; some of his work is also discussed in other chapters of this book). In that context, the work by the late contract Anthropologist Nancy Sullivan should also be emphasized and how modern PNG society gets the 'shorter end of the stick' in the modern policies of mining and fisheries (see related chapters of this book).

For a long time, PNG actually attracted many Anthropology scholars from all over the world, of course. A complete overview is virtually impossible to provide. Many of them are not well-known and their data are missing or ignored. A good example for those ones is Italian Anthropologist Lamberto Loria (https://www.berose.fr/article1755.html?lang=fr). The deep human experiences and the amount and locations traveled by those scholars remains mind-boggling. It's fair to say that for easily over a century, for the discipline of Anthropology there had been vast amounts of 'contact' experiences with 'uncontacted' tribes offering huge amount of research and knowledge gains. But what came from it?

This has wider repercussions, and in the global mind, PNG still remains a feast for Anthropologists and their reviewers (see also West, 2006; Golub, 2014 for PNG and conservation). The Western World has a hard time to grasp PNG either way. Anthropology can be used to show mining impacts, or the advent of the internet. Anthropology certainly shows the dynamics of human society over time (e.g. Foster,

2006, 2008). In PNG decorative stone implements now give way to plastic and metal; what does that indicate though for human society?

Traditional Anthropology has turned the world upside down though for PNG. It's nothing unusual that 'uncontacted tribes' got studied and then mining and natural resource extraction broke loose right after. Missionary work also happened in parallel. Examples are presented in Golup (2014), see also Stewart and Strathern (2002), or examples in Gillison (1993, 2002) for Crater Mountain (West, 2006; Mack, 2014); more exist. It comes almost as a template to study those topics for western scholars and Anthropologists.

Colonial perspectives of an onlooker are dominating the discourse presented. And the vast majority of Anthropology work done in PNG comes from male eyes for a 'professional' male audience (see Gillison, 1993 for a critique and female perspective and research subject in PNG). That's because the dominating colonial rulers are Atlantic-centric—with an Australian spin (see example in Habbia, 1973)—and the practitioners at that time just hardly changed; Anthropology became a culture set in stone with a now set and fixed lense. It's difficult to change from that.

And just until recently many classic Anthropologists never really acknowledged that for PNG as a problem (Rooney & Papoutsaki, 2004). Once more PNG fell victim to a small clique, as can be seen in many aspects. With few exceptions, PNG itself has no dominant Anthropology platform, e.g. university department with high-level public funding and international reading. PNG's Anthropology is still much dominated by effluent westerners; and China and India as emerging world powers are widely not participating in such a science. Anthropology in PNG turned into a niche business and shows a science and policy fiefdome! A classic approach to achieve this is the Anthropology and collection trips of Oxford, Cambridge, Harvard and rich private funders, e.g. the 11 Archbold expeditions for the American Museum (essentially funded—in big part—by private money from Standardoil and such mind-sets; Beehler & Laman, 2020), or see the work by some private donor NGOs, e.g. via Microsoft. All of those suffer from a vast under-analysis of their materials and thus easily carry a certain bias.

Often such biases are ignored to this very day and/or get presented in a very different light (e.g. 'art,' see also with artifacts and sources in Beehler & Laman, 2020). Seen from that angle, PNG is actually a prime place to understand Anthropologists, their institutions, their funders and practitioners/students themselves, while PNG citizens are still the tools to do so (West, 2006 as a modern conservation example; it studies PNG citizens and NGO employees alike, similar to a human experiment on both ends). Such an Anthropology keeps primarily the perspective of an onlooker, a passive not-interfering describer without a meaningful research design for inference and application. Essentially it sets it up for the reader being excited, shaking the head and being 'befuzzled' with PNG, but by PNG itself being reduced to nothing but exotic and remote. Arguably, and as it makes for a big discussion item (e.g. Lahood, 2007), a passive onlooker can never get into the intrinsic fabrics of the study subject. What is to be, and truly has been learned and concluded from such a research? This is hardly different for many biodiversity scholars, e.g. Flannery (1994), Richards (2018), Beehler and Laman (2020), or Laman and Scholes

(2012) for Birds of Paradise, from specimen and museum collections, or perhaps from the human zoos that traveled around the world in the early 1900s (Putnam, 2012) showcasing exotic humans. Instead, being embedded in the subject provides a deeper insight but is seen as biased style. So who is right?

In journalism to be embedded is widely accepted though (see public record on Gonzo journalism by H. Thompson: https://en.wikipedia.org/wiki/Gonzo_journalism). As a matter of fact, it was argued to be the only style that provides authenticity and 'truth.'

The key questions still remain: science and Anthropology done how and for what? What platform is to be used. And what is the progress, what is helped and how? Who gains and how done best for the wider public good, the global community overall?

Arguably, some major changes in the understanding and documentation of PNG came from its citizens themselves, such as shown in books like Kiki & Cheshire (1969), Wilson (2019) with an Australian-PNG perspective, and then Chan (2016). While those are narratives, carry no research design, statistics and science from institutions, they come from a lifetime experience. It adds insights that outsiders can virtually never obtain. Can it get better than that?

Other very relevant PNG society documentations and contributions of course come from non-Anthropologists, see for instance Flannery (2000) for field biologists. This is to say, many sources, and way beyond Anthropology—including media and media projects (e.g. the Judith Neilson Institute with The Guardian: https://jninstitute.org/project/guardian-australia-pacific-editor/)—exist for PNG and for the wider context (Suzuki, 1993, 1996; Diamond 2011b; Harari, 2015). Essential Anthropological research items were provided from people who never studied Anthropology, or do not care for it much, e.g. for PNG see Diamond (2011a, 2011b),B. Narokobi (in Bablis, 2010) or Harari (2015).

26.3 From Anthropology to Paid Consultancy and Commercial Messages?

Anthropology is a science and costs money; and in the western world the sciences are usually carried out on an institutional platform (e.g. a university or governmental agency done by its citizens and tax-funded). They happen under public ethics rules and agreed terms and payments. Anthropology for instance is not a 'narrow science tunnel' (it should not be due to its complexity; see Mead, 1967 for PNG for instance) so it now consists of genetics, paleontology, social science, mapping, etc.

With a consistent lack of money though in those platforms, science shapes itself in the money world. It moves into neoliberalism, consulting and thus caters a direct dependence into funding (see an example for reality situations and outcomes in such type of institutions https://worldnewsera.com/news/science/max-planck-archaeology-director-removed-after-alleged-bullying/). Anthropology also links up closely with associated disciplines of fieldwork and wider field ecology. Those have often already made the wide switch to funders, e.g. on the topic of birds and rapid biodiversity assessments see for instance Richards and Whitmore (2015), Richards (2018) for oil companies and similar funding support (Beehler & Laman, 2020 for mining). Much of the funding is not even exposed, and can come in some but crucial helicopter rides or field camp and staff use plus their local insights. Keeping the critical voice, independent inquiry and lack of bias in Anthropology and in the sciences and its platforms remain the question of the day.

In Anthropology, and PNG, this is a real-world question. Work in Stewart and Strathern (2005) shows it, and mine-related Anthropology studies show us many deep examples of that subject (e.g. Golup, 2014).

26.4 Should One be Anthropology-Phobic?

The overall deep human history on earth is perhaps 300,000 years old, with humans from Siberia being related via DNA with PNG (Beehler & Laman, 2020). Immigration into PNG happened likely in several waves, and many aspects of PNG society and the Sahul remain unknown (e.g. Flannery, 2002 for details). PNG still allows us to get a picture of humanity, and Anthropology can help to do so; done by trained experts and with a wider context

Within that, Anthropology has no monopoly though on human society, on knowledge, on science, or on PNG and sustainability or conservation; far from it. Anthropology is not for the rich and wealthy, nor to be practiced just by them, or used for their entertainment. Anthropology is by everybody and for everybody, serving the wider public good; not? And it needs to be understood that way. Simply based on the track record, traditional Anthropologists have little authority in PNG, or on relevant topics like economics, sustainability and wealth of a society (e.g. Gaffney & Harrison, 1994; Daly & Farley, 2010). Anthropologists are rarely asked for modern politic and legal questions, certainly not in PNG and for sustainability or man-made CO_2 release. Instead see the work by Sullivan (2015) on CEOs and their power and understanding of PNG for real-world decisions (also outlined in Chap. 2 of this book; and Flannery, 2000). This is modern Anthropology, but far from the traditional concepts and not much found in the discipline still, it's hardly taught.

It's not surprising then that many of the major food and farming questions actually come from outside of Anthropological theory, e.g. Madison (1997), Ostrom et al. (2002), Conner (2009), Daly and Farley (2010). Equally, sustainable governance is not well featured in Anthropology (see for islands in the Pacific Rim Dublin & Tanaka, 2015) nor does a good consensus exist (e.g. Raymond, 2007; Narukobi, 1983; Gulop, 2014 on the 'Melanesian Way'). Relevant conservation management theory in major cultures and economies happens much without Anthropology, e.g. Ostrom et al. (2002) or Organ et al. (2012) for an example. Anthropology is widely absent from the major biodiversity, health and sustainability questions of our time—including much of the Machine Learning and AI applications (Humphries et al., 2018), 'the cloud', pandemics and climate change policy—see also Wroe et al. (2013).

While Anthropologists document the decay of languages and cultures, they do little about it for betterment; it's even hardly applied correctly as those language entities are in heavy flux and no real associations or lines or boundaries can be drawn on a map in the first place (*sensu* Gulop, 2014). As a matter of fact, many tribes in PNG had no name for themselves or for their languages. The naming and subsequent association to people and sites—classifications—were essentially just CREATED in response to outside influences and by the western desire to do so, to describe all and everything; regardless of being meaningful or not (Golup, 2014 for details and applications for mining impact studies).

And so, the reality stated by Beehler and Laman (2020) remains: One weak culture gets lost and a strong one forms and stays; it's a process and goes on and off (Short, 2010). As PNG sustainability is now after 47,000 years on the decline, what really caused it and why? Nobody still knows, despite Anthropologists having studied 'uncontacted tribes' for over a century and still are doing it in various shapes and forms (see details in Flannery, 2002; Diamond, 2011b; Acemoglu & Robinson, 2013; Harari, 2015).

And globally as well as in PNG, there is now openly the ongoing dispute of landownership tenure versus privatization and land developers and their land grabs, this process destroys the usual lifestyle (Baraka, 2001; Beehler & Laman, 2020). One may study it, but it's to be resolved with good guidance. The Special Agricultural Business Leases (SABLS) show the conflict well (ATBC, 2011; Beehler & Latam, 2020, p. 347). Anthropology eventually has dealt little with land ownership resolutions, with an equal wealth distribution of the society on a finite space (typical example found here for Crater Mountain: https://www.google.com/books/edition/Anthropology_and_Consultancy/1AGt1K5BNP0C?hl=en&gbpv=1&dq=crater+mountain+photographer+papua+new+guinea&pg=PA65&printsec=frontcover). In the land-associated discipline of Ecological Economics, Anthropology is widely absent, e.g. for rent seeking and sustainability, or how to roll it out globally (Daly & Farley, 2010). And that is certainly true for PNG. And so, what does this promote and tell us, for Anthropology as a discipline? Capitalism, or a good life for all (Hardin, 1998; Ostrom et al., 2002 for the wider common good)? It's Bandura (2007) then being true for a wide section of an entire discipline: Selective disengagement and *laissez-faire*, or worse.

A similar and related example deals in PNG with governmental patrols and village courts to pacify 'uncontacted' peoples, e.g. as done by Australia til the 1970s approved by the U.N. and world community (Beehler & Latam, 2020); most rural villages and society were affected. How could that go on for that long without driving the science record *en masse* and while Anthropologists studied in the villages firsthand for over a century? Silence dominates.

26.5 Human Diversity Beats Anthropology?

The human diversity of PNG within is vast and evolving; it's unique (see Cox 2011 for an example of a money making scam scheme to Christianize finances, U-Vistract Financial Systems, based on the village model though as the central scheme of PNG society) and much wider than found in most western nations, if not even anywhere in the world. Already the density of tribes and their languages (easily over 700; Kulik, 2019). PNG is host to 20% of the world's languages spoken, but despite decades of intense study, many carries now often less than 2000 speakers, sometimes 100, or they are on the verge of extinction. How can and should one study a language, e.g. for areas, people and tongues that do not name themselves even? Whatever we know, we still know little about those languages and they quickly disappear, so do their speakers and the associated cultures and landscapes. Within PNG, the Sepik-Ramu region stands out as one of the highest language diversity in PNG (Swadling & Hide, 2005). As the case elsewhere in the world, e.g. Himalayas (for overview see Regmi & Huettmann, 2020 and citations within), this diversity is also associated with terrain ruggedness and remote valleys. Still, the linguistically so diverse Ramu and Sepik regions are ones with a vast industrial footprint also (details in chapters of this book and citations within).

But it's precisely here where the actual membership and the spatial footprint of tribes, clans and their languages and study get very soft. That's because a strict association of a clan or tribe can often not be made well. Humans do mix, clear distinctions usually do not exist, they are fluent over time, and they even overlap. It's often pointless then to draw such maps or to associate territories and properties, as done in The Western World and in such a science (see Golup, 2014 for examples in PNG).

Tribal governance is still the major item in PNG and drives major decisions and daily life—on the ground (Walton & Jackson, 2020; see Flannery 2000 for examples). Local bartering—not globalization—sits at the core of the PNG lifestyle, namely using betel nut and pigs (not money; Sullivan, 2007; see Golup, 2014 for many uncertainties in 'pay day' payouts) as a currency in many regions of PNG and to settle money disputes. Grassroots gardening remains the lifestyle added with other subsistence activities like fishing and hunting (Baraka, 2001). Not all tribes in PNG use betel to trade, but bartering items, from mussels, over pigs to minerals, and 'mustard' with white lime/ash powder is commonly found (see Fig. 26.4) (Figs. 26.1, 26.2, 26.3 and 26.5).

Most tribes are also still somewhat nomadic, a concept that the western mind cannot grasp well with, e.g. for being registered and when paying local or federal taxes for healthcare and pension fund, or for dividend payouts. Nomads are not easy to administer in a western-style work shift of 9 am til 5 pm with a weekend and a 40 h work week (= the world most Western people live in); they hardly pay tax (tax for what service anyways?). Often the nomads are to be settled and to comply with the administration, which happens to destroy the nomadic culture and sustainability. In PNG, being nomadic and settle into new and 'better' areas is a frequent event, based

Fig. 26.1 Rugby, a favorite sport in Papua New Guinea and beyond

Fig. 26.2 Pig, a famous study object and subject of dispute in Papua New Guinea society

Fig. 26.3 Modern anthropology can deal with markets and demands, such as coffee 'Arabica' telecoupled with U.S. and Europe markets (Note that telecoupling research sits with Ecologists, and some Economists, but virtually not with Anthropologists, yet)

Fig. 26.4 "Mustard" often used with a white powder (grinded ash from reefs) and widely appreciated and consumed in PNG

on warfare and peacemaking for neighboring groups (see Golup, 2014 for real-world examples around the Panguna mine in Bougainville).

For the curious mind on those matters, already just simply spending some time at the airport in Port Moresby (POM) will trigger diverse and conflicting views of the Anthropogenic human diversity spectrum (see also Hayan, 1990). Just when looking at the tourists that come to PNG will achieve that; apart of some diving tourists and

Fig. 26.5 Old ways of bartering found a new object: plastic bottles and containers

Ecotourists. The bulk of the tourists is covering the global diversity from middle class and upper class, transgender over religious families with many children, over to obese, religious right-wingers and to missionaries as well as surf dudes, oil drillers, miners, engineers, businessmen, bankers (nowadays females presenting the western world view of their feminism), crooks, coffee buyers over to Japanese, Chinese, Phillipines, as well as Eastern European ship hands, and of course, various Australians. One will see among the PNG tourists and travelers, Chinese businessmen, American oil seekers as well as diplomats and a few lost biologists from the rich west, mixed up with a few students and scholars. It easily makes for a 'human zoo,' popped up with traveling church choirs, rugby teams and Melanesian Bob Marley look-alikes, rasta females and many crying babies. Human society can hardly be more diverse than that, and Port Moresby (POM) airport has it all.

The human diversity and range of people at POM airport easily stand as a characteristic for the globalization macrocosm itself. They meet, mix, mingle and travel all in PNG and through PNG in space and in time. So how to study it, what languages, and why and for whom (Rooney & Papoutsaki, 2004)? Anthropology can provide the answer, but does it?

The airport snapshot shows us essentially a run-through of PNG time during globalization. It's perhaps like 47,000 years of human civilization ran by in just a few minutes of observation. That's because for most of the PNG society and its time, there were very little relevant changes. Dramatic and substantial changes just came last two hundred years, and then specifically after WW2. The remnants are easily seen in actors passing by at the airport. Individual people and PNG citizen have witnessed such dramatic changes (see reported details for instance in Kiki & Cheshire, 1969; Chan, 2016; Wilsoon, 2019).

26.6 With Anthropology toward a Better World?

So which human society and culture fares best? While this question is widely posed throughout PNG's western history it is virtually impossible to answer. It's clear that the Australia set up, supervises and dominates PNG at large as 'superior'; and all as per U.N. mandate (Baraka, 2001 for overview); and thus they won that game. But still, already the PNG Rugby Club beat the Australian national team twice. PNG developed into a sports nation…one would think (Foster, 2006; West, 2014 for surfing and diving sports and tourism). Competitive sports in PNG were introduced by colonists and their missionaries. In PNG those have taken on a style of their own, with games often inspiring not only team but also village or district loyalty (https://www.britannica.com/place/Papua-New-Guinea/Daily-life-and-social-customs). PNG has participated in Olympic Games but focuses most on the Commonwealth Games. However, instead of the noble sports scheme of the ancient Greece, or UK knights, it's a trend now where the tropical nations provide the (human) raw material, eager athletes, which are 'going to market' abroad (sports entertainment serving industrial society). It's similar to a muscle market, just see it in soccer and Olympics, and it was already equaled with a modern form of slavery (Rollason, 2010; Hunter-Rainey & Rodriguez, 2011; Milenković, 2019). PNG is now into sports, and the Anthropology of sports remains a wide topic (Blanchard, 1995; see Basketball in U.S. or FIFA and Qatar Football World Cup: Thuer, 2017).

In the meantime, the Anthropology topic is certainly a question of sustainable society: how done, how assessed and what good management? One would assume Anthropology can provide an answer, but as intense debates on the topic shows us, the PNG lifestyle is still not universally accepted—hardly globally known and acknowledged—as a governance success story (Naboroko, 1983; widely disputed in Golup, 2014 though)! And one will find little Anthropology or conservation research published that promotes it, certainly the NGO conservation does not (examples in Mack, 2014; West, 2006).

"The next focus of power to fall under the spell of this little gang"(Gray, 1996).

Anthropology is just a small aspect of science and of the western and global society. So it has clearly issues of its own to deal with, e.g. Willis (2011), and to resolve. And it's not so clear how that really is achieved in the discipline, or at the university campus and associated agencies?

Textbox 1: An Assessment of Modern Economy, Culture and Globalization when compared with Papua New Guinea: What did Anthropology fail to detect, collect and to infer, and why not setting an embargo until improved?

Clearly, we are living in a global crisis of the environment, of biodiversity, of social unrest and of economy. What is working? In the meantime, the sciences feature usually high budgets, Anthropology Departments are part of that university system, often with endowed funds even. So how come that we have so few Anthropologists in charge, in the lead, and with a good outcome? Should the study of humans not result into a role model, a well-run human society?

With PNG being of such relevance to Anthropology and its core questions over time, highly funded and with field collections for over 100 years, why has that not solved the human problems?

It's clear that Anthropology by virtually all the nations done in PNG did not provide that progress desired. PNG is not in the best shape, nor do citizens benefit from Anthropology or is Anthropology really sustainable in itself. At best, Anthropology is not effective.

From that it's urgent to realize that something is not adding up here, but what is it? Nobody really knows and more study is needed?

Fieldwork in Anthropology typically requires permits, e.g. by PNG and western nations that are all concerned about PNG and its people. But until real value is provided, and inherent problems are fixed, a research embargo cannot be imposed?

Textbox 2: Is Anthropology in a crisis with itself, with Oxford/Cambridge. Harvard, with its donor ethics and with Jarred Diamond?

Leading universities and institutions—including the Vatican—were involved in collection expeditions in Papua New Guinea, ongoing for over 100 years; it reflects on either and was seen for a long time as 'good style.' Arguably much has happened ever since with Anthropology, now moving perhaps finally into a more social science and holistic modern direction trying to shed its past. But still, the recent work by Diamond (2011a, 2011b)—arguably not a trained Anthropologist—on PNG and the world brought a massive change into the Anthropology perspective. Similar can be said for T. Flannery's work (e.g. 2000, 2002; also not a trained Anthropologist). While too late, changes had to happen in the worldview. It then also resulted into deeper discussions within the community, as well as 'positioning' of different camps and new books on such topics (Acemoglu & Robinson, 2013; Diamond, 2011b; Harari, 2014). Such a debate is important and healthy resulting into new approaches, e.g. indigenization (e.g. Whyte, 2017 for Alaska). But clearly, many things of the past are now discussed, reviewed and revised. While this benefits Anthropology also, universities, the sciences, are now getting more interdisciplinary and digital online. Many traditional research and scientific disciplines are revisiting and revising themselves: adjust or get erased as being obsolete. Departments of

Latin, Ancient Greece or Frequency Statistics are facing such realities for a long time now, more is to come.

For people interested in the subject of change, how entire disciplines adjusted, changed and flipped, one just has to follow their textbooks throughout their editions as a documented fast paradigm and baseline shift over time. With that, it also becomes clear that human education has changed and earlier degrees might become almost worthless. Education and science have now a very high turnover rate, which puts doubt on leadership, because those leaders are usually trained in the 'old world,'. And humans, and when leading, have a tendency to not adjust fast and stick with what they were told; staying on the safe side as conservative as one can. It's here where PNG can help us for a better and more updated reality, whereas Anthropology has not achieved that, really.

Textbox 3: The cannibals and savages…are everywhere

While widely referred to, and linked with, with the topic of Papua New Guinea (PNG), cannibalism—the predatory eating of human flesh—is not a pleasant topic for me to write about, or to inquire. But on the good side, cannibalism is a thing of the past, and this issue is not isolated to PNG whatsoever. It's commonly found in the human history worldwide. In most earlier wars with starvation, and prisoners of wars for instance, it was reported (e.g. German WW2 https://www.spiegel.de/geschichte/kriegsgefangenschaft-zwei-expertinnen-ueber-den-stand-der-forschung-a-805f2ff8-0002-0001-0000-000204216427).

As stated in Beehler and Laman (2020) in Christianity, worldwide Jesus Christ's human body is consumed in the form of a dry cookie-type during weekly services, with a blessed zip of a red wine (~to represent human blood). Sources like Richardson (2005) tend to be very quite about those details, but boost the other side of the coin.

While many unconfirmed rumors exist, e.g. Hoffmann (2015), the last act of cannibalism in PNG was reported from the 1930s. Missionaries and Australia outlawed it early on, and it was essentially gone in PNG by the 1960s, helped by Australian efforts. In the mid-1960s, there were cases reported of the Kuru disease, a disease of prions (e.g. alsoarcheologically reported to occur in Neanderthals and when eating the human brain-occurring in human-to-human transmissions). But in PNG this was tracked down as a holy kinship practice at funerals with the Fore people. This tribe ate parts of the funeral bodies to avoid the soul and body of their family member instead of being eaten by worms in the dirt. It was mostly left to females as a good host for that spirit, and they gave some snacks to their kids. With an incubation time the Kuru disease developed. In that region it actually killed over 10,000 people. By now, it's ended. Whereas scorchery is still ongoing and uncontrolled.

Such stories contribute to the fear factor, and the tourist attraction to PNG, e.g. Silverman (2012).

Instead, the brutal treatment of fellow humans is very commonly found throughout societies and the world. The use of DNA studies can track those details. A good hint seems to sit in the genetics of Iceland; two-thirds of females seem to come from gaelic stock, e.g. Ireland or Scotland. Just 1/3 is 'truly' scandinavian. It's opposite in the male DNA, which likely shows enslavement and robbery of the female body by those vikings. Similar stories were found for Vikings (https://www.stern.de/panorama/wissen/wikinger-als-sklavenhaendler--die-dunkle-seite-des-nordischen-volkes-30355876.html#:~:text=Die%20fiese%20Seite%20der%20Wikinger&text=Insbesondere%20exotisch%20aussehenden%20Sklaven%20%E2%80%93%20sprich,Spanien%20bis%20nach%20%C3%84gypten%20erstreckte). They had app. 10% of slaves in their population, and treated them badly (as judged by the archeological record and graveyards). Slaves that looked 'exotic,' e.g. blue eyes were transferred and received wide-distance trading due to wide demand in the Old World

Why that is so underdocumented and poorly acknowledged for the colonial nations? It might perhaps well have to do with the Aryan view in the sciences and Germany/Central Europe at the time. Early dominating Anthropology—as practiced at the royal courts and with such funders and mindsets—was a lot driven by such views, and the public was misinformed ever since.

26.7 Some Main Tribes of PNG

What are the main tribes of PNG? This question is not much answerable really because there is a lot of flux, with much tribal action happening in Port Moresby also. Using language groups might help instead. Motu is a main language. Other main languages of operation are Melanesian (Tok) Pisin and English. But then there are over 700 languages and those are usually linked to tribes locally. One can thus easily list over 700 tribes, but they vary in size and details. Humans move, mix and mingle; they are on the move. The exact tribe number is not answerable, but also somewhat meaningless of a bookkeeping question (see also Gulop, 2014). Tribes tend not to name themselves (and what language?) and their resource needs on the land are overlapping and vary over time; boundaries are hard to draw. It's not an easy math, hardly relevant

26.8 What is a Tribe Anyways?

Perhaps one can describe a tribe most appropriately as a loose and moving village unit, with several main clans and families within. Tribes are usually not very complex in the high-level hierarchy; they are loose family bands, with members being distributed and several places. Kinship matters, and so do their words (language) as a defining feature. Strictly speaking, such tribes vary in time and space. I remember being in a 'virgin rainforest,' but then I was told by locals that a bigger tribe lived here some decades ago with over 100 people doing farming, slash and burn. They all died and had left, likely due to an unknown disease, linked to 'bad ghosts.' The place grew over, and it looked quickly again like pure virgin rainforest. And because until recently village PNG does not really use concrete or cement structure, it's all organic and the buildings on stilts and footprints leave minor marks behind. Nature takes its circle. Houses in the PNG bush tend not to last longer than 7 years, but get constantly improved and fixed if the location is 'right.' A house is a living thing then. Traditional PNG societies have very little individual and material wealth, just common wealth (Nakorobi, 1983; Baraka, 2011). Generally, wealth was measured in pigs, and only a few people had those. They used them to indebt and impress on others, paying back for peace. With that, fire pits and gardening traces might be the only tracks still found as leftovers. It makes Anthropologists and Palaeobiologists ponder and argue for decades (for New Guinea see Beehler & Laman, 2020 and citations within). This widely differs from what the western civilization would leave behind. Sustainability, the coming and going of human settlements and existence, is the forte of PNG.

26.9 Human Diversity in Nature: Examples from PNG

I think most people will agree with isolation being the main cause for finding separate genetic variation (= random drift within a confined population, or specific local adaptations). So, then, if there is no mixing within that group, it will be somewhat distinct. Isolation can be facilitated by geography, e.g. islands or mountains; PNG has steep valleys and rugged terrain making travel—and thus exchange—difficult. I think this is widely seen as a driver in most mountain regions and why human diversity evolved here so strongly. Arguably, there is a link with the environment and language, and it can be circular in regard to diversity. Language barriers avoid genetic mixing, just like language and culture do.

When it comes to mammals—and thus humans—subspecies and races have been discussed for a long time in the science record, also and specifically in early Anthropology. However, the current wider perspective is, we are all humans, and the race theory was ill-footed, and essentially human races and their study is not pursued. That puts a limit to some PNG study efforts.

PNG is not really a stand-alone island case. It's part of the Sahul region. And that allows to generalize from Australia, as done by Flannery (2002). If that is correct, one can easily relate back to app 65,000 years ago, perhaps as far as 74,000 years for the region (Beehler & Laman, 2020). That's specifically so because PNG people were a lot on the move, and moved all over, on land and at sea. It's a society in flux and always has been. The meaning of a single site is usually less relevant. Doing archeological digs there for inference suffers from errors of omissions and commissions; a typical but poorly acknowledged problem in archeology and in biogeography interpretations!

In PNG, an easy connection can already be made with the Lapita Culture that came from Asia c. 3600 years ago (Beehler & Lamaon, 2020 and citations within); it likely brought the pig with them, presumably also fisheries techniques and other things like specific timber.

The potatoes arrived in the highlands app. 1600 years ago (Beehler & Laman, 2020 and citations within), also with new subspecies brought by the Portuguese colonial powers app. 400 years ago onwards ('English potatoe'; now all done with genetic clones).

Another change happened with the Casuarina tree for timber, arriving in PNG c. 1200 years ago. It somewhat changed some of the forest set up. Planting timber in the forest turns it more into a farming area, a modified virgin forest with human domestication aspects. Much of PNG and its 'virgin' forest has such features, e.g. the flower and food plants grown along trails.

Another major change occurred in PNG's landscapes when the dog arrived c. 2000 years ago to PNG; it likely wiped out the thialine and some tree kangaroos and stressed many ground-living birds.

The problem with those items remains in the fact that a few paleobiologists and their (DNA) analysis machines tell us reasonable-sounding narratives based on a few digs, caves and artifacts, but hardly with a control study or research design, instead based on microsample sizes. Much uncertainty remains from that.

26.10 What is and What Maintains the Human Diversity: Examples from PNG

This might be an even bigger and more relevant question, considering genetic diversity is a healthy metric for viable human life; is it not? Not everybody will agree.

In PNG, the mixing of people is a fact of live, but so is the insulation, e.g. in the valleys and islands, and now the opening up due to globalization and getting lost. Which one is better, and at what time scales, and what is a good mix and balance, in PNG and beyond? Nobody truly knows.

While males seem to dominate, patrilineal and matrilineal societies exist both in PNG. The Wantok system ('one talk') dominates in 'the bush,' but it means that the local tongue is the driver. Another strong PNG feature is the Payback System, it involves public compensation ceremonies and ritual to settle any debts and problems (Gillison, 2002 for examples). Pigs are central in that discussion.

Bride price is still a major expectation in much of the PNG society, and payments take the form of pigs, cassowary and cash. Payment of a bride price is app. 6000$ and a few pigs; some of that is returned over time, which can create disputes. Considering a change of wealth in PNG society, the bride price and policies changed also; a study feast for Anthropologists and social scientists: What are humans worth? While monogamy is promoted by the churches, polygamy is common in PNG and wealthy men can have several wives and children with them. Missionaries are widely against such concepts but achieved little progress on that matter, e.g. Richardson (2005).

As a fundamental concept in PNG, complex land tenure systems are the rule (Baraka, 2001). As a matter of fact, rural land is rarely sold or bought; such transactions tend to be always a source of conflict.

To provide some additional examples of the diversity of live and their associated cultures in PNG: PNG has one of the highest rainfalls in the world. But then, on the Boesa Island there is a freshwater shortage, only vulcanoe rock and little freshwater but PNG citizens get along (Cousteau & Richards, 1999). Those landscape- and climate-driven concepts manifest themselves in many outcomes; a few more are presented here:

- PNG is for instance famous for its **iconic carvings**, such as masks, spirit boards, decorative canoe prows and paddles, and the Ceremonial Draggers. And then, there are the headdresses, many featuring Birds of Prey, and the Yam Storehouses (some examples shown in Cousteau & Richards, 1999; Beehler & Laman, 2020).
- **Trophy Heads** are found in PNG, but are actually nothing special in the tropical world, as also found in Borneo (Gingging, 2007), North and Northeast of New Zealand and Australia as well as Amazonian region etc.
- The **Wigmans** in Kindau have reached a certain world fame; they are quite unique.
- Another fascinating PNG feature is the **Firedance** on Baining, New Britain Island, where people are walking on coals, which is done since deep time (Cousteau & Richards, 1999). While people have done such things in several places of the world, e.g. India, they are such cases found in PNG at just a few sites.
- PNG also uses **mumus** (earth oven food-sharing events) and such type of family and community celebrations. This leaves the entire cooking, health and food concept open for a discussion (not done here). But Melanesia is based on healthy food and live (Narokobi, 1983); and PNG hosts big feasts, for friends and neighbors. It's similar to potlucks held in the Pacific Northwest, which totally stands in conflict of capitalism, direct wealth accumulation (~hoarding) and such theories. Instead, in mumus one gives away wealth and gifts/presents.
- People in PNG are usually very much concerned about the afterlife, for them and the relatives. The **transition of the soul** toward the next generation remains an essential concern. It's more of a concern than living in the present, e.g. (Lohmann, 2005). PNG has many examples that the ancient remains are kept, e.g. Lake Kutubu (Cousteau & Richards, 1999, p. 104). Some tribes use mummies in some of its funeral practices (e.g. Becket, 2021). Other tribes have ceremonial burial cliffs, e.g. in Kukukuku (Beehler & Laman, 2020, p. 265).

- A rather unique example for fishery practices in PNG is found on New Ireland with **shark calling**. It's based on using rattles and predances on a canoe (Cousteau & Richards, 1999, p. 196) and represents a sustainable practice (as sharks had not been overharvested, historically).
- In the Sepik region, the village of Kambaramba presents itself as a unique **stilt village** in the swamp lake. It's located in a Sepik flood zone at a lake and presents a trading hub (Cousteau & Richards, 1999 for details).
- A classic example of culture and architecture is shown in the sacred house, the **Haus Tambaran**, which comes with sacred ceremonies and artwork, Sepik River.
- The Kiriwina Island (**Trobriand Islands**) have been studied for long time for their unique features. Anthropologist investigated for instance their sexual moves and dances (Cousteau & Richards, 1999). It's a typical example for islands and their own dynamics and evolution, but as well as connectedness to other areas (widely different from PNG mainland, etc.)
- Then, as part of initiation rites widely found in PNG, there are **Scarification procedures** to join a secret group of men, e.g. done in Sepik. It goes back to the concept that a human been swallowed by a crocodile as a baby and spit out, reborn as 'crocodile men.' It's a crocodile cult and religion.
- Other parts of the **crocodile worship** in the Sepik river require crocodiles caught by hand, and then consumed with spirits and powers transferring to the hunter.
- PNG's **pottery** is of global fame and collected and studied worldwide. In PNG, it's widely used in good sophistication, e.g. for sago storage. The Aibom people at Sepik show good examples of that skill (Cousteau & Richards, 1999, p. 141).
- Perhaps easily overlooked, but a good example of locally adjusted culture is represented in **smoking and unique tobacco pipes;** for New Guinea see for instance Haddon (1946).
- The **use of non-timber products** is outstanding in PNG. For instance, mud masks consist of a burlap structure and dried mud. It was traditionally used to disguise the killing and fighting of warriors. Non-timber use of the forest widely outcompetes in PNG the timber use.
- Another generic concept found in PNG consists of **gift making**, e.g. in daily life via tobacco and a piece of a newspaper (Hayan, 1990) and through betel nuts. It's an easy way to make peace in PNG and its highly diverse tribes. On a village level, pigs get used for that.
- A major feature of PNG life might be its described **brutality**. It can be found throughout much of the written record. Matthiessen (1987), Gillison (1993), Flannery (2000) report frequent dispute, warfare, rape and mutilation, e.g. a finger cutoff as a sign of mourning for a dead family member (Matthiesen, 1987). Golup (2014) reports on beheadings and skulls cut open by machetes linked to mining disputes.

26.11 Some Modern Aspects of Anthropology in Papua New Guinea and Beyond

Arguably the fast transitions of societies, from ancient to industrial and globalization remain the fascination of the day for PNG, with mining at the core of those questions. Works like Mead (1935, 1967), Flannery (2000, 2002), and specifically PNG documentations by Kiki & Cheshire (1969), Chan (2016), Wilson (2019) provide fascinating, albeit less research-solid, insights. It's an Anthropology by the people themselves; the study subjects document themselves. One may ask though why the lack of Anthropological training and of science rigor in those descriptions matter so much as we know all relevant aspects and impacts of industrialization already?

For myself, I am probably less interested in knowing how the TV impacts PNG life (Wilson, 2019; we know for long time that TVs can bias and brainwash people, e.g. Afrin, 2021). And I am equally not much interested in mobile phone impacts (Watson, 2011), as we know mobile phones are not sustainable and highly commercial and contribute to a slave-type commercial society (Upreti & Singh, 2017). Arguably, the Internet is a feast for Anthropologists to study in PNG also (e.g. Dynson et al., 2006) but what really changes, and what does it mean?

Instead one will be more interested in knowing who is responsible for mining pushing beyond the law, for money grabbing from relatively intact and well-to-do societies, and who keeps denying science? We know some of those already, e.g. Henton and Flower (2007), and others are listed by Golup (2014), Ludlam (2021; for Australian sources and applications).

Still, the real question remains what is a good model for moving forward with mankind, for PNG and how to implement it and with a solid governance for all people?

I see little help with that in the Anthropological literature, or in Anthropological field research or Anthroplogical Institutes, certainly not from PNG studies done there. The examples from Acemoglu and Richards (2012), Harari (2014) are quite heartbreaking in their interpretation (contrast with Daly & Faley, 2010; Laman & Scholes, 2012; Beehler & Laman, 2020) remain in the descriptive world, and the works by Flannery (2002), and Diamond (2011a, 2011b) indeed started great discussions but are still lacking a good policy and environmental outcome and message, or to deal with the direct actors in PNG, Australia and the international scene. Thus, this book and its chapters for a try of more progress. Simply adding Ecological Economics and its applications (Czech, 2000) to that discussion will be progress and make it more concrete and real, for Anthropology and PNG—a finite space.

The discipline of Anthropology has been utterly disappointing on those aspects; it's not leading the global society, certainly not for humanity or for sustainability, or climate change. I had hoped better from the investigators, the discipline of Anthropology and institutions, including over 100 years of frontier fieldwork and from the leading research agencies of Cambridge (Murray & White, 1981), Oxford, Havard and similar ones with world-leading means, infrastructures and large public and private budgets, e.g. The Smithsonian. What an opportunity it has been?

With that, looking at PNG the good outcomes of such an Anthropology, of science itself, remain marginal. Instead, widely popular figures like the late Bob Marley continue to be the guidance of the day for many PNG citizens (Toynbee, 2007). And all of this stays embedded in a tribal system, similar to what was ruling for over 47,000 years. Modern PNG shows us no other.

26.12 What is the Wider Picture Regarding the Anthropology Question? What Can and What Does the West Really Learn from PNG?

Like elsewhere, many people see a wider underlying question in PNG's human diversity: and that is the widely debated topic of genetic set up, race, human evolution, and whether all humans are the same, and whether they differ and how they differ? Or more directly, are there superior races, e.g. based on life style and intelligence? And is PNG coined as 'fuzzy wuzzies' (~savages ; Levi-Strauss, Mead, 1967) inferior or ancient and original? While nowadays nobody would likely admit it, that's a major question in many of the early colonial explorations and descriptions by the western world. Many people and their funders had jumped on that question of 'the wild' underlying and indirectly—but with no good outcome. This is a question that goes back to old roots and whether black, white, the West, Asia, etc., are different from each other and who is better; a topic that I do not believe much in. Personally, I assume all humans are 'same' and more pressing questions are to be pursued, and resolved well among us. Ideally, one wants to know how to get a long well and how to live well?

And then there is the hotly debated topic of human DNA and who owns it (e.g. https://www.nytimes.com/1995/11/27/business/patents-recent-patent-papua-new-guinea-tribe-s-cell-line-prompts-outrage-charges.html), and whether and how one can summarize it from just a few village samples across human history even? PNG plays a big question in that.

In my view, the real questions in PNG are still not widely celebrated and shown and resolved. For one, it has to do with the feudal society, knights and lack thereof. What are its impacts, beyond what Ostrom et al. (2002) described? Related to that, how does the public common good do under such a framework? And from that, how sustainable is it, and at what scale? PNG has species extinctions, but those are local and not global, hardly PNG-wide.

Also, PNG allows to see a window into the human past, that is true for land and sea, and includes the geology and the universe; the cosmologies (Gillison, 1993). But knowing those details, how does that help for progress and crisis that the human society is facing these days?

26.13 Anthropology of the West Used to Study and Treat Yourself Instead: Using PNG also as the Control/Baseline. Something Really went Wrong with the Global Industrialized 'Modern' Society

Simply looking at the metrics and facts, the western world is pretty sick, 'in the head' and in the body. In Canada, and similar for the U.S., where per-capita and national wealth is very high, easily 50% of the citizens use drugs, antidepressants and worse, to get by on a daily basis. The Opioid crisis is for real (Kertesz & Gordon, 2019), so is Obesity and Diabetes crisis, or pandemics. How 'nice' can the paradise and wealth get?

The amount and severity of injuries and days missing at work in western nations are highly driven by industrial work itself (chronic neck and back pain as a national disease there), by the employment and the capitalistic need 'to make money.' '*No pain no gain,*' as the common saying there states; a rather strange concept for human life and if well-being is a goal. And this is not only for physical injuries, but also for psychological ones. It's easily seen in the fact that even 'social' nations of the west suffer from mental health problems on a massive scale, e.g. suicide rates in Sweden, Iceland and Canada (Klerman, 1986). And clearly, the vote for a political party tends to depend on the employment set up and subsequent salary. While not often admitted, the western world is a pure 'money only' world, greed dominates, and so are its sciences and institutions. Much of that money comes from natural resources, and from the tropics. The PNG money often flows and stays abroad. PNG is one of the dirty money pits for the western society (see Revkin, 1994 for Amazonia, see Eichstaedt, 2016 for Congo; see Huettmann, 2020 and citations within for China and its Yellow Sea as the kitchen sink for globalization).

It's easily to see in the overemphasize of 'holidays' in the western world, such as Xmas, Easter, Birthdays and Valentines. In the ancient world of PNG, those commercial things have no meaning. In modern times, such western-imposed cultural celebrations create much deaths in PNG (due to drinking and other reasons).

Arguably, the biggest issue with the western world and globalization might be its vast unsustainability, including its inherent massive warfare and microaggressions in the own society spilling over . It's a topic that is virtually not covered in traditional Anthropology and for PNG. Compared to PNG, the western world is virtually free of 'taboos'; individual freedom and wrongly understood economic growth rules instead (e.g. Czech, 2000). While laws exist, those let economic growth unfold virtually unconstrained (see Taber & Payne, 2003 for examples that in sustainability efforts industry always tends to win and nature pays the prize).

And for sure, the highly developed society—The West and Japan—has a deep history of initiated global warfares and atrocities; that is certainly true for the land of PNG and can be seen there (Cousteau & Richards, 1999 for many examples within). Many of the western nations, certainly the richest ones, still engage in a type of world war and global destruction today. From that, it's not a big surprise that slavery was very common in the western society itself (see for Ireland and Medivial times see Fontaine, 2021; Rusev, 2013 for modern times such as human trafficking).

The wide failure, and the deep divides within, can be read in work from Cockburn (2013) and for Ireland let's say, (O'Toole, 2009). One finds them also in the heart of the EU (e.g. Cunliffe, 2021). But beyond that, there is not so much Anthropology literature on the western misadventure in Greenland, or Cuba for instance. It certainly exists for East Africa and in the UK approach to the Massai (Hughes, 2006) and the German colonial enterprise failure (Firth, 1972; Smith, 2012). Many more can be named.

The failures of the western model are specifically obvious in the environmental sector, which often precludes other problems on the humanity, social and economic front; Czech, 2000). A typical example can be seen there in the poor performance of migratory species (Gallo Orsi & Orhun, 2008), starting with fur seals, e.g. crashed in Alaska around the 1900, and then the North American birds (overharvest for plumage) and fish (straddling stocks, including totally overfished tuna, herring etc; see chapters within this book).

In comparison, gardening PNG operates widely sustainable and has done so for easily over 95% of its time. A typical example is found with the nest mounds of the megapods. Those can be harvested sustainably in the forest for their eggs. Another example is clam gardens in the coast. PNG once had a sustainable clam gardening practice. Clams were part of a community ownership and harvested in a sustainable fashion with 30 clams per garden (Cousteau & Richards, 1999, p. 213). The *Tridacna* clams had an assigned ownership and care for their harvest; they weigh up to 500 pounds. Due to the western cash economy people then overharvested and break down the ancient system. Another reason is that education is NOT free and families now need to accumulate 'cash' to pay the school for their children. This is all done now on the cost of the clams, family structures and of the oceans, to get an industrial education.

Nowadays, when the cash crop economy dominates, gift giving and feasts need money and are not sustainable anymore (Baraka, 2001; Golup, 2014). An ancient sustainability system broke down with a subsidized short-term and failing scheme.

And while the demand for primitive art from Sepik is on the rise, worldwide, there is little hardwood left, and now cavers use stained softwood, etc. (Cousteau & Richards, 1999). It's the wider framework that drives the landscape and its sustainability.

PNG has a reputation for being quite a brutal nation, e.g. Richardson (2005), The Guardian (2019a, 2019b) (see chapters in this book). PNG might have a lot of warfare, tribal, but the WW2 battles—an outside war—fought on PNG grounds like Kokoda trail and Rabaul were similarly atrocious and much worse for the destruction, contamination and outcomes and scales (Cousteau & Richards, 1999, p. 106 for an example).

But in terms of accused brutality, one does not have to go far in the European 'civilization' to find such records, e.g. the 100 years war between England and France, or the 30 years war and the Westphalian Peace Treaties in Central Europe 1648, and the almost frequent 50-year re-occurrence of warfare in the EU landscape and history (a pattern that holds to this very day, even in the otherwise relative peaceful Coldwar period (a 40 years or so outlier in the EU history), where conflicts around Basque and North Ireland were essentially international wars). Peace and such a culture are not a specific forte of the western world, certainly not the EU or the North American one.

One may go beyond that and just look at police brutality; as a widely accused PNG feature (Gosarevsky et al., 2019) but instead found in most parts of the western world, certainly in the dominating U.S. And it does not end there, leaders themselves are very brutal also, as can be seen in the biography of Helmut Kohl—German chancellor of 16 years—dealing with his family and competitors (Kohl, 2011). More of those leaders can be named. And then the historical record is full of uttermost brutality, from leaders like https://en.wikipedia.org/wiki/Gilles_de_Rais to heroes like Ancient Greece philosophers (Husby, 2009) and Mongolian great leaders, e.g. Djinghis Khan (Stone, 2017).

Now, whoever thinks that PNG is alone in this situation should be well aware that virtually all leaders, and most celebrities, of the western world—from poor to rich nations—get commonly confronted with death threats, if not even hostage-taking and assassination. There is a long list of that in the 'modern' past and society at large It is its history. A. Lincoln, J. F. Kennedy, RAF hostage-taking of German embassy in Sweden, Olympics Munich, etc. It's also found in Nicaragua, or with the Nepal Supreme Court Judge ("*Supreme Court judge shot dead in Nepal capital*"; https://www.bbc.com/news/world-asia-18275891) and the Nepal Royal family (Gregson, 2002). So why can PNG not have it, just like many other nations? Why gets PNG singled out?

As the outcry over the killing in PNG shows, it's NOT a new thing as such but on the rise. And while the mining industry actually blames the mobile phones and the internet, modern life (https://www.theguardian.com/world/2019/jul/23/the-karida-massacre-the-start-of-a-new-era-of-tribal-violence-in-papua-new-guinea), in reality, those go hand-in-hand and feed off each other (see Golup, 2014 for sociological details). Isn't mining justified by providing for mobile phones, batteries and such? It's clear that mining blames modern times, a society it actually always claims to create! Mining would be sustainable and provide progress and a modern society?

There are many aspects where The Western World cannot really compete with other nations. For instance, the English language, with its 26 character alphabet, is tiny and a major constraint and does not allow to express things well and locally, beyond the area it comes from (compare for instance with Chinese, app. 1000 character types and coinciding with Megabiodiversity; or PNG with its locally adapted languages, mostly actually unwritten as there is no need to do so). Using English as a global unifier or a diplomacy platform must seem absurd for such instances, just as it appears to most rural PNG citizens I know of. Many details they know of can hardly be expressed in English, certainly not with just 26 characters. Needless to say, the effort that is taken to teach children those 26 western characters remains mind-boggling to anybody who just looks at it with an open mind.

26.14 Anthropology of Money and Mining, Anybody? Research Done thus far by the West and for the West (but not for PNG, Hardly with PNG or with Elite Anthropologists)

Where does Anthropology stand on all of those topics? The answer is also not well-published. And the concept of "*Hey I am an Anthropologist from Oxford, Cambridge and Harvard and I study people for the Smithsonian in a remote jungle village in Papua New Guinea*" likely remains as outdated as it now gets these days. One may easily ask then what has Anthropology really achieved, and what is it good for, after all?

PNG has moved on and is in modern times; it's much more than the framed discussions of the noble savage (Levis-Strauss, 1966), the noble savage ecologist (Raymond, 2007) or the Melanesian way (Narokobi, 1983), or mining impacts being bad or good (Golup, 2014). That's the reason why antiquated approaches on top-down Anthropology by so-called elite institutions and their experts fail us all so much; it's a mismatch (the mismatch comes from the scientist and scholar that has not realized in what world they are already living).

A core question to society, the Anthropology of Sustainability, of Economic Growth, of handling finite space over time, is widely not what made the discipline of traditional Anthropology famous, of the Anthropology of PNG. Anthropology serves PNG little, to this very day

Like with any of the sciences, Anthropology only happens in the funded national institutions and such frameworks and by such people. It sets the questions, and that's precisely where much of the traditional Anthropology fails itself, as the institutions are utterly flawed, biased, political, often ineffective and unsustainable. The constant chase for funding makes that clear. The idea that a big National Science Foundation (NSF) and National Institutes of Health (NIH) grant would cover your cost comes with a massive flaw: already the hosting institutions of such grants hardly cover infrastructure costs, healthcare and printing, field work costs, or long-term work; certainly not tenured salaries with a good pension and child care and paid maternity leaf. And then, neither NSF nor NIH are free of strings attached. Instead they are to serve the donor (which is not always the taxpayer) and follow institutional constraints (e.g. in the U.S. ethics rules are to be followed, including the provision of receipts, annual reports, or outhouse access for its fieldworkers making relevant fieldwork in PNG and the bush next to impossible).

Emphasizing that the western world is not a global failure and not an insult to most people (= the majority people, those who make less than 4$ a day), is like saying '*the Titanic is just experiencing some delays*'. Why can Anthropologists not study their own society and framework, admitting its flaws first and foremost, e.g. the massive failures of mining, oil and gas exploration, and of flying to the moon and Mars, or warfare? Once this is agreed on, one can progress fine. No need to be in PNG. Anthropology and the western critique and progress remain odd bedfellows

(Eriksen et al., 2015 for neoliberalism as a dominating business model). Western universities and their funders are not good at that and must improve.

PNG remains to be pushed violently into unsustainability and destruction, natural resource extraction concepts—mining, gas and oil, forestry, farming and fishery, beyond others—and one may easily study the Anthropology of Mining and Terror (Ballard & Banks, 2003; Gulop, 2014). But the outcome is quite clear anyways: Mining is not really sustainable, impacts cannot be marginalized or divorced from it, it has not good footprint in PNG or anywhere, comes with much brutality and a curse instead, and it creates violence and many other clean up problems unresolved (Kirsch, 2014; Golup, 2014 for examples). So what is learned from science, from Anthropology, from its funding institutions, and why not stated and improved? It should halt til improved for a better outcome.

References

Acemoglu, D., & Robinson, J. A. (2013). *Why nations fail: The origins of power, prosperity, and poverty paperback.*

Afrin, S. (2021). A postmodern cultural study of media manipulation and capital market expansion. *Postmodernism Problems, 11*(3), 270–289.

ATBC. (2011). The urgent need to curtail 'special agricultural and business leases' in Papua New Guinea. In *Resolution, association for tropical biology and conservation.*

Bablis, G. (2010). "A Melanesian Icon-Professor Bernard Mullu Narokobi (Ca 1940–2010)." 236–257.

Ballard, C., & Banks, G. (2003). Resource wars: The anthropology of mining. *Annual Review of Anthropology, 32*, 287–313.

Bandura, A. (2007). Impeding ecological sustainability through selective moral disengagement. *International Journal of Innovation and Sustainable Development, 2*(1), 8–35.

Baraka, R. (2001). The meanings of independence in Post-Colonial Papua New Guinea. *A plan B research paper submitted to the faculty of the center for Pacific Islands studies in the candidacy for the degree of Masters of Arts. School of Hawaii*, Asia and Pacific Studies. The University Of Hawai'i At Manoa.

Beehler, B., & Laman T.(2020). *New Guinea: Nature and Culture of Earth's Grandest Island.* Princeton University Press.

Beckett, R. G. (2021). Smoked bodies of Papua New Guinea. In *The handbook of mummy studies: New frontiers in scientific and cultural perspectives* (pp. 983–1008). Springer Singapore.

Blanchard, K. (1995). *The anthropology of sport: An introduction.* ABC-CLIO.

Chan, J, (2016). *Playing the Game: Life and Politics in Papua New Guinea.* University of Queensland Press.

Cockburn, A. (2013). *A colossal wreck: A road trip through political scandal.* Verso Publishers.

Conner, C. D. (2009). *A people's history of science: Miners, midwives, and low mechanicks.* Hachette.

Cox, J. (2011). *Deception and Disillusionment: Fast Money Schemes in Papua New Guinea.* Universty of Melbourne.

Cousteau, J. M., & Richards, M. (1999). *Cousteau's Papua New Guinea journey.* Henry Abraham's Books.

Cunliffe, B. (2021). *Bretons and Brittons fight for identity.* Oxford University Press.

Czech, B. (2000). *Shoveling fuel for a runaway train: Errant economists, shameful spenders, and a plan to stop them all.* Univ of California Press.

Daly, H., & Farley, J. (2010). *Ecological economics: Principles and applications* (2nd ed.) Island Press.
De Waal, F. (2016). *Are we smart enough to know how smart animals are?* WW Norton & Company Publishers.
Diamond, J. (2011a). *Guns, germs and steel: The fate of human societies*. Random House.
Diamond, J. (2011b). *Collapse: How societies choose to fail or succeed* (Revised ed.). Penguin Books.
Dublin, D., & Tanaka, N. (2015). Comparison between Satoyama and the Ainu way of life. *Kawistara Jurnal Sosial Dan Humaniora, Universitas Gadjah Mada, Indonesia, 5*, 221–328.
Dynson, L. E., Hendriks, M., & Grant, S. (Eds.) (2006). *Information technology and indigenous people*. IGI Global.
Eichstaedt, P. (2016). *Consuming the Congo: War and conflict minerals in the world's deadliest place diamonds*. Laurence Hills Publisher.
Eriksen, T. H., Laidlaw, J., Mair, J., Martin, K., & Venkatesan, S. (2015). The concept of neoliberalism has become an obstacle to the anthropological understanding of the twenty-first century. *Journal of the Royal Anthropological Institute, 21*(4), 911–923.
Firth, S. G. (1972). *The New Guinea Company, 1885–1899: A case of unprofitable imperialism*. Adelaide University.
Fontaine, J. M. (2021). The archaeology of slave trading in Viking-Age Britain and Ireland: A methodological approach. In *The archaeology of slavery in early medieval Northern Europe* (pp. 53–63). Springer.
Foster, R. J. (2006). From Trobriand cricket to rugby nation: The mission of sport in Papua New Guinea. *The International Journal of the History of Sport, 23*, 739–758.
Foster, R. (2008). *Coca-globalization: Following soft drinks from New York to New Guinea*. Springer.
Flannery, T. (2000). *Throwim Way Leg: Tree-kangaroos, Possums, and Penis Gourds--on the Track of Unknown Mammals in Wildest New Guinea*. Grove Press.
Flannery, T. (2002). *The Future Eaters: An Ecological History of the Australasian Lands and People*. Grove Press.
Forsyth, M., & Eves, R. (2015). *Talking it through: Responses to sorcery and witchcraft beliefs and practices in Melanesia*. Australian National University Press (ANU).
Gaffney, M., & Harrison, F. (1994). *The corruption of economics (Georgist Paradigm Series)*. Shepheard-Walwyn.
Gallo Orsi, U., & Orhun, C. (Compilers) (2008). Review of the implementation and the effectiveness of 15 single species action plans for migratory waterbird species. AEWA Technical Series No. 30. Bonn, Germany.
Gillison, G. (1993). *Between culture and fantasy: A New Guinea highlands mythology*. University of Chicago Press.
Gillison, D. (2002). *New Guinea ceremonies*. Harry N Abrams Inc.
Gingging, F. A. M. (2007). "I lost my head in Borneo": Tourism and the refashioning of the headhunting narrative in Sabah, Malaysia. *Cultural Analysis, 6*, 1–29.
Gray, G. G. (1996). 'The next focus of power to fall under the spell of this little gang': Anthropology and Australia's postwar policy in Papua New Guinea. *War and Society, 14*(2), 101–117.
Gregson, J. (2002). *Massacre at the palace: The doomed royal dynasty of Nepal*. Miramax Publishers.
Gosarevski, S., Hughes, H., & Windybank, S. (2019). Is Papua New Guinea Viable? Political Economic Bulletin 19. https://www.semanticscholar.org/paper/Is-Papua-New-Guinea-viable-Gosarevski-Hughes/6d71c70951c12fa82dea578a36b66923997f3f92. Accessed June 1, 2022
Habbia, H. I. (Ed.). (1973). *Anthropology in Papua New Guinea*. Melbourne University Press.
Haddon, A. C. (1946). Smoking and tobacco pipes in New Guinea. *Philosophical Transactions of the Royal Society of London. Series B, Biological Sciences, 232*(586), 1–278. https://www.jstor.org/stable/pdf/92349.pdf
Harari, Y. N. (2015). *Sapiens: A brief history of Humankind*. Harper Perennial.

Harari, Y. N., & Sapiens, A. (2014). A brief history of humankind. Publish in agreement with The Deborah Harris Agency and the Grayhawk Agency.
Hardin, G. (1998). Extensions of "the tragedy of the commons." *Science, 280*(5364), 682–683.
Hayan, D. M. (1990). *Road through the rain forrest: Living anthropology in Highland Papua New Guinea.* Waveland Press.
Henton, D., & Flower, A. (2007). Mount Kare Gold Rush: Papua New Guinea, 1988–1994. Published by Mt Kare Gold Rush, Brisbane Australia. ISBN 10: 0646482815ISBN 13: 9780646482811.
Hoffmann, C. (2015). *Savage Harvest: A Tale of Cannibals, Colonialism, and Michael Rockefeller's Tragic Quest.* William Morrow Paperbacks; Reprint edition.
Huettmann, F. (2020). The Hindu Kush-Himalaya (HKH) in the global and marine context: Major estuaries, coast-scapes, ocean coupling, seawalls, over 2 billion people and global (food) security. In *Hindu Kush-Himalaya watersheds downhill: Landscape ecology and conservation perspectives* (pp. 61–77). Springer.
Hughes, L. (2006). *Moving the maasai: A colonial misadventure.* Springer.
Humphries, G., Magness, D. R., & Huettmann, F. (2018). *Machine learning for ecology and sustainable natural resource management.* Springer.
Hurley, F. (1924). *Pearls and savages: Adventures in the air, on land and sea—In New Guinea.* GP Putnam's Sons.
Husby, T. K. (2009). Justice and the justification of war in Ancient Greece: Four authors. *Classics Honors Papers*, 1. http://digitalcommons.conncoll.edu/classicshp/1. Accessed August 4, 2022.
Hunter-Rainey, S., & Rodríguez, L. C. (2011, July). The Gilded Cage: Contemporary Slavery in American Professional Sports Teams. *In Proceedings of the International Association for Business and Society* (Vol. 22, pp. 44-51).
Kertesz, S. G., & Gordon, A. J. (2019). A crisis of opioids and the limits of prescription control: United States. *Addiction, 114*(1), 169–180.
Kiki, A. M., & Cheshire, F. W. (1969). *10,000 years in a lifetime.* Cheshire Publishing.
Klerman, G. L. (1986). Evidence for increase in rates of depression in North America and Western Europe in recent decades. In *New results in depression research* (pp. 7–15). Springer.
Kulik, D. (2019). *A death in the rainforest: How a language and a way of life came to an end in Papua New Guinea.* Algonquin Books. 9781616209049
Laman, T., & Scholes, E. (2012). *Birds of paradise: revealing the world's most extraordinary birds.* National Geographic Books.
Le Bellac, M. (2011). *Quantum physics.* Cambridge University Press.
Levi-Strauss, C. (1966). *The savage mind.* University of Chicago Press.
Levit, G. S., & Hossfeld, U. (2020). Ernst Haeckel, Nikolai Miklucho-Maclay and the racial controversy over the Papuans. *Frontiers in Zoology, 17*(1), 1–20.
Lipset, D. (2003). Rereading "Sex and temperament": Margaret mead's Sepik triptych and its ethnographic critics. *Anthropological Quarterly, 76*(4), 693–713.
Lohmann, R. I. (2005). The afterlife of Asabano corpses: Relationships with the deceased in Papua New Guinea. *Ethnology*, 189–206.
Lutton, N. (1981). From Port Moresby to Perth. 30–38.
Ludlam, S, (2021). *Full Circle: Power, Hope and the Return of Nature.* Black Inc.
Madison, M. G. (1997). "Potatoes made of oil": Eugene and Howard Odum and the origins and limits of American agroecology. *Environment and History*, *3*(2), 209–238. http://www.jstor.org/stable/20723041
Matthiessen, P. (1987). *Under the Mountain Wall: A Chronicle of Two Seasons in Stone Age New Guinea.* Penguin Classics.
Mead, M. (1930). *Growing up in New Guinea.* Morrow Publisher.
Mead, M. S. (1935). *Temperament in three primitive societies.* Morrow Publisher.
Mead, M. (1935). *Sex and temperament.* Routledge and Kegan Paul, .
Mead, M. (1967). The rights of primitive peoples: Papua-New Guinea: A crucial instance. *Foreign Affairs, 45*(2), 304–318.

Mead, M. (Ed.). (2002). *Cooperation and competition among primitive peoples* (Vol. 123). Transaction Publishers.
Mack, A. (2014). *Searching for Pekpek: Cassowaries and Conservation in the New Guinea Rainforest*. Cassowary Conservation & Publishing.
Milenković, B. (2019). Sport: A call to freedom or agreement to slavery? *Fizičko Vaspitanje i Sport Kroz Vekove, 6*(1), 55–69.
Miller, M. T., Smith, E. A., & Nadler, D. P. (2016). Debate and discourse: The role of the faculty senate on the modern American campus. *Journal of Higher Education Theory and Practice, 16*(3), 22–24.
Murray, T., & White, J. P. (1981). Cambridge in the bush? Archaeology in Australia and New Guinea. *World Archaeology, 13*, 255–263.
Naess, A. (2009). *The ecology of wisdom: Writings by Arne Naess*. Catapult.
Narokobi, B. (1974). The nobility of village life. *Catalyst: Social Pastorial Magazine for Melanesia, 4*(4), 55–70.
Narokobi, B. (1975). *Foundations for nationhood*. Papua New Guinea Press and Bookshop.
Narokobi, B. (1983). *The melanesian way*. Institute of Papua New Guinea Studies.
Narokobi, B. (2000). Papua New Guinea's self-image: Through a glass darkly. In N. Sullivan (Ed.), *Uncertain paradise: Building and promoting a better Papua New Guinea* (pp. 23–31). Divine Word University Press.
Ohff, H. J. (2008). Empires of enterprise: German and English commercial interests in East New Guinea 1884 to 1914 [Unpublished Doctoral dissertation, University of Adelaide, Australia].
Organ, J. F., Geist, V., Mahoney, S. P., Williams, S., Krausman, P. R., Batcheller, G. R., Decker, T. A., Carmichael, R., Nanjappa, P., Regan, R., Medellin, R. A., Cantu, R., McCabe, R. E., Craven, S., Vecellio, G. M., & Decker, D. J. (2012). The North American model of wildlife conservation. In *The wildlife society technical review* (pp. 12–04). The Wildlife Society, Bethesda.
Ostrom, E. E., Dietz, T. E., Dolšak, N. E., Stern, P. C., Stonich, S. E., & Weber, E. U. (2002). *The drama of the commons*. National Academy Press.
O'Toole, F. (2009). *Ship of fools: How stupidity and corruption sank the Celtic Tiger*. Faber & Faber.
Otto, T. (1999, May). Cargo cults everywhere? In *Anthropological forum* (Vol. 9(1), pp. 83–98). Taylor & Francis Group.
Putnam, W. (2012). "Please Don't Feed the Natives": Human zoos, colonial desire, and bodies on display. In *The environment in french and francophone literature and film* (pp. 55–68). Brill.
Kohl, W. (2011). *Leben oder gelebt werden: Schritte auf dem Weg zur Versöhnung*. Integral Verlag.
Kirsch S. (2014). *Mining Capitalism*. University of California Press.
Raymond, H. (2007). The ecologically noble savage debate. *Annual Review of Anthropology, 36*, 177–190.
Revkin, A. (1994). *The burning season*. Penguin Books.
Richards, S. J., & Whitmore, N. (Eds.). (2015). *A rapid biodiversity assessment of Papua New Guinea's Hindenburg wall region*. Wildlife Conservation Society Papua New Guinea Program.
Richards, S. J. (Ed.) (2018). *Rapid biological assessments of Wau Creek, Uro Creek and Lake Kutubu: Documenting biodiversity values to promote forest conservation in the Kikori River basin, Papua New Guinea*. ExxonMobil PNG Limited.
Richardson, D. (2005). *Peace child: An unforgettable story of primitive jungle treachery in the 20th century*. Regal Publishers.
Robinson, J. A., & Acemoglu, D. (2012). *Why nations fail: The origins of power, prosperity and poverty* (pp. 45–47). London: Profile.
Rollason, W. (2010). *We are playing football: Sport and postcolonial subjectivity, Panapompom, Papua New Guinea*. Cambridge Scholars Publishing.
Rooney, D., & Papoutsaki, E. (2004). Who is research for? An observation from Papua New Guinea.
Rusev, A. (2013). *Human trafficking, border security and related corruption in the EU. Migration and the security sector paper series*.

Salsburg, D. (2001). *The lady tasting tea: How statistics revolutionized science in the twentieth century*. Macmillan.

Sanders, M. E. (2008). Stem, stem education, stemmania. The Technology Teacher, December/January. https://vtechworks.lib.vt.edu/bitstream/handle/10919/51616/STEMmania.pdf?sequence=1&isAllowed=y. Accessed August 3, 2022.

Sanjek, R. (Ed.). (2019). *Fieldnotes: The makings of anthropology*. Cornell University Press.

Short, D. (2010). Cultural genocide and indigenous peoples: A sociological approach. *The International Journal of Human Rights, 14*, 833–848.

Silverman, E. K. (2012). From Cannibal Tours to cargo cult: On the aftermath of tourism in the Sepik River, Papua New Guinea. *Tourist Studies, 12*(2), 109–130.

Smith, W. D. (2012). *The German colonial empire*. UNC Press Books.

Spitz, R. A. (1935). Frühkindliches Erleben und Erwachsenenkultur bei den Primitiven: Bemerkungen zu Margaret Mead "Growing up in New-Guinea." *Imago, 21*(3), 367–387.

Stewart, P. J., & Strathern, A. (2002). *Remaking the world: Myth, mining, and ritual change among the Duna of Papua New Guinea*. Smithsonian Institution.

Stewart, P. J., & Strathern, A. (Eds.). (2005). *Anthropology and consultancy: Issues and debates* (Vol. 1). Berghahn Books. https://www.google.com/books/edition/Anthropology_and_Consultancy/1AGt1K5BNP0C?hl=en&gbpv=1&dq=crater+mountain+photographer+papua+new+guinea&pg=PA65&printsec=frontcover. Accessed August 10, 2022.

Stone, Z. (2017). *Genghis Khan: A biography*. Vij Books India Pvt Ltd.

Sullivan, N. (2007). God's brideprice: Laissez faire religion, and the fear of being left behind in Papua New Guinea. *Contemporary PNG Studies, 6*, 63–91.

Suzuki, D. (1993). *Wisdom of the elders: Sacred native stories of nature*. BoD–Books on Demand.

Suzuki, D. (1996). *"The Suzuki Papers" in power and in Papua New Guinea* (pp. 53–55). Aid/Watch Publication.

Swadling, P., & Hide, R. (2005). Changing landscape and social interaction: Looking at agricultural history from a Sepik-Ramu perspective. In *Papuan pasts: Cultural, linguistic and biological histories of Papuan-speaking peoples*. Pacific Linguistics.

Taber, R. D., & Payne, N. F. (2003). *Wildlife, conservation, and human welfare: a United States and Canadian perspective*. Malabar, Florida: Krieger Publishing Company.

The Guardian. (1998). Blood money, 21st January 1998. https://www.theguardian.com/science/1998/jan/21/genetics. Accessed December 1, 2021.

The Guardian. (2016). Papua New Guinea's prime minister, Peter O'Neill, resigns 26th May 2016. https://www.theguardian.com/world/2019/may/26/papua-new-guineas-prime-minister-peter-oneill-resigns. Accessed May 10, 2022.

The Guardian. (2019a). 'I will eventually get killed': Meet Bryan Kramer, Papua New Guinea's anti-corruption tsar, 10th August 2019a. https://www.theguardian.com/world/2019a/aug/11/i-will-eventually-get-killed-meet-bryan-kramer-papua-new-guineas-anti-corruption-tsar. Accessed January 15, 2022.

The Guardian. (2019b). The Karida massacre: Fears of a new era of tribal violence in Papua New Guinea, 22 July 2019b. https://www.theguardian.com/world/2019b/jul/23/the-karida-massacre-the-start-of-a-new-era-of-tribal-violence-in-papua-new-guinea. Accessed December 12, 2021.

Thuer, D. (2017). Qatar 2022, the World Cup of modern slavery. Is FIFA playing fair? [Doctoral dissertation, University of Seville, Spain].

Tutorsky, A. V., Govor, E. V., & Ballard, C. (2019). Miklouho-Maclay's legacy in Russian-and English-language academic research, 1992–2017. *Archaeology, Ethnology and Anthropology of Eurasia, 47*(2), 112–121.

Toynbee, J. (2007). *Bob Marley: Herald of a postcolonial world?* Polity Press.

Upreti, R., & Singh, R. (2017). Transformation of man into smartphone slave. *Indian Journal of Health and Wellbeing, 8*(9), 2–8.

Vetta, A. (2021). Comparison of the scientific developments of first 2 decades of 20th century and 21st century. https://www.academia.edu/49860209/Comparison_of_the_Scientific_Developments_of_First_2_Decades_of_20th_Century_and_21st_Century?email_work_card=title.

Walton, G., & Jackson, D. (2020). *Reciprocity networks, service delivery, and corruption: The wantok system in Papua New Guinea.*
Watson, A. (2011). The mobile phone: The new communication drum of Papua New Guinea [PhD thesis, Queensland University of Technology, Australia]. https://eprints.qut.edu.au/47170/
West, P. (2006). *Conservation is our government now*. Duke University Press.
West, P. (2014). "Such a site for play, this edge": Surfing, tourism, and modernist fantasy in Papua New Guinea. *The Contemporary Pacific*, *26*(2), 411–432.
West, P, (2006). *Conservation is Our Government Now: The Politics of Ecology in Papua New Guinea*. Durham, NC: Duke University Press.
White, F. (1930). The rise and fall of the German colonial empire, 1884–1918. *Journal of the Royal Institute of International Affairs*, 695–696.
Wilkinson, M. (2013). Testing the null hypothesis: The forgotten legacy of Karl Popper? *Journal of Sports Sciences, 31*(9), 919–920.
Wilson, M. (2019). *A true child of Papua New Guinea: Memoir of a life in two worlds*. MacFarland Publishers.
Willis, P. (2011, August 15). *The west's ongoing theft of indigenous knowledge*. openDemocracy. https://www.opendemocracy.net/en/wests-ongoing-theft-of-indigenous-knowledge/. Accessed August 10, 2022.
Wohlleben, P. (2016). *The hidden life of trees: What they feel, how they communicate—Discoveries from a secret world* (Vol. 1). Greystone Books.
Wroe, S., Field, J. H., Archer, M., Grayson, D. K., Price, G. J., Louys, J., Mooney, S. D., et al. (2013). Climate change frames debate over the extinction of megafauna in Sahul (Pleistocene Australia-New Guinea). *Proceedings of the National Academy of Sciences, 110*(22), 8777–8781.
Whyte, K. (2017). Indigenous climate change studies: Indigenizing futures, decolonizing the Anthropocene. *English Language Notes, 55*(1), 153–162.

Chapter 27
Use of Over 100 Open-Access GIS Layers ('Big Data') and Their Analysis and Message for Papua New Guinea with Data Mining ('Boosting') to Overcome Data Gaps for a Robust Inference: Human Footprints (~Roads) of the Last 50 years Destroy a Civilization that is Over 47,000 years Old

Abstract Despite international mega-mining projects, a U.N. mandate, The World Bank support and entire centuries of colonial expeditions, explorations and surveys—on land, in the subsurface, at sea and via remote sensing—geographic information system (GIS) data for Papua New Guinea (PNG) remain few and coarse. Coastal and marine GIS data and compilations are widely missing. Here 82 predictors are used from an overall pool of compiled 100+ open-access GIS layer set from the author that is made publicly available for anybody and free to use for PNG and beyond. This data set is provided in an Open Source GIS format for QGIS, R and similar (ASCII grids, shapefiles, geoTIFF) with ISO-compliant metadata. Using this 'Big Data' for PNG, for the first time a basic analysis of GIS data quality overlays, human footprint clustering (unsupervised; hierarchical clustering hclust) and data mining using machine learning (supervised; boosted regression trees, Salford Predictive Miner Minitab-SPM 8.3) is shown to overcome 10% of PNG-wide data gaps as a case study allowing to infer and to elaborate on the different types of footprints and to make first basic statements from the large amount of quantified predictors as baseline information. While this study remains a somewhat incomplete underestimate of human impacts, the road proximity, e.g. linked with mining and ports, remains the overarching predictor for modern human footprint impacts overruling all ecological features in an otherwise described holistic fashion. Beyond biogeography this work can open up many more detailed questions about the PNG environment on a 1 km scale. While many of those layers are dynamic and might get updated over time, here a first template is shown how such data can be used, with a workflow and for a baseline to work from and what to focus on. Road construction presents the biggest predictor for human footprints and is part of Economic Growth policies that leave PNG citizens and their ancient lifestyle behind. It is hoped that this research triggers more intense work on the issue of publicly owned habitat and ancient land tenure and can help to address (industrial) impacts and resolve disputes in peaceful and mutual ways benefitting PNG and the public wider good, for Melanesia and elsewhere

F. Huettmann, *Globalization and Papua New Guinea: Ancient Wilderness, Paradise, Introduced Terror and Hell*, https://doi.org/10.1007/978-3-031-20262-9_27

Keywords Online Open-access data · Open-source geographic information systems (GIS) · Data mining Machine Learing (Boosting) · Remote sensing · Landscape change models · Data inference

27.1 Introduction

Beyond the ongoing physical impact, the internet/www culminates industrial efforts and kept bringing pervasive changes to the world, specifically it creates (western) bias and impacted the tropical and poor regions in a negative fashion where many people cannot participate: They have no internet or cannot afford it, they have no computer or mobile device, and they do not speak English. Data also play a key role in that discussion.

Papua New Guinea (PNG) is one of those areas in the world, and data are lacking also. Despite centuries of international aid and exploration PNG (e.g. Diamond 2011) is at the bottom of the 'information age' and cannot really join the information society. It does not make PNG a relevant actor in the "New World," and it cannot operate as a nation (as expressed by Gosarevski et al., 2019 but ignoring the WWW and data as a crucial resource). Development aid and their NGOs have left this strategic focus area for PNG widely unfunded and open, continuing to leave the PNG nation and its people simply in the 'dark ages.' and with much biased information.While this is not new for PNG nor new for the developers as a strategy, it remains still shocking for the reality perspective and carried colonial patterns.

According to the New South Wales State Library (2015) mapping and subsequent inventory of PNG remained difficult until the 1970s. Till then, large areas of PNG had not been explored (Beehler & Laman, 2020). Just a few occasional surveys were done leaving the geography and details of the mountainous highlands and offshore islands unexplored for map making; the large German colonial efforts for instance did not result into available and relevant maps for PNG and the world (land and sea) to be used now. This missing knowledge was not really surprising given the combination of rugged inaccessible mountains, dense jungle cover, wet tropical climate and, in the early days, hostile and warlike tribes. However, despite a UN oversight, a combined effort to provide a better internationally shared mapping scheme with shared and documented data policies for PNG never was really done.

Only when aerial photography became more prevalent in the 1950s, it was possible to begin mapping the whole country. Remote Sensing supported that effort. However, continual cloud cover disturbed efforts to photograph many highland areas, until cloud-penetrating radar imagery became available in 1971. The advent of free and open access Landsat remote sensing in the 1980s improved the picture but clouds were again a problem (see Yen et al. 2005 for overview and example).

Data sources were sometimes compiled for PNG but are hardly organized, in a good format overall, or made publicly available (e.g. Faith et al., 2000). Some changes were brought by online portals, e.g. satellite image webportals, DIVA-GIS (http://www.diva-gis.org/gdata), GBIF.org, and Bioclim (https://www.worldclim.org/data/bioclim.html) that helped to set up a basic digital infrastructure. Further, research by Huettmann (2020) and others allowed to discuss, share, use and improve the compiled and publicly provided open-access data for PNG with anybody interested. While this presents progress, it still lacks much of the applications possible (see for examples Huettmann, 2015; Huettmann et al., 2018).

Here I am trying to present those best-available data sets, to point to data gaps, and to subsequently showcase some data mining analysis and Big Data explorations as a first template to work from. I am using a pool of over 100 open-access open-source GIS and remote sensing layers for PNG, which we made publicly available earlier (Sriram and Huettmann unpublished). They were compiled from initial open-access online sources and our processed and modeled ones. More details on sources and an application can be seen in Huettmann (2020).

PNG has a human history of over 47,000 years (e.g. Flannery 2002, Diamond 2011). Using machine learning and a wider pool of GIS data in a holistic fashion, this research aims to look at the human footprint and specifically, it tries to find drivers and explanations for the 'modern' human footprint. It aims to set up a template for what can be done with such 'big datasets' in an otherwise complex and remote land-tenure ownership landscape that generally lacks digital data and information access.

27.2 Methods

27.2.1 GIS Data and Handling

The raster datasets for terrestrial PNG presented by Huettmann (2020) were obtained as ACII grids, shapefiles and primarily geoTIFFs and imported into ArcGIS and QGIS. They are based on a geographic projection of WGS84 and using decimal degrees (latitude, longitude). Out of the over 100 GIS layers available, it was decided to use 82 data sets for this analysis.

I then created a point lattice for PNG in QGIS using vector research tools as a shapefile with a 0.2° spacing. It resulted in 941 lattice point locations for terrestrial PNG (Fig. 27.1).

Using 82 GIS layers and a 0.2° lattice for Papua New Guinea, those were imported and overlaid in QGIS and then assessed for subsequent hierarchical cluster correlations (non-supervised classification) and for explaining their human footprint (supervised classification).

Formula 1 Human Footprint 1 vs Predictors 1 − 100

$$(predictor1 + predictor2 + predictor3 + ...)$$

This allows us to better relate to the findings and identify major clusters and see whether they are related to a wider ecological predictor and patterns that might exist which have not been explored or found yet ('data mining').

The predictor overlay for 941 lattice points in PNG yielded 843 points free of overlap consistency errors (−9999). Those overlay errors occurred for island and coastal sites. It shows that the predictor layers for PNG are not covering all of PNG

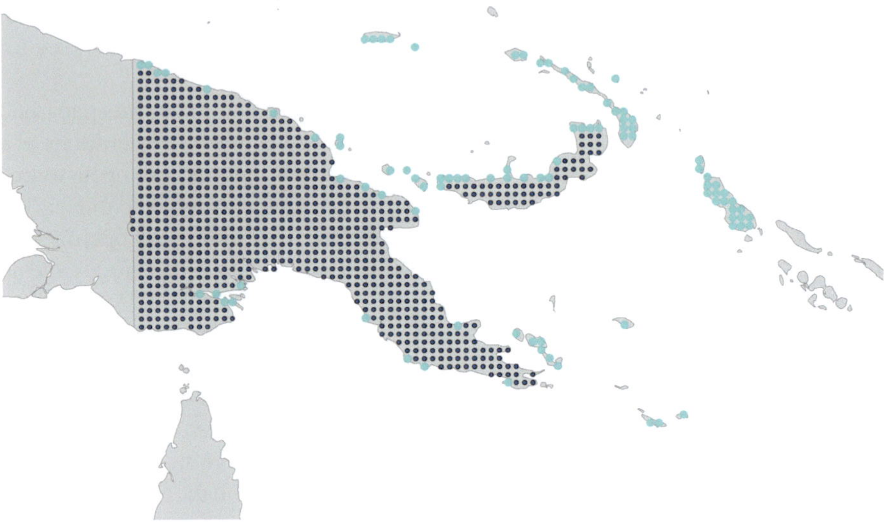

Fig. 27.1 Papua New Guinea study area and point lattice (the highlighted lattice points in 'blue' show data gap errors; c. 10% of PNG points. Note that most of Bougainville is suffering from such problems)

and its complex geography on a 1 km pixel scale; data errors occur, app. 10% of the PNG landscape data is affected and must be cleaned out.

27.2.2 Unsupervised Classification and Supervised Classification

The unsupervised classification allows to assess the data cube of 103 predictors. Here I used the varclust method in R based on the HMisc package (https://rdrr.io/cran/Hmisc/man/varclus.html). Varclus allows to not only find the correlations but also to see the clusters within. It's a quick and easy way of looking and finding structure in data sets and in an otherwise complex GIS data cube.

For the supervised classification, a boosted regression tree was run using formula 1. It allows to assess for the human footprint and what its drivers are, rank them and show the actual nature of the correlation (expressed as a partial dependence plot). This was implemented in Minitab-SPM software, TreeNet vers 8.3 (https://www.minitab.com/en-us/products/spm/).

The supervised classification is using the 'raw' datasets, which have 10% of a spatial overlay error and data aps. This is possible to analyze with the TreeNet implementation of boosted regressions in Minitab-SPM. It's an elegant way to overcome data gaps (~imputation) in complex data cubes (Grillo et al., 2022; see also Jiao et al., 2016 for 'data cloning').

Fig. 27.2 Fieldwork in Papua New Guinea (PNG) on remote trails in the remote bush with some handheld high-tech (Global Positionining System GPS) and ground-truthing remote sensing images and geographic information system GIS data layers

27.3 Results

The GIS data overlays show in the point lattice that PNG is not well mapped on a 1 km scale for quality data; app. 10% of the landscape data carry no values (−9999) for 82 predictors out of the larger pool. In a traditional analysis, this would leave out a large and relevant section of PNG. However, Minitab-SPM can handle such complex data and the supervised classification is based on this approach (for examples see Grillo et al., 2022). As shown in Fig. 27.2, thus far it remains crucial for PNG to ground-truth and compares GIS layers with reality.

As shown in Fig. 27.3, the unsupervised classification shows several tight clusters of environmental data, namely the mean temperature cluster across months, the monthly cloud data, the monthly solar radiation data, and some precipitation and bioclim data. Their correlations are beyond 80% each. Noteworthy is the tight correlation cluster of the human footprint with roads. Arguably roads drive the human footprint. Another tight correlation is found between proximity to coast and proximity to rivers; app. 50% in each case.

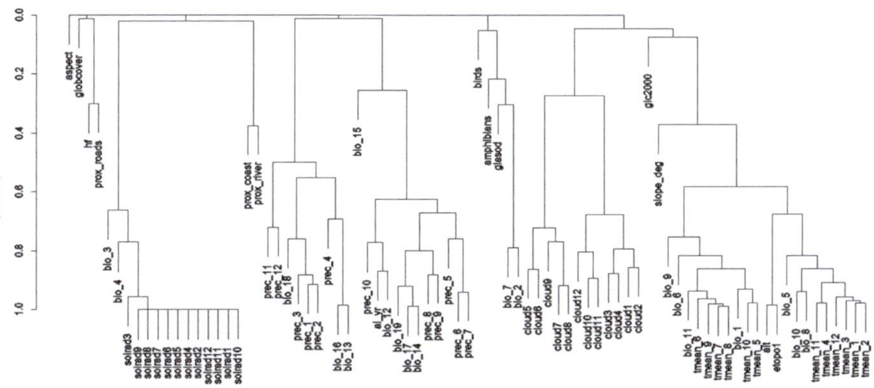

Fig. 27.3 Hierarchical clustering (Varclus) of the GIS data cube using 82 open-access predictors for Papua New Guinea based on 843 point lattice locations

There are just a few independent predictors, lowly correlated, e.g. aspect, slope, landcover predictors, birds and amphibians. In a parametric view, those can likely be perceived as 'independent.'

Out of the 82 predictors used, the supervised classification to explain the human footprint obtained a good model fit and a gains curve of app. 75%. The model was primarily driven by proximity to roads, proximity to coast, two landcover layers and proximity to rivers (see Table 27.1 for top 5 predictors). While 'landcover' layers ranked relatively high, they are of less relevance here for the wider picture. That has two prime reasons: (1) the actual categories of the two landcover predictors all showed a relatively high and equal human footprint index (details not shown here), and (2) categorical predictors can get ranked higher due to ML algorithms like boosting being 'greedy' grabbing higher variance predictors due to the category containers (Friedman, 2001). Landcover GIS layers should not be overestimated, or overinterpreted, accordingly. That is specifically the case when processes are to be estimated and more holistic approaches matter. Further, the ecological factors like average monthly temperature, bioclim, clouds, solar radiation and precipitation all played a lower-ranked relevance for explaining the human footprint index. Likely that is because a road can nowadays be built anywhere, overruling any of the place-based and ecological predictors other than mining, e.g. gold and copper resources. Arguably, the human footprint models are centered around industrial activities the most, including port locations (Figs. 27.4 and 27.5).

The result for the supervised classification can be summarized in Fig. 27.6 by showing the top 2 predictors. It indicates that locations close to roads and close to coast show the highest impact. Road constructions have been done and promoted for PNG since 1940s and still get widely promoted (see Beehler & Laman, 2020). They are a preferred means of the international aid community for 'development,' e.g. Japan, and got strongly implemented by the Australian government after WW2 all over PNG (Fig. 27.7).

Because the human footprint layer for PNG describes a relatively new landscape modification based mostly on recent destruction as part of industrialization and 'modernization,' it summarizes primarily the impact after WW2, centered around roads as promoted by outside actors and nations. The link between mining and road development is very obvious and can be seen for instance in gold mining areas like Wau and the Markham Valley with devastating impacts (e.g. Beer & Church, 2019).

Table 27.1 Importance rank of the top 5 predictors (out of 82 global predictors; Huettmann, 2020) for the human footprint index in Papua New Guinea based on the TreeNet algorithm (Boosting)

Predictor	Importance rank in percent
Proximity to roads	100
Proximity to coast	84
Global landcover GLOBCOVER	80
Proximity to river	78
Global landcover GLC2000	71

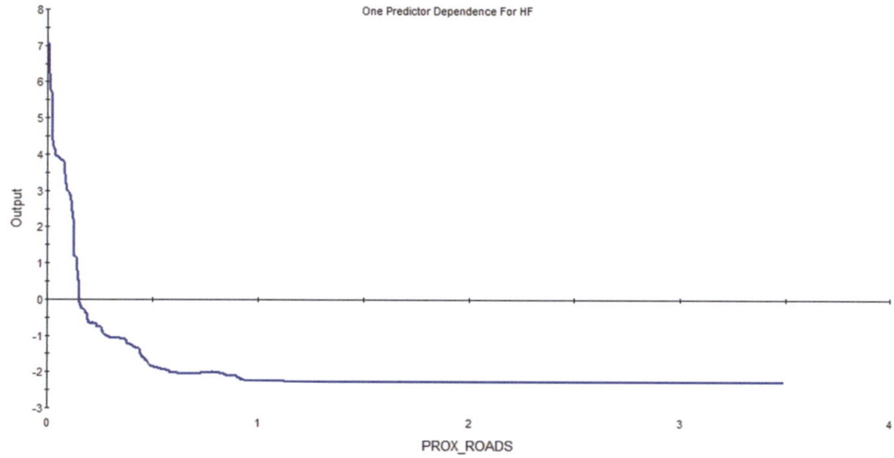

Fig. 27.4 Relationship between human footprint index and proximity to roads; the closer to roads the higher is the human footprint

While roads were clearly shown as a major impact, initially PNG was heavily developed by airplanes and through airstrips. That relates specifically to early (gold) mining, developed primarily through planes (details explained in Sinclair, 1978). The actual impact of airports was actually not well reflected in the GIS layers, yet. It was therefore found, overall, that the human footprint layer used in this study somewhat underestimates the human footprint in PNG. That is also because forest landscape modifications, e.g. planting of timber and fruit along ancient trails, are not well-accounted for, nor are overgrown forests considered, e.g. from ancient settlements,

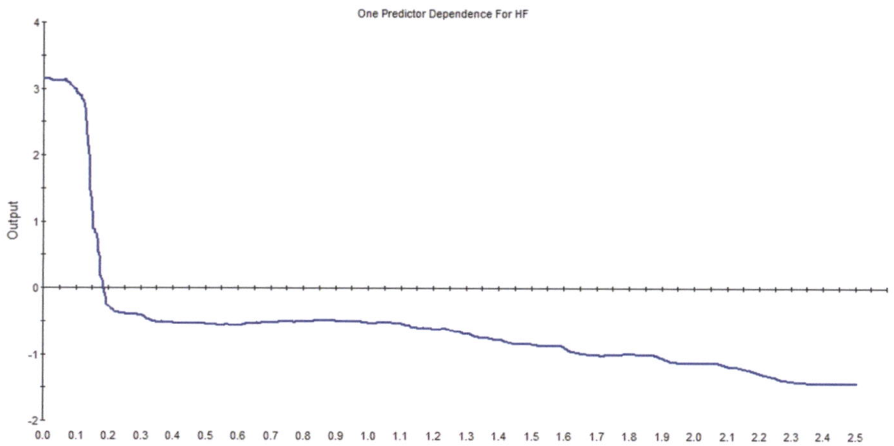

Fig. 27.5 Relationship between human footprint index and proximity to coast; the closer to the coast the higher is the human footprint

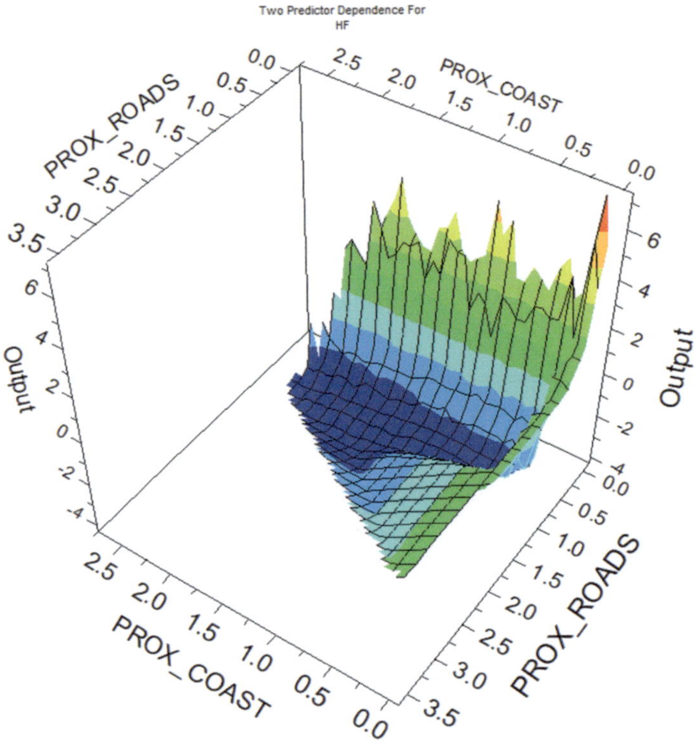

Fig. 27.6 Relationship between human footprint index and proximity to roads and proximity to coast. This figure exemplifies a colonial-style impact, the biggest human footprint for port and access regions from abroad (see red peak). PNG is a typical resource provider for abroad

gardens or man-made fires. Overall, this study must be seen as an underestimate of reality and more study is needed.

27.4 Discussion

This research presents the best publicly available open access GIS data information for terrestrial PNG. It further is the first to use 'Big Data' and to data mine them (for concept and overcoming data gaps compare also with Huettmann et al., 2018; Jiao et al., 2016). The coastal and ocean aspects of PNG still await a data compilation and an assessment.

The unsupervised classifications of the predictor set show a few independent predictors besides several highly correlated and clustered predictors. While this is typical in ecological studies here the unique quantified correlational structure is presented for PNG and its human footprint predictors allowing to capture the 'status quo' quantitatively

Fig. 27.7 Weekly frequented airstrips like shown here are not well mapped and modeled yet for Papua New Guinea but include quite a dramatic human footprint with invasive spcies, diseases, the overall human industrial enterprise including socio-economics and in a wide radius, e.g. easily over 30km; it's part of an urbanization process unfolding globally

Fig. 27.8 Ecological health metrics such as water, the harmony of life, is difficult to express in just a few GIS layers, whereas proxy predictors are more powerful and quite holistic instead

The supervised classification model on those predictors shows roads and their proximities as a prime predictor for the human footprint index, also linked with coastal proximities. Additional predictors are proximity to rivers, as well as two land-cover layers. It points toward industrial impacts with roads as the core arteries, e.g. for mining, while ancient impacts, e.g. in the highlands or forest modifications and brought by cultural immigration waves, are not so well-captured. Those 47,000 years old landscape modifications remain relatively benign and sustainable though, when compared to the most recent 60 years old road construction efforts in PNG, e.g. related to late-colonial, world war and neocolonial efforts.

The results are still presenting underestimates, e.g. airstrips are not well known and mapped (but see basic list as public information with https://en.wikipedia.org/wiki/List_of_airports_in_Papua_New_Guinea). There are also different types of human footprints, beyond roads. Modern forestry is among those but not well mapped and shared yet for impacts. A coherent mining map is also missing for PNG and not readily available, despite it's vast impact for over a century. Human settlements are additional sources of a human impact, and 'in the bush' often now centered around rectangular airstrips, but also poorly mapped. Similar can be said for invasive species (information widely missing for PNG). The sixth impact is climate change despite its relevance for tropical and coastal nations PNG has poor climate layers to work from. While individual data points easily exist, a cohesive PNG-wide GIS map for those layers is widely absent, not publicly available in good GIS formats nor accessible for the global audience, or for the PNG citizens (but see work forthcoming by Steiner and Huettmann for a compiled set of over 132 open-access GIS layers globally available).

Despite those technical and governance shortcomings of the data, from a sustainability aspect, the most worrisome issue likely remains wilderness loss, old-growth forest loss, contamination and decay of the social well-being for PNG. Along similar lines, invasive species are not well addressed—or mapped—nor are hunting impacts (Flannery, 2002 for hunting dogs). All of those questions stand in big contrast to highly funded species collections and many research efforts from abroad still dominating the PNG science and research enterprise, e.g. expressed and promoted in Beehler and Laman (2020) or Novotny and Basset (1999), Sam et al. (2020). It often simply has the wrong focus and does not serve effectively relevant sustainability questions for PNG and beyond with little relevant data to use!

When adding a more holistic and synergetic climate perspective, it likely shows that the human footprint is actually massive in PNG and reaching into even the most remote corners of PNG either way (*sensu* Flannery, 2002).

This research shows that humans - e.g. from abroad - currently widely underestimate the impact they have on PNG, and likely on the earth. Noteworthy is that the ocean impact is not even fully included here (see Halpern et al., 2008; Cousteau & Richards, 1999 for local descriptions) but it should be addressed making the findings even more complete and serious.

The findings of this work further expose and quantify the sustainability failure of the ongoing global governance, international aid and Australian policy model for PNG destroying a viable sustainability concept that evolved over millennia (see chapters in this book). Australia for instance promoted widespread road and bridge construction in PNG since the 1970s onwards while those regions remained widely very poor but carry a vast human footprint as the main driver of wilderness destruction

Overall, this work confirms that a modern sustainable culture is not yet achieved, not in modern PNG, nor likely in the wider region ('Australian arc') and its development policies, and the world (e.g. Rich, 1994; Stiglitz, 2003). Implications from this study remain mind-boggling when it comes to the western failure to actually acknowledge this problem, to quantify it, to create reliable GIS layers, share data open-access across for all actors with the global community, to manage it well, and to promote and to set up a human culture that achieves a good relationship with nature, Mother Earth and its people. PNG did this though in its own culture for most its 47,000 years long history.

Appendix

List of 82 datasets used in the analysis, more details on datasets, names and sources in Huettmann (2020) and Sriram and Huettmann (unpublished); see also Huettmann et al. (2018), and Steiner and Huettmann (in review) for an updated GIS data set. The data are available from the author and can also be obtained from a subsequent Google Drive. This data set is dynamic and receives updates and extensions.

"alt": altitude
"ai_yr": avian influenza prediction (Herrick et al.)
"amphibians": global amphibian diversity
"aspect": aspect in degrees
"bio_1-19": 19 bioclim variables
"birds": global bird diversity
"cloud1-12": monthly cloud coverage
"etopo1": etopo 1 altitude
"glasod": glasod layer
"glc2000": landcover 2000
"globcover": global landcover
"hf" and "hii": human footprint indeces
"prec_1-12": monthly average precipitation
"prox_coast": proximity to coast
"prox_river": proximity to rivers
"prox_roads": proximity to roads
"slope_deg": slope in degrees
"solrad1-12": monthly solar radiation
"tmean_1-12": monthly mean temperature

References

Adams, V. M., Dimitrova, N., Possingham, H. P., Allan, J. R., Kuempel, C. D., Peterson, N., & Tulloch, V. J. (2021). Scheduling incremental actions to build a comprehensive national protected area network for Papua New Guinea. *Conservation Science and Practice, 3*(2), e354.

Beehler, B., & Laman, T. (2020). *New Guinea: Nature and culture of Earth's grandest island*. Princeton University Press.

Beer, B., & Church, W. (2019). Roads to inequality: Infrastructure and historically grown regional differences in the Markham Valley, Papua New Guinea. *Oceania, 89*(1), 2–19.

Cousteau. J.M., & Richards, M. (1999). *Cousteau's Papua New Guinea Journey*. Henry Abraham's Books.

Diamond, J. (2011). *Guns, Germs and Steel: The Fate of Human Societies*. Random House.

Faith, D. P., Nix, H. A., Margules, C. R., Hutchinson, M. F., Walker, P. A., West, J., & Natera, G. (2000). The BioRap biodiversity assessment and planning study for Papua New Guinea. *Pacific Conservation Biology, 6*, 279–288.

Flannery, T. (2002). *The future eaters: An ecological history of the Australasian lands and people*. Grove Press.

Friedman, J. H. (2001). Greedy function approximation: A gradient boosting machine. *Annals of Statistics*, 1189–1232.

Gosarevski, S., Hughes, H., & Windybank, S. (2019). Is Papua New Guinea viable? *Political Economic Bulletin, 19*. https://www.semanticscholar.org/paper/Is-Papua-New-Guinea-viable-Gosarevski-Hughes/6d71c70951c12fa82dea578a36b66923997f3f92. Accessed June 1, 2022.

Green, A., Smith, S. E., Lipsett-Moore, G., Groves, C., Peterson, N., Sheppard, S., & Bualia, L. (2009). Designing a resilient network of marine protected areas for Kimbe Bay, Papua New Guinea. *Oryx, 43*(4), 488–498.

Grillo, M., Huettmann, F., Guglielmo, L., & Schiaparelli, S. (2022). First three-dimensional quantification of planktic food chain lower levels (copepods) for the Ross Sea region marine protected area (RSRMPA), Antarctica: Using FAIR-inspired legacy data with machine learning, and open source GIS. Diversity Ross Sea Special Issue.

Halpern, B. S., Walbridge, S., Selkoe, K. A., Kappel, C. V., Micheli, F., d'Agrosa, C., Bruno, J. F., et al. (2008). A global map of human impact on marine ecosystems. *Science, 319*(5865), 948–952.

Huettmann, F. (2015). On the relevance and moral impediment of digital data management, data sharing, and public open access and open source code in (tropical) research: The Rio convention revisited towards mega science and best professional research practices. In F. Huettmann (Ed.), *Central American biodiversity: Conservation, ecology, and a sustainable future* (pp. 391–418). Springer.

Huettmann, F. (2020). Chapter 24—Investigating Matschie's tree kangaroos with 'modern' methods: Digital workflows, big data project infrastructure, and mandated approaches for a holistic conservation governance. In D. L. P. Valentine, J. Blessington, & K. Schwartzet (Eds.), *Tree kangaroos: Science and conservation* (pp. 379–391). Academic.

Huettmann, F., Mi, C., & Guo, Y. (2018). 'Batteries' in machine learning: A first experimental assessment of inference for Siberian crane breeding grounds in the Russian high arctic based on 'shaving' 74 predictors. In G. Humphries, D. R. Magness, & F. Huettmann (Eds.), *Machine learning for ecology and sustainable natural resource management* (pp. 163–184).

Jiao, S., Huettmann, F., Guo, Y., Li, X., & Ouyang, Y. (2016). Advanced long-term bird banding and climate data mining in spring confirm passerine population declines for the Northeast Chinese-Russian flyway. *Global and Planetary Change, 144 C*, 17–33. https://doi.org/10.1016/j.gloplacha.2016.06.015

Novotny, V. (1999). Body size and host plant specialization: A relationship from a community of herbivorous insects on Ficus from Papua New Guinea. *Journal of Tropical Ecology, 15*(3), 315–328.

New South Wales State Library. (2015). Papua New Guinea and world wars, 15th September 2015. https://www2.sl.nsw.gov.au/archive/discover_collections/society_art/png/wars/index.html. Accessed July 10, 2022.

Rich, B. (1994). *Mortgaging the earth: The World Bank, environmental impoverishment and the crisis of development*. Earthscan Publishers.

Sam, K., Koane, B., Sam, L., Mrazova, A., Segar, S., Volf, M., Moos, M., Simek, P., Sisol, M., & Novotny, V. (2020). Insect herbivory and herbivores of Ficus species along a rain forest elevational gradient in Papua New Guinea. *Biotropica, 52*(2), 263–276.

Sinclair, J. (1978). *Wings of gold: How the aeroplane developed New Guinea*. Pacific Publications. ISBN 0-85807-033-2.

Stiglitz, J. (2003). *Globalization and its discontent*. Norton Paperback.

Yen, P., Ziegler, S., Huettmann, F., & Onyeahialam, A. I. (2005) Change Detection of Forest and Habitat Resources from 1973 to 2001 in Bach Ma National Park, Vietnam, Using Remote Sensing Imagery. International Forestry Review 7(1):1-8.

Chapter 28
Why Are There no Squirrels in Papua New Guinea? Insights from Predictive Models for a Vacant Ecological Niche of Global Relevance

Abstract Papua New Guinea (PNG) is a biogeography textbook example for species distributions, ecology, evolution and human co-evolution. It is located in the tropical region near the equator and driven by the Wallace Line and the Weber Line; neither mammals nor birds or many fish species have really crossed them. Primates as well as squirrels are curiously absent in PNG, despite PNG featuring one of the largest virgin forest covers and biodiversity. Squirrels occur almost all over the world but not in PNG. Here we infer from a global predictive open-access species distribution model (SDM) for squirrel species of the world for PNG the occurrence of potential but unconfirmed seven Southeast Asian squirrel species. We then overlaid this predicted ecological niche map with a proxy-squirrel species from nearby Australia, the sugar glider (*Petaurus breviceps*) to better understand why and where this predicted ecological niche is vacant. In the absence of an archeological record for squirrels in PNG, this research remains inconclusive whether squirrels have occurred in PNG and how climate and/or humans have affected them. Model-predicting absence in SDM is an important but widely overlooked subject for better inference. While more research is needed here we provided progress through a first analysis and make all data publically available for further inquiry.

Keywords Old world squirrel · Ecological niche inference · Vacant niche · Biodiversity · Papua New Guinea, sugar glider (Petaurus breviceps)

28.1 Introduction

Papua New Guinea (PNG) is a global species hotspot, but less so for mammals (Flannery, 1990, 1995). As a megadiversity nation, PNG has just app 289 mammals (Flannery, 1990, 1995) and it lacks primates and squirrels for instance. However, it is the only place on earth (except perhaps for the close-by neighbor Australia) that features the full set of mammal clades: monotremes (egg-laying mammals), marsupials (pouch-raising mammals) as well as placentals (mammals with a placenta as essentially found worldwide and often seen, wrongly though, as the quintessential mammals) (Kennedy, 1992). Many mammals are endemic just to PNG.

F. Huettmann, *Globalization and Papua New Guinea: Ancient Wilderness, Paradise, Introduced Terror and Hell*, https://doi.org/10.1007/978-3-031-20262-9_28

PNG is located south at the Wallace Line and is also affected by the Weber Line. Because PNG has many islands, it is a prime example for biogeography (Wilson & MacArthur, 2001) with classic studies, e.g. Diamond (1973) for birds; see Beehler and Laman (2020) and citations within for wider overview of New Guinea.

Still, the absence of many mammals, tree-living monkeys as well as squirrels, is peculiar for such a diverse, complex and widely forested island landmass (see also Lien, 1999 for Australia as part of the wider Sahul region). PNG has the largest block of virgin old-growth forests left in the Pacific, e.g. Beehler and Laman (2020). While squirrels are essentially found all over the world, except for Antarctica, deserts and high arctic regions, the Wallace Line and the consistent presence of humans for over 47,000 years might have affected such a PNG 'squirrel gap' (see for Wallace Line details Golson, 1971; Van Oosterzee, 1997). Likely this absence of squirrels is compensated by *dendrolagus* tree kangaroos (Huettmann, 2020) and a variety of other closely related canopy and tree-living species such as gliders that are endemic to that region (Flannery, 1990, 2002). On an Australasian scale that should include the family Petauridae, and specifically Squirrel gliders (genus Petaurus) (Cremona et al., 2021; Menkhorst et al., 1988; Osborne & Christidis, 2001; Ziegler, 1981).

Based on best-available squirrel species data in the world, and using over 130 global open-access GIS environmental predictor layers, here we present the first in-depth model prediction inference and assessment for PNG. It's based on a zoom-in version of the global layer model by Steiner and Huettmann (unpublished), and we try to assess here where the potential squirrel habitat would be in PNG using powerful machine learning algorithms, and what such models consist of for its predictors, species and niche compensation.

28.2 Methods

28.2.1 Global SDMs and Species Selection

Most of the world's squirrel species are poorly studied and conserved. They received little public attention; that is specifically the case for the tropical region where most squirrels are found. In order to identify possible squirrel niches in Papua New Guinea, in this study, we performed Species Distribution Models (SDMs) and Species Distribution Forecasts (SDFs) for the year 2100 (results presented elsewhere with the authors) using machine learning algorithms supported by the software MaxEnt (version 3.4.4 https://biodiversityinformatics.amnh.org/open_source/maxent/). The SDMs and SDFs are based on 132 and 7 environmental predictors, respectively. An overview of all included predictors, their descriptions, and with their corresponding reference source can be found in Table 28.1. This table has been reproduced from Steiner and Huettmann (2021) (Table 3.2). This table illustrates all 132 environmental predictors used for the SDMs and also contains the information and references for the ones used in the SDFs.

The global species distribution model serves as a robust approach to obtain a rapid assessment scenario for more fine-tuned assessments. A first Global SDM for

Table 28.1 List of species models for SDMs in Papua New Guinea, see Flannery (1990) for details

Scientific species name	Common species name	Known distribution in New Guinea
Exilisciurus concinnus	Philippine pygmy squirrel	None (Found in Philippines)
Hyosciurus ileile	Lowland long-nosed squirrel	None (Found in Sulawesi)
Lariscus insignis	Three-striped ground squirrel	None (Found in Southeast Asia: Indonesia, Malaysia and Thailand)
Lariscus Niobe	Niobe ground squirrel	None (Found in Sumatra and Java)
Petaurista elegans	Spotted giant flying squirrel	None (Found in Southeast Asia, Central China and Himalayas)
Ratufa bicolor	Black giant squirrel	None (Found all over Southeast Asia)
Rubrisciurus rubriventer	Sulawesi giant squirrel	None (Found in Sulawesi)
Petaurus breviceps	Sugar glider	Undocumented for PNG but outside (Australia)

Taxonomic Serial Numbers (TSNs) for those species are with Steiner and Huettmann (in prep). One species (sugar glider) is just added as a proxy for assessment known to occur in PNG

over 230 species was done by Steiner and Huettmann (in prep) following methods outlined by Elith et al. (2006), Guisan and Zimmermann (2000), Humphries et al. (2018).

The SDMs that represent the occurrence of Southeast Asian inhabiting squirrel species represent their predicted occurrence approximately for the years 2010 until 2020 (based on the reference year of the predictors). Because we used 132 global environmental predictors, those models are assumed to be the most accurate and modern ones globally available for the included species. The species that have been included in this study are Sulawesi giant squirrel (*Rubrisciurus rubriventer*), black giant squirrel (*Ratufa bicolor*), spotted giant flying squirrel (*Petaurista elegans*), three-striped ground squirrel (*Lariscus insignis*), Niobe ground squirrel (*Lariscus niobe*), lowland long-nosed squirrel (*Hyosciurus ileile*), and Philippine pygmy squirrel (*Exilisciurus concinnus*). The occurrence points of these species that have been used for the SDMs and SDFs have been downloaded from www.GBIF.org (using this download https://doi.org/10.15468/dl.665b59) and can be accessed in Appendix 28.1.

For details about how to perform such analysis in MaxEnt, please see Appendix 3.1 in Steiner and Huettmann (2021).

After performing the models in MaxEnt, the output rasters have been imported into ArcGIS and open source QGIS, where they have been standardized and their symbology has been modified in order to represent the desired study area as seen in the results section. Also, all the output files of the SDMs can be found in Appendix 28.2.

28.2.2 PNG Zoom-In

Based on the global models at hand, we then used ArcGIS and created a specific PNG focus view from the global squirrel model for the species mentioned in Table 28.1. That was done with zoom and clip in GIS. For context, we include New Guinea and part of Northern Australia also.

28.2.3 PNG Model Assessment and Validation

The Global SDMs carry robust and commonly used model assessment methods and metrics. Details are presented with Steiner and Huettmann (in prep). In order to summarize these above-mentioned models, we also created an overview table, where the distribution range shifts have been summarized which can be found in the results section.

To validate our models, in the absence of relevant PNG data, we also overlaid the species distribution hotspots and coldspots of the included species with other small mammal species as proxy species that are known to currently inhabit Papua New Guinea, e.g. the sugar glider (*Petaurus breviceps*). This has been done to identify wether the possible niche hotspots of squirrels are already taken by Papua New Guinea inhabiting species. Technically, this has been achieved by overlaying the final SDMs with the recorded occurrence points of this species download from www.GBIF.org (using the following download link (09 July 2022) GBIF Occurrence Download https://doi.org/10.15468/dl.4ec3e2); a cleaned GIS shapefile dataset for New Guinea resulted into 632 data points. This can help us to infer whether the ecological niche of squirrels is used by other animals, and by 'sister' taxa. The sugar glider from nearby Australia is one species that might come close to it (see Discussion and subsequent figures), living in such habitats like shown for PNG in Fig. 28.1).

28.3 Results

28.3.1 PNG Prediction of Squirrels

SDMs have run well in MaxEnt, and corresponding Roc graphs can be found in the Appendix 28.2. These have also been used to assess whether overfitting of the models occurred due to the high number of predictors. By assessing these, and the list of the predictive contribution of the predictors, we were able to conclude that no major overfitting issue occurred when running the SDMs. Using a zoom-in of the world squirrel SDM, we found that PNG features some predicted squirrel occurrences. Those are located in clustered hotspot locations. Most of those are in the interior of

Fig. 28.1 Habitat shot of an ancient old-growth forest at the Huon Peninsula, Papua New Guinea. Surprisingly, this habitat has left a vacant niche ecological niche for monkeys as well as squirrels (photo by FH) while tree kangaroos occur

PNG and some are at coastal locations, namely the Port Moresby area and isolated patches of New Britain and Bougainville.

Figure 28.2 illustrates one hotspot region of the species Philippine pygmy squirrel (*Exilisciurus concinnus*) close to the Deception Bay of PNG, and another hotspot can be identified in the Bismarck range/Central Range of PNG.

Figure 28.3 illustrates one hotspot region of the species lowland long-nosed squirrel (*Hyosciurus ileile*) in the Sarawaget Range in the northeast of PNG and another hotspot in the region around the Goroka Mountain.

Figure 28.4 illustrates multiple hotspot regions of the species. Three-striped ground squirrel (*Lariscus insignis*) across the country of PNG. They are located continental except for some hotspots on the northern coast and Manus Island.

Figure 28.5 illustrates one hotspot region of the species Niobe ground squirrel (*Lariscus niobe*) in the Ramu River basin near Misinki (PNG), and one smaller hotspot region close to the Habuna Volcano on the New Britain island in a region fully covered by plantations (likely palm trees according to satellite imagery).

Figure 28.6 illustrates multiple hotspot regions of the species spotted giant flying squirrel (*Petaurista elegans*) in the continental center of PNG, as well as the Sarawaget and Owen Stanley Range.

Figure 28.7 illustrates two main hotspot regions of the species Black giant squirrel (*Ratufa bicolor*) in the northeastern part of the Bismarck range, as well as in the Sarawaget Range.

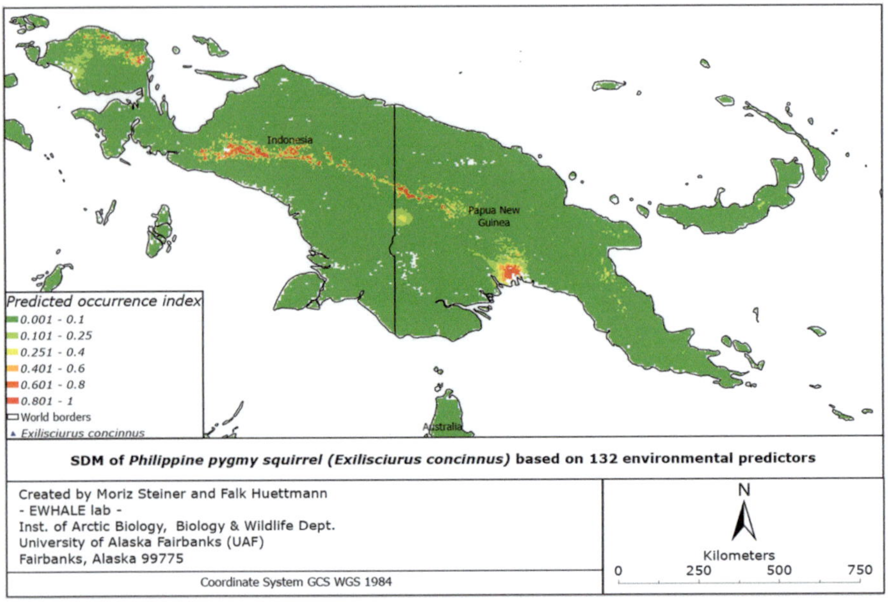

Fig. 28.2 SDM of Philippine pygmy squirrel (*Exilisciurus concinnus*) based on 132 environmental predictors

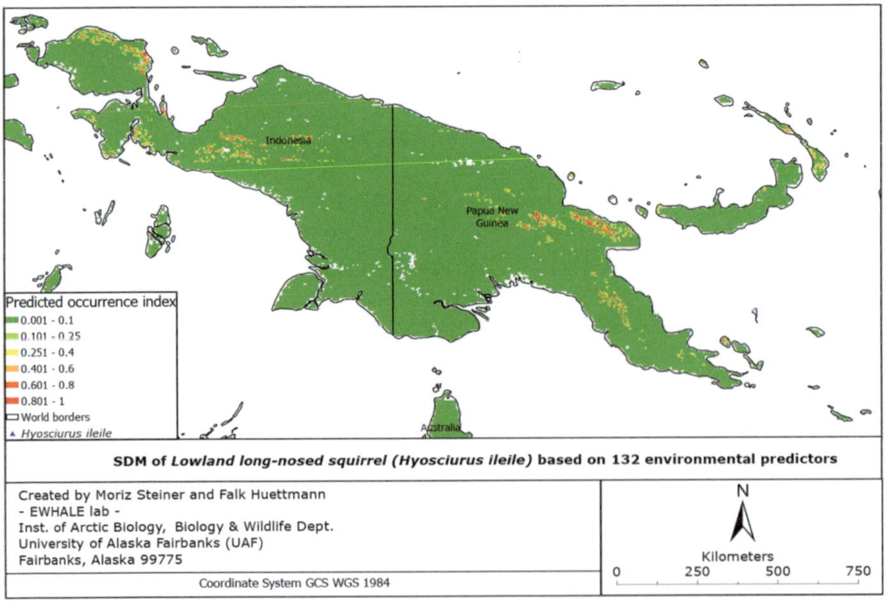

Fig. 28.3 SDM of lowland long-nosed squirrel (*Hyosciurus ileile*), based on 132 environmental predictors , a), b), c), d) and e) SDF of lowland long-nosed squirrel (*Hyosciurus ileile*) based on 7 BioClim predictors

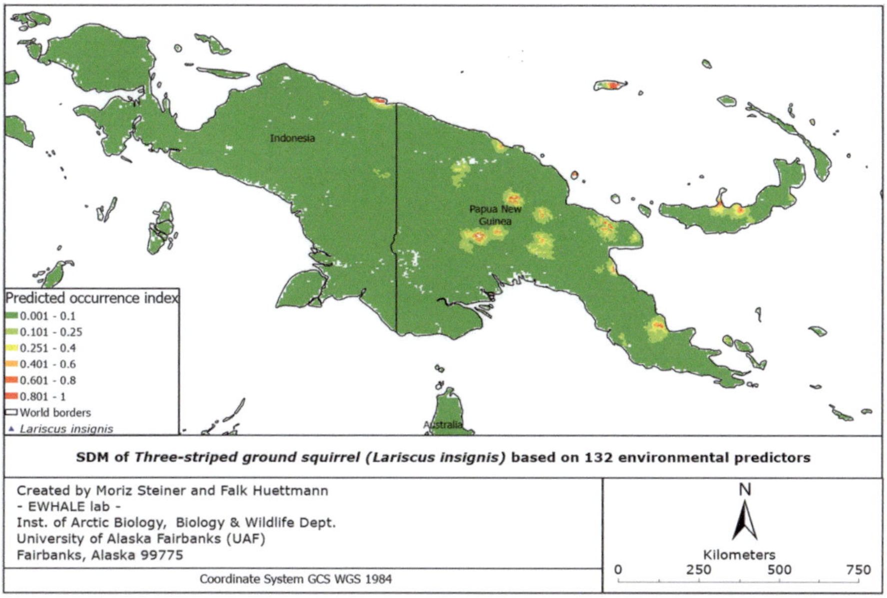

Fig. 28.4 SDM of three-striped ground squirrel (*Lariscus insignis*), based on 132 environmental predictors

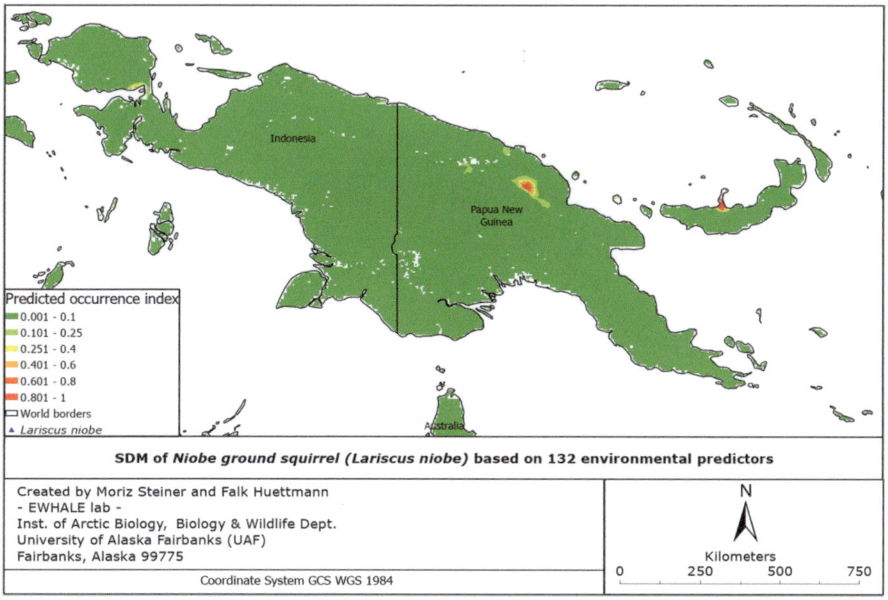

Fig. 28.5 SDM of Niobe ground squirrel (*Lariscus niobe*), based on 132 environmental predictors

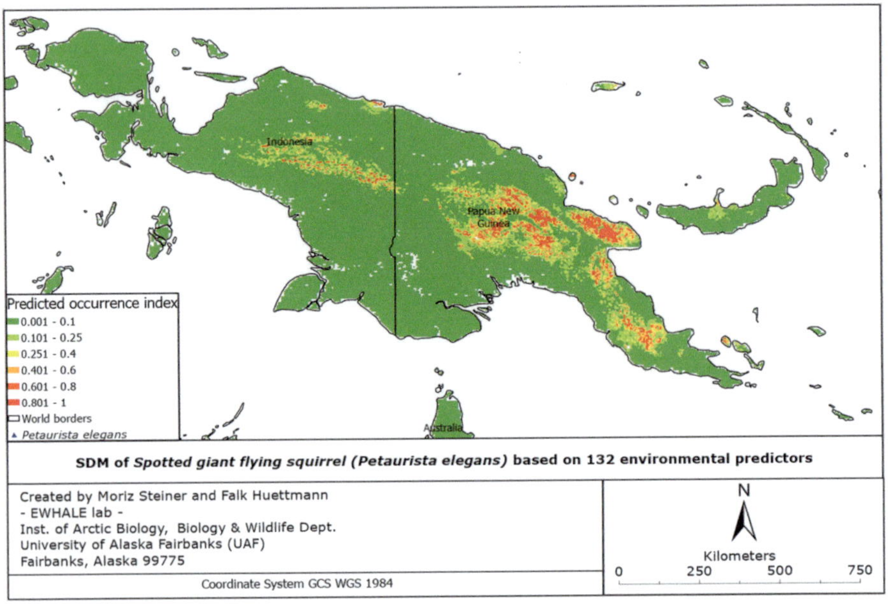

Fig. 28.6 SDM of spotted giant flying squirrel (*Petaurista elegans*), based on 132 environmental predictors

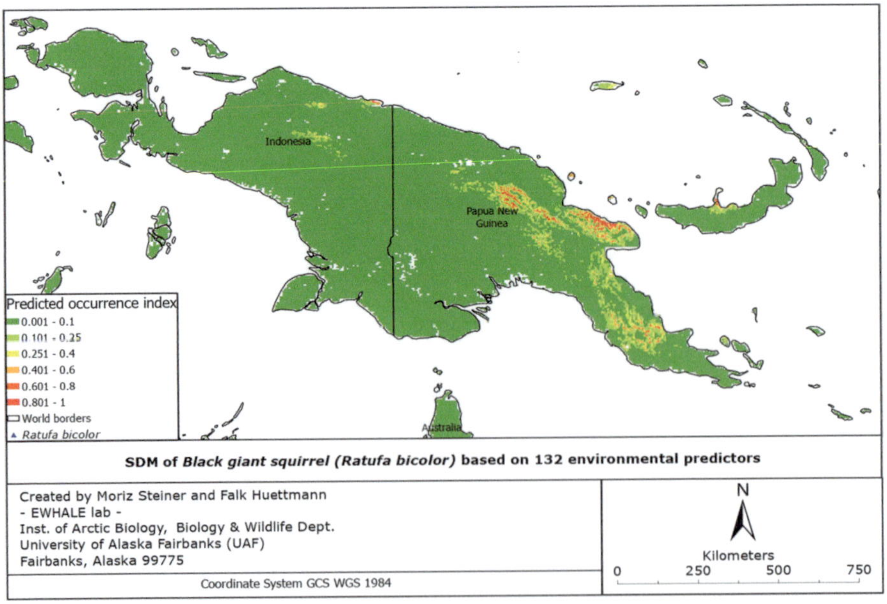

Fig. 28.7 SDM of black giant squirrel (*Ratufa bicolor*), based on 132 environmental predictors

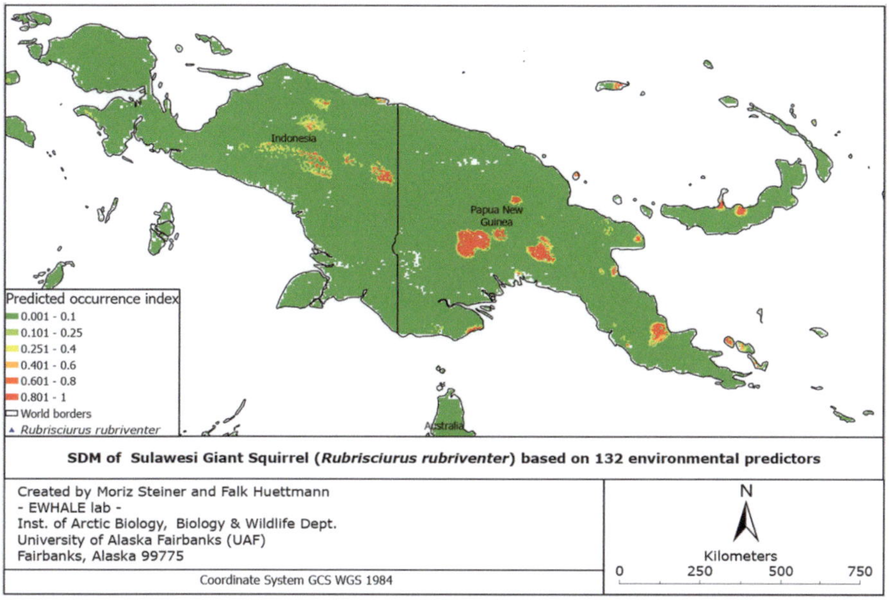

Fig. 28.8 SDM of Sulawesi giant squirrel (*Rubrisciurus rubriventer*), based on 132 environmental predictors

Figure 28.8 illustrates multiple hotspot regions of the species Sulawesi giant squirrel (*Rubrisciurus rubriventer*) across the country of PNG. The main ones can be found in the region of the Southern Highlands and Chimbu (Simbu). Another major one can be found in the area around Kawawoki Mission.

Overall, we find that coastal areas and islands received less squirrel predictions than the mainland PNG. Within mainland PNG we see most predictions in mountain ranges like Mt. Hagen and the Huon Peninsula. These areas are known to cover extensive forests and steep slopes.

28.3.2 SDM Assessment and the use of a Proxy Species

There are virtually no real squirrel species or survey data for PNG to assess the models. Other than plain absences the model assessments show that the Global Models overestimate squirrel occurrences (which is not the case for the global assessments overall; details in Steiner and Huettmann unpublished). In order to further validate these models and understand whether other small mammal species occupy the predicted squirrel niches, here we present a figure illustrating the hot and coldspots of the species included in this study. This figure has been created by simply merging the SDMs of all seven species included in the study using ArcGIS.

In this Fig. 28.9, we can observe that the current modeled cumulative squirrel hotspots can be found almost exclusively in the Middle Range of PNG in the mountains with clusters in the remote middle parts. Whereas, the coldspots can be found

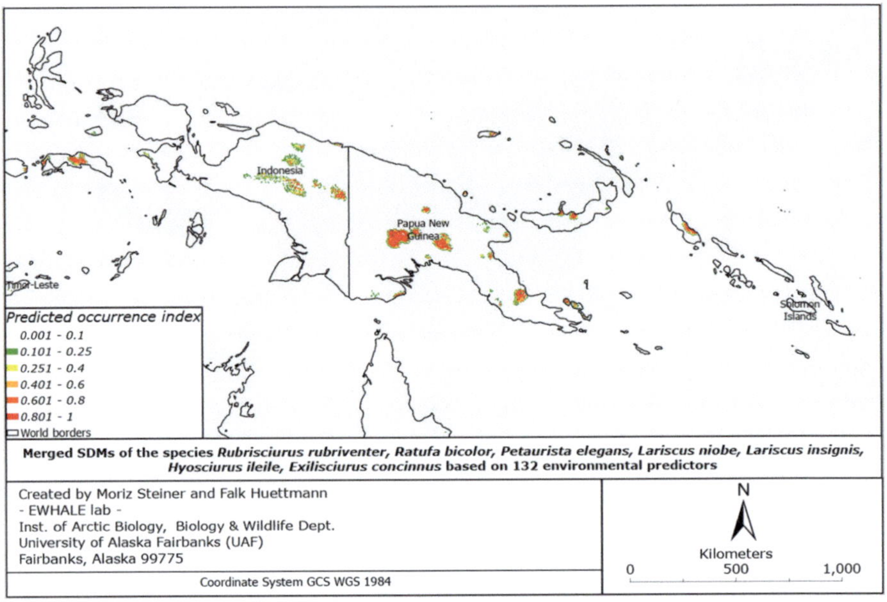

Fig. 28.9 Predicted cumulative squirrel hot and coldspots in PNG

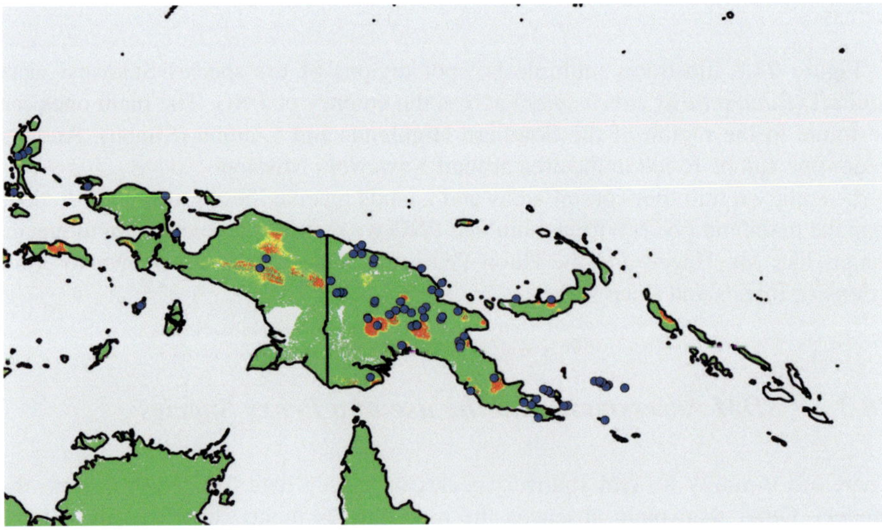

Fig. 28.10 Predicted squirrel hot-and coldspots in PNG with occurrence points of PNG-inhabiting small mammals, the sugar glider

in most of PNG islands and coastal regions including low-elevation areas, except for a few sites in New Britain and Bougainville.

Subsequently, we have overlaid the occurrence points of the squirrel proxy species sugar glider (*Petaurus breviceps*) with Fig. 28.10. The different colors of occurrence points indicate the different species of small mammals that have been recorded in Papua New Guinea and uploaded onto the www.GBIF.org database.

The assessment with proxy-squirrel species, the sugar glider, shows generically a weak overlap with the combined model. It matches generally 'best' in the elevational ranges but still includes many errors of omission and many errors of commission. It does not relate well with the species clusters but shows overlaps and misses. However, it relates visually well with some individual squirrel SDMs, such as the spotted giant flying squirrel and the lowland long-nosed squirrel. Sugar gliders are a species consisting of subspecies (for PNG it is *papuanus, flavidus and tafa*) and their habitat specifics are not so well known yet.

28.4 Discussion

The leading reference on mammals by Flannery (1990) shows no squirrels for Papua New Guinea. No squirrels are described to occur in New Guinea neither (Beehler & Laman, 2020) nor were squirrels ever introduced there to our knowledge. However, our global squirrel SDM work indicates the potential habitat occurrence for such species in PNG. These predicted areas are located in mainland PNG, along mountain ranges, in the Port Moresby area and few locations in New Britain and Bougainville.

Arguably, SDMs are rapid assessment tools. They work by correlations and predict a relative index of occurrence (RIO); Humphries et al. (2010). They can stand for habitats present at the pixel predicted which would potentially indicate the squirrel species. SDMs done for high-resolution canopy work for squirrel-associated species are very rare, thus far (see Robolt & Huettmann, 2021). Primarily that is due to lack of data sharing among biologists and foresters alike (Huettmann 2020).

In the case for PNG, SDMs can be puzzling and because squirrels are absent there while the global models predict them though. It can mean several things. (a) The Global SDMs are overpredicting, (b) squirrel habitat (=the ecological niche) does occur in PNG, (c) squirrels were there initially but got eliminated, (d) others reasons unknown. All of those arguments can be based on existing citations and research work; better conclusions await to be drawn with more data available.

We think that the SDMs show informative and good model assessment metrics and are biologically meaningful overall, e.g. for data used and quantifications. We further think that the ecological niche does occur in PNG for them, but that it is 'vacant'. Other sister taxa, such as gliders might be able to use that ecological niche space, but our assessment shows no strong indication for it, other than the mountain range habitats across New Guinea overall. Further resource allocation to such research sectors is necessary to better understand these pristine ecosystems. Other species, not included in this study, might occupy this identified squirrel niche in PNG. Concepts of 'landscapes of fear' and disease impacts remain also unstudied for this species range.

As found here, the relevance of mountains, elevation, is well-described for ecological niche work, as well as for PNG, e.g. Diamond (1973 and summaries in Beehler & Laman, 2020; Beehler & Alonso, 2001). These authors also presented various explanations for biogeographic patterns observed in PNG, such as skipping habitats and ranges. On a PNG scale, this might apply for squirrels and can be put to a test in a quantitative fashion using GIS data and models made available here.

It is nothing unusual to find vacant ecological niches. Such cases usually indicate extinction events, e.g. caused by harsh environmental factors or by humans. Another option would be a taxonomic mis-classification and sister taxa, e.g. gliders, as those 'gaps' can be very insightful for phylogeny and evolution. The actual model-prediction of absences is rarely done, and here we show such a model with first results; it should complement 'presence models'.

We are not aware of the archeological record for PNG showing any squirrel species, yet. So we do not think that squirrels were overhunted or became extinct through human pressures (O'Connell & Allen, 2007). This is also the reason why we used proxy species here to better understand why and where this predicted ecological niche is vacant. Arguably, this proxy species and its close relatives face a pretty grim outlook for the future since many are already endangered and face significant habitat fragmentation and eradication (Ashman and Ward 2022; Barber-Meyer 2007; Knipler et al. 2021; Suckling 1984). Squirrels would likely face similar issues and fates.

Despite decades of assessments and research, e.g. Flannery (1990), Alcorn (1993), Faith et al. (2000), Diamond (2011), Beehler and Laman (2020), Beehler and Alonso (2001) finding digital data for PNG remains a difficult task, specifically for mammals in PNG (Huettmann, 2020). We hope that this work can help to trigger and fine-tune more data and well-described information on PNG habitats and species, *sensu* Huettmann (2015) *and help to move towards a more modern profession of mammologists, biologists and conservationists.*

Here we showed the wide absence of squirrels and squirrel predictions for PNG. We hope it can facilitate more research and answers on the issue of the squirrel gap in PNG.

Appendix

Appendix 28.1: Species SDM data for seven species—including the additive model, as TIFF rasters.

Appendix 28.2: Environmental Predictors (rasters, available from Steiner and Huettmann in prep).

References

Alcorn, J. B. (1993). *Papua New Guinea needs assessment. A USAID-funded consortium of world wildlife fund*. The Nature Conservancy, and World Resources Institute Washington, D.C. with the Government of Papua New Guinea Department of Environment and Conservation, Boroko, Papua New Guinea. https://www.sprep.org/attachments/VirLib/PNG/png-conservation-needs-assessment-alcorn-1993.pdf. Accessed June 2, 2022.

Ashman, K., Ward, M. (2022). Extinction risk for Australia's iconic glider. Science, 378(6620), 608. https://doi.org/10.1126/science.adf1013.

Barber-Meyer, S. M. (2007). Photopollution impacts on the nocturnal behaviour of the sugar glider (Petaurus breviceps). Pacific Conservation Biology, 13(3), 171-176.

Beehler, B., & Alonso, L. E. (2001). Southern New Ireland, Papua New Guinea: A biodiversity assessment. *RAP Bulletin of Biological Assessment, 21*, 1–101.

Beehler, B., & Laman, T. (2020). *New Guinea: Nature and culture of Earth's grandest island*. Princeton University Press.

Boucher, O., Servonnat, J., Albright, A. L., Aumont, O., Balkanski, Y., Bastrikov, V., Vuichard, N., et al. (2020). Presentation and evaluation of the IPSL-CM6A-LR climate model. *Journal of Advances in Modeling Earth Systems, 12*(7), e2019MS002010.

Cremona, T., Baker, A. M., Cooper, S. J., Montague-Drake, R., Stobo-Wilson, A. M., & Carthew, S. M. (2021). Integrative taxonomic investigation of Petaurus breviceps (Marsupialia: Petauridae) reveals three distinct species. *Zoological Journal of the Linnean Society, 191*(2), 503–527.

Diamond, J. M. (1973). Distributional ecology of New Guinea birds. *Science, 179*, 759–769.

Diamond, J. M. (2011). *Guns, germs and steel: The fate of human societies*. Random House.

Elith, J., Graham, C., & NCEAS Working Group. (2006). Novel methods improve prediction of species' distributions from occurrence data. *Ecography, 29*, 129–151.

Faith, D. P., Nix, H. A., Margules, C. R., Hutchinson, M. F., Walker, P. A., West, J. J., Natera, G., et al. (2000). The BioRap biodiversity assessment and planning study for Papua New Guinea. *Pacific Conservation Biology, 6*, 279–288.

Flannery, T. (1990). *Mammals of New Guinea*. The Australian Museum.

Flannery, T. F. (1995). *Mammals of New Guinea*. Reed.

Flannery, T. (2002). *The future eaters: An ecological history of the Australasian lands and people*. Grove Press.

Golson, J. (1971). Both sides of the Wallace line: Australia, New Guinea, and Asian prehistory. *Archaeology and Physical Anthropology in Oceania, 6*(2), 124–144.

Guisan, A., Zimmermann, N.E. (2000). Predictive habitat distribution models in ecology. Ecological modelling, 135(2-3), 147-186.

Huettmann, F. (2015). On the relevance and moral impediment of digital data management, data sharing, and public open access and open source code in tropical) research: The Rio convention revisited towards mega science and best professional research practices. In F. Huettmann (Ed.), *Central American biodiversity: Conservation, ecology, and a sustainable future* (pp. 391–418). Springer.

Huettmann, F. (2020). Chapter 24—Investigating Matschie's tree kangaroos with 'modern' methods: Digital workflows, big data project infrastructure, and mandated approaches for a holistic conservation governance. In D. L. P. Valentine, J. Blessington, & K. Schwartzet (Eds.), *Tree kangaroos: Science and conservation* (pp. 379–391). Academic.

Humphries, G., Magness, D. R., & Huettmann, F. (2018). *Machine learning for ecology and sustainable natural resource management*. Springer.

Kennedy, M. (Ed.). (1992). *Australasian marsupials and monotremes: An action plan for their conservation*. IUCN.

Knipler, M., Dowton, M., & Mikac, K. (2021). A Major Highway Acts to Genetically Structure a Sugar Glider (Petaurus Breviceps) Population.

Lines, W. (1999). *Taming the great south land: A history of the conquest of nature in Australia*. University of Georgia Press.

MacArthur, R. H., Wilson, E. O. (2001). *The theory of island biogeography*. Princeton, N.J: Princeton University Press.

Menkhorst, P. W., Weavers, B. W., & Alexander, J. S. A. (1988). Distribution, habitat and conservation status of the squirrel glider Petaurus-Norfolcensis (Petauridae, Marsupialia) in Victoria. *Wildlife Research, 15*(1), 59–71.

O'Connell, J. F., & Allen, J. (2007). Pre-LGM Sahul (Australia-New Guinea) and the archaeology of early modern humans. In P. Mellars, K. Boyle, O. Bar-Yosef, & C. Stringer (Eds.), *Rethinking the human revolution: New behavioural and biological perspectives on the origin and dispersal of modern humans* (pp. 395–410). McDonald Institute Monographs.

O'Connell, J. F., & Allen, J. (2015). The process, biotic impact, and global implications of the human colonization of Sahul about 47,000 years ago. *Journal of Archaeological Science, 56*, 73–84.

Osborne, M. J., & Christidis, L. (2001). Molecular phylogenetics of Australo-Papuan possums and gliders (family Petauridae). *Molecular Phylogenetics and Evolution, 20*(2), 211–224.

Robold, R., & Huettmann, F. (2021). High-resolution prediction of American red squirrel in interior Alaska: A role model for conservation using open access data, machine learning, GIS and LIDAR. *PEERJ*. https://peerj.com/articles/11830/

Steiner, M., & Huettmann, F. (2021). Justification for a taxonomic conservation update of the rodent genus Tamiasciurus: Addressing marginalization and mis-prioritization of research efforts and conservation *laissez-faire* for a sustainability outlook. *The European Zoological Journal, 88*(1), 86–116. https://doi.org/10.1080/24750263.2020.1857852

Suckling, G. C. (1984). Population ecology of the sugar glider, Petaurus breviceps, in a system of fragmented habitats. Wildlife Research, 11(1), 49-75.

Van Oosterzee, P. (1997) . *Where worlds collide: the Wallace Line*. Cornell University Press.

Ziegler, A. C. (1981). Petaurus abidi, a new species of glider (Marsupialia: Petauridae) from Papua New Guinea. *Australian Mammalogy, 4*, 81–88.

Chapter 29
A SWOT and PESTLE Analysis for PNG Using 132 GIS Layers: Such Data Cannot Lie

Abstract Valid data with quantitative information explicit in space and time or even good proxies, are a rare commodity for Papua New Guinea's resource management, and also in the wider region including Australia. Well-curated geographic information system (GIS) layers can help to overcome this problem. Using the existing open-access data available for PNG will help for better spatial decisions. Use of SWAT and PESTLE analysis is common in the disciplines outside of ecology and conservation. However, they can prove as a good tool for progress helping to find and see problems in the habitat and species ranges for instance. Such an approach is tried experimentally and compared here for the best globally available open-access data set for PNG. While this approach leaves much to be desired, it sets the stage where data work—big data—can actually go for PNG and similar cases worldwide, and what data sources are sufficient and/or to be improved.

Keywords Papua New Guinea (PNG) · SWOT · PESTLE · GIS layers · Open-access data · Globalization · Sustainable development

29.1 Introduction

Papua New Guinea (PNG)—the nation, the culture and its species diversity with the underlying geography—remains of global importance (Diamond, 2011; chapters in this book). Many relevant issues in PNG can be named that are either not studied at all, not assessed well, or not done to satisfaction and with a good outcome. In the majority of cases, conservation management still takes a shot in the dark with many uncertainties. PNG is one of the poorest nations in the world, dominated from the outside—specifically by Australia—and relevant conservation research is widely not, or under-, funded. Conservation simply does not rank high on most agendas, certainly not in the tropics or in rich and adjacent nations (Laurance et al., 2011) (Table 29.1).

For over 300 years mankind lives in the so-called 'age of enlightenment' though using science and reason (and its associated administration), with gigantic annual

F. Huettmann, *Globalization and Papua New Guinea: Ancient Wilderness, Paradise, Introduced Terror and Hell*, https://doi.org/10.1007/978-3-031-20262-9_29

Table 29.1 Known shortcomings for an effective conservation management in Papua New Guinea

Conservation management topic	Details	Citation	Comment
Complex land tenure	The Melanesian way	Narokobi (1983)	This sits at the core of PNG and Melanesian resources
Planning	Municipal plans, mining plans, national plans are widely incomplete	Beehler and Laman (2020)	Good planning and vision are essential
Poverty	PNG includes one of the poorest regions in the world	Various official and online sources	PNG was an Australian protectorate and colonial nation for over a century but this situation did not allow for PNG to 'prosper'
Social unrest and violence	PNG is frequently rated as unsafe	Salak (2001)	Many unsafe places exist in the world, by now
Laws specific to the environment	Ecological management needs laws and a framework to operate under	Narokobi (1975, 1983)	The framework will shape the outcome
Laws specific for wildlife habitat management	Like above	NA	Classic topics in the discipline of landscape ecology
Local, national and international alignment	Locations and nations do not agree with each other on how to manage the environment, even if it's the same habitat or taxon	Adamowicz (2016)	A typical failure for many colonial nations
Taxonomic disagreement	Large taxonomic disagreement even for the same species, etc	See chapters in this book	If the taxonomy cannot be agreed on, and while alternatives are not pursued, most western approaches cannot really proceed

research budgets, e.g. CERN in the EU with c. 1 billion EUR (http://home.cern), or the National Science Foundation NSF U.S. with c. $8.5 billion (National Science Foundation Centers—NSF https://www.nsf.gov › about › budget › pdf). China currently tries to top U.S. and western research efforts. Research includes data and data collection; as a matter of fact, PNG has received for several centuries major data collection efforts, but where are the data (e.g. Dornelaas et al., 2018; for PNG details and examples see Huettmann, 2020)? And even if data exist, it's not exactly clear what the found evidence and alleged impact really is, e.g. Hannah et al., 2021). Western science is quite a difficult and convoluted platform to work from for progress and

Fig. 29.1 Water, how can it ever be assessed and expressed in a quantitative fashion, e.g. for estuaries, across borders, and the clouds (evapotranspiration) and atmosphere?

when major aspects of ecological and sustainable integrity are in global decay (see an example provided in Figs. 29.1, 29.2, 29.3, 29.4, 29.5 and 29.6).

And while large economic models exist to 'manage' and to apply a science-based management theory to topics of real-world concern, this actually is not applied much to nature (Daley & Farley, 2011) or PNG yet. It also widely fails on global sustainability questions, e.g. climate change (Stern, 2008). Data are widely absent and/or not suitable, hardly assessed for validity. It is a holistic problem because if humans continue to capitalize on nature and its resources, including nations like PNG, but none of the currently highly funded projects will be so useful—hardly needed—for humanity if their surrounding environment is naturally inhabitable for humans in the first place (or of quality is sacrificed). The wider and deeper connections matter for modernity (e.g. Hutchinson 1957 and Carlson 2012; see Buckland et al. 2015 for quantitative wildlife conservation implications)

And although we are living in the information age, (essential) data on PNG are even less solid to use and to obtain, e.g. online. It's not unusual to find that those data are simply incorrect, confusing (Steiner & Huettmann, 2021 for taxonomy), carry no confidence measure, or are completely missing. But in the meanwhile, proxies or relevant correlated data sets do exist and can be used for habitat and similar assessments in order to get at those questions. Over 100 of those data are found online and readily available (Sririam & Huettmann, unpublished, Huettmann, 2020 for a PNG application). And an even larger set was compiled by FH and co-authors consisting of 132 GIS layers available for good use and wider assessment (Steiner & Huettmann, unpublished).

Fig. 29.2 Ocean, beach and habitat pollution ongoing: see the details in subsequent Figs. 29.3, 29.4, 29.5 and til 29.6

Further, management has developed steadily, and many new tools are used and applied in those textbooks. However, the wildlife management textbook (Silvy, 2020), a core reference and now in its 8th edition, has still not implemented much those concepts and less so applies them, nor do many agencies for governance and policy. At minimum, an experimental use of those methods should be explored.

Here, I am looking at a basic but powerful and promoted textbook management tool SWOT (Madsen, 2016), to assess the status of PNG through its environmental data available for such approach. SWOT looks at just four aspects, Strength, Weakness, Opportunities, and Threats. It is a basic but widely used and understood method to get a baseline for a *status quo*, a decision, and a situation; usually applied though to business and competitive assessments. Essentially, it's a rapid-assessment and parsimonious tool, a situational summary template, based on a two-by-two matrix and well known to have weaknesses (e.g. Omer, 2018). While widely applied, a well-done SWOT analysis can be somewhat holistic, in reality, it is often pretty coarse and leaves shortcomings. However, it can be informative, helps to find and assess gaps, and adds new insights.

Therefore, additional and more inclusive and finer-detailed approaches have been proposed in Management theory to be more holistic and meaningful. And so here, I go one step further beyond SWOT and also look at a PESTLE (Political, Economic,

Fig. 29.3 Plastic pollution—plastic bottle from a soft drink—that can often be found on Papua New Guinea's beaches and is not included in the SWOT and PESTLE analysis

Social, Technological, Legal, and Environmental) analysis for PNG and its environmental data. In the absence of good management data in modern times, and to start the process—in an experimental form—here, I base this assessment on the best-available GIS habitat data for PNG, in order to assess whether the conclusions and findings match the SWOT analysis findings across those management tools for these species.

Overall, such specific use of the SWOT and PESTLE approach is part of the social sciences as well as in business and economy, and it equals more a meta-analysis and a trend assessment rather than traditional quantitative and 'precise' science following a hypothesis testing path. Still, SWOT and PESTLE are standard in any planning and for business and administration, and get applied worldwide across disciplines.

Here, I pursue this approach as an experiment overall because PNG and its environment and data are not in the mainstream research agenda nor managed so well; a pro-active approach—as mandated by U.N.—is actually widely missing. In addition, the discipline of wildlife management widely falls short in promoting and using modern management theory and such metrics with all best-available data (Silvy, 2020), while the Anthropocene moves into a global environmental crisis (Hamilton et al., 2015; Houston, 2013). One hopes by using widely used business analysis tools to get new insights and a wider buy-in.

Fig. 29.4 Plastic pollution—leftovers of a flip flop and styrofoam piece—that can frequently be found on Papua New Guinea's beaches. Those are also not addressed well in the SWOT and PESTLE analysis

Finally, this concept I pursue here has also been applied with the conservation management of squirrels (Steiner & Huettmann, unpublished) and it is followed here. It is meant to be an experimental use and extension applied to PNG. If found useful, It could help to become a standard assessment for national and environmental fields and their underlying data qualities and uses.

29.2 Methods

29.2.1 SWOT Analysis

Here I used the pool of 132 GIS habitat layers compiled by Steiner and Huettmann (unpublished; see Table 29.2). Those layers were classified as multidimensional habitat assessments and impacts for Papua New Guinea according to the four SWOT aspects of strength, weakness, opportunities, and threats (Madsen, 2016).

In the SWOT analysis, I asked the question whether the content of those GIS habitat layers meets the four SWOT questions for a PNG conservation management. For instance, is the current 'landcover' in the Anthropocene a threat or gain for PNG?

Fig. 29.5 Another plastic pollution—plastic rope and net leftovers—that can be found on Papua New Guinea's beaches and that is not addressed well in the SWOT and PESTLE analysis

29.2.2 PESTLE Analysis

Like in the SWOT analysis, I used the 132 GIS habitat layers and classified their status and meaning as assessments for PNG according to the six PESTLE aspects: Political, Economic, Social, Technological, Legal, and Environmental (Perera, 2017).

In the PESTLE analysis, I asked the question whether the content of those GIS habitat layers meets the six PESTLE questions for squirrel conservation management. For instance, is the current 'landcover' in the Anthropocene political for squirrels?

In order to achieve a generic conclusion, for SWOT and PESTLE, I added up the majority of the metrics for yes vs no.

29.3 Results

29.3.1 SWOT Analysis

As shown in Table 29.3, it was found that most typical factors for PNG conservation management are also emphasized in the SWOT analysis. While factors related to

Fig. 29.6 Plastic pollution—a water drinking bottle—that can be found on Papua New Guinea's beaches is further not addressed well in the SWOT and PESTLE analysis. PNG has one of the highest rainfalls in the world, so what does a plastic bottle really bring new to the table?

weaknesses and opportunities add a new insight and management options, little new really comes to light when acting under the assumption that resources are finite, as emphasized for a long time in Ecological Economics (Darley & Farley, 2011; Farley, 2014; for Steady-State Economics (SSE) see Czech, 2017; Czech & Daly, 2004; Huettmann & Czech, 2007). For PNG this has not been well applied though.

This is fully reflected in the summary columns of each of the four metrics in SWOT. For strengths, the majority of the 132 GIS habitat layers reflect that. For weaknesses, most of the 132 GIS habitat layers show none. The opportunities of the 132 GIS habitat layers are not the majority result for PNG conservation management. The majority of the 132 GIS habitat layers present threats.

29.3.2 *PESTLE Analysis*

From the PESTLE analysis, Tables 29.4a and 29.4b show a more graduated view than the SWOT analysis. While the specific environmental aspect in PESTLE itself adds the environment to the agenda, the other factors are not really new, except perhaps the political and social ones. Still, from a more holistic perspective PESTLE provides a more rounded assessment.

The summary columns of the PESTLE analysis show us this pattern even stronger: The majority of habitat layers show a political aspect, just like the economic column, but not the social metric. The technological aspect has no majority nor the legal one, but certainly, the environmental one has.

29.4 Discussion

Papua New Guinea (PNG) presents a global showcase for humanity. Conservation management stands at its core and it was done for over 47,000 years. The actual conservation management is part of the wider management theory. While a global crisis unfolds in PNG already for over 50 years, in parallel, management theory has evolved a lot in the last 50 years, specifically for business and administration. However, for ecology and conservation—sustainability—there is little of such concepts available or implemented yet. Tools like SWOT and PESTLE are well established outside of wildlife and conservation management but not much within such sectors (e.g. Silvy, 2020). Here, I show an application for SWOT and PESTLE and what those can bring to the table, using PNG and its available data.

By using business and management analysis tools, it allows to look at the best-available GIS habitat data set for PNG and its environmental aspects in a different light. Here, I focus on a wider environmental assessment, namely with an Ecological Economics foundation.

The 132 GIS habitat layers used here are the best publicly available online data sources. Those have not been truly used for assessments of national environmental management (but see Huettmann, 2020 for PNG, and for other applications Boulanger-Lapointe et al., 2022; Huettmann et al., 2018); nor are SWOT and PESTLE used for species conservation management (compare with Silvy, 2020).

Using SWOT and PESTLE tools for data and data interpretation remains subjective because it involves judgment questions of classification and impact. It becomes clear though that those management tools are mostly designed for human and competitive business aspects, not for wildlife or ecology or for (GIS) data, e.g. the economic or social, especially the technology metrics in PESTLE. Still, some new insights were obtained, which will be elaborated more in the following:

It was found that the SWOT analysis of the data at hand was not able to describe the issues of public land tenure commonly found in PNG. Same applies for the habitat issues. But it was found that the PESTLE analysis helped to put more emphasize on the issue of PNG habitats. Namely, the specific focus on environmental metrics makes it a more holistic tool and better suited for ecological and species management questions.

However, neither SWOT nor PESTLE showed a good outlook or differed in the acknowledged meta-analysis for PNG's conservation management, that is specifically the case when using Ecological Economics as the platform to work from.

In terms of data and data qualities, it becomes clear that specific socio-economic data are needed, but are widely missing in the open-access arena. Secondly, neither SWOT nor PESTLE deal well with data accuracy or confidence, or pixel resolutions; those are essential questions for any data though. It further shows the limits of SWOT and PESTLE.

With these two widely-used management tools, here a very basis for the western world's management theory is laid out for PNG's conservation. It then becomes the task of follow-up studies and research to apply more ecologically adequate management tools for PNG and generally the ecological sector (see Alcorn et al., 1993, Faith et al., 2000; Beehler & Alonso, 2001, and follow-up studies). What is widely missing still is an approach—and tool—that can take PNG and its realities better into account, e.g. tribal management. Possible follow-up research can include Analytical Tools for Environmental Design and Management in a Systems Perspective (Wrisberg et al., 2002), Environmental Management Tools (https://www.sciencedirect.com/topics/earth-and-planetary-sciences/environmental-management-tool), and similar more advanced tools while data quality, data meaning and data access and Q&A must receive other priorities for modern PNG, its wilderness and sustainable management.

Appendices

Appendix 29.1 List of 132 GIS Data Set and Their Attributes

See Table 29.2.

Appendix 29.2 List of 132 GIS Data Set and Their SWOT Values

See Table 29.3.

Appendix 29.3 List of 132 GIS Data Set and Their PESTLE Values

See Tables 29.4a and 29.4b.

Table 29.2 List of 132 GIS predictors

Predictor	Source	Explanation	Citation
BIO1_2_5min - BIO19_2_5min; tmin1 - tmin12; tmax1 - tmax12; tavg1 - tavg12; srad1 - srad12; prec1 - prec12; Wcaltitude	Worldclim (https://www.worldclim.org/data/worldclim21.html)	These data sets tend to represent most of the climate data utilized for SDMs	Fick and Hijmans (2017)
FAOCC	FAO Geonetwork (http://www.fao.org/geonetwork/)	This predictor represents the global climate classes	
LC12asc2; VE4	Geospatial Information Authority of Japan (https://www.gsi.go.jp/kankyochiri/gm_global_e.html)	These predictors represent the global land cover (LC12asc2), and the global vegetation cover (VE4)	
GlobalRiversProxy2	A Simple Global River Bankfull Width & Depth Database (http://gaia.geosci.unc.edu/rivers/)	This predictor represents proximity to all global mid-size and large rivers	Andreadis et al. (2013)
GlobalBigRivers11	Global major rivers (https://www.arcgis.com/home/item.html?id=44e8358cf83a4b43bc863646cd695945)	This predictor represents all global large rivers	
GlobalCities2	Global Cities (https://hub.arcgis.com/datasets/6996f03a1b364dbab4008d99380370ed_0?geometry=-65.394%2C25.931%2C73.737%2C49.818)	This predictor represents all global cities	
GlobalLakes2	Global Lakes and Wetlands Database (GLWD) (http://www.fao.org/land-water/land/land-governance/land-resources-planning-toolbox/category/details/es/c/1043160/)	This predictor represents all global lakes and wetlands	

(continued)

Table 29.2 (continued)

Predictor	Source	Explanation	Citation
GlobalSnowCoverMonthJan2021_7; FFJan2020_3; FFFeb2020_3; FFMar2020_3; FFMay2020_3,FFJun2020_3; FFJul2020_3; FFAug2020_3; FFSep2020_3; FFOct2020_3; FFNov2020_3; FFJan2021_3	Global Snow Cover and Forest fires (https://neo.sci.gsfc.nasa.gov/view.php?datasetId = MOD10C1_M_SNOW and https://neo.sci.gsfc.nasa.gov/view.php?datasetId=MOD14A1_M_FIRE&year=2020)	These predictors mainly represent the Global snow cover in the month of January, and the forest fire information of nearly all months of 2020 with exception of April and December as these months were not available	
WorldSoil2	Global Soil characteristics map (https://webarchive.iiasa.ac.at/Research/LUC/External-World-soil-database/HTML/HWSD_Data.html?sb=4)	This predictor represents all global soil types and its characteristics	
WorldProtectedAreasMerged4	Global Protected areas. (https://www.protectedplanet.net/en/search-areas?geo_type=region&filters%5Bis_type%5D%5B%5D=terrestrial)	This predictor represents all global protected are merged into one shapefile	UNEP-WCMC and IUCN (2020)
WorldMammaldensity4 WorldRodentDensity3 WorldThreatenedMammalDensity3	Global Mammal density. Proximity maps for the world mammal density, world rodent density, world threatened mammal density (https://biodiversitymapping.org/index.php/mammals/)	These predictors mainly represent global biodiversity densities. In detail, they contain the world mammal density, world rodent density, world bird density and the world's threatened mammal density	Jenkins et al. (2013)

(continued)

Table 29.2 (continued)

Predictor	Source	Explanation	Citation
GlobalBirdDensity2			
GlobalRoadsProxy2	Global Roads—Socio-economic data and applications center (SEDAC)—Data center in NASA's Earth Observatory System Data and Information System (EOSDIS) (https://sedac.ciesin.columbia.edu/data/set/groads-global-roads-open-access-v1/data-download)	This predictor represents the global proximity to all world's roads. Minor roads may not be included	
Human footprint	Human Influence Index (HII). (https://sedac.ciesin.columbia.edu/data/set/wildareas-v2-human-influence-index-geographic/data-download)	This predictor represents the global Human Influence index; it's strongly driven by roads and cities for the index	
WorldSlope1	Slope. (https://scholarworks.alaska.edu/handle/11122/7151)	This predictor represents the global terrestrial and aquatic slope	Sriram and Huettmann (2017)
World_MAX_RH_JAN - World_MAX_RH_DEC; World_MIN_RH_JAN - World_MIN_RH_DEC	Global Monthly Relative Humidity. (http://palebludata.com/?q=data)	This predictor set represents the global maximum and minimum relative humidity for the months January to December of the year 2020	Jones and Wint (2015)

Source Steiner and Huettmann unpublished

Table 29.3 SWOT analysis based on 132 GIS habitat layers (see Appendix 29.1 for details) describing the environment of Papua New Guinea, e.g. as needed for conservation management and ecological niche models (see Huettmann, 2020 for an example)

Predictor (Appendix 29.1)	Strength	Weakness	Opportunity	Threat
BIO1_2_5min - BIO19_2_5min; tmin1 - tmin12; tmax1 - tmax12; tavg1 - tavg12; srad1 - srad12; prec1 - prec12; Wcaltitude	Yes (climate to rely on)	Yes (climate as a bottleneck)	Yes (climate as a buffer)	Yes (bad impacts of man-made climate change and by bad data)
FAOCC	Yes (climate to rely on)	Yes (climate as a bottleneck)	Yes (climate as a buffer)	Yes (bad impacts of man-made climate change)
LC12asc2; VE4	Yes	No	Not really	Yes
GlobalRiversProxy2	Yes	No	Not really	Not really
GlobalBigRivers11	Yes	No	Not really	Not really
GlobalCities2	Little	No	Somewhat	Yes (habitat loss)
GlobalLakes2	Yes	No	Not really	Not really
GlobalSnowCoverMonthJan2021_7; FFJan2020_3; FFFeb2020_3; FFMar2020_3; FFMay2020_3,FFJun2020_3; FFJul2020_3; FFAug2020_3; FFSep2020_3; FFOct2020_3; FFNov2020_3; FFJan2021_3	Not directly	No	No	Yes, loss of water and humidity
WorldSoil2	Yes	Yes (e.g. contamination)	Little	Little (soil loss)
WorldProtectedAreasMerged4	Yes (protection)	Little	Little	No
WorldMammaldensity4	Little (hybridization)	No	No	Little (disease transmission)
WorldRodentDensity3	No	No	No	No
WorldThreatenedMammalDensity3	No	No	No	No
GlobalBirdDensity2	Yes (prey)	Little	Little	Little (raptors)
GlobalRoadsProxy2	No	Little	No	Yes (road kill)

(continued)

Table 29.3 (continued)

Predictor (Appendix 29.1)	Strength	Weakness	Opportunity	Threat
WorldSlope1	No	No	Little (refuge habitat)	No
World_MAX_RH_JAN - World_MAX_RH_DEC; World_MIN_RH_JAN - World_MIN_RH_DEC	Yes (climate to rely on)	Yes (climate as a bottleneck)	Yes (climate as a buffer)	Yes (bad impacts of man-made climate change)
Majority sum of yes versus no	Majority yes	Majority no	Majority no	Majority yes

Table 29.4a PESTLE analysis based on 132 GIS habitat layers (see Appendix 29.1 for details) describing the environment of Papua New Guinea, e.g. as needed for conservation management and ecological niche models (see Huettmann, 2020 for an example)

Predictor (Appendix 29.1)	Political	Economic	Social
BIO1_2_5min - BIO19_2_5min; tmin1 - tmin12; tmax1 - tmax12; tavg1 - tavg12; srad1 - srad12; prec1 - prec12; Wcaltitude	Yes	Not directly	Not directly
FAOCC	Yes	Not directly	Not directly
LC12asc2; VE4			
GlobalRiversProxy2	Not directly	Not directly	Not directly
GlobalBigRivers11	Not directly	Not directly	Not directly
GlobalCities2	Yes	Yes	Not directly
GlobalLakes2	Not directly	Not directly	Not directly
GlobalSnowCoverMonthJan2021_7; FFJan2020_3; FFFeb2020_3; FFMar2020_3; FFMay2020_3,FFJun2020_3; FFJul2020_3; FFAug2020_3; FFSep2020_3; FFOct2020_3; FFNov2020_3; FFJan2021_3	Not directly	Not directly	Not directly
WorldSoil2	Yes	Yes	Not directly
WorldProtectedAreasMerged4	Yes	Yes	Yes
WorldMammaldensity4	Not directly	Not directly	Not directly
WorldRodentDensity3	Not directly	Not directly	Not directly
WorldThreatenedMammalDensity3	Not directly	Not directly	Not directly
GlobalBirdDensity2	Not directly	Not directly	Not directly
GlobalRoadsProxy2	Yes	Yes	Not directly
WorldSlope1	No	Not directly	Not directly

Table 29.4b PESTLE analysis based on 132 GIS habitat layers (see Appendix 29.1 for details) describing the environment of Papua New Guinea, e.g. as needed for conservation management and ecological niche models (see Huettmann, 2020 for an example)

Predictor (Appendix 29.1)	Technology	Legal	Environmental
BIO1_2_5min - BIO19_2_5min; tmin1 - tmin12; tmax1 - tmax12; tavg1 - tavg12; srad1 - srad12; prec1 - prec12; Wcaltitude	Not directly	Yes	Yes
FAOCC	Not directly	Yes	Yes
LC12asc2; VE4	Not directly	Yes	Yes
GlobalRiversProxy2	Not directly	Not directly	Not directly
GlobalBigRivers11	Not directly	Not directly	Not directly
GlobalCities2	Not directly	Yes	Yes
GlobalLakes2	Not directly	Not directly	Not directly
GlobalSnowCoverMonthJan2021_7; FFJan2020_3; FFFeb2020_3; FFMar2020_3; FFMay2020_3,FFJun2020_3; FFJul2020_3; FFAug2020_3; FFSep2020_3; FFOct2020_3; FFNov2020_3; FFJan2021_3	Not directly	Not directly	Not directly
WorldSoil2	Not directly	Yes	Yes
WorldProtectedAreasMerged4	Not directly	Yes	Yes
WorldMammaldensity4	Not directly	No	Yes
WorldRodentDensity3	Not directly	No	Yes
WorldThreatenedMammalDensity3	Not directly	Not directly	Yes
GlobalBirdDensity2	Not directly	Not directly	Yes
GlobalRoadsProxy2	Not directly	Yes	Yes
WorldSlope1	Not directly	No	Yes

References

Adamowicz, W. L. (2016). Economic analysis and species at risk: Lessons learned and future challenges. *Canadian Journal of Agricultural Economics/Revue Canadienne D'agroeconomic, 64*(1), 21–32.

Alcorn, J. B. (1993). Papua New Guinea Needs assessment. A USAID-funded Consortium of World Wildlife Fund. The Nature Conservancy, and World Resources Institute Washington, D.C. with the Government of Papua New Guinea Department of Environment and Conservation, Boroko, Papua New Guinea. https://www.sprep.org/attachments/VirLib/PNG/png-conservation-needs-assessment-alcorn-1993.pdf. Accessed July 3, 2022.

Andreadis, K. M., Schumann, G. J. P., & Pavelsky, T. (2013). A simple global river bankfull width and depth database. *Water Resources Research, 49*(10), 7164–7168.

Beehler, B., & Alonso, L. E. (2001). Southern New Ireland, Papua New Guinea: A biodiversity assessment. *RAP Bulletin of Biological Assessment, 21*, 1–101.

Beehler, B., & Laman, T. (2020). *New Guinea: Nature and culture of Earth's Grandest Island.* Princeton University Press.

Boulanger-Lapointe N., Ágústsdóttir, K., Barrio, I. C., Defourneaux, M., Finnsdóttir, R., Jónsdóttir, I. S., Marteinsdóttir, B., Mitchell, C., Möller, M., Nielsen, Ó. K., Sigfússon, A. Þ., Þórisson, S.

G., & Huettmann, F. (2022). Understanding herbivore species coexistence in changing rangeland ecosystems: First high resolution national open-source and open-access ensemble models for Iceland. *Journal of the Total Environment* (in press).
Buckland, S. T., Rexstad, E. A., Marques, T. A., & Oedekoven, C. S. (2015). *Distance sampling: Methods and applications* (Vol. 431). Springer.
Carlson, D. J. (2012). IPY 2007–2008: Where threads of the double helix and sputnik intertwine. In F. Huettmann (Ed.), *Protection of the three poles*. Springer. https://doi.org/10.1007/978-4-431-54006-9_2
Czech, B., & Daly, H. E. (2004). In my opinion: The steady state economy—what it is, entails, and connotes. *Wildlife Society Bulletin, 32*(2), 598–605.
Czech, B. (2017). The steady state economy. In *Routledge handbook of ecological economics* (pp. 467–476). Routledge.
Daly, H. E., & Farley, J. (2011). *Ecological economics: Principles and applications*. Island press.
Diamond, J. (2011). *Guns, germs and steel: The fate of human societies*. Random House.
Dornelas, M. L. H., Antão, F. M. et al. (2018). BioTIME: A database of biodiversity time series for the Anthropocene. *Global Ecol Biogeogr* 1–26. https://doi.org/10.1111/geb.12729
Faith, D. P., Nix, H. A., Margules, C. R., Hutchinson, M. F., Walker, P. A., West, J., & Natera, G. (2000). The BioRap biodiversity assessment and planning study for Papua New Guinea. *Pacific Conservation Biology, 6*, 279–288.
Farley, J. (2014). Steady state economics. In *Degrowth* (pp. 77–80). Routledge.
Fick, S. E., & Hijmans, R. J. (2017). WorldClim 2: New 1 km spatial resolution climate surfaces for global land areas. *International Journal of Climatology, 37*(12), 4302–4315.
Hamilton, C., Gemenne, F., & Bonneuil, C. (Eds.). (2015). *The Anthropocene and the global environmental crisis: Rethinking modernity in a new epoch*. Routledge.
Hannah, S., Wauchope, T. A., Geldmann, J., Johnston, A., Simmons, B. I., Sutherland, W. J., & Jones, J. P. G. (2021). Evaluating impact using time-series data. *Trends in Ecology & Evolution 36*(3), 196–205. https://doi.org/10.1016/j.tree.2020.11.001
Houston, D. (2013). Crisis is where we live: Environmental justice for the anthropocene. *Globalizations, 10*, 439–450.
Huettmann, F., Mi, C, & Guo, Y. (2018). 'Batteries' in machine learning: A first experimental assessment of inference for Siberian crane breeding grounds in the Russian High Arctic based on 'shaving' 74 predictors. In G. Humphries, D. R. Magness, & F. Huettmann (Eds.). *Machine learning for ecology and sustainable natural resource management* (pp. 163–184).
Huettmann, F. (2020). Chapter 24—Investigating Matschie's tree kangaroos with 'modern' methods: Digital workflows, big data project infrastructure, and mandated approaches for a holistic conservation governance. In D. L. P. Valentine, J. Blessington, & K, Schwartz (Eds.), *Tree kangaroos: Science and conservation* (pp. 379–391). Academic Press.
Hutchinson, G. (1957). Concluding remarks. *Cold Spring Harbor Symposia on Quantitative Biology, 22*, 415–427.
Jenkins, C. N., Pimm, S. L., & Joppa, L. N. (2013). Global patterns of terrestrial vertebrate diversity and conservation. *Proceedings of the National Academy of Sciences, 110*(28), E2602–E2610.
Jones, P., & Wint, W. (2015). Data set produced by Waen Associates for Environmental Research Group Oxford, Limited, funded by the International Research Consortium on Dengue Risk Assessment, Management and Surveillance (IDAMS), European Union's Seventh Framework Programme for research, technological development and demonstration under grant agreement no 281803.
Laurance, W. F., Kakul, T., Keenan, R. J., Sayer, J., Passingan, S., Clements, G. R., Villegas, F., & Sodhi, N. S. (2011). Predatory corporations, failing governance, and the fate of forests in Papua New Guinea. *Conservation Letters, 4*, 95–100. https://doi.org/10.1111/j.1755-263X.2010.00156.x
Madsen, D. Ø. (2016). SWOT analysis: A management fashion perspective. *International Journal of Business Research, 16*(1), 39–56.
Narokobi, B. (1975). *Foundations for nationhood*. Papua New Guinea Press and Bookshop.

Narokobi, B. (1983). *The Melanesian Way*. Institute of Papua New Guinea Studies, Boroko, Papua New Guinea.

Omer, S. K. (2018). Swot analysis; the tool of organizations stability (KFC) as a case study. *Journal of Process Management. New Technologies, 6*(4).

Salak, K., (2001). *Four corners: One Worman Solo's Journey into the heart of Papua New Guinea.* Counterpoint Publisher.

Silvy, N. J. (Ed.). (2020). *The wildlife techniques manual: Volume 1: Research. Volume 2: Management* (Vol. 1). JHU Press.

Sriram, S., & Huettmann, F. (2017). A global model of predicted peregrine falcon (Falco peregrinus) distribution with open source GIS code and 104 open access layers for use by the global public. *Earth System Science Data Discussions*, 1–39.

Steiner, M., & Huettmann, F. (2021). Justification for a taxonomic conservation update of the rodent genus Tamiasciurus: Addressing marginalization and mis-prioritization of research efforts and conservation laissez-faire for a sustainability outlook. *The European Zoological Journal, 88*(1), 86–116.

Stern, N. (2008). The economics of climate change. *American Economic Review 98,* 1–37.

UNEP-WCMC and IUCN. (2020). Protected planet: The World Database on Protected Areas (WDPA) [Online], September 2020, Cambridge, UK: UNEP-WCMC and IUCN. Available at: www.protectedplanet.net.

Wrisberg, N., de Haes, H. A. U., Triebswetter, U., Eder, P., & Clift, R. (Eds.). (2002). *Analytical tools for environmental design and management in a systems perspective: The combined use of analytical tools* (Vol. 10). Springer Science & Business Media.

Chapter 30
Some Very Easy Lessons from Papua New Guinea's "*Garden of Abundance*" for the Wider Global Good: Indigenous, Tribal and Marxist Views, Bob Marley as the Public Role Model, Destructive Neoliberalism, Reality and Understanding Globalization Failures Are not Rocket Science

Not only have the oil, gas and mining industries not helped the poorest people in developing countries They have often made them worse off. Scores of recent academic studies and many of the bank's own studies confirmed our findings that countries which rely primarily on extractive industries tend to have higher levels of poverty, child morbidity and mortality Civil war, Corruption and totalitarianism than those with more diversified economies. Does this mean extractive industries can never play a positive role in a nation's economy? No It simply means that the only evidence of such a positive role we could find took place after a country's democratic governance had developed to such a degree that the poorest could see some of the benefits. before the fundamental building blocks of good governance a free press a functioning judiciary Respect for human rights free and fair elections and so on are put in place The development of these industries only aggravates the situation for the poorest
*Dr Emil Salim, Chair of the World Bank Extractive Industry Review (UK times 16th June 2004)**

Abstract While the nation of Papua New Guinea features some low-end economic and modern metrics, it should not be branded as a failure, or dismissed. Instead, PNG is deep global humanity and it has shown remarkable and relatively strong contributions to local, regional and global sustainability and governance public resources. That is, a concept that is utterly missing in 'The West' and with colonial powers and their follow-up entities, in the U.N. and their underlying business approach. PNG is not really a hierarchical or feudal society, and such concepts somewhat entered into communistic and social ideals - theory of Marxism-attributed to ancient societies such as Africa or PNG. In the modern sense, sharing resources on a finite space remains a major skill that is theorized but not well achieved in real live. These

* Cited in N. Sullivan, https://www.academia.edu/4492757/Introduction_to_The_Prospects_for_Sustainable_Development_in_Papua_New_Guinea (accessed 3rd July 2022)

F. Huettmann, *Globalization and Papua New Guinea: Ancient Wilderness, Paradise, Introduced Terror and Hell*, https://doi.org/10.1007/978-3-031-20262-9_30

topics are discussed here, showing some simple conclusions and the generic failure of essential western thought to natural resources and the (tropical) world and wilderness.

Keywords Papua New Guinea (PNG) · Public good · Land rent · Georgism · Marxism · Globalization · Sustainable development

30.1 Introduction

Most times, economic metrics of globalization look down upon Papua New Guinea (PNG) and its culture and people (e.g. metrics presented in Gosarevski et al., 2019). PNG is given a baby place on the international agenda and table of power nations (Miller, 1973) and that has never really changed. Instead, it gets virtually forgotten that capitalism underlying globalization has failed so much, globally. Biodiversity and conservation fail accordingly; it cannot be any other; the wilderness loss in PNG shows it clearly (e.g. Perez-Haemmerle et al., 2022). But as an alternative, PNG has still much to offer for the global society. PNG runs a sustainable concept for over 47,000 years fine. So why not learning from it? PNG does not need more Australian and English-speaking teachers, it should send out its own PNG teachers instead for teaching the world, and its policy advisors. It's worthy to note that the western world is widely positioned on the other side of the spectrum. So what went wrong?

Allowing for a PNG perspective onto the western world, and putting it into context with PNG—as per metrics—a center piece of global sustainability will help to set the record straight and to correct some misconceptions starting with colonialism, the European-centric perspective of the few royal courts[1] and globalization as we know it.

30.1.1 Are Ancient Societies (of PNG) Primitive?

This is an old discussion in the scientific literature and beyond, e.g. Levi-Straus (1966), Mead (1967) or Narokobi (1983), and it can be put to a rest. I think it becomes quickly clear that PNG is far from primitive,[2] trivial or mundane. PNG is as sophisticated as other nations and beyond. That's due to the complexities and what

[1] The European Royal Courts and their advisors created a global *havoc* in the last history of the earth. They did so repeatedly. But those courts are far and few really; compared to the millions of citizens it's a small group. They consist of the French Crown, the British Empire, the Spanish Crown, the Portuguese split-off, the Dutch, the Austrian-Hungarian Empire, the Belgium Royals, the Danish, the Swedish and the Russian Tsar. Most of them were very closely related, due to historic intermarriage for power reasons. It' thus a backwardish European power clique that was in charge that really drove colonialism and global events and outcome. French was their language of diplomacy. Those groups and structures are still major drivers on the global scene, e.g. in the UN, the British Dominions, in Antarctica, and former colonial nations. Even places like Alaska and Canada owe their existence and layout, borders and environmental set up + socio-economic crisis to those inter-royal disputes, which mostly were about money and vanity.

[2] A term used by Mead, Levi-Strauss and others; see literature here and in other chapters of this book.

its people have to go through and to do, in real live. Diamond (2011a) had shown well that skills like planting and gardening knowledge are well entrenched and well-known with indigenous people, certainly in PNG. In tropical nations, the species diversity is very high, and having a good ID and taxonomic skill level is essential for living there. Knowing what to eat and how to prepare it is crucial. Every child learns and practices those things early on within their own living environment. Usually in PNG that's the garden added with bushmeat and 'fruits from the wild.' Hard to starve that way. As a matter of fact, most PMNG people are not starving and due to the rural female and child family labor as a backup net. Also, treatment of plants for safe consumption or medical use is an inherent part of the PNG lifestyle 'in the bush,' as practiced and fine-tuned for millennia. Adding local fishery practices adds to that sustainable lifestyle. When compared with the average urban citizen, a rural PNG person is not only more informed, but also more fit. PNG people can usually walk in rough terrain and walk very long distances and they do in very remote and steep terrains (see Figs. 30.1, 30.2 and 30.3 of travel in rugged terrain and valleys that easy can take 3 days to cross for the 3 km distance covered on a map) (for income, livelihood and realities see also Figs. 30.4, 30.5, 30.6, 30.7, 30.8, 30.9, 30.10, 30.11, 30.12 and 30.13).

As stated by Riley (2004): "*In the absence of a strong national identity in post-colonial creations like Papua New Guinea or Solomon Islands, most people's primary*

Fig. 30.1 Female labor, a global item of dispute that is virtually and brutally left unaccounted for in western economic models and metrics

Fig. 30.2 The photo speaks for itself: modern technology happing in a tribal market economy governance. A blend of modernity trying in a dynamic way to make the best of all worlds offered

loyalty is to their clan, tribe and wantok ('one talk') groups. In competing with each other for resources, these ethno-political units increasingly act like interest groups in developed societies, attempting to divert potential public goods towards the private enrichment of their members alone. Such 'rent seeking' is a pervasive aspect of contemporary political and economic life in Melanesia, precisely because of the way societal fragmentation enables small ethnic collectives to be mobilised , monitored and enforced to acquire and then monopolise control of actual and potential public goods. Micro-ethnic identities thus become the key social structure for facilitating the formation of distributive coalitions, which are then used to divert potentially productive public goods towards the benefit of their group alone" (for more on this, see Reilly, 2004*).*

Fig. 30.3 Rugged terrain and valleys, PNG wilderness

Of course, in the family business world the concept of cronyism and corruption lingers around, see for instance accusations by the own PNG minister in charge of his agency (The Guardian, 2020) and election problems (for a direct Australian link see The Guardian, 2022a). But which nation does not feature such concepts and can offer better examples? Those things are widely known for many nations, see for instance the former colonial power of Holland: now called a Narco-state (The Guardian, 2019a).

Everybody who ever was in PNG will agree that PNG culture, society, biodiversity, land- and seascape indeed can serve as an inspiration, wisdom and guidance for the people and the world, just as expressed by Cousteau and Richards (1999, p. 221). PNG remains a global inspiration, a gift to humanity, and of relevance for the global society.

30.2 Are Ancient Societies (of PNG) 'Primitive' and the Original Communist Life Forms?

One may perhaps argue - as done for some parts of Africa - that the ancient indigenous lifestyle in the wild, unfettered by the nobles, royals and mass warfare, is essentially

Fig. 30.4 Slingshots for Bushmeat as the income of the day, globally. Kids play a big role in that

not feudal and thus 'pure' communism: Everybody is equal, public land tenureship, small group hierarchies (Beehler & Laman, 2020), group decision-making (e.g. by warriors, soldiers = "soviets"), almost a complete lack of money (betel nut and mussels instead); maintenance of the wider common good, and nature and holistic views are favored (Lenin for instance supported many of those concepts, see Shtilmark, 2003 for Russian protected area support).

A lot of ideology is wrapped around this concept of native indigenous societies as human role models (e.g. Narokobi, 1974), for public rent making (Gaffney & Harrison, 1994) and for communism. But arguably, this scheme has found entry into Karl Marx, Marxism, and was introduced there into communism (bottom up, from the people, 'grass roots') and its parties worldwide. And it has worked very well, for millennia and for millions of people; let's just face it. It's true for humanity for over 98% of its existence; well, until colonialism and specifically the last 50 years.

Fig. 30.5 Treekangaroos in a hotel cage: how unsustainable may it be? Many mammals get poached and sold; tourists are in awe

Now, this is true for African indigenous societies, as well as for many in Latin America (see also Brockett, 2019 for Central America) and many places in Asia too (see alone the China experience; lead by a communist party and movement as a world power).

For PNG, it became a certain feature to go by for the entire new nation, e.g. Narokobi (1975, 1983). Melanesia—and PNG—are in a public land ownership (Baraka, 2001; Kinch, 2020). That's even more so true for the ocean. It makes for a classic example of the wider public good and fails the initial work by Hardin (1968) but confirms the role of the framework, such as presented in Ostrom (1990).

But a certain Georgist Economy view remains similar to the basic PNG approach to land, a rent that's sustainable with a good life from around that. Traditional PNG societies have very little individual and material wealth, primarily just common wealth. Houses only lasted 7 years or so; people moved around. Wealth was actually measured in pigs, and only a few people had those. They used them to indebt and impress on others to pacify others and to settle earlier disputes. A complex land tenure dominated and that was embedded in those concepts.

Fig. 30.6 Megapod eggs, a sustainable harvest of the mounds in the forest is possible and done for millennia

30.3 Neoliberalism: An Awful Idea that Works Virtually Nowhere; Certainly not in PNG

Killing was on the order of the day in parts of PNG and what Australia tried to settle; so the narrative says (Matthiessen, 1987; Nelson, 2016). Missionaries claimed the same; but then what did those western religions create instead and left behind!? By now, PNG is the most heavily Christianized area in the world (Beehler & Laman, 2020, p. 325) but it features one of the worst crime metrics in parallel with a modern society described as in decay and education metrics at the bottom of the ranks. Over 1000 ancient cosmologies were actively dismantled in that process. It's indeed an understatement and without justice when Beehler and Laman (2020, p. 326) call it a crime against traditional humankind. And governmental patrols to pacify uncontacted peoples were actually done till the 1970s (Beehler & Latam, 2020) and it affected

Fig. 30.7 An ATM in the remote bush (see the satellite phone link to check your bank account). The author had no problem to withdraw internationally larger amounts of Kina that way, mind any security

villages and society of PNG alike. How can PNG be an equal-partnered nation and society in the global world then?

The initial idea is that modernity was needed to move PNG forward. Australian gold miners, happily pursuing their own feat, jumped in and imposed their own power and employment regime in PNG becoming millionaires in that very game. This happened under a certain rule of the League of Nations and later the UN mandate to put Australia in the lead for PNG (Baraka, 2001). This also created a new group of people, workers and half-breeds that have quite a tough place in PNG society (see Wilson, 2016 for an example) and in Australia. While some tribal fights were stopped, it shifted and it set up another form of abuse, gold mining and nationwide economic growth policies from a western regime that essentially started from European prisoners of the UK and ended in the rascals of POM while leaving wilderness in decay with some gold sold abroad. Crime is record high in PNG. Rascalism does not come from nowhere but has a 'modern' context. It's a direct outflow of how PNG was treated and administered and how the rural PNG society was dismantled when in contact by The West. That's the recent trend coming from globalization and which is widely based on neoliberalism and similar polices, as taught in Western curriculars

Neoliberalism is just a strain of capitalism (MacLaran & Kelly, 2014). It's not really a new concept, and it came in waves upon us. On a wider platform, neoliberalism started as early as the 1970s in the U.S. and became global in the 1980s, with more radical efforts ever since. By now, it's a zombied concept of capitalism

Fig. 30.8 Simple bartering on regional markets has worked well for millennia and was sustainable, why changing it with an imbalanced globalization?

(Springer, 2016). Arguably it's a revised form of an economy, whereas Ecological Economics was much closer to what PNG did for the last 99% of its existence: accept the limits (Daley & Farley, 2010). But Ecological Economics was never applied really, as of yet. While the capitalistic approach to PNG is now called neoliberalism, it actually is not liberal whatsoever, nor is it new/neo. Instead, neoliberalism is an awful old conservative concept in disguise. It promotes a hard-core and ruthless growth at all costs, fully inconsiderate of all else. Privatization and growth for the sake of growth all just try to rip the profit for a few; it's here where a 'small gang' drives it once more (Gray, 1996). The people and the environment—the ones who cannot defend themselves against it—pay the costs. One may state that neoliberalism does not really make money, it just takes it from the public. It's a certain form of public theft and artificial redistribution (Ostrom, 1990). It rakes what generations have built and worked for with their lives and for centuries. It's easy to see if one sticks with the basic concept where money comes from and how created, the Georgist paradigm and laws of themodynamics.

Who believes that neoliberalism works, globally? It only can work if you are the beneficiary helped politically and with the subsequent media and PR. And why is there not more resistance against it, if it is so unpopular and not working? Well, clearly, that's because the powers in favor of neoliberalism also drive the news and the control instruments: money, power, law and police and the military. If one controls that, anybody be turned into a 'sheep' and then can be fed alternative facts: the

Fig. 30.9 Sustainable life is possible

concept of a trickle-down economy: exploit nature, use nature and see the money accumulating upwards with some coming down to you too, eventually.

Has neoliberalism worked? Not really. Is that effective for the people? 100% not. A review of neoliberalism will show no other, from Thatcher's England to railroad privatization to flying to the Mars, fast-growing tree plantations, biofuel or the latest try out: Climate Change mitigation (MacLaren & Kelly, 2014; Springer, 2016). What works?

PNG has witnessed and seen unfolding many neoliberal schemes. One can see them in the mining sector, in the LNG and in natural resource sector now, including education, fisheries, Ramu Sugar, palm oil plants, and tree plantations for carbon sequestrations. It's an endless chase of dreams delivered by fluent experts and CEOs but which leaves the locals behind, and neoliberalism has no good place in PNG, hardly anywhere else (Erikson et al., 2015). No wonder that tribalism remains resilient in PNG (see Hyndman, 1998 for Melanesian resistance).

30.4 The Reality and Globalization Failures in PNG and Elsewhere

Papua New Guinea has poor modern economic metrics; that is true for its money, as it is for per capita income, wealth distribution, and for most things that come from

Fig. 30.10 Ornamental plants along ancient trails species of begonia and *noli tangere* (touch-me-not)

it: administration (The Guardian, 2020), health, public schooling, research output, union performance and status of PNG wilderness. But in parallel PNG is kind of world-leading in resource extractions! PNG in its classic image and Wantok system shows violence and abuse (examples shown in Matthiesen, 1987; Nelson, 1980; West, 2006, etc.). But PNG has helped the world many times. A small example is seen with the people from Boga Boga who rescued the war plane that drowned at Cape Vogel (details in Cousteau & Richards, 1999, p. 110). As a matter of fact, app. 55,000 PNGs locals got conscripted to join the war against evil forces ('Fuzzy Wuzzy Angels'). And during WW2, there were app. 20,000 tons of bombs thrown on Rabaul, the PNG land; one of the largest bombing drop ever done in WW2. And yes, like in most wars there also was 'friendly fire, further killing many innocent PNG citizens (Cousteau & Richards, 1999). PNG is a collateral of the world. And still, PNG contributed to the

Fig. 30.11 Modern concepts of food and meat production pushed onto Papua New Guinea are far from sustainable

win against the Japanese for a 'free world' (Cousteau & Richards, 1999 p. 127). And to this day, PNG remains to play a stable albeit crucial aspect in the 'Pacific Theatre'; that's the PNG support for Australia and for the U.S. alike.

But then, there is also the cultural contribution that PNG makes for the world. The world-famous singsing (Cousteau & Richards, 1999) was not invented in the EU. And it was not the high western society that can harvest needlefish, simply done by a harvested spider web and attached with a self-made kite to attract the fish which then entangles in that web (Cousteau & Richards, 1999). Such sophistication and schemes are PNG made…and sustainable. PNG inspires for its ingenuity.

When compared to PNG as the baseline, there actually is a widespread misunderstanding and mis-judgement what the western world is and what it can do, and what it does, and even for what it stands for. What really is the *status-quo*, what is the vision and what has it been? As an example, Germany itself, as one of the colonial nations is described in the mainstream conservative news outlet DIE WELT as a 'Dysfunct Nation.' Either those nations in charge have no good performance and no vision, or at least not a good one, and one that works, or one that sticks for long. Many metrics show us no other: western education debacle, human right non-achievements, food contamination and air pollution. Cockburn (2013) and Heinsohn (2021) have shown it succinctly for world-leading nations dominating PNG.

Fig. 30.12 Who needs electro junk?

And so, to move forward in good terms, 'going native' in full is not realistic (see Majumder, 2021a, 2021b for a paradigm shift towards indigenous views)? Changes are happening either way and become mandatory. At least, instead of Neoliberalism the Preferential Trade Agreements (PTAs) could allow for a more free and fair trade for everybody, with all benefits of what trade has to offer (Croker, 2021), if it then is such a great system. Arguably, we will not move there any time soon as nature is modified in concrete.

While PNG is a Dominion in the British Commonwealth, schools like Oxford and Cambridge, and their scholars play a major role. But arguably, Rhodes's scholars are not leading the world, nor in a good way. As a matter of fact, they struggle now with themselves in retrospect as they realize they lived a colonial aspect and lived off others, off entire nations and continents (The Guardian, 2016, 2021).

Fig 30.13 Wilderness put into concrete; when can it break free again ?

Textbox 1: Something went Entirely Wrong in the Western World: Education Bias and Debt, Industrialization and Overcommercialization, Planned Obsolescence, Ocean Management, Misuse of Power, Cronyism and Corruption, Façade Democracy and Rule of Law by Supreme Court Judges

Remember globalization and its western dominance, as exemplified by the international corporations, the political grip on the World Trade Organization (WTO) and The World Bank (TWB, 'The Bank')? Well, while currently declining, it's hard to argue that the western world is not in charge of, and affects, virtually all relevant aspects of human life, its ecological processes, and the atmosphere. Clearly, it's still the dominating power in the universe, highly destructive but self-approved with laws, legislations, courts driven by supreme court judges, agencies and deciders and participants to keep itself in place, media included (see Cockburn, 2013 for an example of the richest nation in the world). Education, Science and Policy play a major role in this, while the actual students are in debt (those pay the real bills of such a system, it's intergenerational). PNG is just a smaller collateral in that poor game, so are most poor nations and their people.

The crux now is that those in charge are—relatively speaking—just a few—a tightly linked clique–, whereas the majority are the poor, the ones in tropical nations and other disadvantaged areas. The majority of people earn less than 4$ a day and have little say in the global form of democracy

While it's likely a surprise why the majority—not their union—is not speaking up, and grabbing democracy and governance for betterment, it's obviously not happening and has not happened for long time. Instead, in charge are the banks and their organizations, e.g. the international group of central banks (e.g. Banks for International Settlement; public record shown here:https://en.wikipedia.org/wiki/Bank_for_International_Settlements#:~:text=The%20Bank%20for%20International%20Settlements,a%20bank%20for%20central%20banks%22).

All of this happens in parallel while people are losing it, the environment is decaying, wilderness is lost and the atmosphere gets increasingly polluted. Now, whose fault is that?

According to Cousteau and Richards (1999), during colonial times app 5000 Europeans and 2000 Asians were in PNG; but that changed with WW2 and when 300,000 Japanese settled on Rabaul, with app. an equal amount of Allied Forces namely from the U.S. and Australians. PNG got overrun from the outside, as a battlefield of global relevance. PNG did not start the war. While the war ended eventually and numbers of foreigners reduced, the general influx and mixing actually increased and many problems remained in PNG.

For instance, there actually is an existing international collaboration between Papua New Guinea and many African countries. Papua New Guinea is part of the African, Caribbean and Pacific (ACP) forum (http://www.acp.int/), and so there is a thriving community of Africans who live and work in the country of PNG, often in remote bush hubs.

A typical problem of the new PNG and with associated globalization—all caused and driven by The West—is easily seen in the Sepik river ecosystem. The Sepik is infested with aquatic weed *Salvina molesta* from Amazon, now found there all over (Cousteau & Richards, 1999, p. 136). The ancient deep Sepik received destructive global forestry schemes, crocodile hunts and croque farming, as well as fisheries, tourism and major mining efforts; what is left of the wild Sepik ?

The world is changing rapidly; the economic power now widely sits in the Pacific, not in the Atlantic anymore. For the wider Pacific Rim, Gina Raimondo, U.S Commerce Secretary, announced the US administration's plan to establish an "Indo-Pacific economic framework" (Menkes, 2021). That might well be similar to a NAFTA free trade zone but can involve the U.S., Australia, China, Russia, Malaysia, Indonesia, New Zealand as well as Singapore, Hongkong, Taiwan and …PNG. A global powerhouse for resources and markets but how governed and sustainable!

Rozzi et al. (2015) show us a fully reasonable and widely supported view how ecology and mankind should operate. This message comes straight out of one of the largest Ecological Societies in the world. But thus far, the western ideas have lost their teeth against PNG and how it operates. The western world is great in making and maintaining stereotypes, political narratives and shallow gospels, now boosted with technology (see Rothfels, 2021). The modern world—as started by the western society and then transferred into Asia (funny enough dominated by China with a Marxist paradigm from Germany, England and Russia (= Marx, Engels, Lenin))—runs now into a massive nuclear and environmental crisis (Helfand, 2017). There is no real need to look at Amazonia and its management to understand that globalization and the tropics and wilderness areas are no good friends (The Guardian, 2022b). New views emerge and re-emerge worldwide that blend globalization with sustainability (e.g. Schmithausen, 2000) focusing more on spiritual, tribal and indigenous aspects, e.g. Suzuki (1993), Majumder (2021a, 2021b). But with globalization being so much in trouble, PNG still stands, and it stands tall; despite all the western notions, universities, institutions and NGOs. Like shown elsewhere on the global scene (see Mennonite farming done and applicable virtually worldwide; Loewen, 2021), most of PNG will likely stay robust for centuries to come. Much evidence shows no other, like Anthropologists finding for instance that the culture on Trobriands remarkable unchanged in 1971, despite even the WW2 interruptions (Cousteau & Richards, 1999, p. 118). PNG there went back to what it was used to and its Kula rituals. Indigenous views run much of the world (see Yunkaporta, 2019).

For those readers who find that adventurous, already the ancient food species now become world saviors (e.g. https://www.frontiersin.org/articles/10.3389/fsufs.2020.596237/full) because those have a well-proven track record and resistance against any stressors during the last 47,000 years, and beyond. PNG is one of the world's species cradles of farming and food production, and such a culture! Those species require a maintenance which comes with deep and ancient human knowledge and expertise; much is left in PNG still!

Textbox 2: Female labor and expertise…keeps the PNG society and economy alive but is virtually not accounted or compensated for !

It's by now a well-known item that female labor and expertise – besides many other contributors – is not well accounted for in western metrics like GDP or for taxation rates. In an economy that has no citizen registration, lives widely in rural areas and the bush, that prints its currency abroad (Australia) and which exports many of its goods and resources, a major mismatch must easily occur in the national value appreciation accounting scheme and business model. Already the age limits for the labor force, as set in western nation by law and education, hardly apply in PNG, a pension scheme is virtually absent, and the Wantok scheme widely rules and a nomadic lifestyle occurs. That means, females still get married in arranged fashions, and males can have many wives, and many children. This makes the economic accounting even worse, doubles or triples the error, because many females can contribute to a household, and it's not an 18 or 21 years age limit. One must add here the aspects of currency, which is not the Kina as the single metric, but even more the betel nut, potatoes and fish, pigs for bride prizes and other bartering styles. Social disputes might not get settled 'in cash' but by power and brutality adding a further skew to any accounting and money flow, as practiced and assumed in the west. While western economy will easily meet its nemesis in PNG, female labor is not compensated and rewarded well. Still, while PNG is often described by its male brutality, it's the female part that holds PNG together but which is not well acknowledged, or promoted even. Try to find that in a tourist guide or correctly quantified by an economist...

References

Baraka, R. (2001). *The Meanings of Independence in Post-Colonial Papua New Guinea*. A Plan B Research Paper Submitted to the Faculty of the Center For Pacific Islands Studies in the Candidacy for the Degree of Masters of Arts. School Of Hawaii, Asia and Pacific Studies. The University Of Hawai'i At Manoa.

Brockett, C. D. (2019). *Land, power, and poverty: Agrarian transformation and political conflict in Central America*. Routledge.

Beehler, B., & Lamann, T. (2020). *New Guinea: Nature and Culture of Earth's Grandest Island*. Princeton University Press.

Cockburn, A. (2013). A colossal wreck: A road trip through political scandal, corruption and American culture. Verso Books.

Crocker J. (2021). The U.S. should pursue preferential trade agreements (PTAs). *Academia Letters*, Article 2258. https://doi.org/10.20935/AL2258

Cousteau, J.M. & Richards, M.(1999). *Cousteau's Papua New Guinea Journey*. Henry Abraham's Books.

Daly, H. & Farley, J. (2010). *Ecological Economics: Principles and Applications*. (2nd ed.). Island Press.

Diamond, J. (2011a). *Guns, Germs and Steel: The Fate of Human Societies*. Random House.

Eriksen, T. H., Laidlaw, J., Mair, J., Martin, K., & Venkatesan, S. (2015). The concept of neoliberalism has become an obstacle to the anthropological understanding of the twenty first century. *Journal of the Royal Anthropological Institute, 21*(4), 911–923.

Gaffney, M., & Harrison, F. (1994). *The corruption of economics (Georgist Paradigm Series)*. Shepheard-Walwyn.

Gosarevski, S., Hughes, H., & Windybank, S. (2019). Is Papua New Guinea viable? Political Economic Bulletin 19. https://www.semanticscholar.org/paper/Is-Papua-New-Guinea-viable-Gosarevski-Hughes/6d71c70951c12fa82dea578a36b66923997f3f92. Accessed June 1, 2022

Gray, G. G. (1996). 'The next focus of power to fall under the spell of this little gang': Anthropology and Australia's postwar policy in Papua New Guinea. *War & Society, 14*(2), 101–117.

Hardin, G. (1968). *The Tragedy of the Commons Science, 162*(3859), 1243–1248.

Heinsohn, G. (2021). The disastrous tenth century: Cataclysm and collapse. https://doi.org/10.20935/AL2958

Helfand, I. (2017). Climate disruption and global famine: Nuclear weapons impact on the environment. *Civil Society and Disarmament* (pp. 16–25). https://doi.org/10.18356/7769c0b4-en

Huettmann, F. (2015). If leaders like Alexander Von Humboldt, Ronald Reagan, and Xi Jinping Would Have been environmental justice radicals using public internet and drones: A short political

ecology of Central America with regard to global agenda setting in 2014. In: F. Huettmann (Ed.), *Central American biodiversity: Conservation, ecology, and a sustainable future* (pp. 25–76). Springer.
Hyndman, D. (1998). Melanesian resistance to ecocide and ethnocide: Transnational mining projects and the fourth world on the Island of New Guinea. In H. J. Bodley (Ed.), *Tribal peoples and development and issues: A global overview* (pp. 281–298). Mayfield Publishing Company.
Kinch, J. (2020). Changing lives and livelihoods: Culture, capitalism and contestation over marine resources in Island Melanesia. School of Archaeology and Anthropology, Research School of Humanities and the Arts College of Arts and Social Sciences, Australian National University, Canberra.
Levi-Strauss, C. (1966). *The savage mind*. University of Chicago Press.
Loewen, R. (2021). *Mennonite farmers: A global history of place and sustainability*. John Hopkins University Press.
MacLaran, A., & Kelly, S. (2014). Neoliberalism: The rise of a bad idea. In A. MacLaran, & S. Kelly (Eds.), Neoliberal urban policy and the transformation of the city. Palgrave Macmillan. https://doi.org/10.1057/9781137377050_1
Majumder, N. (2021a). Paradigm war: A distinct indigenous vision. *Academia Letters*, Article 926. https://doi.org/10.20935/AL926
Majumder, N. (2021b). Paradigm war: A distinct indigenous vision. *Academia Letters*, Article 926. https://doi.org/10.20935/AL926
Matthiessen, P. (1987). Under the mountain wall: A chronicle of two seasons in stone age New Guinea. Penguin Classics.
Mead, M. (1967). The rights of primitive peoples: Papua-New Guinea: A crucial instance. *Foreign Affairs, 45*(2), 304–318.
Menkes, J. (2021). Free trade area without free trade agreement. *Academia Letters*. https://doi.org/10.20935/AL4149. Accessed May 25, 2021
Miller, J. D. B. (1973). Papua New Guinea in world politics. *Australian Outlook, 27*, 191–201.
Narokobi, B. (1974). The nobility of village life. *Catalyst: Social Pastorial Magazine for Melanesia, 4*(4), 55–70.
Narokobi, B. (1975). *Foundations for nationhood*. Papua New Guinea Press and Bookshop.
Narokobi, B. (1983). *The melanesian way*. Institute of Papua New Guinea Studies.
Nelson, H. (2016). Black, white and gold: Goldmining in Papua New Guinea 1878–1930. ANU Press, Google Books. https://www.google.com/books/edition/Black_White_and_Gold/8M0PDQAAQBAJ?hl=en&gbpv=1&printsec=frontcover
Nelson, H. (1982). *Taim Bilong Masta, The Australian involvement in Papua New Guinea Australian Broadcasting Commission*, Sydney . ISBN 9780642975669, 0642975663. Google Books https://www.google.com/books/edition/Taim_Bilong_Masta/y2MhAAAAMAAJ?hl=en
Ostrom, E. (1990). *Governing the commons: The evolution of institutions for collective action*. Cambridge University Press.
Perez-Haemmerle, K.-V., Moon, K., Venegas-Li, R., Maxwell, S., Simmonds, J. S., Venter, O., Garnett, S. T., Possingham, H. P., & Watson, J. E. M. (2022). Wilderness forms and their implications for global environmental policy and conservation. *Conservation Biology*, e13875. https://doi.org/10.1111/cobi.13875
Reilly, B. (2004). State functioning and state failure in the South Pacific. *Australian Journal of International Affairs, 58*, 479–493.
Rothfels, N. (2021). *Elephant trails: A history of animals and culture*. John Hopkins University.
Rozzi, R., Chapin III, F. S., Callicott, J. B., Pickett, S. T., Power, M. E., Armesto, J. J., & May Jr, R. H. (Eds.). (2015). *Earth stewardship: Linking ecology and ethics in theory and practice* (Vol. 2). Springer.
Schmithausen, L. (2000). Buddhism and the ethics of nature—Some remarks. *The Eastern Buddhist, 32*, 26–78.
Shtil'mark, F. R. (2003). *History of the Russian zapovedniks, 1895–1995*. Russian Nature Press.

Springer, S. (2016). *The discourse of neoliberalism: An anatomy of a powerful idea.* Rowman & Littlefield.
Suzuki, D. (1993). Wisdom of the elders: Sacred native stories of nature. BoD–Books on Demand.
The Guardian. (2016). Rhodes scholars: can you justify taking the money of the 'godfather of apartheid'? 1st February 2016. https://www.theguardian.com/commentisfree/2016/feb/02/rhodes-scholars-can-you-justify-taking-the-money-of-the-godfather-of-apartheid. Accessed May 7, 2022.
The Guardian. (2019a). 'We live in a narco-state': Murder of Dutch lawyer prompts fear and fury. 19th September 2019a. https://www.theguardian.com/world/2019a/sep/19/netherlands-narco-state-dutch-lawyer-murder-fear-fury. Accessed May 7, 2022.
The Guardian. (2019b). 'I will eventually get killed': Meet Bryan Kramer, Papua New Guinea's anti-corruption tsar. 10th August 2019b. https://www.theguardian.com/world/2019b/aug/11/i-will-eventually-get-killed-meet-bryan-kramer-papua-new-guineas-anti-corruption-tsar. Accessed January 15, 2022.
The Guardian. (2020). Papua New Guinea police accused of gun running and drug smuggling by own minister. 18th Sept 2020. https://www.theguardian.com/world/2020/sep/18/papua-new-guinea-police-accused-of-gun-running-and-drug-smuggling-by-own-minister. Accessed November 28, 2021.
The Guardian. (2021). 'Colonialism had never really ended': My life in the shadow of Cecil Rhodes. 14th January2021 https://www.theguardian.com/news/2021/jan/14/rhodes-must-fall-oxford-colonialism-zimbabwe-simukai-chigudu. Accessed January 4, 2022.
The Guardian. (2022a). PNG prime minister denies connection to suitcase full of cash found as voting starts. 6th July 2022a. https://www.theguardian.com/world/2022a/jul/06/png-prime-minister-denies-connection-to-suitcase-full-of-cash-found-as-voting-starts. Accessed July 6, 2022.
The Guardian. (2022b). Climate crisis: Amazon rainforest tipping point is looming, data shows 7th March 2022b. https://www.theguardian.com/environment/2022b/mar/07/climate-crisis-amazon-rainforest-tipping-point. Accessed March 9, 2022.
West, P. (2006). *Conservation is our government now.* Duke University Press.
Wilson, M. (2019). *A True Child of Papua New Guinea: Memoir of a Life in Two Worlds.* McFarland.
Yunkaporta, T. (2019). *Sand talk: How indigenous thinking can save the world.* Text Publishing.

Chapter 31
Bringing It All Home: Where on Earth is Papua New Guinea, Its Society, Their Environment and Well-Being Heading? Summarized Examples of Western Failure and 'Forwards to the Roots,' in Your Own Garden Without Social Engineering!

During a rainfall and heavy fog in the mountains, on a muddy forest road somewhere in remote PNG, I see walking umbrellas towards me coming out of nowhere. It was utterly scurrile; stuff for a movie. Who is under these fast-moving colorful umbrellas, how many people will be there, and will they mug you ? But instead they either just passed by, or stated 'monning' in the nicest and most friendly way imaginable; it was a travel group consisting of a male, females and kids; a wilderness frontier experience 'in the bush'.
Source unknown.
The Convergence of Worlds: Or simply Convergence, for short. What exactly IS Convergence? A convergence occurs when objects or beings from different regions of the multiverse instantaneously cross vast distances and fuse together momentarily – usually only for a zeptosecond before the merged matter reverts back to its normal state. But this convergence is unique– as these planets have remained meshed together. A new reality was cobbled together in the blink of an eye, as mysterious forces brought pieces of four disparate worlds together. The citizens of Eemia, Ossuary, Numina, and the Immensity had never before been aware of each other – and in an instant they were pulled apart from their homeworlds and thrust together to carve out a new future. The cosmic immigrant story of the Convergence was born.*
Quote from Q.Dot art exhibition Denver Colorado 'The Convergence of Worlds' https://www.convergencestation.com/about-qdot

Abstract Papua New Guinea (PNG)'s culture is over 47,000 years old and resilient. The modern nation construct of that PNG is just less than 60 years old featuring a 'blink' in PNG's overall history but it was quite a devastating experience for its people, biodiversity and wilderness. PNG is relatively free of leftist or Marxist ideologies, but it is a control site showing that the 'modern' form of development has not resulted into a wider, higher and more equal distribution of wealth while the closest

F. Huettmann, *Globalization and Papua New Guinea: Ancient Wilderness, Paradise, Introduced Terror and Hell*, https://doi.org/10.1007/978-3-031-20262-9_31

Fig. 31.1 Forward is the way ahead

neighbor (Australia and to a lesser extent New Zealand) as well as colonial and global powers still are cashing in from PNG, e.g. mining, oil and gas, fisheries, timber, human resources. The science is only partly done for PNG but a quite ruthless extraction, collection and policy industry was allowed to operate in PNG to this very day, benefitting from views that are outdated, incomplete and self-serving for those nations. Based on a topical summary, here those details are acknowledged, and then, a synthesis outlook is presented to move forward for PNG in reality terms while sorcery, bush life and tribalism ('Wantok') are ongoing and might contribute to resilience and sustainability in the absence of any other and better options at hand.

Keywords Papua New Guinea (PNG) · Marxism · Globalization · Sustainable development · Global governance · Generational justice · Environmental justice

History will not be so kind to the ruthless colonialism, industrialization and globalization times; it's only a very short period in the earth's history but one that created much havoc and death in just 300 years or less; and the last 60 years were specifically impactful (Flannery, 2002) (Figs. 31.1, 31.2, 31.3, 31.4, 31.5, 31.6, 31.7, 31.8, 31.9 and 31.10).

This book and its chapters provide a powerful set of details and messages, for PNG and beyond (Table 31.1 and 31.2).

Fig. 31.2 However, it looks like the way should be robust as an ancient trail used for millennia

There are many non-believers and deniers while facts are clear and obvious. Once the deniers are established—as powerholders in an institution, agency, funding apartus, media or otherwise (as widely found on the PNG topics)—such people and their followers don't like to move and adjust their views anymore, or their associated income streams and associated routines. A bad culture is set and rolls out. That's specifically so if other humans and elsewhere in remote and exotic places can cater their lifestyles and pay the bills for them; out of mind out of sight with window dressing. As shown in many of the topical assessments of this book, PNG plays a role for many of such people, a clique and their outcomes. That clique, or gang (Gray, 1996) then stays in the driver seat and tries to steer the entire enterprise and feeds the media accordingly. It's the history of PNG and needs an improvement.

While those people do not perceive themselves that way ('nice' and 'friendly', 'what's the problem?', 'we work so hard'), many biologists and conservation people can be part of such a 'game' (*sensu* Chan, 2016). This group can easily include

Fig. 31.3 Celebrate your heritage, Papua New Guinea style as done here in a village center in the remote bush (which happens to be often centered around a rectangular airstrip with cut lawn)

biodiversity lovers, specimen collectors, many entrenched contractors, amateurs, and many other uncritical foreign travelers who go to PNG 'for fun.' It's a concept out of the wealthy colonial world and mind, it's ongoing and might get worse: 'watching the third world' tourism.

The world—as it is currently designed and runs—is in a poor lock-in. It's in a crisis state—socially, environmentally and economically. Like a bad machinery, it's a cloned template that one can easily watch in action—in terms of behavior, economy, social fabric and the environment. And it's difficult to break out from for nations, people and societies. As shown in this book, the 'cloning' and entrenchment of the scheme is done through the western religion, through development aid, loans, via schooling, by endless but mindless apparatchniks, by the tax system, with employment schemes, and with the media; supported by the so-called free press and social media online with their 'influencers' including fashion models. Anything that brings you into a depending position will then be exploited there. No mercy. With a UN

Fig. 31.4 Bartering, a way to stay afloat for most people in PNG and the world

mandate, it's a certain predatory system operating on a global level cemented by the security council of just a few nations but their permanent seats in th U.N. at large. PNG is predetermined and predefined within that path (see Premdas, 1976 for PNG's suggested role to accept; Baraka, 2001 for a discussion). The role of pandemics made that also very clear; the recent COVID outbreak is not really a PNG-originated disease but it affects PNG in many ways, e.g. lockdowns, travel restrictions, cargo restrictions and missing products and goods. Just like many other nations, PNG citizens are a recipient of action, but not much an actor. No wonder than that PNG remains often so passive. Australia is not an equal neighbor nation in that regard, an unequal partner instead with many faulty concepts. PNG is embedded in, and serving, globalization, but hardly can do more. It's still a one-way street but should be more mutual and coming from bottom-up, shared governance and consent. That's what tribal management can achieve, in a way, now also using Facebook and similar (see Watson, 2011 for mobile phones as the new 'drum').

PNG has relative little human population and much intact rainforest and oceans; use that as a good starting point on the world scene. Like in many places of the world, national parks are not compatible with PNG's land ownership. The existing national park in New Guinea is reported as poorly managed, designed around mines, and people who live in them are hardly aware it's a protected area or what that is to mean (Beehler & Laman, 2020). The conservation prioritization workshops done by academia and NGO leading to find and protect biodiversity hotspots on land or

Fig. 31.5 Papua New Guinea is alive, modern and well in its own terms, as shown by this mural

sea often suffer from similar problems. As PNG is part of a wider landmass and ecosystem, New Guinea has two governments to deal with it, none work well. It's hardly known though.

The J. Nelson Institute states it as a crucial truism (https://jninstitute.org/project/guardian-australia-pacific-editor/):

> *The Pacific receives minimal coverage at a time when the region is increasingly important for geopolitical, military and security reasons. Critical strategic, social, environmental, crime and business stories are not being told.*

So what are our options, really?

Alternatives to the current *status quo* exist, one of them should be to practice more of Ecological Economics ('Steady State Economics') approaches for PNG (see for Australian's missed opportunities Greasley et al., 2017). PNG will not be a world leader any time soon, nor will PNG and its people be given a fair chance in the global

Fig. 31.6 Water flows freely around rocks in Papua New Guinea

community, be it in the UN, in UNESCO, with The World Bank, in Climate Change agreements or elsewhere. Other nations will not consider the fate of PNG, much. If lucky, PNG can make its way through with as little further bruises as possible by providing resources to Australia and China and other dominant nations. That's what Michael Somare set up PNG with (Somare, 1975), and it hardly was changed.

Arguably, the big institutions, nations and organizations are essentially using PNG and failing its people. This book has provided some assessment and metrics on that matter, and one hopes that the *status quo* gets accepted as a problem and now improved, toward a better world order (Stiglitz, 2003) benefitting wilderness conservation and mankind more holistically. What's wrong with it ?

But like expressed by Cousteau & Richards (1999, p. 221) PNG can serve as an inspiration, wisdom and guidance for the people, for humanity and for the world. PNG remains a gift to the world! While this message was totally forgotten by Laman and Scholes (2012) and Beehler and Laman (2020), any other than war, the solution can be as simple as gardening ("grass roots"), local food production and such a (family/social) life. One can easily extend from that Melanesian way (Narokobi, 1983). It happens to match what the elders of the world communicate (https://www.wisdomweavers.world/).

Let's agree that many forecasters on a global scale did not get it right (see example in Friedman, 2010). The economic forecasters get it wrong most of the time, while politicians mysteriously still get their ear and act on them…just to learn how wrong it all was again. Ludlam (2021) showed us a full circle to go by. Along those lines, one may suggest once more the Global Wisdom of the Elders (Suzuki, 1993).

Fig. 31.7 Papua New Guinean village beauty

If the forecasters remain consistently wrong on one thing, then it is that they overrate modernity, including education and reason, in what it is and in what it offers and where it goes. Arguably, mining in PNG left virtually no good progress to the society in PNG, in Melanesia or anywhere else. Science got ignored. For most parts, the ecology aspects, telecoupling as well as Mother Earth views proved equally or more relevant, but are still widely pushed aside…just to be surprised and reminded again in shock that those are in the driver seat.

Tribal views, and often affiliated with Mother Earth, are equally ignored and not taken serious yet, but should be. They usually are still not part of any model view and equation (while far from perfect, New Zealand has taken a lead on that issue, but that's not on the agenda for Australia or for PNG at all)

Providing a reality view where PNG is heading is not much possible yet, but several scenarios can be outlined for the next 100 years:

1. PNG would break down in full
2. PNG would become fully controlled by Australia
3. PNG would become a China-dominated resource provider
4. PNG would become a U.S. and Australian military-base
5. PNG would become a widely recognized indigenous leader

Fig. 31.8 Reflecting in the wild about the way ahead for Papua New Guinea and the world

Textbox 1: Total Denial Still Dominates: Insights from the public across many years

Despite its name, modern Papua New Guinea (PNG) is not like an independent western nation (Baraka, 2001). Many PNG citizens can vote, but it often does not have a real impact or progress for them. When talking with the public over the years—in PNG and outside—it's pretty clear to me. It's the one single large message I can confirm. Metrics show that clearly also. In the meantime, such details get widely ignored and denied—even by experts working on the ground—while the 'business as usual' goes on, and that is, the (neo) colonial business. There are many actors in it, and they come from many angles and disciplines, but eventually, whatever those actors do and engage in, they widely deny the suffering and reality, and not much changes and not for the betterment. It's those facts that are undeniable!

Textbox 2: Failure of the West: Policies and Legal Concepts

Over the last 60 years, after the WW2, the western world has essentially set up, designed and fine-tuned most international policies and created a certain

Fig. 31.9 Clean natural water in a wild landscape remains the essence of life, anywhere

world order, as cemented by its institutions, laws, courts, policies and financial incentives. This is widely seen in the U.N. and its Security Council with permanent seats for just a few nations. It's put in concrete that way for the world. In such a structure, nobody really can break out or add new concepts and insights. The Cold War was lost by Russia in the 1980s onwards, and until recently, the West got the collateral. China now pushes more into that global sphere, so does India. Whereas most tropical nations are small, and those had to see and watch, and being run over by those events. They had to deal with the subsequent policies, e.g. Washington Consensus (https://en.wikipedia.org/wiki/Washington_Consensus). It looked like the great western success and take-over eventually. But the wider social and environmental aspects of that global regime kept popping up, repeatedly and increasingly. Laws and policies got more stringent but failed with maintaining a sustainable governance while resources got used up. And while subsequent conflicts increased, so did the refugees and many resulting problems, e.g. having a stable but consuming middle class. Beyond wars and inequities, many refugees are due to climate change. While the strongholds of The West—wealth, economy and health—started to decay on a finite space, all other problem areas popped up to the

Fig. 31.10 A hand-woven bilum made of tree fiber; when compared to plastics, its organic and flexible sophistication can be the vessel that keeps PNG sustainable and together, as a nation and a global entity

surface even more. The western enterprise became more complicated than anticipated.

There are many aspects now that show up and present the case of global governance—its laws and policies—as a serious problem on most relevant aspects: economy, social and environmental.

Culture and human actions 'in the remote bush' cannot really be governed through words or policies. It is here where the (urbanized) western world has hardly acknowledged the problems nor a good answer to solve them and presents no vision. They are not in the bush or in the wild. The western courts and practitioners of law are usually not seeing it as a problem. It's primarily the next generations that will keep paying those costs though.

Table 31.1 Summary and conclusion from this book

Conclusion	Chapters	Comments and thoughts
Papua New Guinea offers deep sustainability insights to inspire and to learn from	Chaps. 1,3, 4, 6, 7, 10, 15, 16, 17, 20, 22, 26, 27, 28	Mother Earth, holistic approaches and Cosmologies prevail
PNG as a biodiversity cradle	Chaps. 3, 6, 10, 14, 15	A natural resource
Modern PNG is widely controlled from the outside	Chaps. 9, 11, 12, 13, 15, 17, 18, 22, 24, 25, 26	Neocolonialism dominates
PNG problems	Chaps. 2, 11, 12, 13, 25, 26	Over 100 years of cultural science and explorations have helped little
PNG operates on a finite space	Chaps. 1, 6, 12, 13	A classic example of Ecological Economics, Steady State Economy
Man-made climate change as a massive impact	Chaps. 8, 10, 17	Major changes for PNG and the world fabirc
A good PNG way to move forward	Chaps. 1, 27, 30	Melanesian way, open access baseline data, and beyond

Textbox 3: Failure of the West: Migratory Species as a Deal Breaker for Governance Theory

Migratory species of birds are managed internationally. It's to assure avoiding their decline and extinction. Well, most shorebirds are on the decline and populations have been lost. The fate of boreal passerines is hardly any different. So what went wrong? Migratory birds are managed through a flyway policy. North America is rather strong on those efforts due to its Migratory BIrd Act (MBA). The EU follows, whereas nations like China or Russia, or the African continent are much weaker on that aspect, if there really is any law and means at all to address the issue. Still, most nations, including the richer ones, have virtually uncontrolled poaching cases. Based on the annual reviews and feedback received on the matter, official and unofficial (see Huettmann, 2015; Yong et al., 2021; https://www2.cifor.org/bushmeat/), there is no good outlook or control whatsoever. That's equal to what many nations in the tropics have done but they had not caused such extinctions or such declines. A good example is the

Spoon-billed Sandpiper, the Far Eastern Curlew or Red Knot. Those species were traditionally pursued, perhaps, but it had not that global-scale decline we see now, e.g. as caused by industrialization and such nations. Culprits are clear and our Governance does not stop it, rather vice versa!

Table 31.2 Example list of major or relevant professional societies with large budgets and governmental and industrial support in the region, and their (missing) link with a brief justification in regards to PNG and its people and realities

Professional society	Mission	Membership	Ecological link with PNG	Issue unresolved for PNG
Pacific Seabird Group (PSG)	The Pacific Seabird Group (PSG) is a society of professional seabird researchers and managers dedicated to the study and conservation of seabirds and their environment	U.S. Canada, Mexico and some international	(Migratory) Seabirds, management, oceans, focus on entire Pacific and islands	Seabird and habitat conservation, no statement on bad economic growth impacts, no open-access digital data promotion, attempts for a good sustainable and holistic governance
American Ornithological Society (AOS; formerly American Ornithological Union AOU)	The mission of the American Ornithological Society is to advance the scientific understanding of birds, to enrich ornithology as a profession, and to promote a rigorous scientific basis for the conservation of birds	U.S., Canada, Mexico and worldwide	(Migratory) birds, e.g. biogeography, specimen, egg collection, authoritative taxonomy, bird catching	Bird and habitat conservation, bird and egg collections (e.g. for Birds of Paradise BoP), no meaningful stable taxonomies, no open-access digital data promotion, economic growth impacts not addressed

(continued)

Table 31.2 (continued)

Professional society	Mission	Membership	Ecological link with PNG	Issue unresolved for PNG
American Geophysical Union (AGU)	To support and inspire a global community of individuals and organizations interested in advancing discovery in Earth and space sciences and its benefit for humanity and the environment	U.S., Canada, and worldwide	Sea level rise, geology, atmosphere, data sharing	CO_2, climate change, ocean acidification, sea level rise
British Ornithologist's Club	To advance education in the subject of ornithology for the benefit of the public by promoting scientific discussion between members and others interested in ornithology and to facilitate the dissemination of scientific information concerned with ornithology, and in particular avian systematics, taxonomy and distribution	Primarily UK, some EU	Birds	Birds, habitats and impacts of mining, oil and gas, private data policies, no statement on bad economic growth impacts, no open-access digital data and metadata promotion
Society of Conservation Biology (SCB)	A global community of conservation professionals. The society's mission is to advance the science and practice of conserving Earth's biological diversity	Worldwide	Conservation	Mining, oil and gas, Palm Oil Plantations, poverty, economic growth ignorance, no open-access digital data and metadata promotion

(continued)

Table 31.2 (continued)

Professional society	Mission	Membership	Ecological link with PNG	Issue unresolved for PNG
Association for Tropical Biology and Conservation (ATBC)	International scientific and professional organization whose mission is to promote research, education, and communication about the world's tropical ecosystems	Worldwide	Tropical Ecology, Forests, Conservation	An effective sustainable culture and wilderness maintenance, no forest inventory promotion, no open-access digital data and metadata promotion; modern governance, indigenous and poverty questions ignored
Orchid Society of Papua New Guinea	NA	PNG and Australia	Source for many endemic orchid species	Essential questions of species trade, any relevant forest and habitat loss efforts, mining impacts, conservation management plan, any open-access digital data sharing or transparency efforts
Australasian Wader Study Group (AWSG)	A special interest group of BirdLife Australia to coordinate and focus studies on waders, or shorebirds, in Australia and throughout their migration routes in the Asia Pacific	Australia, and some Asia and North America	PNG is an inherent part of the shorebird flyway and conservation concept	Shorebird distributions and abundances with confidence estimates, end poaching, wetland and river conservation performance, little open-access digital data-sharing and metadata promotion, effective policy link

Textbox 4: Failure of the West: Example of Education as a requirement 'to play'

When the PISA report (https://www.oecd.org/pisa/) came out on the national student assessments of the education system, many nations were in deep shock. Germany was not really among the top 10, the U.S. was much worse than Costa Rica, and industrial nations performed poorly overall. Still, some peculiar western nations like Finland and Switzerland performed very good and consistently according to the Pisa study; so did Scandinavia overall. 'Hooray' for those, one might say. But in reality, the small nations educate just a relatively small amount of people and they influence marginal groups; whereas the future leaders that come out of such a 'great' education system are hardly globally relevant. They play no role and it does not affect the many billion people educated less. The vast majority of people and nations are simply left in the dark, essentially somewhat uneducated thus uninformed about the western world and how decisions are made; they live in the tropics and make less than $4. It's a global dilemma of epic/biblical proportions. One way out is public and community education and such colleges. Elite Ivy League schools cannot achieve that or leave a good legacy. Open enrollment and free tuition should be the rule for motivated individuals of the society. But such views are still far from reality still.

In the meantime, applying and repeating the traditional schooling and education policies in PNG with a one-sided industrial focus benefiting Australia will fail. It has hardly worked anywhere else, so why now in PNG in a global context of a world crisis?

Textbox 5: Failure of the West: Industrialization rotten at the core

The mass production process—industrialization—has substantial repercussions and impacts, affecting resources, markets, society and well-being at large, the world and future generations. It needs a supply chain, and where do those root products still come from while the actual processing also needs energy and resources then leaving behind much waste? By now, the industrial impacts are indeed 'universe-wide,' e.g. see the space waste, and how industrial society affects plate tectonics and earth spin, including space travel and satellite placement. But also on earth the industrialization left major marks socially, economically and environmentally. Who for instance wants to be a black smith anymore, or who works as a leather tainter, or a hatter? Already with the advent of the loom, and its industrialization process society went into *havoc*, as well as sheep keeping and grazing and the commons overall. The loom and wool production affected colonies in India and the British Commonwealth worldwide (nowadays, India has a larger GDP and economy than the UK; Australia and New Zealand are essentially independent).

The industrialization process produces and accumulates unnaturally high concentrations and high densities of goods, e.g. concentrated meat and milk, eggs and cheese...or chemicals and plastics. In nature, none of those exist in that density, e.g. poultry farms or feed lots for cattle, or grazing impacts. In nature, cows do not produce milk daily, nor do chicken lay eggs every day. Plastics are virtually absent. The growth rates of calves to produce veal are as artificial as big chicken are (McKenna, 2017; see also Singer, 1973). It will result into invasive species, diseases and pandemics, e.g. zoonotic disease spread and spill-overs such as rabies, malaria, avian influenza or corona, besides others (e.g. diseases from air pollutants and contaminations). The mitigation and avoidance costs for such diseases are easily in the billions, affecting the profit and life quality of mankind and the economy at large.

And then there are labor costs; arguably dealt with by unions (several of those exist, not all achieve well or are efficient for their missions, as the globe's poverty shows us well). The costs of such labor disputes employ an entire army of people, including judges, attorneys and employees.

Further, there are specific health costs that come with the industrial society, labor and its holiday scheme. Those can be accidents at the workplace, of which there are many. But then, there are also long-term damages due to wider contamination etc., e.g. asbestos in the building (EU Brussels had such a case in its administrative offices; FH pers. com) or handling them, e.g. agent orange. But there are also more long-term psycho-somatic problems, from neck and spine problems to psychological diseases, namely depression and subsequent addiction. And no need to mention ergonomics for their impacts! While alcohol plays a big role in much of this, other drugs come to play also people seek for mitigation, such as cocaine (found in parliaments and with leaders, in clinics, with the public and with celebrities; annual police and news reports show that clearly).

Then, there are social costs, e.g. having to pay for a day care even for infants because the parents cannot afford to go off work. The costs of negligence are extremely high affecting entire generations. There are also hidden costs of industrialization, for instance the effects of a required education to obtain a qualifying job in industry. Nowadays, one goes to school till aged 25, often even longer. In the chemical industry, having a PhD is a common entry requirement, and only having a business degree in parallel will land you the job there with a lab or corporation. Ideally, having a legal degree in addition provides for a job; none of those things come with a young age.

Of course, the industrial society needs a road system, a policing apparatus, usually military support (as seen with the oil, gas and mining apparatus) a tax system and its administration.

The climate change impacts have been discussed for a long time and go back to early research Ladurie (1976) and others, now widely shown by David Suzuki's daughter, Greta Thunberg, etc. The science record on that goes through the roof by now.

In sum, costs of industrialization are incredibly high and pervasive; they now affect virtually all aspects of the universe, including chemical cycles, biological processes, the ocean and deep sea, and the atmosphere reaching into the universe. All beings are affected; most cost-and-benefit analysis are flawed as they exclude those 'true' costs. It's the Anthropocene, and with over 7 billion people on the rise, where will it ever end? PNG is not really part of that and went a different way.

Textbox 6: Failure of the West: The prescribed Economic Model as a governance scheme *ad absurdum*

Remember the textbook example of an Economy, the circular model of consumers and producers with an oversight agency? Makes all sense:consumers buy the product for a cheap prize, and producers provided it for financial gain, done with some good governance oversight. Well, reality is, not all citizens—the majority—do participate in this model, they hardly can, others do not even want to join. If one takes all those numbers together, globally, the majority of people is not in that model at all. One may agree that a registered buyer, let's say via a tax number or with Amazon.com, is needed for such an economy. But how many people pay tax, or are registered? It's not the vast majority of people in the world, often not even in a nation; squatters are a good example, the unempowered such as the unemployed, pensioners and lower casts are others. And one may easily go further and ask about gender participation in that economy? Arguably, it's male-oriented, with female labor either fully excluded, or paid less. The contributions of children and pensioners are not included neither; another vast section of the society. The vast amount of mankind is not participating in that model scheme of the economy!

It has been discussed for a long time, but got widely ignored on purpose, that the market should actually never be in charge anyways. Even the market promoters agree for a wider market framework required. A so-called open market will simply make rich people richer and impose brutal terror on the consumer and providers and the others (= majority). We see those things with 'planned obsolence' and lack of liability and impact studies, as an example. Who wants that?

Then, the entire focus on coins, money, for live interactions is widely misperceived. For the largest currency in the world, the U.S. Dollar, remember its move away from the gold standard? Any dollar bill says '*In God we Trust*'; well, not all people trust in god. As a matter of fact, the bitcoin currencies have already started to challenge and outcompete the initial coin monopoly. Then, one must understand that the U.S. Dollar, as the global currency, is controlled in the U.S. but actually used and valued, worldwide: a money market much bigger than the U.S. economy itself. How can a central bank keep a currency stable, when most of the currency is used or valued outside, and often gets manipulated there? Russia for instance uses for many of its domestic transactions the U.S. Dollar. And China has tried for a long time to use and buy U.S. Dollar so that it works in their favor (China has been the strongest 'hoarder' of U.S. Dollars in the world; at some stage, it was $600 per Chinese citizen). And then, there is money and values not traded at the legitimate (western) stock markets, Dubai, Qatar, Hongkong and Pakistan etc are among those, more exist. Even many London assets unlikely would be legal and allowed in New York's stock market due to the strong U.S. ethics laws. This, among many others of the money flows in existence (e.g Caribbean tax heavens), is another big proportion of the money value. Needless to say, the biggest farming cash crop likely is…drugs, namely cocaine, heroin and marihuana (now legal in parts of the U.S. and on the rise with farmers worldwide). While that might be true globally, it is also true for smaller markets, and islands, showing classic economic theory as widely absurd and not applicable. I have worked on such islands in the world where that is obvious. And PNG is known to be affected by it full scale. For instance, in the Caribbean, they even tried to drop the smaller coins (cents, etc.) for many islands. That's for a large island region: simply drop the pocket money for many nations and economies with one of the poorest. What an economic model would that be, and benefit from it in regards to people's wealth and business? Clearly, the use of metal coins and paper money and their maintenance is a national loss, with some coin units and paper not worth what they actually state or are traded/exchanged for.

And then, for places like PNG, the Kina matters but there are other currencies in PNG and its tribal Wantok system: Betel Nut for instance. In POM and its criminal gangs, it's often a stolen SUV (c.$15,000 worth per unit). And of course, for more official measures, in PNG the pig actually remains the currency of choice (value c. $1000 to $5000 U.S.). So what economic model are we now understanding, studying, advising and really following, and how to fix the 'PNG economy' and its currency, always running after the Australian Dollar by design in a globalized economy?

Textbox 7: Professional Societies come to the rescue, for the money or for the funeral? Missing and inefficient conservation and sustainability examples for Papua New Guinea and the Pacific (Seabirds) within dominating natural resource professions

For a good sustainability outcome, Papua New Guinea (PNG) is widely ignored by many conservation and professional societies in 'The West.' That is also true for professionally dominating ones in the region, e.g. professional societies from the EU and the U.S. with large funding and political support and with authoritative media outlets (peer-reviewed literature, twitter campaigns, Facebook and so on) to affect public opinion, funders and self-justification. However, PNG is widely used by those societies instead for their own purposes as they see fit; it's totally ignored any other. This situation equals a colonial approach of exploitation, and it is affecting conservation in a bad way—in PNG and outside. It's difficult to argue that this would be 'professional' or that it matches a well-meant intent of their mission statement and mandate of such societies, or that it is truly approved by the membership, or by that profession respectively. Being tax-exempt just adds fire to the fuel. Instead it exemplifies a fully institutionalized ongoing western culture and attitude, made worse by funding, annual conferences media promoting those professional societies and their 'jobs,' while PNG is once more at a loss.

One can easily see the sustainability ignorance of this topic in the socio-economic and environmental performance metrics of PNG of the last decades, and also in the fact that PNG is not directly part of many national and international professional societies from abroad. PNG does not appear as an actor there, has widely underrepresented members, and has no real chance even to join and 'to play' in those professions and their societies. However, due to PNG's global uniqueness, in professional societies PNG frequently serves for delivering many examples, arguments and ideologies for them; it even makes money for those societies and professions. e.g. in their commercial journals, international conferences and in contracts, while PNG itself remains one of the poorest regions in the world with a decaying sustainability and wilderness record. Rock solid.

A typical example can already be found with seabed mining, where it's widely acknowledged and known by experts of the world that mining the seafloor is harmful to the environment at large; the acclaimed and hoped-for triple-down economy—the blue economy—does simply not happen for all members of society, and the natural resource curse hits instead. Seabed mining certainly will affect the seafloor, ocean water qualities, coral reefs, some major fisheries as well as marine mammals, and coastal communities and their sophisticated fabric that co-evolved fine otherwise for millennia. However, professional societies devoted to topics of the Pacific biodiversity and sustainability have been widely quite on the matter; an elephant in the room for decades. The seabed mining impacts for seabirds of 'the Pacific' have virtually never been discussed with them, and certainly not for the seabirds of PNG (a location where much of that mining action is to unfold though).

Instead, the mystery and the unknowns—the fascination—of PNG seabirds even got presented in keynote speaker sessions at annual meetings of the Pacific Seabird Group (PSG, https://pacificseabirdgroup.org/), involving New Zealand, UK, U.S. and NGO actors (but not PNG or PNG matters). In the meantime, the PSG stayed totally clear of any statement regarding the bad impacts of seabed mining, or of economic growth for that matter (Huettmann & Czech, 2006). While the PSG is promoting, and involved, in many seabird issues in China, Japan, Russia, Peru, Hawaii and Mexico, a relevant PNG conservation plays mysterioulsy no role on the agenda. Useable seabird science data on the topic are not shown or shared well, or get promoted. It must be noted that PSG is heavily drawing on managers, governance members as well as contractors in powerful roles in that society and in setting their agenda. But PNG is not part of it, never really was

But there is no need to go as far as seabed mining. Already impacts of land mining and oil and gas exploration and extraction—some of those are the largest operations in the world, in the British Dominion, or in the Pacific—have bluntly been ignored for impacts and get sidelined by professional societies and their respected internationally peer-reviewed outlets, e.g. Woxvold et al.

(2019). Watersheds, rivers and estuaries clearly are getting polluted for decades (Cousteau & Richardson, 1999; Kirsch, 2014) negatively affecting the pristine ocean ecosystem at large! The notion of 'Mining Melanesia'—that's a large-scale mining effort in the vast region of Melanesia—is virtually uncovered in most professional societies for the region.

One may then continue with palm oil plantations in PNG, which the Society of Conservation Biology (SCB, https://conbio.org/) has virtually not effectively addressed whatsoever (SCB has some conservation efforts on palm oil plantations in SE Asia and holds professional meetings in Malaysia, etc., but the assigned budget is not high, and even there it has not resolved major conservation problems like Orang Utang deaths, decay of indigenous people livelihoods, massive pristine forest loss but the parallel extension of palm oil plantation areas. SCB does not deal effectively with economic growth nor with PNG really).

Further, most societies of professional foresters and their educational institutions are not concerned about PNG; even when the timber of those nations gets frequently imported from tropical nations to 50% or more (a typical figure in the world trade scheme these days, but widely mis-labeled and not well unerstood neither). There is perhaps the exception of the Association for Tropical Biology and Conservation (https://tropicalbiology.org/) that has indeed taken a stand on the farming land property problems in PNG—Special Agricultural and Business Leases (SABLS)—in PNG (ATBC, 2011) and even with Australian expert leads; but then, what came from it? It's essentially another paper tiger with a lip service, but it lacks the true effort for a good change, certainly for PNG. Certainly most Australian experts and other (museum) taxonomists do not stop it, if even trying. Economic Growth as a problem, and the bad western influence, narrow science and global development aid—culminating in UN's Aichi Targets and Sustainable Development Goals (SDGs)—is not even well acknowledged as a problem. So one can easily ask what's professional about that, in a society, and why tax excempt?

Birds and their habitats—as a core issue for PNG—are not addressed by most professional societies with a PNG focus for their conservation. That is easily seen with Birds of Paradise (poached, traded in vast numbers, not protected, hardly assessed, unresolved taxonomies even, and collected), but also includes other essential topics of ornithology, such as parrots, or specimen collection and egg collection. Those are core schemes for PNG, and none are addressed there, while all experts know it. It's not hypocrite?. As a matter of fact, Beehler and Laman (2020) stated boldly that for New Guinea—the landmass of two nations, Indonesia and PNG—PNG is widely chosen by scientists for easier permits (see also Mack, 2006 for examples)! Those specimen collection and egg collection efforts in PNG are not based on sustainable management quotas, not enforced, hardly documented, but have been ongoing for centuries in full force and are currently not much mitigated, with no end in sight. In the year 2022, PNG simply keeps serving the world as an exploitation reservoir for 'exotic' species and their parts to be put in the museum and collection basements, *ex-officio* (referred to by Beehler & Laman, 2020 as '*cabinet of curiosities*'; see Graham et al., 2004 for new directions, Huettmann, 2020a, 2020b for overview. Check for instance with GBIF.org for a data release assessment of progress, e.g. Germany and its colonial data and specimen holdings, Stresemann, 1923 for PNG, not really made well available online; note that many publicly funded German museums are in support of the tropical ornithological society https://www.tropenornithologie.de/ as a conglomerate of amateurs and some professionals but where open-access and professional transparent research are ignored/not promoted; ethics statements are hard to come by or position statements on such topics, including climate change).

Using birds again as a representative example, many professional societies and their outlets and publications use PNG as they need, e.g. for publications and to make a case for them. PNG co-authors are far and few there though; it's primarily still a white man's game (see Textboxes 9 and 10 for more details). And when it comes to PNG conservation, habitats and their people, PNG is left by itself, and just delivered to a few outside actors, e.g. from Australia, the EU, the U.S. and Asia. Typical examples are found for Australian bird societies that have left the PNG widely untouched and unhelped while many species are on the vast decline facing even extinction, e.g. curlews. Some Australian and New Zealand ornithologists take instead great pride focusing on North and South Korea, and eastern China, but leaving adjacent PNG simply behind (details for Australasian Wader Study Group found here: https://awsg.org.au/).

And as the largest issue ongoing with the Anthropocene, man-made CO_2 release—climate change—remains widely not addressed for PNG. For instance, the American Geophysical Union (AGU, https://www.agu.org/)—as one of the largest professional society in the world featuring over 100,000 members and frequently expressing deep public concern on the climate topic and on the Pacific region (with the world-famous Hawaiian'-measured Keeling curve at the center) has virtually no effort on PNG, or a relevant session or relevant participation from PNG. It does though feature the Hindu Kush-Himalayas frequently, as well as Antarctica, the Greenland ice sheet and Alaska and Russia, but PNG is a wide gap in that professional society.

All of this matters because islands—as the case shows for adjacent Hawaii (the place with one of the largest socio-economic and conservation problems in the U.S.)—are the prime regions of conservation and humanity concern, e.g. through sea level rise, coastal communities, finite space, refugees, remoteness and endemic species. Virtually, all island environments can easily be destroyed by globalization, and many have been already (Czech & Krausman, 1997; the infamous case of the extinct Dodo comes from Mauritius, an island location not that far away from PNG eventually). PNG is no exception but actually the 'grandest' and highest island of the world (Beehler & Laman, 2020).

Ignoring and using PNG so bluntly, uncritically—professionally—for many decades consistently comes with a cost to the global community.

While ethics might be promoted, in reality, professional societies have no solid and sustainable business model allowing for relevant ethics; already the metrcs on EID (Equity, Diversity and Inclusion) speak to that problem fine. The ethics that such a wide-spread 'professionalism' entails is neoliberal and it tries to balance gender and race issues perhaps, but falls already widely short on equity, fairness/equity and diversity (EID), and certainly on PNG and any global sustainability. That is true globally, and even in wealthy nations. Such professions and institutions—their unions included—make it worse and lead an ecological spiral to the ground, a global environmental bankruptcy. Those type of professional societies and NGOs are stand-alone entities, often tax-exempt—but they are 100% not independent. They are often supported and driven by governmental support staff, e.g. as editors and board members, with a university and public participation as 'officers,' and receive large amounts of industrial and other private support and money, such as from mining, oil and gas industries, and the zoos and museums. PNG features many of those as prime actors. This set up is not well-addressing sustainability issues though and results into ignorant and abusive outcomes, and many are easily perceived as 'perverse' to the environment, the profession and for a professional conduct, or for training future decision-makers and for PNG.

It should be noted that neither Russia, China or India actually has many dominant professional societies on the matter, or ones that consider PNG and which can express opinions and advice on the profession, or on sustainability or PNG for that matter. The notion of 'professionalism'—often also expressed as a 'certification' scheme—sits primarily with 'The Neoliberal West' -also adopted by China as a nice top-down governance metric to enforce on without much freedeom—but arguably it fails PNG and its people. Being 'professional' is part of a relatively recent approach in 'The West' and is widely neoliberal and remains rather harmful. It has no relevant track record.

Overall, just like with many Non-Governmental Organizations (NGOs, Fowler, 1996; see West, 2006; Mack, 2014 for a PNG example), such professional societies are highly ineffective for protection and conservation, or for what they promote and stand for as a 'Profession.' Professional societies are not really 'professional' and serve themselves the most, e.g. by raising funds in various forms (done through memberships, donations, PR, etc.). A large chunk of their efforts is devoted to that task: how to stay afloat. Arguably, the improvements brought forward through such professional societies are HIGHLY ineffective for conservation and sustainability, certainly for PNG and its people and habitats.

PNG is virtually not served through professional societies, certainly not by the major western ones. Instead, with professional societies hanging so much on fund raising for conservation and their own struggle for survival they mostly just can promote the *status quo,* if even that. The conservation and sustainability gain that professional societies brought for PNG are miniscule, at best. In reality, and as can be seen with forestry, mining or seabirds, many of those 'professions' achieve the net destruction of PNG, its wilderness and its society. The later though is a community of world relevance that instead operated for over 47,000 years sustainably; it had its own sustainable professional conduct that is now in vast destruction.

Textbox 8: Funded curators and curation problems as the serious bottleneck for Papua New Guinea artefacts, media PR and the global public: Dominating colonialism still to be overcome in 2023 for a sustainable 'One World'

The industrial nations and globalization relies for its ongoing functioning on many resources and inspirations from abroad. It also needs entertainment to keep the workforce fluent, entertained and happy. For that reason virtually all big nations feature large entertainment parks and such an industry, entities and subsidized 'sin cities' to cater those needs. Exotic resources—the 'unknown'—rank high in that regard to keep curiosity, attraction and the industrial life of globalization afloat, despite its unsustainability. It shows off resources from abroad, often with a national and ideological pride in an internationally competitive framework, but remains widely unsustainable, flawed and does so on the costs of others, e.g. perceived human underlings and less powerful nations who have to pay the cost.

The dominating cultures achieve that—in part—through 'fascinating' exhibitions and collections that are to be 'curated' for value and promoted to large audiences. That is, they are to show some narrow good-looking aspects to 'inform', but hardly present the reality and holistic context (e.g. Lodder, 2007). Seeing the full reality would not be 'entertaining', fun or it detracts. That's the job that curators do, and what 'curated' collections provide (see for global influence: https://www.artnews.com/list/art-news/artists/top-curators-shaping-art-world-today-1234593112/gabi-ngcobo-berlin-biennale/). The job description of curators, and the debates surrounding such a profession center widely around that. Curators receive funds to do their work, and those funds can be highly political, one-sided, dubious, intransparent and not representative. Many examples for this exist throughout the history of the industrialized world, e.g. Boakes et al. (2010).

Based on complex realities, curators essentially are employed to offer a lens to specific details but emphasize from that only the ones of their interest and of their employer. It's a PR job with curators as the well-spoken 'hired gun'. Entire disciplines and topics are affected, and shaped that way (see Gower et al., 2019 for mammal fossils, Saumarez Smith, 2022 for the development of western art and their museums). A wide discussion now exist on the bias in curators, in ethical funding of exhibitions, and in institutions overall, namely art, photography and museums (see for instance Lightfoot, 1989; The Guardian, 2018; see https://www.nbcnews.com/id/wbna18789206 for The Smithsonian allegedly altering warming exhibit).

It's easy to see this bias when it comes to Papua New Guinea (PNG). Most people of 'the west' are arguably not well informed about deep subjects such as wilderness or PNG and its realities. Most of the PNG subjects presented still center around PNG's rawness, and 'exotic' lifestyle in the South Pacific. It's a widely portrayed template and stereotype, usually accompanied with Birds of Paradise, beaches, ancient tribes, half-nakedness and PNG's richness, e.g. gold (see Beehler & Latam, 2020, p. 43 and 238/239 for U.S. examples; compare with Gillison, 2002). Using the internet does not make it better (Hamilton & Hopwood, 2022).

Exhibitions also carry national biases and this can easily be seen in how PNG is portrayed in Australia, with Sydney as one of the biggest collection holders and exhibition spaces for PNG over time (e.g. https://rb.gy/osof3e). Topics like mining problems, socio-economic destruction, poverty, wild specimen collection and ruthless profiteering—or climate change and human immigration- are widely ignored, for more than a century. It's not convenient any other...even offending? While the public gets sold dumb through curators and such efforts.

While attention to this problem of global proportions is on the rise (e.g. Knoefel, 2021), thus far, this template has received little solution, certainly not for PNG and its people. How much longer will it take for solving a problem that should have been resolved centuries ago and with all people living happily together on earth as the specified aim (Globalization states it but does not achieve whatsoever)?

Textbox 9: Indigenous, Hinduist & Buddhist concepts, ecology, steady-state-economics, and conservation management all can converge in a more holistic and sustainable approach to living well with nature and when using open-data sharing and modern research approaches

A unifying view (by M. Steiner and F. Huettmann)

The world is in a wider crisis: ecologically, socially, politically, and economically (e.g. Chapin et al. 2011a, b; Chapin et al. 2022; Rozzi et al. 2015). While this situation affects most aspects of the global processes, it specifically involves the 'western' world as the dominating culture of the Anthropocene (Diamond 2011a, b; Huettmann 2017).

According to the Cambridge Dictionary (July 2021), the word "indigenous" expresses the meaning of "naturally existing in a place or country rather than arriving from another place". However defined, indigenous species can be seen as the ones that inhabit a place or country for several thousands of years, or millennia rather than arriving very recently from another place. For the human species, the definition is a bit more complex. Saugestad (2001) proposed four criteria: first come, non-dominance, cultural difference, and self-ascribed. Such concepts shall be considered when we talk about "indigenous people". Their stewardship has dominated Mother Earth for over 99% of human civilization, over 60,000 years (Suzuki 1993; Diamond 2011; Dawson 2016; Huettmann 2020a). If at all, it has just resulted in benign impacts, and no such incidences as world wars, nuclear or industrial contamination, or microplastics, including no man-made global climate change or excessive sea floor decay.

Shamanism, Hinduism, and Buddhism are among the world's oldest spiritual practices and religions promoting holistic concepts—stewardship- linked with the wider universe, 'all is one' and those have equally left no deeper negative impacts on Mother Earth but provided guidance and better sustainability and wealth than the current 'regime', e.g. Lipner (2011). The deeper definitions of Ecology promote no other, e.g. Naess 1989). Along similar lines, ecological economics, steady-state-economics, as part of modern reasoning, is defined by H. Daly, B. Czech, etc. (see Center of the Advancement of the Steady State Economy https://steadystate.org/) as originating from "ecological economics" with roots in classical economics, most notably the "stationary state" (Czech and Daly 2004, 2014; Daly 1974; Daly et al. 2007; see Gaffney 1994 for some roots and flaws of a recently designed economics that rules the world currently). A steady-state economy is often discussed in the context of economic growth, consumption, and the impacts of economic growth on ecological integrity, environmental protection, and economic sustainability. Whereas a wider ecology perspective does exist and is officially taught – even in the western world—and is widely published, e.g. Naess (1989), Madison (1997). Nature conservation has actually been defined by Carl F. Jordan as "biological conservation as being a philosophy of managing the environment in a manner that does not despoil, exhaust or extinguish" (Jordan 1995). The former US president Theodore Roosevelt—as part of a globally dominant leadership and culture for decades—commented on nature conservation as follows "The conservation of natural resources is the fundamental problem. Unless we solve that problem, it will avail us little to solve all others" (Roosevelt 1907). Aldo Leopold, John Muir, and many others have argued along similar lines (Fox 1985; Leopold 1933; Naess 1989). There is an entire economy school system related to that concept: The Georgoists (see Gaffney 1994, and Wenzer 1997 and references within). It can be observed that all of these topics are closely related to, and fundamentally embedded in nature and its management and sustainability. These schemes—environment, social, and economics—have been clearly identified and acknowledged in the past to be of crucial importance, and much research has already been conducted on them. But there have barely been any connections made among them, and especially not among all together in synergy. Those views are widely missing but deserve more attention, as the main scheme of action, including policy and diplomacy (Buckley 2021 for Buddhism views of the Dalai Lama); world peace, and subsequent sustainability can be in reach. Instead, disconnected, and parsimonious views, as widely promoted by the western world, e.g. Burnham and Anderson 2004), act counterproductively for global sustainability and deviate from the more ancient human approaches to nature. It also does not provide a wider buy-in across nations, Bob Marley for instance, and his message of World Peace reaches easily billions of people across continents; a feat that many economic schemes or politicians never achieved. Even the Sustainable Development Goals (SDGs) by the U.N. are not well accepted, have been missed for decades, and get widely debated for their meaningless directions (Huettman and Young 2022, Acharya et al. in press) Therefore, here we are attempting to briefly outline the problem and better address these issues, and present suggested ways forward, with an additional discussion on the potential benefits and challenges for governments and industries to approach these topics with a focus on more holistic modern concepts and opportunities, including digital options, the world wide web, and open data initiatives available to any citizen of the world.

Indigenous people are known to live close to, and inherently with, nature (Baver 2020; Katewa 2009), Shamanism, Hinduism, and Buddhism promote similar concepts (Buckley 2021; Huang 2020; Okafor and Stella 2018; Schmithausen 2000). Steady-state economics (SSE) is likewise highly interconnected with nature as the definition above indicates ("...impacts of economic growth on ecological integrity, environmental protection, and economic sustainability"). Additionally, SSE supports the ground-baring ecological "trophic theory of money", which states that all money (a common medium for economics, e.g. costs, taxes, and payments) is originating from natural resources, namely agriculture (Czech 2019). Together with water and soil, and photosynthesis, it's 'the sun' that actually sits at the core of this concept. Farming can be sustainable, and Odum in Madison (1996), and Czech 2019) also mention, that agriculture, as the very basis (Producers), must grow in order to sustain and increase "the primary consumers", which is in the case of the trophic theory of money indicates "heavy manufacturing". "Heavy manufacturing" in turn must grow in order to sustain "secondary consumers" or "light manufacturing" (examples of this are biomass, hydroelectricity, geothermal power, wind power, etc.). This means that all the consumerism and thus, human consumption is based upon agriculture, a natural resource, and 'land rent'

(Gaffney 1994). Or in other terms, nature presents the foundation for human well-being, any financial business in a market, and its very existence (Czech 2019). Without growth in the agricultural sector, there cannot be any meaningful growth in any other sector, also not in the industrial sectors, higher up in the trophic levels. (many book-keeping tricks exist trying to show otherwise) Claiming any other must be a fallacy, e.g. the Internet Economy (as wrongly promoted per Nobel Laureate Paul Romer, see details in Choi and Yi 2001), most investment bubbles (such as Hedge funds) or mysterious money creation (e.g. bitcoins), and investment schemes (national debt trading or debt for nature swaps among those; the famous Credite Suisse Bank case among them). Entire books have been written on those problems, e.g. (Rich 1994, Fine and Milonakis 2009, and citations within). Lastly, nature conservation is primarily about conserving nature, and thus, all about nature as the resource – the basis for any meaningful lifestyle and trustworthy economy (Naess 1989; Suzuki 1993). It can be observed that all the three main topics (indigenous people wisdom, SSE, and nature conservation) of this study are strongly interlinked with nature and one can hardly survive sustainably without a healthy interaction among themselves (it can be compared to the famous "Three-Legged Stool" metaphor, removing one leg (in this case integrity of nature), the whole stool will collapse).

As mentioned above, without agriculture (= food production from the soil fueled by the sun), an economy cannot really grow or maintain sustainable stability. The land and the planet overall are finite but fueled by the sun energy; the wider universe matters (Buckley 2021), those are the laws of thermodynamics (Kümmel et al. 2010) and widely acknowledged in any wider and long-term world views, e.g. Suzuki (1993), and Buckley (2021). So, whenever nature and arable land are depleted or destroyed, less agriculture can be practiced, which in turn has negative influences on the higher trophic levels, economy incluyded. One may try to insert fuel and energy, e.g. via fertilizers produced by oil (Madison 1996) but that takes from the climate cycles and affects climate change; as per SSE theory. Thus, when a "modern" economy, which is often just being approached in a capitalistic extractive and technological manner, destroys nature and arable land, itself and its stakeholders are destroyed in that very process. In light of this, it must be realized that there is a sustainable limit for everything, agricultural production, our consumerism, and especially economic growth. And exactly these limits are outlined by the SSE, for us to respect them and act in favor of a more sustainable, limited, and stable lifestyle (Daly 1974; Madison 1996). This very principle has been understood and taught already in many indigenous groups for millennia (Suzuki 1993; Nirmal and Rocheleau 2019); the western world has known it since the infamous Club of Rome report (https://www.clubofrome.org/). In addition, indigenous people are known to be very good natural resource managers (e.g. Baver 2020), where nature and the wider universe are certainly part, if not the very center of resource management. Conclusively, this leads to sustainable nature conservation.

However, if such information and details are not shared and acknowledged, its impacts can only remain local and cannot reach/ penetrate the western or global world. By now, the human footprint has extended into the atmosphere even; the wider universe and the planetary fabric is affected. Once efforts are invested in learning this knowledge from indigenous people on how to manage the local resources and nature more sustainably, such practices can be applied on a wider scale. Wider applications can be scaled up even to regional and continental extents if synergized well among different indigenous groups. Such efforts to collect the knowledge about all the applied practices would be highly valuable and indeed a leap forward toward a sustainable steady-state economy (Kunkel and Daly 2008 for a critical review and transition concept), and more generally towards a promising future with the increasing human population on earth. From this, it can be concluded that the more data and information is collected, and shared with the wider public good, the more beneficial it would be for the wider society and the sustainable well-being of our future and nature.

Therefore, we focus in the upcoming sections briefly on some major benefits and challenges for governments and industries concerning this topic of learning information for the management of natural resources from indigenous peoples, and how it connects with SSE and nature conservation, as seen from an open-data initiative perspective for the wider common good.

a) Governmental and well-constrained industrial benefits of contributing to holistic and open data initiatives to achieve synergy

Linking ancient well-proven views with modernity offers many options and it is likely the only good and mutually agreeable outcome for promising progress (e.g. Cullis and Suzuki 2010). Like the addition of science, the internet, and its options, e.g. data sharing and cloud computing, remain essential in this effort (Carlson 2012). The main expected benefits for governments and industries in the involvement of open data initiatives regarding information collection of indigenous people's resources and nature management are manifold. Briefly, they include the prevention of future impacts on the environment and living conditions caused by the unsustainable economy, the establishment of higher social apprise for indigenous and virtually all people groups, the appreciation and utilization of their knowledge, eradication of possible misconceptions of indigenous people's culture, and practices by sharing and spreading information about them and for inquiries. This is mainly achieved by contributing to, and promoting, open data initiatives, including metadata to describe data. Also, making the indigenous groups and all people feel more appreciated within their territory and the global community, increasing the conservation success of endemic species, by respecting SSE theories/practices and including/consulting and acknowledging indigenous people is assumed to beneficially impact all included parties. Lastly, possible business collaborations between industries and indigenous people would also most likely result in beneficial relations. All these topics will now be shortly discussed individually and how they contribute to the aim of this study and the overall synergy. It follows what most groups and paradigms promoted for decades and brings them together, e.g. Naess (1989), Buckley (2021), Chapin et al. (2011a, b), Collins and Suzuki (2010).

To begin with, the prevention of future impacts on the environment caused by an unsustainable economy has a major benefit for governments and industries as they can continue to exist and execute such practices. Any other way has no good future, certainly not with 10 billion people on earth, and for taxpayers as they run out of resources, thus funds to be tax and their governance that needs it (e.g. see Czech 2000). The world has already most people earning less than $4 a day, so where will this end unless addressed better? Once governments and industries invest both efforts and financial resources in collecting this valuable knowledge and learning from indigenous people (in collaboration with them; see for instance Native People Federation for the Arctic and www.CAFF.org), it can be shared and utilized to prevent future negative impacts on the environment caused by an unsustainable economy and insufficient knowledge.

Next, we will discuss in this section—in short—the establishment of higher social apprise for indigenous groups and the appreciation and utilization of their knowledge. By providing information and by sharing these about the cultures, practices, and traditions of indigenous groups it is expected that the wide society can better understand the lifestyle and traditions of these indigenous groups. From this better understanding, it is expected that the commonly misunderstood inferior social values (*"primitive"*) for indigenous people can actually be raised to, and understood as, socio-economic equity (Dockery 2010). Such increased socio-economic values and the spread of knowledge about the indigenous' traditional culture of sustainability would most likely be admired and supported by the wider informed society, and thus elevate the indigenous' perception of Mother Earth within the society as well. Similar statements can be discussed for Shamans, Hindu, and Buddhist beliefs, and several other religions, old or new (see Dublin and Tanaka 2015 for an example in Japan). One can therefore easily include other religions in this like Christianity, Islam, etc. as well, but in the mainstream, the latter two have currently no main focus on the wider holistic and universal aspects of global sustainability and harmony (see exceptions with Francis of Assisi - Kim (2017) or Liberation Theology; Marx (2008). As a matter of fact, many of those beliefs have already fully embraced the internet and its concepts (e.g. the Vatican https://www.vatican.va/content/vatican/it.html). Industries and especially the government can benefit from such a change as the well-being of its citizens is increased and possible misconceptions can be eradicated. It is another way of dealing with, and improving, poverty. The distribution of wealth and socialism as major items of relevance for mankind sparked world wars and many national-scale revolutions bringing down emperors, tsars and royalty. Additionally, the corresponding governmental entities can utilize the newly obtained knowledge on how to manage natural resources more sustainably and thereby decrease the country-wide environmental impacts.

The next point to focus on is making the indigenous sustainability groups feel more appreciated within their territory/ country and participatory; national parks do not really achieve that or need to be improved on those aspects (see Beehler and Laman 2020 for examples in New Guinea). This goes hand in hand with the previous point of discussion. Once this socio-economic equity for indigenous groups has been reached, it is expected that indigenous peoples can feel more appreciated and acknowledged within their territories/ countries. This again benefits the government because its citizens' well-being is better established.

The next discussion point focuses on the increase of conservation success of a wider harmonic lifestyle in the world, with nature fully included. For instance, endemic species can be considered by respecting SSE theories/ practices and including/consulting indigenous people about their ancient, but successful management practices. Food production should be pursued in a way not to stand in broad conflict with species and wildlife (e.g. the North American Model of Wildlife Conservation Management; Organ et al. (2012); see wild rice production for an example Hauan (2015)). Currently, the conservation of species and especially endemic ones are not yet successful enough to call it "well under control", as can be seen with the apparently current mass extinction. Modern farming tends to result in a direct loss of wilderness, habitats, and species. A more specific example is the broad marginalization of more than 95% of all global squirrel species (approx. 300 species) (Steiner and Huettmann in prep). Another major issue in conservation is still the lack of open-access data sharing, a combined effort from governments, institutions, industries, and citizens toward more successful and sustainable species conservation is necessary (Huang et al. 2012). It can be evidently observed that species conservation is not yet successful enough when according to the World Wide Fund for Nature (WWF) approximately 10,000 species become extinct every year (WWF, 2021), and scientists call modern times "*a period of mass extinction*" (Briggs 2017; Ceballos et al. 2015; Elewa 2008; McCallum 2021; Wake and Vredenburg 2008). Who wants such modernity then?

In contrast to this mainly western-managed domain, indigenous groups have managed nature and wildlife for centuries without such major extinction rates (Rogers 2018, see Ayers 2010 for indigenous "Gardens of Abundance' even in the Arctic regions). It has even been documented that nature conservation data has been modified and not shared accordingly in the western world, in order to present statuses that are not aligned with nature to receive increased governmental funding (Egan et al. 1998). Therefore, it is of major importance to consult local indigenous groups on how they managed nature and wildlife in the past and present to better preserve our prestigious species on this planet. Once this information has been collected, it must be published, with the good consent of the indigenous sources, to the public and policymakers with a call for change. Industries and governments can highly benefit from this as their country and its wildlife, and nature within will be managed more selectively based on century-old valuable knowledge. This is expected to lead to more successful species conservation, benefiting not only governments and industries, but also the citizens, human well-being, and most importantly nature.

Lastly, possible business collaborations among industries, spiritual groups, and indigenous people are being discussed for a Mother Earth business (see for instance Collins and Suzuki 2010, Chapin et al. 2011a, b) which aim to favor all parties equally. When it comes to sustainable policies, indigenous people are quite often "let alone" with the execution and continuation of their traditions (Beltran 2000), and thus, support from non-indigenous businesses is often poor (but see a viable link in Native Corporations as practiced in Alaska for instance, https://www.ncai.org/tribal-directory/alaska-native-corporations). The possible establishment of business relationships between indigenous groups and non-indigenous industries could benefit both parties significantly. Further examples of such beneficial collaborations can be the supply of materials for traditional clothing, education of children, technology, affordable food, health services, etc. (Dyson et al. 2006; Dyson et al. 2015; Zinck and Marmion 2011). These possible business relationships are however only possible if there is an exchange of information happening and the industries are actively involved in open data initiatives.

These upper-mentioned benefits can become relatively challenging to achieve them in some cases (e.g. because of unwillingness to collaborate from one of the involved parties). Therefore, to bridge this gap, the challenges for governments and industries in the involvement of open data initiatives will be discussed in the upcoming section.

b) Governmental and industrial challenges of contributing to modern and holistic efforts including open data initiatives using the World Wide Web (www)

The main expected challenges for governments and industries include providing respectful and adequate legislation for indigenous and spiritual approaches and their promoters. This is for each group and their legislative country, moving from an initially capitalistic-oriented economy towards a SSE without negatively influencing local businesses too heavily. Additionally, it shall not influence the unwillingness to change from the industry's side due to a possible reduction of the production of their goods following SSE approach, and unwillingness to collaborate from the indigenous people's side. Instead, we have no other choices than to collaborate.

To begin with, one major challenge for governments seems to be to provide unbiased lobbying as well as adequate legislation for indigenous and sustainability groups (Hunter 2009). Arguably, western law is not the only way to govern and it is certainly not one that is sustainable, or native to indigenous people (the latter are often organized in tribal management and their legislation and policies, including a widely found female-domination; an example of the discrepancies and challenges found in Wilson (2019)). The challenge might be caused by a lack of knowledge from the western governmental side about the indigenous groups, and thus, miscommunication is likely to occur. Well-designed Open Data initiatives could resolve some of such knowledge gaps and understanding. An additional challenge regarding this issue is that a mediating entity/ party might be missing which could aid to bridge this gap. An example of such a mediator could be a member of an indigenous group, e.g. an elder from other tribes, employed by the government, that is in charge of bridging this gap.

The role of industry and employment awaits more details but arguably, the governmental approach sets the stage to achieve this ‘best’, e.g. Huettmann (2015).

Another challenge to be focused on is moving from a capitalistic-oriented economy to a mildly-fluctuating SSE with the least negative impacts on local and regional societies. Governments tend to focus mainly on economic growth without considering its negative impacts (Cozzolino 2018). However, this is highly unsustainable (Sargen 2018), and it is suggested to move more toward a steady-state economy. But in order to pursue a SSE, moving away from growth, a period of degrowth is required to reach the sustainable steady-state (Daly 1974). It will involve a mildly fluctuating economy and a certain downsizing of priorities and sectors while the wider whole can remain intact fine. Neoliberal governments and entities will probably not be in favor of this, as the degrowth of the economy is perceived as negative by the wider uninformed western society and would result in fewer votes in the following legislative period. Additionally, industries will most likely be against it as it would result in a period of degrowth, and thus, decreased profit for businesses. Also, a decreased production of each industry’s goods will be considered highly negative. This needs trust with a well-thought-out transition plan to overcome misconceptions and bottlenecks. The challenge thus is, to handle this period of transition, or degrowth, sustainably and with good leadership, to assure the least possible impact on local businesses and the remaining parties. The only way of facilitating this transition period is to fully inform all participating parties and stakeholders. This is likely best achieved by the involvement of the government and the industries in open data initiatives and information sharing. Because the better all stakeholders are informed, the fewer issues and unpredicted scenarios are expected to be encountered. Together with the above-mentioned points, changing the public opinion/ narrative around degrowth, can be achieved by involving universities and youth-led organizations (e.g. AIESEC, The Wildlife Society student chapters)/ events and opening discussions in academia too.

In reality, COVID pandemics have achieved some of it benefitting some conservation aspects and might offer a (painful) but now available platform to start out from. It should be fully noticed that assumed growth rates of the western world are usually far from great. What has been coined as a recession offers actually great opportunities to move into a better world and to rebuild in a more sustainable way. Many opportunities remained unused though.

Overall, a new culture has to emerge either way – and to be taught at our leading institutions - that overcomes those problems and one which is based on a culture of holistic respect towards Mother Earth and the universe.

The last discussion point focuses on indigenous and spiritual sustainability groups that might be unwilling to collaborate with governments and industries and prefer to be "left alone" (e.g. Gerstner, 2019). Here, the challenge is to convince the indigenous groups that collaborations between them, governments, and industries can be beneficial if the latter two parties respect the indigenous traditions and treat them equally, and fair, as the rest of society. Achieving this trust from those groups might be challenging since major distrust in information sharing with the western world has been experienced due to unacceptable actions in the past (see Flanney 2002 for the bad 'contact' experience in Australia of the last remaining 'wild' aborigenees). There, indigenous knowledge has for instance been utilized and essentially stolen from indigenous groups without any form of credit, even a now acknowledged genocide widely occurred requiring deep healing (Hinton et al. 2014; Short 2010; Willis 2011). Therefore, in order to regain this trust from indigenous groups, credit should be given to the source of the knowledge, and thus to them and their achievements as part of world civilization (sensu Suzuki 1993). Additionally, it is important to openly share the information on what is happening with the information, and how it is used. This information should of course not be used against indigenous groups or any human. This is only possible through a good culture, and with the active involvement of governments and industries in sustainable open data initiatives. In some cases, it is also possible that not all information can or is allowed to be shared. In such cases, a justification and an exception must be made where the information should be kept private and local to protect the involved individuals (e.g. privacy information, or indigenous details that are tribe/ group-specific confidential information). However, the more often data can be shared for better decision-making, it should be. This aims to inform the wider society better, to eradicate misconceptions, hate, and prejudgment.

Conclusively, challenges will be encountered in this process and the approach we promote, however, as long as the involvement of governments and industries in open data initiatives is as high as possible, the outcome is expected to be highly beneficial. It's the only way forward. Thus, it is crucial to work together towards a more sustainable nature and resource management as well as increased social well-being.

Conclusion

The world is facing a resource, sustainability, social, economic, and related governance crisis. For resolution and reaching more holistic and better governance, the involvement of industries and governments in open data initiatives can have major benefits for themselves, the wider society, as well as the indigenous groups, the world, and human well-being overall. Additionally, by consulting indigenous sustainability groups about their local resource management practices, major progress can be achieved towards a sustainable steady-state economy and a more successful future of nature conservation. However, this should only be done by treating the indigenous groups with respect and as part of the wider global community and providing them with good credit for sharing their knowledge. Once this is done, it is suggested for governments and industries to transition and heavily involve themselves in open data initiatives to yield benefits for the wider society, indigenous and spiritual people, the environment, and generally, nature and the universe in which all are embedded.

Textbox 10: Being a "Waldlaeufer" in Papua New Guinea: Some Personal Afterthoughts for the future

All my life, I walked in deep forests, in treed landscapes, and worked about forests and their wildlife and ecological services. I tried to have a live with and in forests and around forests every day. For that, I also studied 'forestry' with institutions and projects all over the world to be able to do so. Deep ancient old-growth forests I dealt with in Russia, in Canada, in Norway, in Africa, China, Hindu Kush-Himalaya, Tasmania and Australia, and Patagonia, and PNG of course. One can easily connect with forestry people anywhere—in the boreal zone, in temperate Europe or in the tropics regardless. Forest people will understand. PNG is such a case where as a '*Waldlaeufer*' on ancient trails you can easily see many deep connections with people, souls, villages, species, the atmosphere and the universe at large. Just try it!

Being a "*Waldlaeufer*" refers to the German word and underlying concept of walking (~"laufen") in the forest (~"Wald"). In New Age terms that might be equal to '*forest bathing,*' but as a concept it's widely spread and recognized since humankind, see for instance "Walking with Wolf" for Mennonites in Costa Rica (Chornook & Guindon, 2008). Most 'Bird Watchers' are '*Waldlaeufer.*' And anti-cancer and other generic but deeper health benefits of walking in the woods are described in the medical literature, and beyond (e.g. Wohlleben, 2016). Mongolians and Tibetans—the moving healers and shamans—who tend to be nomadic, have similar concepts for their steppe and mountain landscapes, so do the Athabascans in interior Alaska, or the Tuareg tribes in the Sahara desert etc. Moving and walking, throughout the landscape and its vegetation - shrubs and trees - is an effort humans do traditionally engage in, and it's healthy and balanced, shaping mind and soul. It provides a spiritual basis for your doings (e.g. Huettmann, 2020a, 2020b). Use skis, a mushing sled, dogs, a bike or a horse, but the overall effect will remain. It sums all what humans can do on earth, and leave little footprints in the process. It's as sustainable as it can get for us. Most people in PNG I met are sophisticated "*Waldlaeufer*'; they know the bush and can operate in it fine since childhood. It's their home, like it's our all home. It's where everything sits and starts...and ends.

I have done a lot of 'field work' (a North American term that reflects the forest connection poorly though) in forests, and in prime forests of this world to get closer to the heart of ecology, of nature, or Mother Earth and the universe, and the people.

Many people will understand those links, but few actually make it their life's work. While it's a privilege, it's a devotion and can be burden. That's because it's difficult to relay those concepts to urbanized people and burocrats and ones that are now so industrial and corporate, and who come from such institutions directed and trained. In such a regime one gets loosely linked with similar people working on 'wildlife,' 'game,' or now, 'conservation' to fit a category. But people like us are actually many, as I am a tree-lover, a tree-addict and a tree-hugger, but we are on our own. I openly admit that I feel similar for oceans, for mountains, for the atmosphere, and for most landscapes with wilderness and species. Old-growth forests, and relatively, untouched wilderness—anywhere on the planet—do have the same connection. Tree-lovers are to speak their minds (Wohlleben, 2019; Ludlam, 2021), and they always have. It's hard to have good democracy—and world peace—without trees, or without 'Waldlaeufer'

PNG and its metrics—its biological and forestry facts—remain very appealing to us, but often those are now very shocking. Still, PNG remains as a gift to mankind. But the way how PNG is treated, how wilderness and its people are treated, remains something that is hard to fathom. Being a Waldlaeufer or not, it makes one sick to the stomach to see the poor reality.

This book and work is to change it for those tree people and their relatives, all of them!

31.1 PNG People and Nature Will Come First, for a Long Time

The people of PNG are the asset, so is the environment. They are relevant curators of the world, also for PNG (see Steiner 2011, James et al. 2012 for approach; compare with National Strategic Plan Taskforce 2011). For whatever happens in PNG, it will be with those factors in mind as the driver seat. The tribal structure around PNG—Wantok—will not go away any time soon. For a social license for industry to operate well in PNG, it is invaluable to have and will be strengthened as a requirement (see Steiner, 2012, 2013 for proposed UN Civil Oversight requirement and applied to PNG seabed mining). Instead, modernity—as we know it—is currently on the edge though and could well crumble. Implications are unclear but the tribal PNG life has been with us for over 47,000 years, the Melanesian way. We may assume it's one of the most and well-proven concepts out there and will last. Now, will there be mobile phones within that? Yes, a new form of technology and availability is forming and moving with many features 'blinking in and out' in between. It likely will be a blend where many styles and concepts come to reality and where it can meet nature and sustainability (Majumder, 2021). But certainly it can only achieve what Mother Nature allows; we are all embedded within that. Eventually, nature and wilderness combined is the only way forward for operating in the universe and space ship earth (= Mother Earth; Carlson, 2012; Rozzi et al., 2015; Schmithausen, 2000; Suzuki, 1993); not only for nature conservation (Hannah et al., 2020) but also human happiness, well-being and people, as well as for the economy (see Gaffney & Harrison, 1994 for the Georgists approach of a land rent).

In the meantime, PNG will stay if the gods allow and are not angered; one is to stay in the ecological boundary of the planet (sensu Huettmann, 2020a, 2020b).

Hope and inspiration? Thus far we have only concerns though (Hunter, 2009).

> Only love can stop war (The Guardian, 2022).

A hope would be to have better models, better data and thus a better inference and subsequent better governance and society with a much better sustainability action and outlook. All else what SDG and Aichi targets promise us (contrast with Robinson et al., 2022).

The road is always where you find it.

> "*A hungry mob is an angry mob*"
>
> Bob Marley and The Wailers song "*Them Belly Full (But We Hungry)*"

References

Acharya, K. P., Chand, R., Huettmann, F., & Ghimire, T. R. (in press). Rabies elimination: Is it feasible without considering wildlife? *Tropical Diseases*.

Anderson, D., & Burnham, K. (2004). *Model selection and multi-model inference* (Vol. 63, 2nd ed., p. 10). Springer.

Ayers, H. (2010). *Arctic gardens: Voices from an abundant land*. Arctic Voices Publisher.
Baraka, R. (2001). The meanings of independence in Post-Colonial Papua New Guinea. *A plan B research paper submitted to the faculty of the center for Pacific Islands studies in the candidacy for the degree of Masters of Arts. School of Hawaii*, Asia and Pacific Studies. The University Of Hawai'i At Manoa.
Baver, S. (2020). Nature conservation, extractivist conflicts, and indigenous rights in the Americas. *Global Environmental Politics, 20*(4), 192–197.
Beehler, B., & Laman, T. (2020). *New Guinea: Nature and culture of Earth's grandest island*. Princeton University Press.
Beltran, J. (Ed.) (2000). *Indigenous and traditional peoples and protected areas: Principles, guidelines and case studies* (Vol. 4). IUCN-The World Conservation Union.
Briggs, J. C. (2017). Emergence of a sixth mass extinction? *Biological Journal of the Linnean Society, 122*(2), 243–248.
Boakes, E. H., McGowan, P. J., Fuller, R. A., Chang-qing, D., Clark, N. E., O'Connor, K., & Mace, G. M. (2010). Distorted views of biodiversity: Spatial and temporal bias in species occurrence data. *PLoS Biology, 8*(6), e1000385.
Buckley, M. (Ed.) (and The Xivth Dalai Lama Tenzin Gyatso) (2021). *This fragile planet: His holiness the Dalai Lama on environment*. Sumeru Press.
Cousteau. J.M., & Richards, M. (1999). *Cousteau's Papua New Guinea Journey*. Henry Abraham's Books.
Carlson, D. J. (2012). IPY 2007–2008: Where threads of the double helix and sputnik intertwine. In *Protection of the three poles* (pp. 35–50). Springer.
Ceballos, G., Ehrlich, P. R., Barnosky, A. D., García, A., Pringle, R. M., & Palmer, T. M. (2015). Accelerated modern human–induced species losses: Entering the sixth mass extinction. *Science Advances, 1*(5), e1400253.
Chan, J. (2016). *Playing the Game: Life and Politics in Papua New Guinea*. University of Queensland Press.
Chapin, F. S., Power, M. E., Pickett, S. T., Freitag, A., Reynolds, J. A., Jackson, R. B., Bartuska, A., et al. (2011a). Earth stewardship: Science for action to sustain the human-earth system. *Ecosphere, 2*(8), 1–20.
Chapin, F. S., Pickett, S. T., Power, M. E., Jackson, R. B., Carter, D. M., & Duke, C. (2011b). Earth stewardship: A strategy for social–ecological transformation to reverse planetary degradation. *Journal of Environmental Studies and Sciences, 1*(1), 44–53.
Chapin, F. S., Weber, E. U., Bennett, E. M., Biggs, R., van den Bergh, J., Adger, W. N., de Zeeuw, A., et al. (2022). Earth stewardship: Shaping a sustainable future through interacting policy and norm shifts. *Ambio*, 1–14.
Choi, C., & Yi, M. H. (2009). The effect of the Internet on economic growth: Evidence from cross-country panel data. *Economics Letters, 105*, 39–41.
Chornook, K., & Guindon, W. (2008). *Walking with Wolf: Reflections on a life spent protecting the Costa Rican wilderness* (Vol. 7). Wandering Words Press.
Cozzolino, A. (2018). Trumpism as nationalist neoliberalism. A critical enquiry into Donald Trump's political economy. *Interdisciplinary Political Studies, 4*, 47–73.
Cullis, T., & Suzuki, D. (2010). *The declaration of interdependence: A pledge to planet earth*. Greystone Books Ltd.
Czech, B., & Krausman, P. R. (1997). Distribution and causation of species endangerment in the United States. *Science, 277*(5329), 1116–1117.
Czech, B. (2000). *Shoveling fuel for a runaway train*. University of California Press.
Czech, B. (2019). The trophic theory of money: Principles, corollaries, and policy implications. *Journal and Proceedings of the Royal Society of New South Wales, 152*(471/472), 66–81.
Czech, B., & Daly, H. E. (2004). In my opinion: The steady state economy—What it is, entails, and connotes. *Wildlife Society Bulletin, 32*(2), 598–605.
Czech, B., & Daly, H. E. (2014). The steady state economy as the sustainable alternative to economic growth. In *Peak oil, economic growth, and wildlife conservation* (pp. 119–129). Springer.

Daly, H. E. (1974). The economics of the steady state. *The American Economic Review, 64*(2), 15–21.

Daly, H. E., Czech, B., Trauger, D. L., Rees, W. E., Dobson, T., & Trombulak, S. C. (2007). Are we consuming too much: For what? *Conservation Biology, 21*(5), 1359–1362.

Dawson, A. (2016) *Extinction: A radical history*. Or Books.

Diamond, J. (2011a). *Guns, germs and steel: The fate of human societies*. Random House.

Diamond, J. (2011b). *Collapse: How societies choose to fail or succeed* (Revised ed.). Penguin Books.

Dockery, A. M. (2010). Culture and wellbeing: The case of Indigenous Australians. *Social Indicators Research, 99*(2), 315–332.

Dublin, D., & Tanaka, N. (2015). Comparison between Satoyama and the Ainu way of life. *Kawistara Jurnal Sosial Dan Humaniora, Universitas Gadjah Mada, Indonesia, 5*, 221–328.

Dyson, L. E., Grant, S., & Hendriks, M. (Eds.) (2015). *Indigenous people and mobile technologies*. Routledge.

Dyson, L. E., Hendriks, M., & Grant, S. (Eds.) (2006). *Information technology and indigenous people*. IGI Global.

Elewa, A. M. (2008). Current mass extinction. In *Mass extinction* (pp. 191–194). Springer.

Egan, C., Smith, E. A., & McCarter, J. (1998). Contested Arctic: Indigenous peoples, industrial states, and the circumpolar environment. *Anthropologica, 40*(2), 230.

Fine, B., & Milonakis, D. (2009). *From economics imperialism to freakonomics: The shifting boundaries between economics and other social sciences*. Routledge.

Flannery, T. (2002). *The future eaters: An ecological history of the Australasian lands and people*. Grove Press.

Fowler, A. (1996). Demonstrating NGO performance: Problems and possibilities. *Development in practice, 6*(1), 58–65.

Friedman, G. (2010). *The next 100 years: A forecast for the 21st century*. Knopf Doubleday Publishing Group.

Fox, S. R. (1985). *The American conservation movement: John Muir and his legacy*. Univ of Wisconsin Press.

Gaffney, M., & Harrison, F. (1994). *The corruption of economics*. Shepheard-Walwyn.

Gerstner, L. (2019). The right to be left alone? Protecting "Uncontacted" tribes of India and Brazil. *Tul. J. Int'l & Comp. L., 28*, 81.

Gillison, D. (2002). *New Guinea ceremonies*. Harry N Abrams Inc.

Graham, C. H., Ferrier, S., Huettmann, F., Moritz, C., & Peterson, A.T. (2004). New developments in museumbased informatics and applications in biodiversity analysis. *Trends in Ecology and Evolution*, 19: 497–503.

Gray, G. G. (1996). 'The next focus of power to fall under the spell of this little gang': Anthropology and Australia's postwar policy in Papua New Guinea. *War and Society, 14*(2), 101–117.

Greasley, D., McLaughlin, E., Hanley, N., & Oxley, L. (2017). Australia: A land of missed opportunities? *Environment and Development Economics, 22*(6), 674–698.

Gower, G., Fenderson, L. E., Salis, A. T., Helgen, K. M., van Loenen, A. L., Heiniger, H., Cooper, A., et al. (2019). Widespread male sex bias in mammal fossil and museum collections. *Proceedings of the National Academy of Sciences, 116*(38), 19019–19024.

Hannah, L., Roehrdanz, P. R., Marquet, P. A., Enquist, B. J., Midgley, G., Foden, W., Lovett, J. C., Corlett, R. T., Corcoran, D., Butchart, S. H. M., Boyle, B., Feng, X., Maitner, B., Fajardo, J., McGill, B. J., Merow, C., Morueta-Holme, N., Newman, E. A., Park, D. S., … Svenning, J. C. (2020). 30% land conservation and climate action reduces tropical extinction risk by more than 50%. *Ecography, 43*, 943–953. https://doi.org/10.1111/ecog.05166

Hamilton, A. J., & Hopwood, B. (2022). Online Museum Collections as Artifacts: A Case Study of New Guinean Necklaces Illustrating Mathematical Approaches as Diagnostics for Collections Management. *Collections*, 15501906221101188.

Hauan, H. (2015). *Zizania aquatica L., wild rice; an evaluation of cultivation, domestication, and production for use in The United States*. https://conservancy.umn.edu/handle/11299/175836

Hinton, A. L., Woolford, A., & Benvenuto, J. (2014). *Colonial genocide in indigenous North America*. Duke University Press.
Huang, H. (2020). Afterthoughts: Nature, culture, and shamanism in inner Mongolia, PRC. *Envirolab Asia, 3*(3), 2.
Huang, X., Hawkins, B. A., Lei, F., Miller, G. L., Favret, C., Zhang, R., & Qiao, G. (2012). Willing or unwilling to share primary biodiversity data: Results and implications of an international survey. *Conservation Letters, 5*(5), 399–406.
Huettmann, F. and B. Czech (2006) Taking Marine Seabird Conservation Seriously: Towards a Steady State Economy for the Pacific and Beyond. *Pacific Seabirds* 33 Fall 2: 52–54.
Huettmann, F. (2015). On the relevance and moral impediment of digital data management, data sharing, and public open access and open source code in (tropical) research: The rio convention revisited towards mega science and best professional research practices. In: F. Huettmann (Ed.), *Central American biodiversity: Conservation, ecology, and a sustainable future* (pp. 391–418). Springer.
Huettmann, F. (2017). Climate change effects on terrestrial mammals: A review of global impacts of ecological niche decay in selected regions of high mammal importance. *Encyclopedia of the Anthropocene, 2*(2018), 123–130.
Huettmann, F. (2020a). Spirituality beats it all: A quick conservation overview, self-organization and the great value of (indigenous) religions for Hindu Kush-Himalaya landscapes, its geo-parks, species, ecological processes and watersheds. In *Hindu Kush-Himalaya watersheds downhill: Landscape ecology and conservation perspectives* (pp. 283–311). Springer.
Huettmann, F. (2020b). Settling the terror of your mind with the deities: About fear, anxiety, inherent chaos and self-doubt in Hind Kush-Himalaya expeditions and associated research. In G. R. Regmi, & F. Huettmann (Eds.), *Hindu Kush-Himalaya watersheds downhill: Landscape ecology and conservation perspectives* (pp. 313–328). Springer Gland.
Huettmann, F., & Young, B. (2022). The so-called modern 'sustainable forestry' destroys wilderness, old-growth forest landscapes and ecological services worldwide: A short first-hand review and global narrative on the use of 'growth-and-yield' as a destructive and even impossible goal. In *Forest dynamics and conservation* (pp. 53–82). Springer.
Hunter, B. (2009). Indigenous social exclusion: Insights and challenges for the concept of social inclusion. *Family Matters, 82*, 52–61.
Kunkel, B., & Daly, H. (2008). Economies of scale: An interview with H. Daly. *New Left Review, 109*, 81–104.
James, P., Nadarajah, Y., Haive, K., Stead, V. (2012). sustainable communities, sustainable development OTHER PATHS FOR PAPUA NEW GUINEA University of Hawaii.
Jordan, C. F. (1995). *Conservation: Replacing quantity with quality as a goal for global management*. Wiley.
Katewa, S. S. (2009). Indigenous people and forests: Perspectives of an ethnobotanical study from Rajasthan (India). In *Herbal drugs: Ethnomedicine to modern medicine* (pp. 33–56). Springer.
Knoefel, U. (2021). Debatte über koloniale Raubkunst: Schluss mit der Verharmlosung. Der Spiegel 28th May 2021. https://www.spiegel.de/kultur/koloniale-raubkunst-schluss-mit-der-verharmlosung-kommentar-a-338f7afd-8f91-4da2-98e6-272cec7cf2e1. Accessed July 1, 2021.
Kim, Y. (2017). St. Francis of Assisi and ecological conversion. *Catholic Theology and Thought, 78*, 47–89.
Kirsch S. (2014). *Mining Capitalism*. University of California Press.
Kümmel, R., Ayres, R. U., & Lindenberger, D. (2010). *Thermodynamic laws, economic methods and the productive power of energy*. De Gruyter.
Laman, T., & Scholes, E. (2012). *Birds of paradise: revealing the world's most extraordinary birds*. National Geographic Books.
Ladurie, E. L. R. (1976). *The peasants of Languedoc* (Vol. 635(6)). University of Illinois Press.
Leopold, A. (1933). The conservation ethic. *Journal of Forestry, 31*(6), 634–643.
Lightfoot, D. T. (1989). Contemporary art-world bias in regard to display holography: New York City. *Leonardo*, (pp. 419–423).
Lipner, J. (2022). The Oxford history of Hinduism: Modern Hinduism. *Religion, 52*(1), 156–162. https://doi.org/10.1080/0048721X.2021.2000315

Lodder, C. (2007). Dream Factory Communism: The Visual Culture of the Stalin Period/Traumfabrik Kommunismus: Die visuelle Kultur der Stalinzeit. Ed. Boris Groys and Max Hollein. Ostfildem, Germany: Hatje Cantz Verlag, 2003. 464 pp. Notes. Bibliography. Chronology. Plates. Photographs. $55.00, hard bound. *Slavic Review, 66*(3), 571–573.

Ludlam, S. (2021). *Full circle: Power*. Black Inc.

Mack, A. (2014). *Searching for Pekpek: Cassowaries and Conservation in the New Guinea Rainforest*. Cassowary Conservation & Publishing, LLC.

Madison, M. G. (1997). "Potatoes Made of Oil": Eugene and Howard Odum and the origins and limits of American agroecology. *Environment and History, 3*(2), 209–238. http://www.jstor.org/stable/20723041

Majumder, N. (2021). Paradigm war: A distinct indigenous vision. *Academia Letters*, Article 926. https://doi.org/10.20935/AL926.

Marx, R. (2008). *Das Kapital*. Pattloch Verlag.

McCallum, M. L. (2021). Turtle biodiversity losses suggest coming sixth mass extinction. *Biodiversity and Conservation, 30*(5), 1257–1275.

McKenna, M. (2017). *Big chicken: The incredible story of how antibiotics created modern agriculture and changed the way the world eats*. National Geographic Books.

Naess, A. (1989). Ecology, community and lifestyle: Outline of an ecosophy. translated by David Rothenberg. Cambridge University Press; reprint. https://doi.org/10.1017/CBO9780511525599

Narokobi, B. (1983). *The Melanesian Way*. Institute of Papua New Guinea Studies.

National Strategic Plan Taskforce, (2011). PAPUA NEW GUINEA VISION 2050. Port Moresby PNG.

Nirmal, P., & Rocheleau, D. (2019). Decolonizing degrowth in the post-development convergence: Questions, experiences, and proposals from two Indigenous territories. *Environment and Planning E: Nature and Space, 2*(3), 465–492.

Okafor, J. O., & Stella, O. (2018). Hinduism and ecology: Its relevance and importance. *Fahsanu Journal, 1*(1).

Organ, J. F., Geist, V., Mahoney, S. P., Williams, S., Krausman, P. R., Batcheller, G. R., Rentz, T., et al. (2012). The North American model of wildlife conservation. *The Wildlife Society Technical Review, 12*(04).

Ostrom, E. (2009). Engaging with impossibilities and possibilities. In R. Kanbur & K. Basu (Eds.), *Arguments for a better world: Essays in honor of Amartya Sen | Volume II: Society, institutions and development* (pp. 522–541). Oxford University Press.

Ostrom, E., Chang, C., Pennington, M., & Tarko, V. (2012). *The future of the commons beyond market failure and government regulation*. The Institute of Economic Affairs.

Premdas, R. (1976). Toward a Papua New Guinea foreign policy: Constraints and choice. *Australian Journal of International Affairs, 30*, 263–279.

Rich, B. (1994). *Mortgaging the earth: The world bank*. Earthscan Publishers New York.

Robinson Sa, E., Dolan, B. M., et al. (2022). A new framework for rapidly assessing national adaptation policies: An application to small island developing states in the Atlantic and Indian Oceans. *Regional Environmental Change, 22*, 37. https://doi.org/10.1007/s10113-021-01855-2

Rogers, C. (2018). Investing in indigenous communities is most efficient way to protect forests, report finds. Mongabay. https://news.mongabay.com/2018/07/investing-in-indigenous-communities-most-efficient-way-to-protect-forests-report-finds/. Accessed July 2, 2018.

Roosevelt, T. (1907). Address to the Deep Waterway Convention Memphis, TN.

Rozzi, R., Chapin, F. S., Callicott, J. B., Pickett, S. T., Power, M. E., Armesto, J. J., & May Jr, R. H. (Eds.) (2015). *Earth stewardship: Linking ecology and ethics in theory and practice* (Vol. 2). Springer.

Sargen, N. P. (2018). *Investing in the trump era, how economic policies impact financial markets*. Springer Link. https://doi.org/10.1007/978-3-319-76045-2.

Saugestad, S. (2001). The inconvenient indigenous: Remote area development in Botswana, donor assistance and the first people of the Kalahari. Nordic Africa Institute.

Saumarez Smith, C. (2022). *The art museum in modern times*. Thames and Hudson Publisher.

Schmithausen, L. (2000). Buddhism and the ethics of nature—Some remarks. *The Eastern Buddhist, 32*(2), 26–78.

Short, D. (2010). Cultural genocide and indigenous peoples: A sociological approach. *The International Journal of Human Rights, 14*(6), 833–848.

Singer, P. (1973). Animal liberation. In *Animal rights* (pp. 7–18). Palgrave Macmillan.

Somare, M. T. (1975). *Sana: An autobiography of Michael Somare*. Niugini Press.

Steiner, R. (2012). Citizens' advisory councils: A prerequisite for socially responsible mining. In *2nd International Conference on Social Responsibility in Mining* (p. 293).

Steiner, R. (2013). Citizens' advisory councils to enhance civil society oversight of resource industries-UNEP Perspectives No. 10. Gland, Switzerland.

Steiner, R. (2011). Citizens' Advisory Councils for Mining in the Pacific Richard Steiner, Professor Oasis Earth.

Stiglitz, J. (2003). *Globalization and its discontent*. Norton Paperback.

Stresemann, E. (1923). Ornithologische Ausbeute im Stromgebiet des Sepik. Ein Beitrag zur Kenntnis der Vogelwelt Neuguineas. Archiv fuer Naturgeschichte ser. A.A. 89 (7)–96 (8) –92.

Suzuki, D. (1993). *Wisdom of the elders: Sacred native stories of nature*. BoD–Books on Demand.

The Guardian. (2018). Why are white curators still running African art collections? 3rd April. https://www.theguardian.com/commentisfree/2018/apr/03/brooklyn-museum-white-curators-african-art-open-letter. Accessed October 10, 2022.

The Guardian. (2022). Only love can stop war: A Northern Cheyenne chief's call to the world, 17th June 2022. https://www.theguardian.com/us-news/2022/jun/17/only-love-can-stop-war-a-call-to-the-world-from-a-northern-cheyenne-chief. Accessed July 7, 2022.

Wake, D. B., & Vredenburg, V. T. (2008). Are we in the midst of the sixth mass extinction? A view from the world of amphibians. *Proceedings of the National Academy of Sciences, 105*(Supplement 1), 11466–11473.

Watson, A. (2011). The mobile phone: The new communication drum of Papua New Guinea [PhD thesis, Queensland University of Technology, Australia]. https://eprints.qut.edu.au/47170/

West, P. (2006). *Conservation is Our Government Now: The Politics of Ecology in Papua New Guinea*. Durham, NC: Duke University Press.

Wenzer, K. C. (1997). Tolstoy's georgist spiritual political economy (1897–1910). *Pennsylvania History: A Journal of mid-Atlantic Studies, 64*(4), 534–542.

Wilson, M. (2019). *A true child of Papua New Guinea: Memoir of a life in two worlds*. McFarlands Publisher.

Willis, P. (2011). The West's ongoing theft of indigenous knowledge. openDemocracy. https://www.opendemocracy.net/en/wests-ongoing-theft-of-indigenous-knowledge/. Accessed August 15, 2011.

Wohlleben, P. (2016). *The hidden life of trees: What they feel, how they communicate—Discoveries from a secret world* (Vol. 1). Greystone Books.

Wohllcbcn, P. (2019). *The Secret Wisdom of Nature: Trees, Animals, and the Extraordinary Balance of All Living Things-—Stories from Science and Observation* (Vol. 3). Greystone Books Ltd.

Wohlleben, P. (2016). *The hidden life of trees: What they feel, how they communicate—Discoveries from a secret world* (Vol. 1). Greystone Books.

Woxvold, I. A., Diamond, J. M., Bishop, K. D., & Legra, L. (2019). Avifauna of the Lake Kutubu Wildlife Management Area, Papua New Guinea. *Bulletin of the British Ornithologists' Club, 139*(3), 266–292.

WWF. World Wide Fund for nature. (2021). How many species are we losing? https://wwf.panda.org/discover/our_focus/biodiversity/biodiversity/. Accessed July 11, 2021.

Yong, D. L., Heim, W., Chowdhury, S. U., Choi, C. Y., Ktitorov, P., Kulikova, O., & Szabo, J. K. (2021). The state of migratory landbirds in the East Asian Flyway: Distributions, threats, and conservation needs. *Frontiers in Ecology and Evolution, 9*, 613172.

Zinck, K., & Marmion, S. (2011). Global focus, local acts: Providing mental health services to indigenous people. *Archives of Psychiatric Nursing, 25*(5), 311–319.

Index

F. Huettmann, *Globalization and Papua New Guinea: Ancient Wilderness, Paradise, Introduced Terror and Hell*, https://doi.org/10.1007/978-3-031-20262-9

N

O

P

W

Y

Z

MIX
Papier aus verantwortungsvollen Quellen
Paper from responsible sources
FSC® C105338

If you have any concerns about our products,
you can contact us on
ProductSafety@springernature.com

In case Publisher is established outside the EU,
the EU authorized representative is:
Springer Nature Customer Service Center GmbH
Europaplatz 3, 69115 Heidelberg, Germany

Printed by Libri Plureos GmbH
in Hamburg, Germany